RENEWALS 458-4574

DRILLING FLUIDS PROCESSING HANDBOOK

DRILLING FLUIDS PROCESSING HANDBOOK

ASME

AMSTERDAM • BOSTON • HEIDELBERG • LONDON • NEW YORK • OXFORD
PARIS • SAN DIEGO • SAN FRANCISCO • SINGAPORE • SYDNEY • TOKYO

Gulf Professional Publishing is an imprint of Elsevier

Gulf Professional Publishing is an imprint of Elsevier
30 Corporate Drive, Suite 400, Burlington, MA 01803, USA
Linacre House, Jordan Hill, Oxford OX2 8DP, UK

Copyright © 2005, Elsevier Inc. All rights reserved.

No part of this publication may be reproduced, stored in a retrieval system, or transmitted in any form or by any means, electronic, mechanical, photocopying, recording, or otherwise, without the prior written permission of the publisher.

Permissions may be sought directly from Elsevier's Science & Technology Rights Department in Oxford, UK: phone: (+44) 1865 843830, fax: (+44) 1865 853333, e-mail: permissions@elsevier.com.uk. You may also complete your request on-line via the Elsevier homepage (http://elsevier.com), by selecting "Customer Support" and then "Obtaining Permissions."

∞ Recognizing the importance of preserving what has been wirtten, Elsevier prints its books on acid-free paper whenever possible.

Library of Congress Cataloging-in-Publication Data
Application submitted.

British Library Cataloguing-in-Publication Data
A catalogue record for this book is available from the British Library.

ISBN 0-7506-7775-9

For information on all Gulf Professional Publishing publications visit our Web site at www.books.elsevier.com

04 05 06 07 08 09   10 9 8 7 6 5 4 3 2 1

Printed in the United States of America

CONTENTS

Biographies xvii

Preface xxiii

1 Historical Perspective and Introduction   1
    1.1   Scope   1
    1.2   Purpose   1
    1.3   Introduction   2
    1.4   Historical Perspective   4
    1.5   Comments   11
    1.6   Waste Management   13

2 Drilling Fluids   15
    2.1   Drilling Fluid Systems   15
        2.1.1   Functions of Drilling Fluids   15
        2.1.2   Types of Drilling Fluids   16
        2.1.3   Drilling Fluid Selection   17
        2.1.4   Separation of Drilled Solids from Drilling Fluids   20
    2.2   Characterization of Solids in Drilling Fluids   25
        2.2.1   Nature of Drilled Solids and Solid Additives   25
        2.2.2   Physical Properties of Solids in Drilling Fluids   26
    2.3   Properties of Drilling Fluids   31
        2.3.1   Rheology   32
    2.4   Hole Cleaning   38
        2.4.1   Detection of Hole-Cleaning Problems   38
        2.4.2   Drilling Elements That Affect Hole Cleaning   40
        2.4.3   Filtration   45
        2.4.4   Rate of Penetration   47
        2.4.5   Shale Inhibition Potential/Wetting Characteristics   51
        2.4.6   Lubricity   52
        2.4.7   Corrosivity   53
        2.4.8   Drilling-Fluid Stability and Maintenance   54

- 2.5 Drilling Fluid Products 54
  - 2.5.1 Colloidal and Fine Solids 54
  - 2.5.2 Macropolymers 55
  - 2.5.3 Conventional Polymers 56
  - 2.5.4 Surface-Active Materials 57
- 2.6 Health, Safety, and Environment and Waste Management 58
  - 2.6.1 Handling Drilling Fluid Products and Cuttings 58
  - 2.6.2 Drilling Fluid Product Compatibility and Storage Guidelines 58
  - 2.6.3 Waste Management and Disposal 62
  - References 66

3 Solids Calculation 69
- 3.1 Procedure for a More Accurate Low-Gravity Solids Determination 70
  - 3.1.1 Sample Calculation 73
- 3.2 Determination of Volume Percentage of Low-Gravity Solids in Water-Based Drilling Fluid 77
- 3.3 Rig-Site Determination of Specific Gravity of Drilled Solids 78

4 Cut Points 81
- 4.1 How to Determine Cut Point Curves 85
- 4.2 Cut Point Data: Shale Shaker Example 90

5 Tank Arrangement 93
- 5.1 Active System 94
  - 5.1.1 Suction and Testing Section 94
  - 5.1.2 Additions Section 95
  - 5.1.3 Removal Section 95
  - 5.1.4 Piping and Equipment Arrangement 96
  - 5.1.5 Equalization 98
  - 5.1.6 Surface Tanks 99
  - 5.1.7 Sand Traps 100
  - 5.1.8 Degasser Suction and Discharge Pit 102
  - 5.1.9 Desander Suction and Discharge Pits 102
  - 5.1.10 Desilter Suction and Discharge Pits (Mud Cleaner/Conditioner) 103
  - 5.1.11 Centrifuge Suction and Discharge Pits 103
- 5.2 Auxiliary Tank System 104
  - 5.2.1 Trip Tank 104
- 5.3 Slug Tank 105
- 5.4 Reserve Tank(s) 105

- 6 Scalping Shakers and Gumbo Removal   107
- 7 Shale Shakers   111
  - 7.1 How a Shale Shaker Screens Fluid   113
  - 7.2 Shaker Description   116
  - 7.3 Shale Shaker Limits   118
    - 7.3.1 Fluid Rheological Properties   119
    - 7.3.2 Fluid Surface Tension   120
    - 7.3.3 Wire Wettability   120
    - 7.3.4 Fluid Density   120
    - 7.3.5 Solids: Type, Size, and Shape   120
    - 7.3.6 Quantity of Solids   121
    - 7.3.7 Hole Cleaning   121
  - 7.4 Shaker Development Summary   121
  - 7.5 Shale Shaker Design   122
    - 7.5.1 Shape of Motion   123
    - 7.5.2 Vibrating Systems   133
    - 7.5.3 Screen Deck Design   134
    - 7.5.4 g Factor   136
    - 7.5.5 Power Systems   140
  - 7.6 Selection of Shale Shakers   143
    - 7.6.1 Selection of Shaker Screens   145
    - 7.6.2 Cost of Removing Drilled Solids   145
    - 7.6.3 Specific Factors   146
  - 7.7 Cascade Systems   148
    - 7.7.1 Separate Unit   150
    - 7.7.2 Integral Unit with Multiple Vibratory Motions   150
    - 7.7.3 Integral Unit with a Single Vibratory Motion   152
    - 7.7.4 Cascade Systems Summary   152
  - 7.8 Dryer Shakers   153
  - 7.9 Shaker User's Guide   154
    - 7.9.1 Installation   155
    - 7.9.2 Operation   156
    - 7.9.3 Maintenance   157
    - 7.9.4 Operating Guidelines   158
  - 7.10 Screen Cloths   159
    - 7.10.1 Common Screen Cloth Weaves   160
    - 7.10.2 Revised API Designation System   167
    - 7.10.3 Screen Identification   174
  - 7.11 Factors Affecting Percentage-Separated Curves   174
    - 7.11.1 Screen Blinding   176
    - 7.11.2 Materials of Construction   177
    - 7.11.3 Screen Panels   178

            7.11.4 Hook-Strip Screens   180
            7.11.5 Bonded Screens   180
            7.11.6 Three-Dimensional Screening Surfaces   180
     7.12 Non-Oilfield Drilling Uses of Shale Shakers   181
            7.12.1 Microtunneling   181
            7.12.2 River Crossing   182
            7.12.3 Road Crossing   182
            7.12.4 Fiber-Optic Cables   182

8   **Settling Pits   183**
     8.1 Settling Rates   183
     8.2 Comparison of Settling Rates of Barite and Low-Gravity Drilled Solids   186
     8.3 Comments   187
     8.4 Bypassing the Shale Shaker   188

9   **Gas Busters, Separators, and Degassers   189**
     9.1 Introduction: General Comments on Gas Cutting   189
     9.2 Shale Shakers and Gas Cutting   192
     9.3 Desanders, Desilters, and Gas Cutting   192
     9.4 Centrifuges and Gas Cutting   193
     9.5 Basic Equipment for Handling Gas-Cut Mud   193
            9.5.1 Gravity Separation   195
            9.5.2 Centrifugal Separation   195
            9.5.3 Impact, Baffle, or Spray Separation   195
            9.5.4 Parallel-Plate and Thin-Film Separation   196
            9.5.5 Vacuum Separation   196
     9.6 Gas Busters   196
     9.7 Separators   197
            9.7.1 Atmospheric Separators   197
            9.7.2 West Texas Separator   198
     9.8 Pressurized Separators   199
            9.8.1 Commercial Separator/Flare Systems   199
            9.8.2 Pressurized, or Closed, Separators: Modified Production Separators   200
            9.8.3 Combination System: Separator and Degasser   202
     9.9 Degassers   202
            9.9.1 Degasser Operations   203
            9.9.2 Degasser Types   205
            9.9.3 Pump Degassers or Atmospheric Degassers   207
            9.9.4 Magna-Vac^{TM} Degasser   207
     9.10 Points About Separators and Separation   209
            References   210

10 Suspension, Agitation, and Mixing of Drilling Fluids 213

- 10.1 Basic Principles of Agitation Equipment 213
- 10.2 Mechanical Agitators 214
  - 10.2.1 Impellers 215
  - 10.2.2 Gearbox 222
  - 10.2.3 Shafts 222
- 10.3 Equipment Sizing and Installation 223
  - 10.3.1 Design Parameters 223
  - 10.3.2 Compartment Shape 226
  - 10.3.3 Tank and Compartment Dimensions 226
  - 10.3.4 Tank Internals 226
  - 10.3.5 Baffles 227
  - 10.3.6 Sizing Agitators 227
  - 10.3.7 Turnover Rate (TOR) 228
- 10.4 Mud Guns 232
  - 10.4.1 High-Pressure Mud Guns 233
  - 10.4.2 Low-Pressure Mud Guns 233
  - 10.4.3 Mud Gun Placement 234
  - 10.4.4 Sizing Mud Gun Systems 235
- 10.5 Pros and Cons of Agitation Equipment 237
  - 10.5.1 Pros of Mechanical Agitators 238
  - 10.5.2 Cons of Mechanical Agitators 238
  - 10.5.3 Pros of Mud Guns 238
  - 10.5.4 Cons of Mud Guns 238
- 10.6 Bernoulli's Principle 239
  - 10.6.1 Relationship of Pressure, Velocity, and Head 240
- 10.7 Mud Hoppers 244
  - 10.7.1 Mud Hopper Installation and Operation 246
  - 10.7.2 Mud Hopper Recommendations 248
  - 10.7.3 Other Shearing Devices 250
- 10.8 Bulk Addition Systems 250
- 10.9 Tank/Pit Use 253
  - 10.9.1 Removal 253
  - 10.9.2 Addition 254
  - 10.9.3 Suction 254
  - 10.9.4 Reserve 255
  - 10.9.5 Discharge 255
  - 10.9.6 Trip Tank 255
  - References 255

11 Hydrocyclones 257

- 11.1 Discharge 261
- 11.2 Hydrocyclone Capacity 265

- 11.3 Hydrocyclone Tanks and Arrangements 266
  - 11.3.1 Desanders 267
  - 11.3.2 Desilters 268
  - 11.3.3 Comparative Operation of Desanders and Desilters 269
  - 11.3.4 Hydrocyclone Feed Header Problems 269
- 11.4 Median (D_{50}) Cut Points 270
  - 11.4.1 Stokes' Law 271
- 11.5 Hydrocyclone Operating Tips 276
- 11.6 Installation 278
- 11.7 Conclusions 279
  - 11.7.1 Errata 281

12 Mud Cleaners 283
- 12.1 History 286
- 12.2 Uses of Mud Cleaners 288
- 12.3 Non-Oilfield Use of Mud Cleaners 291
- 12.4 Location of Mud Cleaners in a Drilling-Fluid System 291
- 12.5 Operating Mud Cleaners 292
- 12.6 Estimating the Ratio of Low-Gravity Solids Volume and Barite Volume in Mud Cleaner Screen Discard 293
- 12.7 Performance 295
- 12.8 Mud Cleaner Economics 297
- 12.9 Accuracy Required for Specific Gravity of Solids 300
- 12.10 Accurate Solids Determination Needed to Properly Identify Mud Cleaner Performance 300
- 12.11 Heavy Drilling Fluids 301

13 Centrifuges 303
- 13.1 Decanting Centrifuges 303
  - 13.1.1 Stokes' Law and Drilling Fluids 308
  - 13.1.2 Separation Curves and Cut Points 308
  - 13.1.3 Drilling-Fluids Solids 310
- 13.2 The Effects of Drilled Solids and Colloidal Barite on Drilling Fluids 311
- 13.3 Centrifugal Solids Separation 313
  - 13.3.1 Centrifuge Installation 316
  - 13.3.2 Centrifuge Applications 316
  - 13.3.3 The Use of Centrifuges with Unweighted Drilling Fluids 317
  - 13.3.4 The Use of Centrifuges with Weighted Drilling Fluids 317

          13.3.5   Running Centrifuges in Series   318
          13.3.6   Centrifuging Drilling Fluids with Costly Liquid
                   Phases   320
          13.3.7   Flocculation Units   320
          13.3.8   Centrifuging Hydrocyclone Underflows   321
          13.3.9   Operating Reminders   321
          13.3.10 Miscellaneous   321
   13.4 Rotary Mud Separator   321
          13.4.1   Problem 1   322
   13.5 Solutions to the Questions in Problem 1   324
          13.5.1   Question 1   324
          13.5.2   Question 2   324
          13.5.3   Question 3   324
          13.5.4   Question 4   325
          13.5.5   Question 5   325
          13.5.6   Question 6   325
          13.5.7   Question 7   325
          13.5.8   Question 8   325
          13.5.9   Question 9   326
          13.5.10 Question 10   326

14 **Use of the Capture Equation to Evaluate the Performance of Mechanical Separation Equipment Used to Process Drilling Fluids   327**
   14.1 Procedure   330
          14.1.1   Collecting Data for the Capture Analysis   330
          14.1.2   Laboratory Analysis   330
   14.2 Applying the Capture Calculation   331
          14.2.1   Case 1: Discarded Solids Report to Underflow   331
          14.2.2   Case 2: Discarded Solids Report to Overflow   331
          14.2.3   Characterizing Removed Solids   331
   14.3 Use of Test Results   332
          14.3.1   Specific Gravity   332
          14.3.2   Particle Size   332
          14.3.3   Economics   333
   14.4 Collection and Use of Supplementary Information   334

15 **Dilution   335**
   15.1 Effect of Porosity   337
   15.2 Removal Efficiency   338
   15.3 Reasons for Drilled-Solids Removal   339
   15.4 Diluting as a Means for Controlling Drilled Solids   340
   15.5 Effect of Solids Removal System Performance   341

- 15.6 Four Examples of the Effect of Solids Removal Equipment Efficiency 342
  - 15.6.1 Example 1 343
  - 15.6.2 Example 2 344
  - 15.6.3 Example 3 346
  - 15.6.4 Example 4 347
  - 15.6.5 Clean Fluid Required to Maintain 4%vol Drilled Solids 347
- 15.7 Solids Removal Equipment Efficiency for Minimum Volume of Drilling Fluid to Dilute Drilled Solids 348
  - 15.7.1 Equation Derivation 349
  - 15.7.2 Discarded Solids 350
- 15.8 Optimum Solids Removal Equipment Efficiency (SREE) 351
- 15.9 Solids Removal Equipment Efficiency in an Unweighted Drilling Fluid from Field Data 354
  - 15.9.1 Excess Drilling Fluid Built 356
- 15.10 Estimating Solids Removal Equipment Efficiency for a Weighted Drilling Fluid 357
  - 15.10.1 Solution 358
  - 15.10.2 Inaccuracy in Calculating Discard Volumes 360
- 15.11 Another Method of Calculating the Dilution Quantity 361
- 15.12 Appendix: American Petroleum Institute Method 361
  - 15.12.1 Drilled Solids Removal Factor 361
  - 15.12.2 Questions 362
- 15.13 A Real-Life Example 362
  - 15.13.1 Exercise 1 362
  - 15.13.2 Exercise 2 364
  - 15.13.3 Exercise 3 365
  - 15.13.4 Exercise 4 365
  - 15.13.5 General Comments 366

16 Waste Management 367
- 16.1 Quantifying Drilling Waste 367
  - 16.1.1 Example 1 368
  - 16.1.2 Example 2 368
  - 16.1.3 Example 3 369
  - 16.1.4 Example 4 370
  - 16.1.5 Example 5 371
  - 16.1.6 Example 6 372
- 16.2 Nature of Drilling Waste 372
- 16.3 Minimizing Drilling Waste 374
  - 16.3.1 Total Fluid Management 375
  - 16.3.2 Environmental Impact Reduction 377

16.4 Offshore Disposal Options   377
  16.4.1 Direct Discharge   378
  16.4.2 Injection   378
  16.4.3 Collection and Transport to Shore   380
  16.4.4 Commercial Disposal   380
16.5 Onshore Disposal Options   382
  16.5.1 Land Application   382
  16.5.2 Burial   386
16.6 Treatment Techniques   391
  16.6.1 Dewatering   391
  16.6.2 Thermal Desorption   395
  16.6.3 Solidification/Stabilization   397
16.7 Equipment Issues   399
  16.7.1 Augers   400
  16.7.2 Vacuums   402
  16.7.3 Cuttings Boxes   403
  16.7.4 Cuttings Dryers   406
  References   412

17 The AC Induction Motor   413
17.1 Introduction to Electrical Theory   413
17.2 Introduction to Electromagnetic Theory   421
17.3 Electric Motors   423
  17.3.1 Rotor Circuits   424
  17.3.2 Stator Circuits   425
17.4 Transformers   427
17.5 Adjustable Speed Drives   429
17.6 Electric Motor Applications on Oil Rigs   432
  17.6.1 Ratings   432
  17.6.2 Energy Losses   433
  17.6.3 Temperature Rise   434
  17.6.4 Voltage   435
17.7 Ambient Temperature   435
17.8 Motor Installation and Troubleshooting   438
17.9 Electric Motor Standards   439
17.10 Enclosure and Frame Designations   441
  17.10.1 Protection Classes Relating to Enclosures   443
17.11 Hazardous Locations   444
17.12 Motors for Hazardous Duty   449
17.13 European Community Directive 94/9/EC   451
17.14 Electric Motors for Shale Shakers   454
17.15 Electric Motors for Centrifuges   459

17.16 Electric Motors for Centrifugal Pumps   459
17.17 Study Questions   460

18 Centrifugal Pumps   465
18.1 Impeller   465
18.2 Casing   467
18.3 Sizing Centrifugal Pumps   470
    18.3.1 Standard Definitions   471
    18.3.2 Head Produces Flow   479
18.4 Reading Pump Curves   480
18.5 Centrifugal Pumps Accelerate Fluid   484
    18.5.1 Cavitation   485
    18.5.2 Entrained Air   486
18.6 Concentric vs Volute Casings   488
    18.6.1 Friction Loss Tables   490
18.7 Centrifugal Pumps and Standard Drilling Equipment   491
    18.7.1 Friction Loss and Elevation Considerations   491
18.8 Net Positive Suction Head   503
    18.8.1 System Head Requirement (SHR) Worksheet   506
    18.8.2 Affinity Laws   506
    18.8.3 Friction Loss Formulas   507
18.9 Recommended Suction Pipe Configurations   508
    18.9.1 Supercharging Mud Pumps   510
    18.9.2 Series Operation   512
    18.9.3 Parallel Operation   513
    18.9.4 Duplicity   513
18.10 Standard Rules for Centrifugal Pumps   513
18.11 Exercises   514
    18.11.1 Exercise 1   514
    18.11.2 Exercise 2: System Head Requirement Worksheet   515
    18.11.3 Exercise 3   517
    18.11.4 Exercise 4   517
18.12 Appendix   518
    18.12.1 Answers to Exercise 1   518
    18.12.2 Answers to Exercise 2: System Head Requirement Worksheet   518
    18.12.3 Answers to Exercise 3   520
    18.12.4 Answers to Exercise 4   520

19 Solids Control in Underbalanced Drilling   521
19.1 Underbalanced Drilling Fundamentals   521
    19.1.1 Underbalanced Drilling Methods   523

19.2 Air/Gas Drilling   523
    19.2.1 Environmental Contamination   524
    19.2.2 Drilling with Natural Gas   525
    19.2.3 Sample Collection While Drilling with Air or Gas   526
    19.2.4 Air or Gas Mist Drilling   527
19.3 Foam Drilling   529
    19.3.1 Disposable Foam Systems   529
    19.3.2 Recyclable Foam Systems   530
    19.3.3 Sample Collection While Drilling with Foam   532
19.4 Liquid/Gas (Gaseated) Systems   532
19.5 Oil Systems, Nitrogen/Diesel Oil, Natural Gas/Oil   535
    19.5.1 Sample Collection with Aerated Systems   535
19.6 Underbalanced Drilling with Conventional Drilling Fluids or Weighted Drilling Fluids   536
19.7 General Comments   537
    19.7.1 Pressurized Closed Separator System   538
19.8 Possible Underbalanced Drilling Solids-Control Problems   539
    19.8.1 Shale   539
    19.8.2 Hydrogen Sulfide Gas   540
    19.8.3 Excess Formation Water   540
    19.8.4 Downhole Fires and Explosions   540
    19.8.5 Very Small Air- or Gas-Drilled Cuttings   541
    19.8.6 Gaseated or Aerated Fluid Surges   541
    19.8.7 Foam Control   542
    19.8.8 Corrosion Control   542
    Suggested Reading   542

20 Smooth Operations   547

20.1 Derrickman's Guidelines   548
    20.1.1 Benefits of Good Drilled-Solids Separations   549
    20.1.2 Tank and Equipment Arrangements   549
    20.1.3 Shale Shakers   550
    20.1.4 Things to Check When Going on Tour   552
    20.1.5 Sand Trap   552
    20.1.6 Degasser   553
    20.1.7 Hydrocyclones   554
    20.1.8 Hydrocyclone Troubleshooting   557
    20.1.9 Mud Cleaners   558
    20.1.10 Centrifuges   560
    20.1.11 Piping to Materials Additions (Mixing) Section   561
20.2 Equipment Guidelines   562
    20.2.1 Surface Systems   562

        20.2.2   Centrifugal Pumps   572
  20.3 Solids Management Checklist   577
        20.3.1   Well Parameters/Deepwater Considerations   577
        20.3.2   Drilling Program   579
        20.3.3   Equipment Capability   579
        20.3.4   Rig Design and Availability   580
        20.3.5   Logistics   580
        20.3.6   Environmental Issues   580
        20.3.7   Economics   581

Appendix   583

Glossary   585

Index   651

BIOGRAPHIES

Bob Barrett received his BSAS from Miami University and his MBA from Northern Kentucky University. From 1997 through 2003 he worked as the Screen Manufacturing Engineer for SWECO, developing manufacturing processes and technologies for industrial vibratory separator screens. Since 2003 he has served as the Senior Development Engineer of oilfield screens for MI-Swaco.

Eugene Bouse has a degree in petroleum engineering from Louisiana State University and has worked in drilling fluids for over 40 years, specializing in solids control for the past 15. He is a past chairman of the American Society of Mechanical Engineers (ASME) Drilling Waste Management Group and is a member of the Society of Petroleum Engineers (SPE) and Mensa.

Brian Carr has his BSME from the University of Louisville's Speed Scientific School. From 1993 through 2001 he worked as the New Product Development Engineer/Engineering Manager for SWECO-Division of Emerson Electric, developing new technologies for industrial and oilfield screens and vibratory separators. Since 2001 he has worked as the Engineering Manager of Shakers, Screens, Hydrocyclones, and Gumbo Removal for M-I SWACO.

Bob DeWolfe has extensive field, technical services, and operational management experience in the energy industry with drilling fluids, solids management, and refinery waste management. He has had specific assignments in Europe, Africa, the former Soviet Union, Southeast Asia, Latin America, and the United States. He is presently working in the Middle East.

Fred Growcock has been serving as Product Applications Team Leader and R&D Advisor for M-I SWACO since early 1999, and, most recently, as director of a U.S. Department of Energy-funded project on the fundamentals of aphron drilling fluids. Prior to joining M-I SWACO, he worked at Brookhaven (NY) and Oak Ridge (TN.) National Labs on coal liquefaction and gasification, and problems related to gas-cooled nuclear reactors; Dowell Schlumberger (Tulsa) on reservoir stimulation, acid corrosion inhibition, and foamed fracturing; and Amoco (Tulsa), now BP, on drilling-fluid development. Fred holds BA/BS degrees in chemistry from the University of Texas at Austin and MS/PhD degrees in physical chemistry from New Mexico State University.

Tim Harvey holds degrees from Oklahoma State University and the University of Florida. He has worked in drilling fluids, solids control, and drilling waste management for over 30 years in the USA, Middle East, West Africa, and the Far East. Tim has served on various API and Drilling Waste Management Group (DWMG) workgroups and is currently based in Kuala Lumpur as Manager of Technical Services for Oiltools International. He is a member of the Society of Petroleum Engineers and is affiliated with American Association of Drilling Engineers (AADE), API and ASME.

Jerry Haston is a graduate of the University of Oklahoma with a degree in petroleum geology. He has spent his entire career, more than 40 years, in oil and gas exploration and production. Most of those years have been drilling-related. Jerry's experience includes international operations world-wide and industry-wide domestic operations. He is currently supervising drilling operations for a major international oil and gas company.

Michael Kargl is a mechanical engineer from Southern Illinois University. Since 1995, he has been the chief engineer for shale shaker electric vibrators at Martin Engineering Company. Prior to that time, he was an engineering group leader at Underwriter's Laboratories Inc. for 13 years, working mostly on design reviews and testing of explosion-proof motors and generators.

Todd H. Lee is the marketing manager for National Oilwell, with 13 years' experience sizing, training, and troubleshooting centrifugal pumps. He has been a business owner and product manager, and has worked in product design of centrifugal pumps.

Bob Line has over 30 years' experience in the oilfield, with major areas of expertise in wellhead and valves, pressure control, subsea controls, drilling instrumentation, and solids control. He spent 7 years in mechanical engineering design, and sales and marketing. For the past 9 years, Bob has been with M-I SWACO and is currently the Global Capital Equipment sales manager. Bob has a BS degree in mechanical engineering from the University of Houston.

Hemu Mehta is one of the founders and current president of KEM-TRON Technologies, Inc. Previously, he was Manager of International Operations for M/I Drilling Fluids. He holds degrees in chemical engineering, petroleum engineering, and international finance. Mr. Mehta's background in chemical engineering and solids control equipment allowed him to help develop today's most advanced dewatering technology. Mr. Mehta grew up in India and has lived in Houston, Texas since 1972 with his wife and two children.

James Merrill has been involved in the design and manufacturing of shale shakers and shale shaker screens for the past 17 years in the petroleum, mining, and utilities business sectors. His vast knowledge of wire cloth and shaker screens has allowed him to solve screen problems around the world. His career has taken him from a roughneck on drilling rigs around the Gulf Coast to Technical Manager of a leading solids control company.

Mark C. Morgan has been the Technical Services Manager for Derrick Equipment Company since 1994. Prior to this he worked as a drilling fluids engineer and seaplane pilot for NL Baroid. Then he worked as a drilling fluids and solids control consultant for 10 years, mainly working offshore of Angola for Texaco. Mark has a BS in professional aviation from Louisiana Tech and a BS from the University of Southwestern Louisiana in petroleum engineering.

Mike Morgenthaler has been involved with the drilling fluids and solids control equipment since 1980, when he joined IMCO Services as a mud engineer. Mike has a degree in mechanical engineering from the University of Texas. Mike is a principal consultant for CUTPOINT, Inc. and specializes in technology for drilling waste management.

Nace S. Peard has over 23 years experience in the oilfield, initially as a drilling engineer with Gulf Oil/Chevron. He later managed exploration and development drilling projects for a large independent oil and gas company. Since early 2000, he has been vice president of sales and marketing for DF Corporation. Nace is a registered PE in Texas, and has a degree in petroleum engineering from Purdue University.

William Piper worked over 20 years with Amoco as a drilling engineer and environmental specialist for the international drilling group. Upon his retirement in 1998, he formed Piper Consulting to continue working in the specialty niche market of environmental affairs relating to the drilling industry. He is recognized as one of the foremost experts in drilling waste management, as well as other environmental issues in drilling. He has been published extensively on the subject of environmental practices in drilling, including an article in the *Encyclopedia of Environmental Analysis and Remediation*. He continues to teach a course in drilling waste management for the ChevronTexaco/BP drilling training alliance and has developed a series of software programs to aid drilling engineers with their environmental issues. Mr. Piper has a BS in chemical engineering from the University of Colorado and is active in the SPE.

Bill Rehm is an underbalanced drilling and completions consultant in Houston, Texas. He, at various times, has been a mud engineer, well control supervisor, president of directional drilling company, R&D manager, and drilling consultant. He has written some 50 papers on such subjects as Solids Control, Well Control, and Underbalanced Drilling.

Mike Richards has worked with Brandt in varying capacities since 1979, including product development manager, engineer, senior project manager, training coordinator, and in technical sales. Many performance parameters of mechanical agitators and mud hoppers were calculated and documented while Mike was manager of Product Development at Brandt. He has a BA degree from the University of Houston.

Leon Robinson has been teaching drilling classes for Petroskills/OGCI since retiring in 1992 from a 39-year career with Exxon Production Research. He earned a BS and MS degree from Clemson, a PhD in engineering physics from N.C. State, the 1985 SPE Drilling Engineering Award, and the 1999 AADE Meritorious Service Award.

Wiley Steen has been involved in drilling and production since the mid-1960s, both domestically and internationally. He lived overseas in both Europe and Southeast Asia from 1970 to 1975 where he served as manager for IMCO Services Division Halliburton. Since 1976, Wiley has worked as an independent consultant in both domestic and international theaters. He was one of the four founding Officers of the SPE Singapore Chapter (serving as secretary), and one of the 12 founders of Houston's AADE Chapter. He has served two terms as Houston AADE president and on the AADE National Board.

Mike Stefanov is currently a Regional Drilling Superintendent for BP and is involved in international deepwater drilling operations. Mike has over 27 years experience managing drilling operations in isolated locations both offshore and on shore, from conception through to maturity in Africa, the Middle East, Far East, Gulf of Mexico (Deepwater), South America and in the Indian Ocean. He has been involued in rig construction and upgrades in Europe, the Far East and North America. Mike began his career in North America with Seismogrph Service Corporation, moving into international operations with Amoco International Oil Company, as a drilling engineer. Mike has worked as a Drilling Foreman, Drilling Supervisor, Drilling Superintendent and Drilling Manager throughout his career. He holds a BS degree in mechanical engineering from the University of Connecticut and is a registered professional engineer in the state of Texas. He has supervised inititiatives such as Technical Limits, HTHP Practices, International Operations' Practices, Floating Rig Construction and Outfitting, and Deepwater Drilling and Testing Practices.

PREFACE

In the early 1970s, the International Association of Drilling Contractors (IADC) formed a committee to study solids control on drilling rigs. After 10 years of work, we published the IADC *Mud Equipment Manual*. The committee started with only six members and ended with about 27. Many members remained on the committee when they changed employers, and employers wanted to stay represented and active on the committee. Others heard about our work and asked to join. The 11 handbooks in the *Mud Equipment Manual* discussed individual components of a drilling fluid system. Each subject was discussed in a series of seminars/conferences presented by local IADC districts all around the United States. Writing the *Manual* was a great education, and the discussions of all components of the material enticed members to remain engaged. After the *Manual* was published, the committee formed a new group, the IADC Rig Instrumentation and Measurements Committee (RIM). RIM was responsible for many publications relating to measurements and information transfer around a drilling rig. After several years of this work, the committee decided that the *Shale Shaker Handbook* needed to be revised; a majority of the committee had been involved in writing the first edition. When it was published, linear motion shale shakers were not available. New technology made the rewrite necessary. After a disagreement on final editing procedures, the committee left the IADC and went in search of a new sponsor.

Culminating a discussion of several offers, the Houston chapter of the American Association of Drilling Engineers (AADE) was selected, and the board of directors at that time enthusiastically welcomed the committee. Many of the original committee members continued to work on the committee and some new members from the AADE joined. The group assembled information and published the AADE *Shale Shaker*

Handbook. Since that book focused primarily on shale shakers, other components of the drilling-fluid system were relegated to a subordinate position. Gulf Publishing Company, which had published the IADC *Mud Equipment Manual*, published the *Shale Shaker Handbook* and retained the copyright for both books.

The committee decided that assigning relative contributions of all of the authors would be too difficult, if not impossible, for purposes of distributing royalty payments equitably. Hence, it requested that Gulf Publishing Company reduce the cost of the books by the amount that would normally be distributed in royalty. The AADE *Shale Shaker Handbook* found a comfortable niche in technology enlightenment. Gulf Publishing sold their book division and now Elsevier owns the copyright to the book.

Immediately after the *Shale Shaker Handbook* was published, the American Petroleum Institute (API) decided to revise RP (Recommended Practices) 13C, "Solids Control," and RP 13E, "Screen Designation." API Subcommittee 13 requested that the committee use the technology discussed in the *Shale Shaker Handbook* to modify the RP. The committee accepted the challenge. While working with the concepts and details available, the committee recognized that some additional technology and new methods needed to be developed. This quest required about five years before the final document was written.

Toward the end of the work on API RP 13C, Elsevier requested that the committee revise the AADE *Shale Shaker Handbook* because supplies were dwindling. The Houston chapter of the AADE, however, was less interested in technology. During the couple of years after the publication of the *Shale Shaker Handbook*, the number of technical committees organized and supported by the Houston chapter of the AADE dwindled from six to one, with only the Deepwater group remaining active. The executive committee of the Houston chapter notified the committee that they would not be a sponsor for the rewrite. Once again the committee went in search of a new sponsor. This time the American Society of Mechanical Engineers (ASME) Petroleum Division enthusiastically assumed that role. This book is, therefore, sponsored by the Petroleum Division of ASME.

Because we had a new sponsor and the need was obvious, the scope of this book was expanded to discuss all aspects of drilling-fluid processing. Most of the people involved with the API work volunteered to write this new book. Several new chapters have been added. Much additional technology has been developed since the AADE *Shale Shaker Handbook* was published. The committee decided that the activity and structure of this group would be slightly different from previous years. Many members are also still actively participating with the API in rewriting API RP 13C relating to solids control. Some felt that they could not spend the necessary time to participate in both committees.

Each chapter of this book was assigned to one or two recognized industry professionals—most volunteered for their assignment. Many of the chapters from the AADE *Shale Shaker Handbook* were modified and brought up to date by one or two volunteers. Some chapters represent completely new material that was missing from the original book. The writer and/or modifiers of the chapters are recognized below the chapter title. However, many of the chapters also benefited from comments and critiques from colleagues. Sometimes the truth is difficult to find in drilling. This committee comprised only authors. In the past, some of the members acted primarily as editors, and this created a situation that, although educational for committee members, increased the writing times for the documents. Each chapter in this book was assigned to three other authors for editing and comments. Streamlining the process allowed this book to be published more rapidly than either of the other two editions. This procedure, however, resulted in some minor variances of technology. The material in each chapter was not read or approved by all of the committee. Industry professionals do not always agree on all aspects of drilling fluid and its processing; some of this different interpretation may appear in the various chapters. A good example of this is the issue of sand traps. In some drilling situations, some of the committee found sand traps to be very beneficial; others had different experiences. Both viewpoints are presented here and, hopefully, with sufficient information so that the benefits expected can be identified from the discussion in this book.

Each chapter retains some of the individuality of the principle author, but an attempt has been made to provide some homogeneity of styles. Sentences have been examined for clarity, accuracy, and their ability to be readily understood. The latter objective is sometimes the most difficult to accomplish. Words may indicate something with clarity but can still be incorrectly interpreted. Surplus words and personal pronouns have been mostly eliminated. Many of the concepts presented in this book have been discussed in depth and the consensus presented here. The oil patch is filled with misconceptions and erroneous "facts." Every attempt has been made to present balanced, accurate science. In some places, duplicate information is provided because the basic technology needs to be understood for each chapter.

The individuals who have written this book represent a superior group of professionals who not only have great knowledge, but also are willing to dedicate many hours of their time to share it with the industry. Many companies were involved in creating this book, and some of the work was done "on company time." However, the authors dedicated many hours of weekend and holiday time to create this book. The committee has never requested funds from any of the sponsors. All of the work (cost of meetings, preparation of manuscripts, etc.) was supported financially by individuals and some of the companies involved. No royalty was ever accepted for any of the books. Many in this unusual group of individuals have worked together on this committee for many years. Several recently joined to further enhance the prestige of this group of authors. As an interesting fact, Bill Rehm joined the IADC committee when it was organized in the early 1970s, and had to resign because of work assignments. However, he has now returned to the group, with great expertise in underbalanced drilling, to enhance this book.

Very few industry committees retain their identity through two or three decades of work. One of the secrets of this group's longevity has been its commitment to developing meaningful, useful products. Another secret has been the learning experience gained from in-depth technical discussions among people interested in facts and willing to argue without involving personalities or ancestry. The authors, who volunteered for this book, have built a firm basis for the technology involved in processing drilling fluid. This technology has evolved through many years of testing, trial and error, and discussions. Some of the history of that technology is captured in the Historical Perspective section of Chapter 1.

Many who started working with the committee in the beginning have retired or have died. All authors owe a debt of gratitude to the pioneers who preceded us. My association with this group has been one of the best learning experiences possible. I have great respect for all of the many talented, professional people who have shared so much knowledge with me. The association with so many brilliant people is deeply cherished. The past 40 years have been remarkable in the great strides made in technology, and this committee has sought to capture this for the primary benefit of the industry.

Leon Robinson
Chairman of the Rewrite Committee, 2004

CHAPTER 1

HISTORICAL PERSPECTIVE AND INTRODUCTION

Leon Robinson
Exxon, retired

1.1 SCOPE

This handbook describes the method and mechanical systems available to control drilled solids in drilling fluids used in oil well drilling. System details permit immediate and practical application both in the planning/design phase and in operations.

1.2 PURPOSE

Good solids-control programs are often ignored because basic principles are not understood. This book explains the fundamentals of good solids control. Adherence to these simple basic principles is financially rewarding.

This American Society of Mechanical Engineers (ASME) textbook/handbook is a revision of the American Association of Drilling Engineers (AADE) *Shale Shaker Handbook*, which was a revision of the International Association of Drilling Contractors (IADC) *Mud Equipment Manual*. Many of the authors of this book were authors of those books as well. Patience, dedication, many long hours of work, and evaluation of the latest technology have been required of all members of this committee. Ten years were required to write the IADC Manual;

7 years were required to write the AADE Handbook; and 2 years were required to write this textbook.

None of the authors of any of the three books have received any compensation for their work and writing. The group was dedicated to providing the drilling industry with the best technology available, and many hours of discussion were frequently required to resolve controversial issues.

1.3 INTRODUCTION

Fallacious arguments persist that drilled solids are beneficial. Drilled solids are evil and insidious. Increases in drilled-solids concentrations generally do not immediately reveal their economic impact. Their detrimental effects are generally not immediately obvious on a drilling rig; so skeptics fail to believe that drilled solids foster the havoc that they truly do. The secret to drilling safely, fast, and under budget is to remove drilled solids. Drilled solids increase drilling costs, damage reservoirs, and create large disposal costs. Specific problems associated with drilled solids are:

- Filtrate damage to formations
- Drilling rate limits
- Hole problems
- Stuck pipe problems
- Lost circulation problems
- Direct drilling-fluid costs
- Increased disposal costs

These bad effects of drilled solids are explored in greater detail here and in the rest of the book. The eradication of these effects is discussed in great detail in this book. The book may be used for planning and designing a drilling-fluid processing system, improving current systems, troubleshooting a system, or improving rig operations. Drilled solids are EVIL, and this is the theme of this Handbook.

The effects of drilled solids on the economics of drilling a well are subtle. Increasing drilled-solids content does not immediately result in disaster on a drilling rig. When a drill bit ceases to drill and torque increases, a driller knows immediately that it is time to pull the bit. When drilled solids increase, the detrimental effects are not immediately apparent. Decreasing drilled solids is analogous to buying insurance for an event that will not happen. Proving that something will not happen—like stuck pipe—is difficult to do. This is somewhat like the story of Salem, who was walking down Main Street snapping his fingers. Friend asks, "Why are you snapping your fingers?" Salem: "Keeps the tigers away." Friend: "There are no tigers on Main Street." Salem: "Yeah, works doesn't it?" No drilling program calls for stuck pipe or fishing jobs even if they are common in an area with a particular drilling rig. The evil effects of drilled solids are real. Acknowledging that fact and preparing to properly handle them at the surface will result in much lower drilling costs.

Good drilled-solids removal procedures start at the drill bit. Cuttings should be removed before another drill bit cutter crushes rock that has already been removed from the formation. These cuttings should be transported to the surface with as little disintegration as possible. In addition to the cuttings produced by the drill bit, slivers or chunks of rock from the well-bore walls also enter the drilling fluid stream. Large drilled solids are easier to remove than small ones. After the cuttings have reached the surface, the correct equipment must be available to handle the appropriate solids loading, and the processing routing must be correct. Surprisingly, after all these years of using drilling fluids, the simple principles of arranging equipment are seldom practiced in the field. Some drilling rigs, particularly offshore ones, have a complex manifold of plumbing in the surface drilling fluid pits. The concept is that any one of the centrifugal pumps can pump from any compartment to any other compartment by adjusting valves. This concept is incorrect and detrimental to proper drilled-solids removal. Generally, arranging the complex routing for correct solids-removal processing is so unobvious that all of the drilling fluid is not processed by the equipment. Also, valves can leak in this system and go undetected for many wells. Better to follow the rule, One pump/one purpose. Add additional plumbing or pumps but do not use solids-removal equipment feed pumps for anything but their stated purpose. This book shows how the equipment works and how it should be plumbed.

While drilling wells, drilling fluid is processed at the surface to remove drilled solids and blend the necessary additives to allow drilling fluid to meet specifications. Drilling-fluid processing systems are described in this book from both a theoretical point of view and practical guidelines. It will be as useful for a student of drilling as for the person on the rig.

Drill bit cuttings and pieces of formation that have sloughed into the well bore (collectively called *drilled solids*) are brought to the surface by the drilling fluid. The fluid flows across a shale shaker before entering the mud pits. Most shale shakers impart a vibratory motion to a wire or plastic mesh screen. This motion allows the drilling fluid to pass through the screen and removes particles larger than the openings in the screen. Usually drilled solids must be maintained at some relatively low concentration. The reason for the need for this control is explained in the next section. The shale shaker is the initial and primary drilled-solids removal device and usually works in conjunction with other solids-removal equipment located downstream.

Solids-control equipment, also called solids-removal equipment or drilled-solids management equipment, is designed to remove drilled solids from a circulating drilling fluid. This equipment includes gumbo removers, scalper shakers, shale shakers, dryer shakers, desanders, desilters, mud cleaners, and centrifuges. These components, in various arrangements, are used to remove specific-size particles from drilling fluid. Knowledge of operating principles of auxiliary equipment, such as agitators, mud guns, mud hoppers, gas busters, degassers, and centrifugal pumps, is necessary to properly process drilling fluid in surface systems. All of this equipment is discussed in this book. However, the best equipment available is insufficient if it processes only a portion of the active drilling fluid coming from the well.

1.4 HISTORICAL PERSPECTIVE

Drilled-solids management has evolved over the years as drilling has become more challenging and environmental concerns have become paramount. Equipment changes and improvements have responded to the necessity to treat more and more expensive drilling fluids. In this context, probably the largest impact on the drilling industry has been the recognition that polymers can make much better drilling fluids than those used heretofore even though they are expensive. Polymer drilling fluids require lower drilled-solids concentration, so superior solids-removal systems were developed to meet those demands. A historical perspective on drilling-fluid management, specifications, solids control, and auxiliary processes, provides a clear and complete picture of the evolution of current equipment

Drilling fluid was used in the mid-1800s in cable tool (percussion) drilling to suspend the cuttings until they were bailed from the drilled hole. (For a discussion of cable tool drilling, see *History of Oil Well Drilling* by J. E. Brantley.) With the advent of rotary drilling in the water-well drilling industry, drilling fluid was well understood to cool the drill bit and to suspend drilled cuttings for removal from the well bore. Clays were being added to the drilling fluid by the 1890s. At the time that Spindletop, near Beaumont, Texas, was discovered in 1901, suspended solids (clay) in the drilling fluid were considered necessary to support the walls of the borehole. With the advent of rotary drilling at Spindletop, cuttings needed to be brought to the surface by the circulating fluid. Water was insufficient, so mud from mud puddles, spiked with some hay, was circulated downhole to bring rock cuttings to the surface. Most of the solids in the circulating system (predominantly clays) resulted from the so-called disaggregation of formations penetrated by the drill bit. The term *disaggregation* was used to describe what happened to the drilled clays. Clays would cause the circulating fluid to thicken, thus increasing the viscosity of the fluid. Some of the formation drilled would not disperse but remain as rock particles of various sizes commonly called *cuttings*.

If the formations penetrated failed to yield sufficient clay in the drilling process, clay was mined on the surface from a nearby source and added to the drilling fluid. These were native muds, created either by so-called mud making formations or, as mentioned, by adding specific materials from a surface source.

Drilling fluid was recirculated and water was added to maintain the best fluid density and viscosity for the specific drilling conditions. Cuttings, or pieces of formation—"small rocks"—that were not dispersed by water, required removal from the drilling fluid in order to continue the drilling operation. At the sole discretion of the driller or tool pusher, a system of pits and ditches were dug on site to separate cuttings from the drilling fluid by gravity settling. This system included a ditch from the well, or possibly a bell nipple, settling pits, and a suction pit from which the "clean" drilling fluid was picked up by the mud pump and recirculated.

Drilling fluid was circulated through these pits, and sometimes a partition was used to accelerate settling of the unwanted sand and cuttings. Frequently, two or three pits would be dug and interconnected with a ditch or channel. Drilling fluid would slowly flow through these earthen pits. Larger drilled solids would settle, and the cleaner fluid would overflow into the next pit. Some time later, steel pits were used with partitions between compartments. These partitions extended to within a foot or two of the bottom of the pit, thereby forcing all of the drilling fluid to move downward under the partition and up again to flow into a ditch to the suction pit. Much of the heavier material settled out, by gravity, in the bottom of the pit. With time, the pits filled with cuttings and the fluid became too thick to pump because of the finely ground cuttings entrained in the drilling fluid. To remedy this problem, the fluid was pumped out of the settling pits to reserve pits to provide room for dilution. Water was added to thin the drilling fluid and drilling continued.

In the late 1920s, drillers started looking at other industries to determine how similar problems were being solved. Ore dressing plants and coal tipples were using fixed bar screens placed on an incline; revolving drum screens; and vibrating screens. The latter two methods were selected for cleaning cuttings from drilling fluids.

The revolving drum, or barrel-type, screens were widely used with the early, low-height substructures. These units could be placed in a ditch or incorporated into the flow line from the well bore. The drilling fluid flowing into the machine turned a paddle wheel that rotated the drum screen through which the drilling fluid flowed. In those days, a coarse screen was 4 to 10 mesh and a fine screen was a 12 mesh. These units were quite popular because no electricity was required and the settling pits did not fill so quickly. Revolving drum units have just about disappeared.

The vibrating screen, or shaker, became the first line of defense in the solids-removal chain and for a long time was the only machine used. Early shakers were generally used in dry sizing applications and went through several modifications to arrive at a basic type and size for drilling. The first modification was a reduction in the size and weight of the unit for transport between locations. The name *shale shaker* was adopted to distinguish between shakers used in mining and shakers used in oil well drilling. This nomenclature was necessary, since both types of shakers were obtained from the same suppliers. The first publication about using a shale shaker in drilling operations, describing a "Vibrating Screen to Clean Mud," was in the *Oil Weekly* of October 17, 1930. The shaker screen was a 30 mesh, 4 by 5 feet, supported by four coil springs.

Prior to the new ISO (International Standards Organization) standard, screens were identified by mesh size. Mesh size was the number of openings per linear inch of screen. Most screens were woven with square openings, so the designation was logical. With ISO nomenclature, the English unit of inches could not be used. In addition to the change in units, a more compelling change was required because of the complexities of the new shaker screens. Screens ceased to be easily described with a simple measurement of openings in either direction.

Screens are now layered to form complex opening patterns and are described with the equivalent opening size in microns and an API (American Petroleum Institute) number (which was formerly the mesh designation). Currently, API 20 to API 50 are considered coarse screens. API 150 to API 325 are fine screens.

The early shale shakers had 4- by 5-feet hook strip screens mounted that were tensioned from the sides with tension bolts. The vibrators were usually mounted above the screens, causing the screens to move with an elliptical motion. The axis of the ellipse pointed toward the vibrator. Since the axis of the ellipse at the feed end pointed toward the discharge end and the axis of the discharge end pointed toward the feed end, these shakers were called *unbalanced elliptical motion shakers*. The screens required a downslope to move cuttings off the screen. Solids at the feed end, particularly with sticky clay discards, would frequently start rolling back uphill instead of falling off the shaker. Screen mesh was limited from about API 20 to API 30 (838 microns to 541 microns). These units were the predominant shakers in the industry until the late 1950s. Even though superseded by circular motion and linear motion shale shakers, the unbalanced elliptical motion shale shakers are still in demand and are still manufactured today.

Research laboratories of large oil companies and began to explore oil well drilling problems. The smaller cuttings, or drilled solids, left in the drilling fluid were discovered to be detrimental to the drilling process. Another ore dressing machine was introduced from the mining industry: the cone classifier. This machine, combined with the concept of a centrifugal separator, taken from the dairy industry, became the hydrocyclone desander, introduced to the industry around 1957. The basic principle of the separation of heavier (and coarser) materials from the drilling fluid lies in the centrifugal action of rotating the volume of solids-laden drilling fluid to the outer limit or periphery of the cone. Application of this centripetal acceleration causes heavier particles to move outward against the walls of the cone. These heavier particles exit the bottom of the cone and the cleaner drilling fluid exits from the top of the cone. The desander ranges in size from 6 to 12 inches in diameter and removes most solids larger than 30 to 60 microns. Desanders have been refined considerably through the use of more abrasion-resistant materials and more accurately defined body geometry. Hydrocyclones are now an integral part of most solids-separation systems today.

After the oilfield desander development, it became apparent that side wall sticking of the drill string on the borehole wall was generally associated with soft, thick filter cakes. Using the already existing desander design, a 4-inch hydrocyclone was introduced in 1962. Results were better than anticipated. Unexpected beneficial results were longer bit life, reduced pump repair costs, increased penetration rates, less lost circulation problems, and lower drilling-fluid costs. These smaller hydrocyclones became known as *desilters*, since they removed solids called silt down to 15 to 30 microns.

The Pioneer Centrifuge Company related a story about the first desilter it installed on a drilling rig (private communication from George Stonewall Ormsby). The bank of 4-inch desilters was mounted on the berm of the duck's nest (the duck's nest was an earthen pit used for storing excess drilling fluid and was usually an area of the reserve pit). The equipment was removing large quantities of drilled solids from an unweighted drilling fluid. After 2 days, however, the rig personnel called to have the equipment picked up because, they said, it was no longer working. When Pioneer arrived at the location, the equipment was completely buried in drilled solids, so that there was no way that more could be removed by the hydrocyclones.

During this period, major oil company research recognized the problems associated with ultra-fines (colloidal) in sizes less than 10 microns. These ultra-fines "tied up," or trapped, large amounts of liquid and created viscosity problems that could be solved only by water additions (dilution). As large cuttings are ground into smaller particles, surface area increases greatly, even though the total cuttings volume does not change. Centrifuges had been used in many industries for years and were adapted to drilling operations in the early 1950s. They were used first on weighted drilling fluids to remove and discard colloidal solids. The heavy slurry containing drilled solids and barite larger than about 10 microns is returned to the drilling fluid system.

In recent years centrifuges have been used in unweighted drilling fluids to remove drilled solids. In these fluids, the heavy slurry containing drilled solids down to around 7 to 10 microns is discarded and the light slurry with solids and chemicals (less than 7 to 10 microns) is returned to the drilling fluid. This application saves expensive liquid phases of drilling fluid. Dilution is minimized, thereby reducing drilling-fluid cost. However, these machines are quite expensive and require a great amount of care.

Unfortunately, many drillers did not believe that these benefits accrued to drilling-fluid systems that were properly arranged to take advantage of them. Mud tanks were, and still are, frequently plumbed incorrectly because of indifference concerning the detrimental effects of drilled solids. These benefits were not really generally accepted until the mid-1980s. Inspection of drilling-fluid processing systems on drilling rigs still reveals that proper plumbing is not well understood or is not a priority.

These hydrocyclones were usually loaded with solids because of the coarse screens on the shale shakers. Removing more of the intermediate-size particles led to the development of the circular motion shale shakers. These "tandem shakers," utilizing two screening surfaces, were introduced in the mid-1960s. Development was slow for these so-called fine screen–high speed shakers for two reasons: First, screen technology was not sufficiently developed for screen strength, so screen life was short. There was not sufficient mass in the screen wires to properly secure the screens without their tearing. Second, the screen basket required greater development expertise than had been required for earlier modifications in drilling-fluid handling equipment.

The tandem shakers had a top screen with larger openings for removal of larger particles and a bottom screen with smaller openings (finer mesh screen) for removal of the smaller particles. Various methods of screen openings were developed, including oblong, or rectangular, openings. These screens removed fine particles and had a high fluid capacity. They could be made of larger wires, so they had greater strength. Layered screens (a fine mesh screen for good solids removal over a coarse mesh screen for strength) were developed. These layered screens were easier to build and had adequate strength for proper tensioning for increased screen life. This development made it possible for the shale shaker to remove particles greater in size than API $80 \times$ API 80 (177 microns).

In the 1970s the mud cleaner was developed. During this period, no shale shaker could handle the full rig flow on an API 200 screen. Desanders and desilters were normally used after the shale shaker; however, they discarded large quantities of barite when used on a weighted drilling fluid—this meant drilled solids larger than an API 80 and the upper limit of the barite size. API specifications currently allow three weight percent of barite larger than 74 microns, which is an API 200 screen. To solve this problem, the underflow from desanders and desilters was presented to a pretensioned API 200 screen on a shaker. Much of the liquid from the underflow of the hydrocyclones and most of the barite passed through an API 200 screen. This was also the first successful oilfield application of a pretensioned fine screen bonded to a rigid frame. Many mud cleaners had screen cleaners, or sliders, beneath the screen to prevent screen blinding. Mud cleaners have also been used with API 250 screens in unweighted drilling fluids that have expensive liquid phases.

A more recent development, introduced in the 1980s, has been the linear motion shale shaker. Linear motion is the best conveying motion to move solids off the screen. Solids can be conveyed uphill out of a pool of liquid as it flows onto the screen from the flow line. Screens with smaller openings, such as API 200 (74 microns), can be used on linear motion shakers, but they could not be used on any of the earlier types of shakers. Developments in screen technology have made it possible for pretensioned screens to be layered and, in some cases, have three-dimensional surfaces.

The latest entry into the shale shaker challenge is a balanced elliptical motion shaker. The motion is similar to an unbalanced elliptical motion shaker except that all axes of vibration are pointed toward the discharge end. The movement of the screen is similar to a linear motion shaker except that the motion makes an ellipse. Solids are transported from a pool of liquid at the feed end of the shaker screen just as they are on a linear motion screen.

When the linear motion shale shakers were introduced, several were frequently arranged in parallel to receive drilling fluid from scalping shakers. Since API 200 screens could be used on these primary shale shakers, mud cleaners were widely considered superfluous, and mud-cleaner use diminished significantly. However, installation of mud cleaners, even with API 150 screens downstream from these linear motion shale shakers, revealed that some removable drilled solids were still in the drilling fluid. In real situations, sufficient drilling fluid bypasses linear motion shale shakers to make mud-cleaner installation economical. In retrospect, since the lower apex discharge of desilters frequently plugs downstream from linear motion shale shakers, this provides proof that all of the large solids are not removed by linear motion shakers.

Emphasis on minimization of liquid discharges for environmental considerations has created techniques to remove liquid from the drilled-solids discard. Since the decanting centrifuge is a very low shear-rate device for the drilling fluid (even though the drilling fluid is rotating at over 15,000 rpm), it can be used to concentrate flocculated and coalesced solids. The light slurry, which is almost a clear stream of water, is returned to the drilling fluid. This has become an important part of the "closed mud" system. Actually, the intent is to eliminate or reduce the quantity of liquid discarded.

A recent innovation for environmental purposes and minimization of liquid discharge is the dryer. The discharge from linear motion shale shakers, desanders, and desilters flows onto another linear motion shaker that has even finer screens than the main shale shakers (as fine as API 450, or 32 microns) and usually has a larger screening surface. The dryer has a closed sump under the screen with a pump installed. Any liquid in the sump is returned to the active system through a centrifuge.

These systems, or combinations of the various items discussed above, meet most environmental requirements and conserve expensive liquid phases. The desirable effect is to reduce the liquid content of the discarded drilled solids so that they can be removed from a location with a dump truck instead of a vacuum truck.

An innovation introduced in the Gulf of Mexico in the 1990s was the gumbo conveyer. Before this was introduced, some drilling rigs would mount stainless steel rods about 2 to 3 inches apart on a downward slope. Gumbo, or large, pliable sticky cuttings, would slide down these rods and be removed from the system. Drilling fluid would easily flow through the openings between the steel rods. At least two versions are currently marketed. One is a chain and the other is a continuous permeable belt. These special conveyors drag gumbo out of the drilling fluid before the drilling fluid encounters a shale shaker. This operation reduces the severe screen loading problems caused by gumbo.

Innovations in drilled-solids removal equipment will probably continue. However, novel, spectacular equipment is useless if it is installed improperly and subjected to poor maintenance and operating procedures. This book concentrates on providing guidelines for practical operations of the surface drilling fluid system.

1.5 COMMENTS

One word of caution is appropriate here. Neophytes in drilling have a tendency to try to minimize the cost of each category of expense on the basis of the misconception that this will minimize the cost of the well. Minimizing individual items will only minimize a total if there is no dependence of variables on other costs. For example, increasing mud weight with drilled solids is cheaper than using barite. The cost savings from not purchasing barite is easy to calculate. The cost of all of the problems that ensue is much more difficult to predict. This is the insidious nature of drilled solids.

Decreasing individual costs to decrease the total cost is somewhat analogous to the accountant with appendicitis who decides to save money by renting a room at a cheap motel and calling a doctor friend rather than going to a hospital for an appendectomy. Room and board might be cheaper, but the net cost of improper care will probably make the decision very costly. Extra costs can be incurred because of inadvisable decisions to cut costs in easily monitored expenses while drilling wells. When line items are independent of each other, minimization of each line item will result in the lowest possible cost. When line items are interconnected, minimization of each line item may be very expensive. Drilled-solids concentrations and trouble costs (or costs of unscheduled events) are very closely intertwined.

One common mistake, usually made with the misconception that the well will be less expensive, is to allow the initial increase in mud weight to occur with drilled solids. Clearly, less money will be spent on the drilling fluid if no weighting agents are added to it. These savings are easily documented. Less revealing, however, will be the additional expenses because of the excessive drilled solids in the drilling fluid. Many of these problems will increase the well cost and have been discussed in the preceding sections.

Another common mistake, usually made while drilling with weighted drilling fluid, is to relate the cost of the weighting agent discarded with the drilled-solids discard. The cost of discarded weighted agents (barite or hematite) can be relatively small compared with the tragedies associated with drilled solids. This is particularly true in the expensive offshore environment. Even in cheaper land drilling, a comparison normally tilts in favor of discarding weighting agents.

Solids-control equipment, properly used, with the correct drilling-fluid selection, will usually result in lower drilling costs. Decisions made for various wells are very dependent on the well depth and drilling-fluid density. Shallow, large-diameter, low-mud-weight wells can tolerate more drilled solids than can deeper, more complicated wells. Each well must be evaluated individually with careful consideration of the risk of problems associated with drilled solids. As a general practice, however, since rigs drill a variety of wells during the course of a year, investing in a proper mud tank arrangement with adequate equipment is wise and frugal.

Yet another common mistake is to believe that different types of drilling-fluid systems will require different mud tank arrangements for solids removal. This is FALSE. Following the guidelines presented in this book will result in a system that will properly remove drilled solids from water-based, oil-based, synthetic-based drilling fluids. The water-based drilling fluids could be dispersed or nondispersed, with or without polymers, or of low or high density or mud weight.

1.6 WASTE MANAGEMENT

Polymer drilling fluids, synthetic oil–based drilling fluids, and other fluids with expensive additives provide a great incentive to use good solids-control procedures. However, minimizing the waste products from these expensive systems will also have a great impact on drilling costs.

Most drilling operations have a targeted drilled-solids concentration. Failure to remove drilled solids with solids-control equipment leads to solids control with dilution. This creates excessive quantities of fluid that must be handled as a waste product. If this fluid must be hauled from the location, the excess fluid becomes a large additional expense. Even if the fluid can be handled at the location, larger quantities of fluid frequently increase cost. This is discussed in depth in Chapter 15 on Dilution.

Smaller quantities of waste products can significantly decrease the cost of a well. Decreasing the quantity of drilling fluid discarded with the drilled solids will decrease the cost of rig-site cleanup. Dilution techniques for controlling drilled-solids concentrations greatly increase the quantity of waste products generated at a rig. This results in an additional expense that adds to the total cost of drilling.

CHAPTER 2

DRILLING FLUIDS

Fred Growcock
M-I SWACO

Tim Harvey
Oiltools, Inc.

2.1 DRILLING FLUID SYSTEMS

2.1.1 Functions of Drilling Fluids

A drilling fluid, or mud, is any fluid that is used in a drilling operation in which that fluid is circulated or pumped from the surface, down the drill string, through the bit, and back to the surface via the annulus. Drilling fluids satisfy many needs in their capacity to do the following [M-I LLC]:

- Suspend cuttings (drilled solids), remove them from the bottom of the hole and the well bore, and release them at the surface
- Control formation pressure and maintain well-bore stability
- Seal permeable formations
- Cool, lubricate, and support the drilling assembly
- Transmit hydraulic energy to tools and bit
- Minimize reservoir damage
- Permit adequate formation evaluation
- Control corrosion

- Facilitate cementing and completion
- Minimize impact on the environment
- Inhibit gas hydrate formation

The most critical function that a drilling fluid performs is to minimize the concentration of cuttings around the drill bit and throughout the well bore. Of course, in so doing, the fluid itself assumes this cuttings burden, and if the cuttings are not removed from the fluid, it very quickly loses its ability to clean the hole and creates thick filter cakes. To enable on-site recycling and reuse of the drilling fluid, cuttings must be continually and efficiently removed.

2.1.2 Types of Drilling Fluids

Drilling fluids are classified according to the type of base fluid and other primary ingredients:

- Gaseous: Air, nitrogen
- Aqueous: Gasified—foam, energized (including aphrons)
    Clay, polymer, emulsion
- Nonaqueous: Oil or synthetic—all oil, invert emulsion

True foams contain at least 70% gas (usually N_2, CO_2, or air) at the surface of the hole, while energized fluids, including aphrons, contain lesser amounts of gas. Aphrons are specially stabilized bubbles that function as a bridging or lost circulation material (LCM) to reduce mud losses to permeable and microfractured formations. Aqueous drilling fluids are generally dubbed water-based muds (WBMs), while non-aqueous drilling fluids (NAFs) are often referred to as oil-based muds (OBMs) or synthetic-based muds (SBMs). OBMs are based on NAFs that are distilled from crude oil; they include diesel, mineral oils, and refined linear paraffins (LPs). SBMs, which are also known as pseudo–oil-based muds, are based on chemical reaction products of common feedstock materials like ethylene; they include olefins, esters, and synthetic LPs.

Detailed classification schemes for liquid drilling fluids are employed that describe the composition of the fluids more precisely. One such classification scheme is shown in Figures 2.1 and 2.2. An even more precise classification scheme is described in Table 2.1, which includes

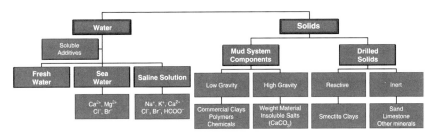

Figure 2.1. Types of Water-Based Muds.

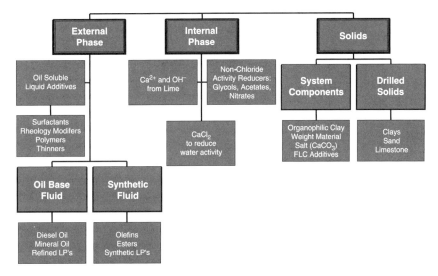

Figure 2.2. Types of Invert-Emulsion Muds.
FLC = ferroelectric liquid crystals.

the mud systems most commonly used today, along with their principal components and general characteristics [adapted from Darley & Gray].

2.1.3 Drilling Fluid Selection

Drilling-fluid costs can constitute a significant fraction of the overall costs of drilling a well. Often the cost is quoted per unit length drilled, which takes into account any problems encountered (and avoided), such as stuck pipe. In many cases, the cost ascribed to the fluid also includes costs associated with solids control/management and waste disposal.

Table 2.1
Classification of Drilling Fluid Systems

Mud Type	Principal Components	General Characteristics
Aqueous		
Simple freshwater	Freshwater	Low cost, onshore applications; fast drilling in stable formations; need space for solids settling, flocculants may be used
Simple seawater	Seawater	Low cost, offshore applications
Spud mud	Bentonite, water	Low cost, surface hole
Saltwater	Seawater, brine or saturated saltwater; saltwater clay, starch, cellulosic polymer	Moderate cost, drilling salt and workovers
Lime or gyp	Fresh or brackish water; bentonite, lime, or gypsum, lignosulfonate	Moderate cost, shale drilling; simple maintenance, high temp. tolerance to salt, anhydrite, cement, drilled solids
Lignite or lignosulfonate (chrome or chrome free)	Fresh or brackish water; bentonite, caustic, lignite or lignosulfonate	Moderate cost, shale drilling; simple maintenance, high temp. tolerance to salt, anhydrite, cement
Potassium	Potassium chloride; acrylic, bio or cellulosic polymer, some bentonite	Moderate cost, hole stability; low tolerance to drilled solids, high pH
Low solids ("nondispersed" when weighted up)	Fresh to high saltwater; polymer, some bentonite	High cost, hole stability; low tolerance to drilled solids, cement and divalent salts

Drilling Fluids

Nonaqueous		
Oil	Weathered (oxidized) crude oil; asphaltic crude, soap, water 2–5%	Moderate cost, low-press well completions and workovers, low-press shallow reservoirs; water used to increase density and cuttings-carrying capacity; strong environmental restrictions may apply
Asphaltic	Diesel oil; asphalt, emulsifiers, water 2–5%	Moderate cost, any applications to 600°F; strong environmental restrictions may apply
Invert emulsion	Diesel, mineral, or low-toxicity mineral oil; emulsifiers, organophilic clay, modified resins, and soaps, 5–40% brine	High cost, any applications to at least 450°F; low maintenance, environmental restrictions
Synthetic	Synthetic hydrocarbons or esters; other products same as invert emulsion	Highest cost, any applications to at least 450°F; low maintenance

Thus, it is just as important to minimize costs associated with these twoaspects as it is to ensure that the drilling fluid fulfills its primary functions [Young & Robinson]. Muds that require special attention and equipment to control the levels and types of solids frequently incur higher costs. Likewise, muds that generate waste fluid and cuttings, which must be hauled off (and perhaps treated) rather than discharged directly into the environment, generally incur higher costs.

Until recently, waste WBMs did not require any treatment and could be discharged directly into the environment. However, a number of components in WBMs are becoming increasingly restricted or prohibited. Chrome-containing materials, such as chrome lignosulfonates, are prohibited in many areas by governmental regulations. Tight restrictions are imposed in many areas on chloride, nitrate, and potassium salts, or, more generally, on the total electrical conductivity of the mud. In the North Sea, use of polyacrylamide polymers, such as partially hydrolyzed polyacrylamide (PHPA), is also severely restricted. OBMs tend to be restricted even more than WBMs, especially offshore, and in many places they can be used only if a zero discharge strategy (sometimes called a *closed loop system*) is adopted [Lal & Thurber].

On the other hand, SBMs often can be discharged directly into the sea if they meet certain toxicity/biodegradability criteria and, in the United States, do not create a sheen; as a result, though SBMs generally incur higher initial costs than OBMs, disposal costs for SBMs tend to be considerably less, which can make them more economical to run.

2.1.4 Separation of Drilled Solids from Drilling Fluids

The types and quantities of solids (insoluble components) present in drilling mud systems play major roles in the fluid's density, viscosity, filter-cake quality/filtration control, and other chemical and mechanical properties. The type of solid and its concentration influences mud and well costs, including factors such as drilling rate, hydraulics, dilution rate, torque and drag, surge and swab pressures, differential sticking, lost circulation, hole stability, and balling of the bit and the bottom-hole assembly. These, in turn, influence the service life of bits, pumps, and other mechanical equipment. Insoluble polymers, clays, and weighting materials are added to drilling mud to achieve various desirable properties.

Drilled solids, consisting of rock and low-yielding clays, are incorporated into the mud continuously while drilling. To a limited extent, they can be tolerated and may even be beneficial. Dispersion of clay-bearing drilled solids creates highly charged colloidal particles (<2 μm) that generate significant viscosity, particularly at low shear rates, which aids in suspension of all solids. If the clays are sodium montmorillonite, the solids will also form thin filter cakes and control filtration (loss of liquid phase) into the drilled formation. Above a concentration of a few weight percent, dispersed drilled solids can generate excessive low-shear-rate and high-shear-rate viscosities, greatly reduced drilling rates, and excessively thick filter cakes. As shown in Figures 2.3 and 2.4, with increasing mud density (increasing concentration of weighting material), the high-shear-rate viscosity (reflected by the plastic viscosity [PV]) rises continuously even as the concentration of drilled solids (low-gravity solids [LGSs]) is reduced. The methylene blue test (MBT) is a measure of the surface activity of the solids in the drilling fluid and serves as a relative measure of the amount of active clays in the system. It does not correspond directly to the concentration of drilled solids, since composition of drilled solids is quite variable. However, it is clear that, in most cases, drilled solids have a much greater effect than barite on viscosity and that the amount of active clays in the drilled

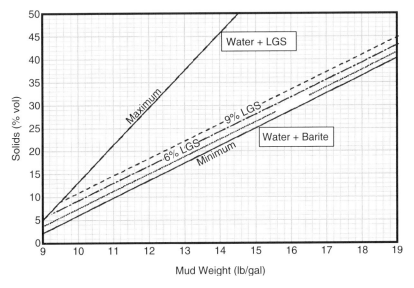

Figure 2.3. Effect of Solids on Mud Weight of Water-Based Muds. (Courtesy of M-I SWACO.)

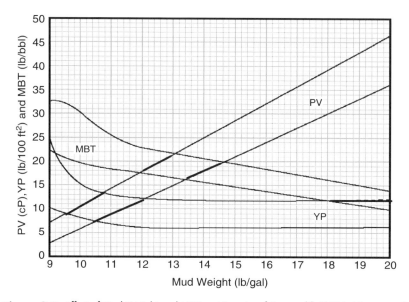

Figure 2.4. Effect of Mud Weight and MBT on Viscosity of Acceptable WBM. (Courtesy of M-I SWACO.)

solids is one of the most important factors. Thus, as mud density is increased, MBT must be reduced so that PV does not reach such a high level that it exceeds pump capacity or causes well-bore stability problems.

As shown in Figure 2.4, increasing the mud density from 10 lb/gal to 18 lb/gal requires that the MBT be reduced by half [M-I LLC]. Different mud densities require different strategies to maintain the concentration of drilled solids within an acceptable range. Whereas low mud densities may require only mud dilution in combination with a simple mechanical separator, high mud densities may require a more complex strategy: (a) chemical treatment to limit dispersion of the drilled solids (e.g., use of a shale inhibitor or deflocculant like lignosulfonate), (b) more frequent dilution of the drilling fluid with base fluid, and (c) more complex solids-removal equipment, such as mud cleaners and centrifuges [Svarovsky]. In either case, solids removal is one of the most important aspects of mud system control, since it has a direct bearing on drilling efficiency and represents an opportunity to reduce overall drilling costs. A diagram of a typical mud circulating system, including various solids-control devices, is shown in Figure 2.5 [M-I LLC].

While some dilution with fresh treated mud is necessary and even desirable, sole reliance on dilution to control buildup of drilled solids in

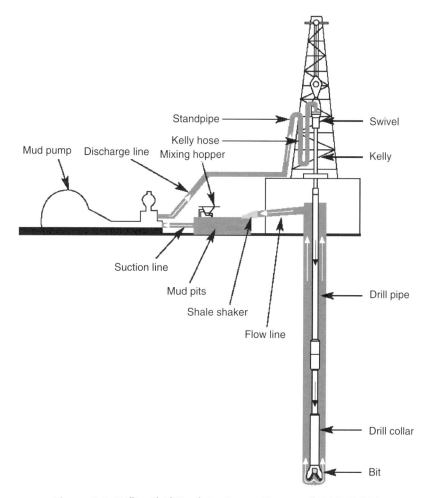

Figure 2.5. Drilling Fluid Circulating System. (Courtesy of M-I SWACO.)

the mud is very costly. The dilution volume required to compensate for contamination of the mud by 1 bbl of drilled solids is given by the following equation:

$$V_{dilution} \text{ (bbl drilling fluid/bbl drilled solids)} = (100 - V_{solids})/V_{solids}$$

where V_{solids} is the volume of drilled solids expressed in volume percentage. As discussed earlier, drilled solids become less tolerable with increasing mud density. For drilling-fluid densities less than 12 lb/gal, $V_{solids} < 5\%$ is desirable, whereas for a density of 18 lb/gal, $V_{solids} < 2$ or

3% is best. When $V_{solids} = 5\%$, the equation above gives $V_{dilution} = 19$ bbl drilling fluid/bbl drilled solids. The cost of this extra drilling fluid (neglecting downhole losses) is the sum of the cost of the drilling fluid itself plus the cost to dispose of it. This dilution cost is generally so high that even a considerable investment in solids-control equipment is more economical.

Solids removal on the rig is accomplished by one or more of the following techniques:

- Screening: Shale shakers, gumbo removal devices
- Hydrocycloning: Desanders, desilters
- Centrifugation: Scalping and decanting centrifuges
- Gravitational settling: Sumps, dewatering units

Often these are accomplished using separate devices, but sometimes these processes are combined, as in the case of the mud cleaner, which is a bank of hydrocyclones mounted over a vibrating screen. Another important hybrid device is the cuttings dryer (also called a rotating shaker), which is a centrifuge fitted with a cone-shaped shaker; this apparatus is used to separate cuttings from NAF-based muds and strip most of the mud from the cuttings' surfaces before disposal. Additional devices can help to enhance solids-removal efficiency. For example, a vacuum or atmospheric degasser is sometimes installed (before any centrifugal pumps, typically between the shakers and desanders) to remove entrained air that can cause pump cavitation and reduction in mud density. Refer to Chapter 5 on Tank Arrangements for more details.

With the advent of closed loop systems, dewatering of WBMs has received strong impetus, and it has been found useful to add a dewatering unit downstream of a conventional solids-control system [Amoco]. Dewatering units usually employ a flocculation tank—with a polymer to flocculate all solids—and settling tanks to generate solids-free liquid that is returned to the active system. Dewatering units reduce waste volume and disposal costs substantially and are most economical when used to process large volumes of expensive drilling fluid.

Solids-control equipment used on a rig is designed to remove *drilled* solids—not all solids—from a drilling fluid. As such, the equipment has to be refined enough to leave desired solids (such as weighting material) behind while taking out drilled solids ranging in size from several millimeters to just a few microns. Although such perfect separation of desired from undesired solids is not possible, the advantages offered by the solids-control equipment far outweigh their limitations. Each device is designed to remove a sufficient quantity and size range of solids. The key to efficient solids control is to use the right combination of equipment for a particular situation, arrange the equipment properly, and ensure that it operates correctly. This, in turn, requires accurate characterization of the drilled solids, along with optimal engineering and maintenance of the drilling fluid.

2.2 CHARACTERIZATION OF SOLIDS IN DRILLING FLUIDS

Selecting, arranging, and operating solids-removal equipment to optimize the drilling-fluid cleaning process require accurate information about the intrinsic nature of the cuttings (drilled solids) and solid additives.

2.2.1 Nature of Drilled Solids and Solid Additives

Particle size, density, shape, and concentration affect virtually every piece of equipment used to separate drilled solids and/or weighting material from the drilling fluid. In the theoretically perfect well, drilled solids reach the surface with the same shape and size that they had when they were created at the drill bit. In reality, cuttings are degraded by physicochemical interaction with the fluid and mechanical interaction with other cuttings, the drill string, and the well bore.

Cuttings hydrate, become soft, and disperse in aqueous fluids and even in invert-emulsion NAFs with excessively low salinity. On the other hand, cuttings may become more brittle than the formation in high-water-phase-salinity NAFs and can be mechanically degraded by the action of the rotating drill string inside the well bore, particularly in deviated, slim-hole, and extended-reach wells. Cuttings are also degraded by mechanical action. Abrasion of the cuttings by other cuttings, by the steel tubulars, and by the walls of the well bore can lead to rapid comminution of the particles. In summary, cuttings recovered at the surface are generally smaller and frequently more rounded than at their moment of creation, depending on the nature of the cuttings themselves and the drilling fluid. Accordingly, the particle size distribution (PSD) seen at the flowline can range from near-original cutting size to submicron-sized particles.

The surface properties of the drilled solids and weighting material, such as stickiness and amount of adsorbed mud, also can play major roles in the efficiency of a rig solids-separation device. Large, dense particles are the easiest to separate using shakers, hydrocyclones, and centrifuges, and the differences in size and density among different types of particles must be well known to design the appropriate piece(s) of equipment for the separation process. Indeed, the optimum efficiency window for each device depends on all four of these parameters: concentration, size, shape, and density. Furthermore, since removal of some—but not all—particles is desirable, characterization of each and every type of particle with respect to those variables is critical. LCM serves as a good example of this. Usually economics dictates removal of large LCM along with cuttings using scalping shakers. Sometimes, however, large concentrations of LCM are required—as much as 50 to 100 ppb—in the circulating system. In such cases, a separate scalping shaker may be installed ahead of the regular battery of shakers to remove the LCM and recycle it back into the mud system [Ali *et al*].

2.2.2 Physical Properties of Solids in Drilling Fluids

Particle sizes in drilling fluids are classified as shown in Table 2.2 [M-I LLC]. PSD is measured using various techniques. For particles > 45 µm diameter, wet sieve analysis is simple, accurate, and fast [API RP 13C]. Alternative methods include the American Petroleum Institute (API) sand test, which provides a measure of the *total* amount of particles > 74 µm diameter [API RP 13B1]; microscopic image analysis, whose size limit at the low end depends on the type of microscope employed; sedimentation, for particles 0.5 to 44 µm diameter [Darley & Gray]; Coulter counter, for particles 0.4 to 1200 µm diameter [API RP 13C]; and laser granulometry (also called laser light scattering, diffraction analysis, and Fraunhoffer diffraction), for particles 1 to 700 µm diameter [API RP 13C].

Table 2.2
Classification of Particles in Drilling Fluids

Category	Size (µm)	Types of Particles
Colloidal	< 2	Bentonite, clays, ultra-fine drilled solids
Silt	2–74	Barite, silt, fine drilled solids
Sand	74–2,000	Sand, drilled solids
Gravel	> 2000	Drilled solids, gravel, cobble

With the Coulter counter, the solids are suspended in a weak electrolyte solution and drawn through a small aperture separating two electrodes, between which a voltage is applied and current flows. As each particle passes through the aperture, it displaces an equal volume of conducting fluid and the impedance between the electrodes increases in a manner that can be correlated with the particle size.

Laser granulometry is rapidly gaining popularity as the method of choice for PSD measurements. In laser granulometry, the solids are dispersed in a transparent liquid and suspended by circulation, if necessary, the slurry may be viscosified with a material like xanthan gum polymer. A beam of light is shone on a sample of the suspended solids, and the intensity versus the angle of the scattered light is analyzed to determine the PSD. Freshwater is used to disperse inert materials like barite. The drilling-fluid base fluid (saltwater, etc.) is used for all other solids (e.g., drilled solids). The sample is diluted to make it sufficiently transparent to obtain accurate readings. The instrument fits the particles to a spherical model to generate a histogram of number of particles versus particle size. For particles that do not fit a spherical model very well, such as plates or rods, calibration with a known PSD of those particles is preferable. Laser granulometry results also depend on the step size chosen—for instance, for step sizes of 5 μm versus 10 μm, using 5 μm will generate two peaks that are each about half the size of a peak generated using 10 μm. If the step size chosen is too large, the reported PSD may miss some of the fine structure of the spectrum; on the other hand, a step size that is too small will generate excessive oscillations and the spectrum will appear to be very "noisy."

Figure 2.6 shows typical laser granulometry PSD curves for feed, liquid effluent (overflow), and solids discharge (underflow) for a field mud processed by a centrifuge. The efficiency of the device may be calculated from these data. PSD curves for each piece of equipment allow a more detailed understanding of what the device is doing and whether the equipment is optimally configured for the fluid being processed. There are calls within the drilling industry now to make laser granulometers standard equipment on critical wells, particularly high temperature/high pressure and extended-reach wells, where the equivalent circulating density (ECD) is likely to exceed the fracture gradient.

Adsorbed mud, as well as swelling and/or dispersion of the cuttings resulting from interaction with the mud, can affect the PSD of cuttings. Comminution (degradation) of drilled solids has a strong impact on rheology and the total amount of mud adsorbed on the solids, inasmuch

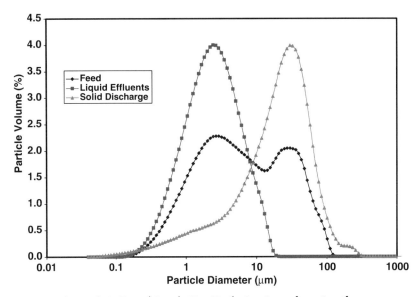

Figure 2.6. Typical Particle Size Distribution Curves for a Centrifuge.

as the forces between the particles and the amount of mud adsorbed on them is proportional to their surface area. Drilled solids generally become comminuted while in the well bore and mud pits, as well as during passage through solids-control devices, through abrasion and chemical interaction with the base fluid. Surface-area increase due to comminution is proportional to the decrease in particle diameter. For example, breaking up a 100-μm-diameter particle into 5-μm particles will increase the total surface area by a factor of 20. Consequently, the amount of mud adsorbed on the solids in this case will increase roughly by a factor of 20 as a direct result of comminution. Low-shear-rate viscosity will also increase significantly with this increase in total surface area, though the relationship is not strictly linear.

Average particle density, also termed "true" or "intrinsic" density, has units of weight/volume. Specific gravity (SG) is the ratio of the density of the material in question to the density of water and is, of course, unitless. Since the density of water is close to 1 g/cm^3 over a wide range of temperature and pressure, the values reported for average particle density and SG are essentially the same. Average particle density should not be confused with bulk density (as often given in the Material Safety Data Sheet), which is a measure of the density of the packaged material. The LeChatelier flask method is the standard for determination of the average particle density of barite and hematite [API 13A]. In this method, one measures the incremental change in volume accompanying the addition of 100 g of the weighting material to a precisely measured volume of kerosene. A more convenient, but less accurate, method for determining density of weighting materials is the air pycnometer [API RP 13I]. Another convenient method, which is rapidly gaining in popularity, is the stereopycnometer [API RP 13I]. In contrast to the air pycnometer, the stereopycnometer is as accurate as the LeChatelier flask method, and it can be used to measure density of any kind of particulate, including drilled cuttings. The stereopycnometer employs Archimedes' principle of fluid displacement (helium, in this case) and the technique of gas expansion [API RP 13I].

Particle shape, partly described by the so-called aspect ratio, is not fully quantifiable. Neither is it possible to incorporate the broad spectrum of particle shapes in drilling fluids into particle-separation mathematical models. At this time, an old simple classification scheme is still used: granule, flake, fiber [Wright].

Concentration of particles in a mud is generally measured using aretort (an automatic portable still). The volume percentage of low-gravity solids (% LGS)—clays, sand, and salt—and the volume percentage of high-gravity solids (% HGS)—weighting material—are calculated from the measured volumes of the distilled fluids and the density of the mud. The calculated % LGS serves as an indicator of the effectiveness of the solids-control equipment on the rig. Occasionally both the overflow and underflow solids from each piece of equipment are reported. Unfortunately, inaccuracies inherent in the retort, combined with the common practice of using an average density for the LGS and an average density for the HGS, can generate considerable uncertainties in % LGS. This is particularly true for low-density fluids, where a slight error in reading the retort will generate misleading—usually high—values of % LGS. However, if the calculated % LGS is below the target limit (typically 5%), and dilution is not considered excessive, the solids-control equipment is considered to be efficient. (Calculation of solids-removal efficiency is presented in Chapter 15 on Dilution.) It should be noted that % LGS includes any clays that are purposefully added to the drilling fluid (for viscosity and filtration control). If a fluid contains 20 lb/bbl bentonite, it already contains 2.2% LGS before it acquires any drilled cuttings; in such fluids, the target limit of % LGS may be somewhat higher than 5%.

Concentration of particles affects mud properties, particularly rheology, which in turn affect the amount of residual mud on drilled solids. For noninteracting particles, the Einstein equation describes the effect of particles on the effective viscosity, μ_e, fairly well:

$$\mu_e = \mu + 2.5\phi$$

where μ is the viscosity of the liquid medium and ϕ is the volume fraction of the inert solids. This effect is independent of particle size, as long as the particles are suspended in the medium. The Einstein equation represents the effect of "inert" particles like barite fairly well, at least until their concentration becomes so great that the particles begin interacting with each other. Most particles in drilling fluids, however, have strong surface charges and interact strongly with each other at any concentration. Since all particles are enveloped by drilling fluid, attractive forces among strongly interacting particles (e.g., clays, drilled solids) generally lead to higher internal friction, hence a higher viscosity. Repulsive forces, such as are generated in muds containing high levels of lignosulfonate or other anionic polymers, will tend to exhibit lower viscosity. Because of these attractive/repulsive forces, strongly interacting particles generate an internal "structure" in a fluid, which manifests itself most clearly at low fluid velocities. Thus, in most drilling fluids, significant deviations from the Einstein equation are the norm, as is discussed in more detail in the next section.

The viscosity of a drilling fluid must be maintained within certain limits to optimize the efficiency of a drilling operation: low-shear-rate viscosity needs to be high enough to transport cuttings out of the hole efficiently and minimize barite sag, while high-shear-rate viscosity needs to be as low as possible to maintain pumpability, remove cuttings from beneath the bit, and minimize ECD of the mud. In an analogous manner, for efficient operation of solids-control devices, the concentration of drilled solids needs to be maintained within a specified range [Amoco]. The upper end (e.g., 5%) is particularly important, but the lower end (typically higher than 0%) is also important for most devices.

Stickiness of cuttings and its effect on the performance of solids-control devices are only beginning to be investigated. Various properties of the mud, along with lithology of the formation being drilled, are known to affect stickiness of particles, especially cuttings [Bradford et al.]. Generally, separation efficiency of any solids-control device decreases with increasing stickiness of the cuttings. Rheology, shale inhibition potential, and lubricity of the mud all can affect the stickiness of particles, which in turn affects performance of solids-control equipment, especially shale shakers. To handle gumbo (very sticky cuttings consisting primarily of young water-sensitive shale), operators will install special gumbo removal devices ahead of the shakers. To aid in conveyance of gumbo, the shaker screens are kept wet with a fine mist and angled horizontally or downward toward the discharge end. Gumbo cannot be transported effectively on a linear motion or balanced elliptical motion screen that is sloped upward.

2.3 PROPERTIES OF DRILLING FLUIDS

Just as the nature of drilling-fluid solids affects the efficiency of solids-control equipment, the nature of the solids also plays an integral role in the properties of drilling fluids, which in turn affect the properties of the solids and the performance of the equipment. This intricate and very complex dynamic relationship among the solids, drilling fluid, and solids-control equipment is represented in Figure 2.7. Any change made to one of these affects the other two, and those in turn affect all three,

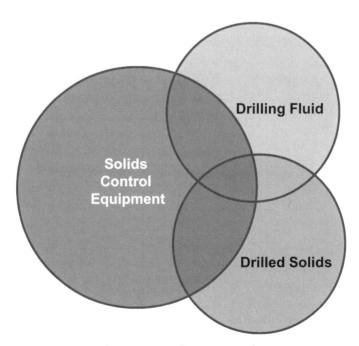

Figure 2.7. Mud Processing Circle.

and so on. To optimize a drilling operation, it is important to understand how the solids affect bulk mud properties, particularly rheology, hole cleaning, filtration, drilling rate (rate of penetration [ROP]), along with surface properties such as shale inhibition potential, lubricity, and wetting characteristics.

2.3.1 Rheology

Rheology is the study of the deformation and flow of matter. Viscosity is a measure of the resistance offered by that matter to a deforming force. Shear dominates most of the viscosity-related aspects of drilling operations. Because of that, shear viscosity (or simply, "viscosity") of drilling fluids is the property that is most commonly monitored and controlled. Retention of drilling fluid on cuttings is thought to be primarily a function of the viscosity of the mud and its wetting characteristics. Drilling fluids with elevated viscosity at *high shear rates* tend to exhibit greater retention of mud on cuttings and reduce the efficiency of high-shear devices like shale shakers [Lundie]. Conversely, elevated viscosity at *low shear rates* reduces the efficiency of low-shear devices like centrifuges, inasmuch as particle settling velocity and separation efficiency are inversely proportional to viscosity. Water or thinners will reduce both of these effects. Also, during procedures that generate large quantities of drilled solids (e.g., reaming), it is important to increase circulation rate and/or reduce drilling rate.

Other rheological properties can also affect how much drilling fluid is retained on cuttings and the interaction of cuttings with each other. Some drilling fluids can exhibit elasticity as well as viscosity. These *viscoelastic* fluids possess some solid-like qualities (elasticity), particularly at low shear rates, along with the usual liquid-like qualities (viscosity). Shear-thinning drilling fluids, such as xanthan gum–based fluids, tend to be viscoelastic and can lower efficiency of low-shear-rate devices like static separation tanks and centrifuges.

Viscoelasticity as discussed above is based on flow in shear. There is another kind of viscoelasticity, however, that is just now receiving some attention: *extensional* viscoelasticity. As the term implies, this property pertains to extensional or elongational flow and has been known to be important in industries in which processing involves squeezing a fluid through an orifice. This property may be important at high fluid flow rates, including flow through the drill bit and possibly in high-throughput solids-control devices. High-molecular-weight (HMW) surface-active polymers, such as PHPA and 2-acrylamido-2-methyl-propane sulfonic acid (AMPS)–acrylamide copolymers, which are used as shale encapsulators, produce high extensional viscosity. Muds with extensional viscosity—especially new muds will tend to "walk off" the shakers. Addition of fine or ultra-fine solids, such as barite or bentonite, will minimize this effect [Growcock, 1997].

Rheology Models

Shear viscosity is defined by the ratio of shear stress (τ) to shear rate (γ):

$$\mu = \tau/\gamma.$$

The traditional unit for viscosity is the Poise (P), or 0.1 Pa-sec (also 1 dyne-sec/cm^2), where Pa = Pascal. Drilling fluids typically have viscosities that are fractions of a Poise, so that the derived unit, the centipoise (cP), is normally used, where 1 cP = 0.01 P = 1 mPa-sec.

For *Newtonian* fluids, such as pure water or oil, viscosity is independent of shear rate. Thus, when the velocity of a Newtonian fluid in a pipe or annulus is increased, there is a corresponding increase in shear stress at the wall, and the effective viscosity is constant and simply called the viscosity [Barnes *et al.*]. Rearranging the viscosity equation gives

$$\tau = \mu \cdot \gamma$$

and plotting τ versus γ will produce a straight line with a slope of μ that intersects the ordinate at zero.

Drilling fluids are *non-Newtonian*, so that viscosity is not independent of shear rate. By convention, the expression used to designate the viscosity of non-Newtonian fluids is

$$\mu_e = \tau/\gamma$$

where μ_e is called the "effective" viscosity, to emphasize that the shear rate at which the viscosity is measured needs to be stipulated.

All commonly used drilling fluids are "shear-thinning," that is, viscosity decreases with increasing shear rate. Various models are used to describe the shear-stress versus the shear-rate behavior of drilling fluids. The most popular are the Bingham Plastic, Power Law, and Herschel-Bulkley. The *Bingham Plastic* model is the simplest. It introduces a nonzero shear stress at zero shear rate:

$$\tau = \mu_p \cdot \gamma + \tau_0$$

or
$$\mu_e = \tau/\gamma = \mu_p + \tau_0/\gamma$$
where μ_p is dubbed the plastic viscosity and τ_0 the yield stress, that is, the stress required to initiate flow. μ_p, is analogous to μ in the Newtonian equation. Thus, with increasing shear rate, τ_0/γ approaches zero and μ_e approaches μ_p. If τ is plotted versus γ, μ_p is the slope and τ_0 is the ordinate intercept. The Bingham Plastic model is the standard viscosity model used throughout the industry, and it can be made to fit high-shear-rate viscosity data reasonably well. μ_p (or its oilfield variant PV) is generally associated with the viscosity of the base fluid and the number, size, and shape of solids in the slurry, while yield stress is associated with the tendency of components to build a shear-resistant.

When fitted to high-shear-rate viscosity measurements (the usual procedure), the Bingham Plastic model overestimates the low-shear-rate viscosity of most drilling fluids. The *Power Law* model (also called the Ostwald de Waele model) goes to the other extreme. The Power Law model can be expressed as follows:
$$\tau = K \cdot \gamma^n \quad \text{or}$$
$$\mu_e = \tau/\gamma = K \cdot \gamma^{n-1}$$
where K is dubbed the consistency and n the flow behavior index. The Power Law model underestimates the low-shear-rate viscosity. Indeed, in this model, the value of τ at zero shear rate is always zero.

To alleviate this problem at low shear rates, the *Herschel-Bulkley* model was invented. It may be thought of as a hybrid between the Power Law and Bingham Plastic models and is essentially the Power Law model with a yield stress [Cheremisinoff]:
$$\tau = K \cdot \gamma^n + \tau_0 \quad \text{or}$$
$$\mu_e = \tau/\gamma = K \cdot \gamma^{n-1} + \tau_0/\gamma.$$
Portraits of the three rheology models are shown in Figure 2.8. For NAFs and clay-based WBMs, the Herschel-Bulkley model works much better than the Bingham Plastic model. For polymer-based WBMs, the Power Law model appears to provide the best fit of the three models; better yet is the Dual Power Law model: one for a low-shear-rate flow regimen (annular flow) and one for a high-shear-rate flow regimen (pipe flow). More discussion is presented in the example below.

Other models have been used too, including the Meter model (also called the Carreau or Krieger-Dougherty model), which describes

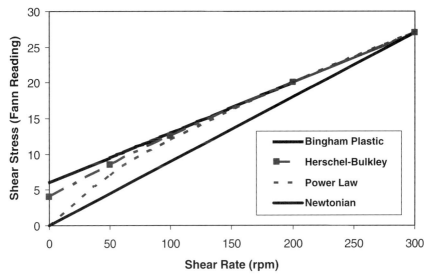

Figure 2-8. Drilling Fluid Rheology Models.

structured particle suspensions well, and the Casson model, which fits OBM data well. However, neither of these models has been widely adopted by the drilling-fluid community.

Measurement of Viscosity

Drilling fluid rheological parameters are usually measured with a concentric cylinder rotary viscometer, like the Fann VG (viscosity grade) meter, which has a geometry that gives the following expression for a fit of the data to the Bingham Plastic model [API RP 13D]:

$$\vartheta_\omega = [(\vartheta_{600} - \vartheta_{300})/(600 - 300)] \cdot \omega + [\vartheta_{300} - (\vartheta_{600} - \vartheta_{300})].$$

This enables calculation of a Fann Reading ϑ_ω (degrees) at a Fann Speed ω (rpm). ϑ_{600} and ϑ_{300} are the Fann Readings at Fann Speeds of 600 and 300 rpm, respectively; $(\vartheta_{600} - \vartheta_{300})/(600-300) = \mu_p$; and $\vartheta_{300} - (\vartheta_{600} - \vartheta_{300}) = \tau_0$. By convention, $(\vartheta_{600} - \vartheta_{300})$ is called the oilfield PV; note that PV $= \mu_p \cdot (600 - 300) = 300\ \mu_p$. Similarly, τ_0 is called the yield point (YP). Thus,

$$\vartheta_\omega = (\text{PV}/300) \cdot \omega + \text{YP}$$

Shear rates in a drill pipe generally encompass the range from 511 to 1022 sec^{-1} (Fann Speeds of 300 to 600 rpm), whereas in the annulus, flows are usually one to two orders of magnitude lower, such as 5.1 to 170 sec^{-1} (Fann Speeds of 3 to 100 rpm). To change the Fann Speed to units of sec^{-1}, ω (rpm) is multiplied by 1.7; to change the Fann Reading to units of dyne/cm^2, ϑ (degrees) is multiplied by 5.11. YP is actually in units of degrees but is usually reported as lb/100 ft^2, since the units are nearly equivalent: 1 degree = 1.067 lb/100 ft^2. Not only is the Bingham Plastic model easy to apply, it is also quite useful for diagnosing drilling-fluid problems. Because electrochemical effects manifest themselves at lower shear rates, YP is a good indicator of contamination by solutes that affect the electrochemical environment. By contrast, PV is a function of the base fluid viscosity and concentration of solids. Thus, PV is a good indicator of contamination by drilled solids.

Application of the Power Law and Herschel-Bulkley Models to Rotary Viscometer Data

The Power Law model may be applied to Fann Readings of viscosity using the expression

$$\text{Fann Reading} = K \cdot (\text{Fann Speed})^n.$$

To obtain representative values of K and n, it is best to fit all of the Fann Readings to the model. A simple statistical technique, such as least squares regression, is quite satisfactory. If a programmable computer is not available but the flow regimen of interest is clearly understood, two Fann Readings will usually suffice to estimate the values of K and n. Naturally, this will lead to weighting of K and n to the Fann Speed range covered by those two Fann Readings. To estimate K and n in the simple Power Law model from two Fann Readings, take the logarithm of both sides and substitute the data for the respective Fann Readings and Speeds:

$$\text{Log (Fann Reading)} = \text{Log } K + n \text{ Log (Fann Speed)}.$$

For example, assume pipe flow and a shear rate of interest covered by the 300 and 600 rpm Fann Readings. If the Fann Readings are 50, 30, 20, and 8 at 600, 300, 100, and 6 rpm, respectively,

$$\text{Log}(50) = \text{Log } K + n \text{ Log}(600)$$

and

$$\text{Log}(30) = \text{Log}\,K + n\,\text{Log}(300).$$

Subtracting the second equation from the first eliminates K and produces one equation for n:

$$n = [\text{Log}(50/30)]/\text{Log}(600/300)] = 0.737.$$

The value of K may be determined by substituting 0.737 for n in one of the equations above:

$$\text{Log}(50) = \text{Log}\,K + (0.737)\text{Log}(600)$$

or

$$K = [(50)/(600)^n] = 0.448.$$

Application of these parameters to the low-shear-rate range is woefully inaccurate, as suggested above. If the Power Law model fitted to the 300 and 600 rpm data is used at, for example, 6 rpm, the calculated Fann Reading is

$$\text{Fann Reading} = 0.448(6\,\text{rpm})^{0.737} = 1.68$$

and the viscosity is

$$\text{Viscosity} = [(\text{Fann Reading})(511)/(\text{Fann Speed})(1.7)] = 84\,\text{cP}.$$

By contrast, the Bingham Plastic model fit gives

$$\begin{aligned}\text{Fann Reading} &= (\text{PV}/300) \cdot \text{Fann Speed} + \text{YP} \\ &= (20/300)(6\,\text{rpm}) + 10 = 10.4\end{aligned}$$

and the viscosity at 6 rpm is

$$\text{Viscosity} = [(\text{Fann Reading})(511)/(\text{Fann Speed})(1.7)] = 521\,\text{cP}.$$

Thus, none of the rheology models should ever be used to extrapolate outside of the range used for the data fit.

The Herschel-Bulkley model is more difficult to apply, since the equation has three unknowns and there is no simple analog solution. Calculation of K, n, and τ_0 is generally carried out with an iterative procedure, which is difficult to do without a programmable computer. An alternative, which works fairly well, is to simply assign to τ_0 the 3-rpm

Fann Reading, subtract that reading from all of the Fann Readings, and fit them to the Power Law model as described above.

2.4 HOLE CLEANING

Good solids control begins with good hole cleaning. One of the primary functions of the drilling fluid is to bring drilled cuttings to the surface in a state that enables the drilling-fluid processing equipment to remove them with ease. To achieve this end, quick and efficient removal of cuttings is essential.

In aqueous-based fluids, when drilled solids become too small to be removed by the solids-control equipment, they are recirculated downhole and dispersed further by a combination of high-pressure shear from the mud pumps, passing through the bit, and the additional exposure to the drilling fluid. The particles become so small that they must be removed via the centrifuge overflow (which discards mud, too) and/or a combination of dilution and chemical treatment. Thus, to minimize mud losses, drilled solids must be removed as early as possible. Figure 2.9 shows a decision tree that can be useful in identifying and solving hole-cleaning problems.

2.4.1 Detection of Hole-Cleaning Problems

Historically, the combination of the necessity to pump or backream out of the hole and a notable absence of cuttings coming over the shale shaker prior to pulling out of the hole has been a reliable indicator of poor hole cleaning. When some cuttings are observed, however, the quantity of cuttings itself does not adequately reflect hole-cleaning efficiency. The *nature* of those cuttings, on the other hand, provides good clues: Good cuttings transport is indicated by sharp edges on the cuttings, whereas smooth and/or small cuttings can indicate poor hole cleaning and/or poor inhibition. With the advent of PWD (pressure while drilling) tools and accurate flow modeling, a number of other indicators have come to light that foreshadow poor hole cleaning and its attendant consequences. Among these are:

- Fluctuating torque
- Tight hole
- Increasing drag on connections
- Increased ECD when initiating drill string rotation

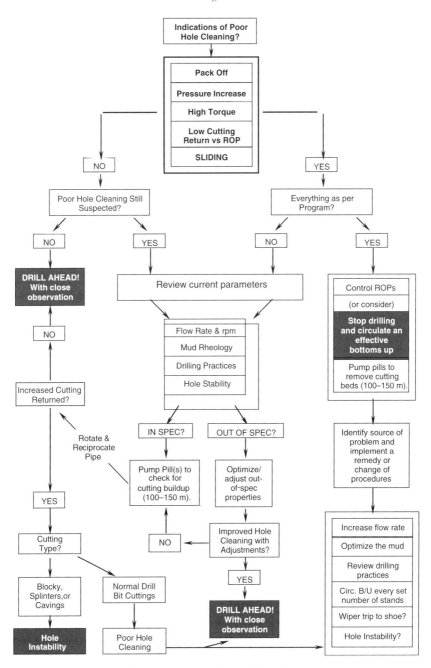

Figure 2-9. Hole-Cleaning Flow Chart.

2.4.2 Drilling Elements That Affect Hole Cleaning

Critical elements that can affect hole cleaning include the following:

- Hole angle of the interval
- Flow rate/annular velocity
- Drilling-fluid rheology
- Drilling-fluid density
- Cutting size, shape, density, and integrity
- ROP
- Drill string rotational rate
- Drill string eccentricity

For a given drilling-fluid density, which is generally determined by well bore stability requirements, the hole may be classified into three hole-cleaning "zones" according to hole angle:

Hole Angle	Critical Parameters (in order of importance)	To Improve Hole Cleaning
Zone I 0°–35°	1. Flow rate/annular velocity	Increase
	2. Rheology (YP, or better yet, K)	Increase
	3. ROP	Decrease
Zone II 35°–65°	1. Flow rate/annular velocity	Increase
	2. Drill string rotational rate	Increase
	3. Rheology (6 rpm/LSRV and PV)	Flatten profile
	4. ROP	Decrease
Zone III 65°+	1. Bit cutter size (PDC)/cutting size	Decrease
	2. Drill string rotational rate	Increase
	3. Flow rate/annular velocity	Increase
	4. Rheology (6 rpm/LSRV and PV)	Flatten profile
	5. ROP	Decrease

LSRV = low-shear-rate viscosity; PDC = polycrystalline diamond compact.

Flow Rate/Annular Velocity

Generally, in near vertical and moderately inclined hole intervals, annular velocity (AV) has the largest impact upon whether a hole can be cleaned of cuttings. However, in extended-reach, high-angle wells (Zone III), AV places third in critical importance, though there is a critical velocity below which a cuttings bed will not form [Gavignet & Sobey]. In practice, the optimum theoretical flow rate may vary from the achievable flow rate. The achievable flow rate is restricted by surface pressure constraints, nozzle selection, use of MWD (measurement while drilling) tools, and allowable ECD. On the other hand, little is gained from very high AVs. Indeed, above 200 ft/min, little improvement in hole cleaning is usually observed, and the primary effect of increasing AV above this level is to increase ECD. In Zone III applications, low-viscosity sweeps—so low that the flow regime in the annulus changes from laminar to turbulent—can be effective. Unfortunately, the volume of fluid required to reach critical velocity for turbulent flow is frequently outside the achievable flow rate for hole sizes larger than 8½-inch and is frequently limited by maximum allowable ECD and/or hole erosion concerns.

Another way to increase AV is to reduce the planned size of the annulus by using larger-OD drill pipe. Not only does a larger pipe generate a smaller annular gap, thereby increasing fluid velocity, it also increases the effect of pipe rotation on hole cleaning. Thus, increasing the OD of drill pipe to 6⅝ inches with 8-inch tool joints has proven to be effective in aiding the cleaning of 8½-inch well bores. A caveat: Although reducing the annular gap can greatly improve hole cleaning, it also makes fishing more difficult; indeed, it violates the rule of thumb that stipulates a 1-inch annular gap for washover shoes.

Rheology

In a vertical hole (Zone I), laminar flow with low PV and elevated YP or low n-value and high K-value (from the Power Law model) will produce a flat viscosity profile and efficiently carry cuttings out of the hole [Walker]. Viscous sweeps and fibrous pills are effective in moving cuttings out of a vertical hole.

In a deviated hole (Zones II and III), cuttings have to travel but a few millimeters before they pile up along the low side of the hole. Consequently, not only do cuttings have to be removed from the well bore, they also have to be prevented from forming beds. Frequently a stabilized cuttings bed is not discovered until resistance is encountered while attempting to pull the drill string out of the hole. Close monitoring of pressure drops within the annulus using PWD tools can provide warning of less than optimal hole cleaning. Increased AV coupled with low PV, elevated low-shear-rate viscosity, and high drill string rpm will generally tend to minimize formation of a cuttings bed. To remove a cuttings bed once it has formed, high-density sweeps of low-viscosity fluid at both high and low shear rates, coupled with pipe rotation, are sometimes effective in cleaning the hole. Viscous sweeps and fibrous pills tend to channel across the top of the drill pipe, which is usually assumed to be lying on the lower side of the hole.

For extended-reach drilling programs, flow loop modeling has generated several rules of thumb for low-shear-rate viscosity to avoid cuttings bed formation. The most popular is the rule that for vertical holes the 6-rpm Fann Reading should be 1.5 to 2.0 times the open-hole diameter [O'Brien and Dobson]. Another rule of thumb specifies a 3-rpm or 6-rpm Fann Reading of at least 10, though 15 to 20 is preferable. However, each drilling fluid has its own rheological characteristics, and these rules of thumb do not guarantee good hole cleaning. If the well to be drilled is considered critical, hole-cleaning modeling by the drilling-fluid service company is a necessity.

NAFs generally provide excellent cuttings integrity and a low coefficient of friction. The latter allows easier rotation and, in extended-reach drilling, more flow around the bottom side of the drill string. As the drill string is rotated faster, it pulls a layer of drilling fluid with it, which in turn disturbs any cuttings on the low side and tends to move them up the hole.

Optimizing the solids-control equipment so as to keep a fluid's drilled-solids content low tends to produce a low PV and a flat rheological profile, thereby improving the ability of the fluid to clean a hole, particularly in extended-reach wells. The fluid is more easily placed into turbulent flow and can access the bottom side of the hole under the drill pipe more easily. In the Herschel-Bulkley model, a moderate K, a low n (highly shear-thinning), and a high τ_o are considered optimal for good hole cleaning.

Carrying Capacity

Only three drilling-fluid parameters are controllable to enhance moving drilled solids from the well bore: AV, density (mud weight [MW]), and viscosity. Examining cuttings discarded from shale shakers in vertical and near-vertical wells during a 10-year period, it was learned that sharp edges on the cuttings resulted when the product of those three parameters was about 400,000 or higher [Robinson]. AV was measured in ft/min, MW in lb/gal, and viscosity (the consistency, K, in the Power Law model) in cP.

When the product of these three parameters was around 200,000, the cuttings were well rounded, indicating grinding during the transport up the well bore. When the product of these parameters was 100,000 or less, the cuttings were small, almost grain sized.

Thus, the term *carrying capacity index* (CCI) was created by dividing the product of these three parameters by 400,000:

$$CCI = (AV) \cdot (MW) \cdot (K)/400,000.$$

To ensure good hole cleaning, CCI should be 1 or greater. This equation applies to well bores up to an angle of 35°, just below the 45° angle of repose of cuttings. The AV chosen for the calculation should be the lowest value encountered (e.g., for offshore operations, probably in the riser).

If the calculation shows that the CCI is too low for adequate cleaning, the equation can be rearranged (assuming CCI = 1) to predict the change in consistency, K, required to bring most of the cuttings to the surface:

$$K = 400,000/(MW) \cdot (AV)$$

Since mud reports still describe the rheology of the drilling fluid in terms of the Bingham Plastic model, a method is needed to readily convert K into PV and YP. The chart given in Figure 2.10 serves well for this purpose. Generally, YP may be adjusted with appropriate additives without changing PV significantly.

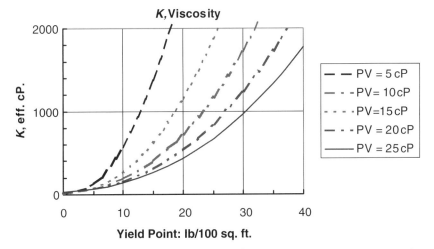

Figure 2.10. Conversion of Bingham Plastic Yield Point to Power Law K [Bourgoyne *et al*].

Example: A vertical well is being drilled with a 9.0 lb/gal drilling fluid circulating at an AV, with PV = 15 cP and YP = 5 lb/100 ft^2. From Figure 2.10, $K = 66$ cP, and from the CCI equation, CCI = 0.07. Clearly, the hole is not being cleaned adequately. Cuttings discarded at the shale shaker would be very small, probably grain size. For a mud of such low density, PV appears to be much too high, very likely the result of comminution of the drilled solids. Solving the equation for the K value needed to give CCI = 1 generates $K = 890$ cP. From Figure 2.10, YP needs to be increased to 22 lb/100 ft^2 if PV remains the same (15 cP). If the drilled solids are not removed, PV will continue to increase as drilled solids are ground into smaller particles. When PV reaches 20 cP, YP will need to be raised to 26 lb/100 ft^2. As PV increases and YP remains constant, K decreases. It is easier to clean the borehole (or transport solids) if PV is low. Low PV can be achieved if drilled solids are removed at the surface.

Cuttings Characteristics

The drier, firmer, and smaller the cuttings, the easier they are to remove from the hole. Small polycrystalline diamond compact (PDC) bit, small cutters on the bit generate small cuttings, which settle out more slowly than large cuttings and are more easily entrained in the annular column of drilling fluid by drill string rotation. As per Stokes' law (see Chapter 13 on Centrifuges), large cuttings will fall out of suspension more rapidly than smaller cuttings, but in high-angle holes, even smaller cuttings may settle and form a cuttings bed. Rounded or agglomerated cuttings are indicative of an extended period of time in the hole and poor hole cleaning.

Rate of Penetration

Preventing cuttings beds in deviated wells is far easier than removing them. Controlling instantaneous ROP is one way to avoid overloading the annulus with cuttings. ROP should always be controlled so as to give the fluid enough time to remove the cuttings intact from the bottom of the hole and minimize spiking of the fluid density in the annulus. The treatment for poor hole cleaning is to reduce ROP, circulate the hole clean, and take steps to optimize hole cleaning. Additional information is provided in the next section.

Pipe Rotation

As pipe rotation rate increases, the pipe drags more fluid with it. In deviated wells, this layer of drilling fluid disrupts cuttings beds that have formed around the pipe while lying on the low side of the hole. Step changes appear to be the norm, occurring in most cases at around 85, 120, and 180 rpm. There is some evidence that above 180 rpm, turbulent flow ensues for many fluids. At these high levels, there seems to be little additional benefit to hole cleaning from increasing pipe rotation any further; most likely this is because cuttings beds cannot form in turbulent flow. During sliding, hole cleaning is minimal and cuttings beds are likely to form. Thus, sliding should be kept to a minimum during any drilling operation. Indeed, this is one of the reasons that rotary steerable tools have become popular.

Drill String Eccentricity

In high-angle wells, the drill string does not remain stable on the bottom of the hole while rotating. The drill string tends to climb the wall of the well bore and fall back, providing additional agitation—though also additional cuttings degradation—while aiding in the removal of cuttings beds on the low side of the hole.

2.4.3 Filtration

When the hydrostatic pressure of the drilling fluid is greater than the pore pressure, drilling fluid invades the formation (spurt loss). Suspended solids attempt to flow in with the liquid fraction, but very quickly particles of the appropriate size (generally one-sixth to one-third the size of pore throats at the well bore) bridge the pores and begin to build a filter cake. In time, finer and finer particles fill the interstices left by the bridging particles and ultimately form such a tight web that only liquid (filtrate) is able to penetrate. Once this filter cake is established, the flow rate of fluid into the formation is dictated by the permeability of the cake. When mud is not being circulated, filter cake grows undisturbed (static fluid loss) and the rate of filtration after the cake is established is proportional to the square root of time. When the mud is being circulated, the filter cake grows to the point at which the shear stress exerted by the mud balances the shear strength of the filter cake (dynamic fluid loss). Under this condition, the cake has a limiting thickness and the rate of filtration after the cake is established is proportional to time. Often, spurt loss is greater under dynamic conditions. Whether static or dynamic, the particles that invaded the formation during the spurt-loss phase may or may not ultimately help to form an internal filter cake, too.

The API Fluid Loss Test (30 min, $\Delta P = 100$ psi through No. 50 Whatman filter paper, ambient temperature) is the standard static filtration test used in the industry; however, because it uses very fine mesh paper as the filter medium, all of the bridging particles are stopped at the surface of the paper and the spurt-loss phase is not simulated properly. Usually this leads to gross underestimates of the spurt loss. A better static filtration test is the PPT, or permeability plugging test, which uses a 1/4-inch-thick ceramic disk of known permeability [API 13B1/API 13B2]. Dynamic filtration, such as in the Fann 90 test, uses a core made of the same ceramic material and simulates shearing of the filter cake by the fluid in the annulus.

For a given pressure and temperature, cake thickness is related to the filtration rate and is a function of the concentration of solids, PSD, and the amount of water retained in the cake. Filtration rate decreases with increasing concentration of solids, but cake thickness increases. Permeability, on the other hand, does not change. Permeability is almost entirely dependent on the proportion and properties of the *colloidal* fraction (<2 μm diameter). Permeability decreases with increasing fraction of colloids and is affected strongly by particle size and shape. A broad distribution of particle sizes is important to attain low permeability. Particles that are flat (e.g., bentonite) can pack very tightly, in contrast to spherical, granular, or needle-shaped particles. On the other hand, some organic macromolecules, such as hydrolyzed starch, are highly deformable and appear to fit well in the interstices of most filter cakes. Similarly, polyelectrolytes like CMC (carboxymethyl cellulose) and PAC (polyanionic cellulose) are large enough to be trapped in the pores of filter cakes. In NAFs, colloidal control of filter cake permeability is achieved with surfactants and water, as well as organophilic clays.

Flocculation causes particles to join together to form a loose, open network. When a drilling fluid is flocculated (e.g., through the addition of salts), the filter cake that it generates at the well bore contains some of that flocculated character, and the rate of filtration increases. Conversely, thinners (deflocculants) like lignosulfonates disperse clay flocs, thereby decreasing cake permeability.

As important as it is to have a substantial colloidal fraction of solids with a broad PSD in the mud, it is equally important that it contain a substantial concentration of bridging particles with a broad PSD. Also critical is the maximum size of bridging particles. Particles about one-sixth to one-third the size of the maximum "pore throat" in a drilling interval suffice, but the fluid must maintain a significant concentration of those particles throughout the interval. The following can serve as a rough idea of the required maximum bridging particle size [Glenn & Slusser]:

Permeability, mD	Maximum Particle Size, μm
100	2
100–1,000	10
1,000–10,000	74 (200 mesh)

A drilling fluid containing particles of sizes ranging up to the requisite maximum should be able to effectively bridge the formation and form filter cake. Above 10 D or in fractures, larger particles are required, and most likely the amounts needed to minimize spurt loss will also increase with the size of the opening. Generally, with increasing concentration of bridging particles, bridging occurs faster and spurt loss declines. For consolidated rock with permeability in the range 100 to 1000 mD, only 1 lb/bbl of 10-mm particles is necessary to prevent mud spurt from invading farther than 1 inch into the rock. On unconsolidated sands of that same permeability, 5 to 30 times that amount may be required. Reservoir drilling fluids typically contain as much as 30 lb/bbl total of acid-soluble bridging materials (usually $CaCO_3$), sized to provide a broad size distribution for all solids in the fluid.

For nonreservoir applications, enough particles of the required size range are usually present in most drilling fluids after cutting just a few feet of rock. However, extensive use of desanders and desilters when drilling unconsolidated sands may deplete these particles, and some bridging material may have to be added back. Likewise, when no drilling is involved (e.g., production repair jobs), bridging particles will need to be added to the fluid.

2.4.4 Rate of Penetration

With higher ROP, both rig time and cost of bits are greatly reduced, and the total drilling operation is less costly, as shown in Figure 2.11.

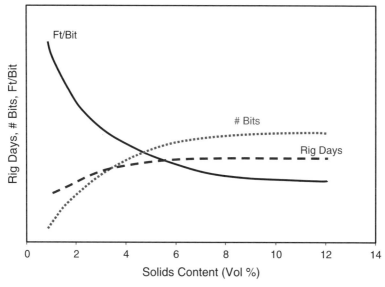

Figure 2.11. Effect of Solids on Drilling Rate.

Drilling-fluid parameters that can affect ROP include:

- Density
- Solids content/solids control
- Filtration
- Rheological profile
- Coefficient of friction/lubricity
- Shale inhibition

Generally, any process that leaves only the desirable solids, lubricants, and rheology modifiers in a drilling fluid will enhance ROP.

Density

The most important mud property affecting ROP is density. ROP decreases as the pressure differential (well-bore pressure minus pore pressure) across the rock face increases. Accordingly, a drilling fluid should be as light as possible while still able to maintain well-bore stability. For a given fluid density, use of weighting materials with a high SG (e.g., hematite or ilmenite instead of barite) can increase ROP, because the volume of solids required to generate that fluid density is less and high-shear-rate viscosity is lower. If well-bore stability is not a factor, a gaseous fluid or one containing entrained gas should be considered first, followed by fluids containing hollow crush-proof beads, then oil or emulsion fluids, followed by invert-emulsion muds, freshwater muds, and brines. On the other hand, special equipment is required to drill with gases or gaseous fluids, and safety risks increase with increasing pore pressure of the formation (see Chapter 19 on Underbalanced Drilling).

Solids Content

The relationship between solids content and drilling rate has been known for many years. Broadly speaking, a low concentration of solids leads to a high ROP, as shown in Figure 2.11. The effect is most pronounced with low-solids drilling fluids, such as clear brine fluids and low-solids nondispersed muds. PSD also affects ROP, indirectly. A wide distribution of particle sizes is required to achieve adequate filtration control, which is necessary for most drilling operations. However, if the concentration of noncolloidal drilled solids can be kept below 4% by weight, ROP can be maintained at a high level [Darley]. When these particles are at such a low concentration, they are not able to form an internal filter cake below the chip (cutting) between successive tooth or cutter strikes, that is, the spurt loss is high, and the pressure differential across the chip remains high (see the following section).

Colloidal solids, which fill the interstices of a filter cake formed by the noncolloidal particles, also reduce ROP, but in a different way. With increasing fraction or total concentration of colloidal solids, the external filter cake on a chip forms more quickly and is less permeable, again reducing the probability of being able to form an internal filter cake. The result is that ROP decreases as dynamic fluid loss decreases (more on this in the following section). For clear-brine polymer drilling fluids, very high ROP is achievable only by removing essentially all of the drilled solids. If a clear brine is infused with enough fine solids to be opaque, ROP will decrease by more than 50%. Nevertheless, in hard rock formations, desanders/desilters or mud cleaners may be able to keep the brine clear, but it is unlikely in younger formations.

Both concentration and PSD of solids also affect the performance of solids-control equipment. For optimal removal of cuttings at the shakers, controlled drilling—limiting the ROP—may be necessary so as not to exceed the operating limits of the pumps and shakers.

Filtration

Historically, the significant reduction in ROP observed during displacing of clear water with clay-based drilling fluid was attributed to chip hold-down pressure (CHDP) [Garnier & van Lingen]. As a tooth from a tricone bit creates a crack in a rock, a vacuum is created under the chip (cutting) unless enough liquid can rush in to fill this incipient crack. Better penetration of the fluid into this crack reduces the pressure drop holding the chip in place, thus facilitating its removal and enabling the tooth to engage fresh rock. For a permeable formation, the fluid to fill the crack can come from within the formation; this is one of the reasons that sandstones generally drill faster than shales. A somewhat different scenario is postulated for PDC bits. Here the argument is that the differential pressure acting across the chip opposes its *initial* dislodgement.

In keeping with CHDP theory, several years of study indicate that ROP increases as density decreases and filtration control is relaxed, regardless of the type of bit. However, ROP does not appear to correlate with static fluid loss, such as is measured with the API Fluid Loss Test. On the other hand, ROP appears to correlate very well with *dynamic* fluid loss. These are tests designed to simulate downhole flow of fluid across the face of the filter cake, leading to continual erosion and production of a constant thickness cake. Thus, as dynamic fluid loss increases, so does ROP. It is essential in these tests to use core from the area to be drilled. As might be expected on the basis of CHDP, if the rock being drilled has very low permeability (in the extreme case, shales with no microfractures), dynamic fluid loss measurements will show very low fluid loss, and consequently ROP will be lower than in permeable rock.

Rheological Profile

Fluids with low viscosity at high shear rates effectively overcome the chip hold-down effect and sweep the hole clean of drilled solids quickly, thereby minimizing regrinding. Often, though not necessarily, PV is also low; the relevant viscosity, however, is the viscosity under the bit, that is, the Fann Reading at high Fann Speed. Generally a formation drills faster the more shear-thinning and flatter the rheological profile of the mud. This again reinforces the advantages of a "clean" drilling fluid.

Lubricity

The ability to turn the drill string, log the well, and run casing in highly convoluted well bores is considered desirable. High lubricity (low coefficient of friction) of the drilling fluid, whether attained mechanically with the addition of glass or polystyrene beads or chemically with the addition of oils or surfactants, enables more accurate control of weight-on-bit and drill string rotation, thereby enhancing ROP.

Shale Inhibition

"Balling" occurs when drilled cuttings are not removed from beneath the bit and they collect between the bit and the true hole bottom (bottom balling) or in the cutters or teeth of the drill bit (bit balling). This effect is most pronounced with hydratable shales (e.g., gumbo) in WBMs. Static and dynamic CHDP conspire with the hydrational and adhesive forces to make the cuttings very soft and sticky. This effect can be reduced by making the mud either more inhibitive or less inhibitive so as to reduce the hydrational and adhesive forces. More on this may be found in the following section.

2.4.5 Shale Inhibition Potential/Wetting Characteristics

Because the continuous phase in NAFs is nonaqueous, cuttings drilled with NAFs do not hydrate, and they are left oil wet and nearly intact. Invert-emulsion NAFs can actually increase cuttings' hardness by osmotically removing water from the cuttings. WBMs, on the other hand, generally are not very efficient at removing water from the cuttings; indeed they may only slow hydration, so that cuttings will still tend to imbibe water, swell, soften, and even disperse. The same phenomenon occurs in the well bore, so that both well-bore stability and cuttings integrity suffer with increased residence time of the mud downhole.

Highly inhibitive WBMs, such as PHPA/glycol in a 20 to 25% solution of NaCl, can remove water from the cuttings, but the cuttings may actually get stickier, depending on how wet they were when generated. The Atterberg limits give a qualitative picture of the effect of removal or imbibition of water on plasticity, or stickiness, of cuttings [Bowles], which may promote bit balling (see Fig. 2.12). Shale-laden cuttings from young formations, such as gumbo-like argillaceous formations of the Gulf of Mexico, tend to be very wet and on the

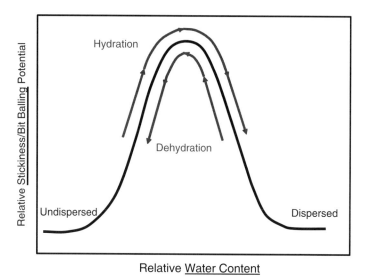

Figure 2.12. Effect of Water Content of Clays and Shales on Stickiness.

downslope right side of the Atterberg curve. Exposure to highly inhibitive WBMs may remove some, but not enough, water and cause the shale to travel left back up the curve to a more sticky condition. Cuttings generated using this kind of mud tend to be stickier than those generated with a less inhibitive mud [Friedheim et al.], so that blinding of shaker screens is a common occurrence. Replacing the fluid with NAF can remedy this problem, but treatment of the WBM with a drilling enhancer (or ROP enhancer) may be more economical. Although WBM treated with a drilling enhancer presents more risk than NAF, it can reduce bit-balling tendency significantly, as well as blinding of screens and other solids-control problems. Most drilling enhancers possess the added virtue of imparting additional lubricity to the fluid and reducing abrasiveness of the cuttings.

2.4.6 Lubricity

A drilling-fluid coefficient of friction that is low (0.1 or less) is generally advantageous, inasmuch as it helps the cuttings to travel as discrete particles over shaker screens. Most mud lubricants will also tend to adsorb onto almost any surface, including the exposed surfaces of the solids-control equipment. A thin film or coating of mud lubricant on those surfaces can help to protect them from corrosion and mitigate adhesion of sticky solids.

2.4.7 Corrosivity

To minimize corrosion of steel tubulars and solids-control devices, control of the responsible agents is a necessity. NAFs effectively prevent corrosion because they are nonconductive and oil-wet the steel surfaces. WBMs, on the other hand, can contain dissolved materials that set up electrochemical cells that ultimately lead to loss of iron from the steel surfaces in contact with the drilling fluid [Bush]. Dissolved O_2 forms rust and pits on the steel surface, and is best controlled by minimizing air entrainment: use only submerged guns in the mud pits; rig all return lines from desanders, etc., to discharge below mud level; and minimize use of the hopper. Keeping the mud at a pH between 9 and 10—with $Ca(OH)_2$ (lime), NaOH (caustic), or MgO—helps greatly to keep the rate of corrosion at an acceptable level. A higher pH is not recommended, particularly in high-temperature wells, because under those conditions the hydroxyl ion becomes very reactive toward clays and polymers. If too much corrosion still occurs, O_2 scavengers such as sodium sulfate (Na_2SO_3) and triazine can be very effective. Less common but also very effective are corrosion inhibitors, such as amines and amine salts, which produce an oily barrier to O_2. The other two primary agents of corrosion are carbon dioxide (CO_2) and hydrogen sulfide (H_2S). Both of these form acids in aqueous drilling fluids. H_2S in particular is a cause for concern because of its high toxicity and its ability to cause hydrogen stress cracking that can lead to fatigue failure of tubulars and solids-control equipment. Again, a high pH can serve as the front line of defense. For high levels of H_2S, though, zinc carbonate, zinc chelate, powdered iron, or magnetite may also be necessary. A mixture recommended by the API for polymer-based WBMs to minimize both corrosion and degradation of polymers by O_2, CO_2, and H_2S consists of MgO, Na_2SO_3 or triazine, and triethanolamine (to sequester iron and remove H_2S/CO_2) [API RP 13C].

It should be noted that dissolved CO_2, O_2, and salts can all accelerate stress cracking and failure of steel hardware, though the effect is most pronounced with H_2S.

Finally, microbes can form corrosive agents, particularly H_2S, via degradation of mud components in the drilling fluid like lignosulfonate or biopolymers. The most effective ways to control microbial corrosion are through use of clean make-up water and a biocide, such as glutaraldehyde or bleach.

2.4.8 Drilling-Fluid Stability and Maintenance

Maintaining the drilling fluid in good condition is essential not only for controlling the mud properties but also to ensure proper operation of solids-control equipment. Vigilance against the effects of contamination and elevated temperatures is particularly important. Invasion of foreign materials, such as water and oils, and thermal degradation of polymers can affect viscosity and filtration properties radically and compromise the performance of some solids-control equipment. Elevated temperatures can also destroy direct and invert-emulsion systems and can cause gelation in clay-based muds, either of which can negatively affect equipment performance. Keeping the mud properties within the design parameters is critical, which requires maintaining the concentrations of mud products and drilled solids at appropriate levels.

2.5 DRILLING FLUID PRODUCTS

Many of the components in drilling fluids can affect the efficiency of solids-control devices. As discussed in the previous section, fluid rheology, shale inhibition potential, wetting characteristics, lubricity, and corrosivity can all affect both the properties of cuttings and the performance of solids-control equipment. Key components that affect those properties include colloidal materials, macropolymers, conventional polymers, and surface-active materials.

2.5.1 Colloidal and Fine Solids

Clay solids (e.g., bentonite, attapulgite) along with clay-laden drilled solids, weighting material, and fine grades of bridging material (e.g., limestone, $CaCO_3$) will all affect the viscosity of a drilling fluid at moderate to high shear rates, thereby increasing the retention of drilling fluid on solids. Although elevated low-shear-rate viscosity is beneficial in many ways (improved hole cleaning, etc.), nothing is gained by having elevated high-shear-rate viscosity. Thinners are often used in both water-based and oil-based muds to reduce high-shear-rate viscosity, especially in high-density muds. Unfortunately, thinners tend to reduce the low-shear-rate viscosity more. One of the best solutions is to minimize the reliance on large quantities of bentonite and other clays for viscosity control and utilize mud systems that provide high yield stress and Bingham-like rheology with low thixotropy, that is, muds with high low-shear-rate viscosity and nonprogressive gels. Examples include xanthan-based polymer muds and mixed metal systems, such as mixed metal hydroxide, mixed metal oxyhydroxide, mixed metal oxide, and mixed metal silicate.

Weighting materials increase the entire rheology profile, and some weighting materials tend to be abrasive not only to the drill string and casing, but also to solids-control devices. Shaker screen life tends to be shorter with weighting materials like hematite, magnetite, and ilmenite, all of which are twice as hard as barite. Alternatives include going to a finer grind of weighting material, treating the system with lubricants, and using high-density brines instead of the standard weighting material. The latter option is risky, though, because the brines tend to be corrosive and may present incompatibility problems with polymers in the drilling fluids, especially acrylamide-based polymers like PHPA.

2.5.2 Macropolymers

Large solids, such as those typically used as LCM, have irregular shapes that are generally classified as flake, fiber, or granule. These are usually removed at the shakers and discarded along with the cuttings. Treatment of whole mud with LCM (rather than 100-bbl pill or slug) requires reintroduction of fresh LCM into the circulating system before pumping it back downhole. At low concentrations (2 to 15 ppb), LCM tends to have little effect on standard mud properties. However, at higher concentrations, LCM will tend to absorb significant amounts of water (especially noticeable in invert-emulsion muds) and increase high-shear-rate viscosities of WBMs and NAFs; a detergent or wetting agent will usually remedy this problem, though addition of base fluid may also be necessary. For applications requiring high concentrations of LCM in the whole mud, a separate set of shakers to remove and recycle the LCM back to the active mud system downstream of the centrifuge may prove economical. High concentrations of LCM will tend to blind shaker screens when the shaker is used to remove both cuttings and LCM, and an additional scalping shaker may be necessary, whether or not the LCM is recycled back to the active mud system.

Like conventional LCM, gilsonite and other asphaltenes will tend to plug shale shaker screens, particularly triple-layer designs; single- or double-layer screens work much better [Amoco].

2.5.3 Conventional Polymers

Highly shear-thinning polymers that are used to provide high viscosity at very low shear rates (e.g., xanthan gum) can be expected to affect the performance of low-shear-rate solids-control devices, such as settling tanks, hydrocyclones, and centrifuges. If these polymers exhibit viscoelasticity, the elastic nature of the polymers at low shear rates may reduce efficiency of these low-shear-rate devices even more. Extensional effects (see the Section 2.3), which can produce a different kind of viscoelasticity, may also be generated by these polymers. In this case, it is usually extensional viscosity, rather than extensional elasticity, that is of most concern. Extensional viscosity manifests itself at high fluid velocities and may be important in any high-throughput solids-control device that contains orifices or openings.

Shale encapsulators are high-molecular-weight (HMW) polymers that serve to inhibit dispersion of large cuttings, but they also tend to flocculate clay fines. Some of the more popular shale encapsulators are PHPA, AMPS-acrylamide copolymers, and HMW polypropylene glycol. Generating a coarser distribution of solids in the mud can improve the efficiency of the solids-control equipment. On the other hand, in the absence of fines, these polymers exhibit extensional viscosity, a property that can cause blinding of shaker screens. They also tend to adsorb strongly on solid surfaces, such as shaker screens, thus exacerbating the problem. Addition of a small quantity (less than 3% by volume) of fine solids (bentonite, barite) to the mud will deplete the highest-molecular-weight fraction of the polymer first, which is the fraction that is mainly responsible for the extensional viscosity and strong adsorption. At the same time, at that low level of solids, shear viscosity (the type of viscosity normally measured) will not be affected very much. Thus, addition of a small quantity of fine solids as soon as possible to a polymer mud will increase shaker screen conductivity and throughput of hydrocyclones and centrifuges.

Another problem that can occur with dry HMW long-chain polymers like PHPA is formation of "fish eyes" and "strings" in the mud that are attributable to improper mixing. When incompletely hydrated and dispersed, the polymer cannot perform as expected and can "gunk up the works," blinding screens and plugging hydrocyclones and centrifuges. The following guidelines can help to prevent such problems:

- Add all mud products after initiation of the solids-control equipment.
- Add HMW dry polymers like PHPA slowly over a full circulation, using a venturi hopper.
- If bentonite is part of the formulation, mix some in with the polymer before adding the latter through the hopper.
- If possible, coat HMW polymers with an oil, or better yet, predisperse the polymer in oil or mutual solvent before adding it to the system. In all cases, consult the product manufacturer for proper addition.

2.5.4 Surface-Active Materials

The surface-active additives in mud products are designed to adsorb on a specific type of substrate: cuttings (shale inhibitors, antiaccretion agents, and oil wetting agents), weighting material and LCM (oil-wetting agents), liquid internal phase (emulsifiers), and drill string/casing/well bore (lubricants, ROP enhancers). However, none of these additives is perfectly selective for its designed target substrate and every one adsorbs on various types of surfaces. Lubricants and ROP enhancers, in particular, will generally adsorb on anything, including shaker screens and the internal surfaces of hydrocyclones and centrifuges. They may accumulate to such an extent that they reduce throughput in these devices. Little can be done to alleviate this other than to avoid overtreatment of the mud with surface-active additives; alternatively, a permanent coating of a low-energy substance (e.g., Teflon^{TM}) on the exposed surfaces of the equipment will discourage adsorption, but that solution can prove expensive. On the other hand, a very thin coating of mud lubricant, ROP enhancer, or wetting agent can actually be beneficial, inhibiting corrosion and solids accumulation in and on solids-control equipment surfaces. Using modest treatment levels will promote formation of relatively thin coatings without forming thick deposits that can reduce solids-equipment performance.

Low-molecular-weight (LMW) shale inhibitors (e.g., various amines) serve as clay intercalators and hydration suppressants and keep cuttings from swelling and dispersing; like shale encapsulators, they can inhibit degradation of drilled solids and make them easier to remove. Thinners and deflocculants, which include lignosulfonates, lignites, polyphosphates, and LMW acrylamides, improve solids-separation efficiency of all the solids-control devices by reducing the overall viscosity of the drilling fluid.

2.6 HEALTH, SAFETY, AND ENVIRONMENT AND WASTE MANAGEMENT

2.6.1 Handling Drilling Fluid Products and Cuttings

Working with drilling fluids can be hazardous. Some drilling-fluid products emit noxious or hazardous vapors that may reach levels that exceed the maximum recommended short-term or long-term safe exposure limits. Some shale and corrosion inhibitors and some oil-base mud emulsifiers tend to produce ammonia or other hazardous volatile amines, particularly in hot areas on a rig. Other products are flammable or combustible (flash point $<140°F$), so that they too must be handled with caution. Thus, proper ventilation is vital in the mud pit areas and around the solids-control equipment.

Various mud products, brines, cleaning agents, solvents, and base oils commonly found on drill rigs are irritating or even hazardous to body tissues. Cuttings may be coated with these materials, too. Consequently, proper protective equipment should be worn for hands, body, and eyes when working around solids-control devices, even though the protective equipment may be inconvenient or uncomfortable.

2.6.2 Drilling Fluid Product Compatibility and Storage Guidelines

Mud products and test reagents can be particularly hazardous when stored improperly. As in any well-run chemistry laboratory, materials on the rig that are chemically incompatible should be stored apart from each other, and preferably in separate spill trays (secondary containment vessels). Some general storage guidelines are given in Table 2.3. Mud products and test reagents are classified into six hazard groups, in decreasing order of hazard risk (priority)—reactive/oxidizer, toxic, flammable, acids and bases, unknown, and nonhazardous—and each group should be segregated from the others. There should be very little or no material on the rig that falls into the reactive/oxidizer category. Acids and bases, though grouped together, should be placed in separate spill trays.

Table 2.3
Hazard Classification of Chemical Reagents and Mud Products
Chemical Segregation Guidelines*

Safe storage practices require that materials be separated according to chemical compatibility and hazard class. The following hazard classes should be used for segregating the waste of decreasing hazard potential. Each hazard class of chemicals should be stored in a separate secondary containment labeled with the hazard class name. The containment vessels for hazard classes containing primarily solids (e.g., nonhazardous materials) should be placed above all others. Priority 3 materials should be isolated from the flammables cabinet. The secondary containment vessel for the oxidizer hazard classes should be made of metal and sit on a metal shelf.

Priority	Hazard Class	Definition	Example
1	Water/air	Materials that are potentially explosive, react violently, or generate toxic vapors when allowed to come in contact with air or water	Acetyl chloride, sodium metal, potassium metal, phosphorus (red and white), inorganic solid peroxides
	Oxidizers, inorganic salts	Specific "listed" inorganic compounds that react vigorously with organic materials and/or reducing agents	Inorganic liquid peroxides, chlorates, perchlorates, persulfates, nitrates, permanganates, bleach
	Oxidizers, inorganic acids (liq)	Inorganic liquids with pH <2 and strong tendency to oxidize organics	Perchloric, pitric, concentrated sulfuric, bromic, hypochlorous
	Oxidizers, organic	Specific "listed" organic compounds that react vigorously with organic materials and/or reducing agents	Organic peroxides
2	Toxic materials, metals	Materials that contain specific "listed" water-soluble or volatile, nonoxidizing/nonreacting metallic compounds that are regulated at levels below a few mg/L	Metals and water-soluble compounds of arsenic, barium, beryllium, cadmium, chromium, copper, lead, mercury, thallium (e.g., chrome lignosulfonate (CLS) and materials contaminated with CLS.

(continued)

Table 2.3
Continued

Priority	Hazard Class	Definition	Example
	Toxic materials, organic reagents	Specific "listed" compounds whose concentrations in wastes are regulated at levels of 0.1 to 200 mg/L	Phenol, biocides, cyanides, propargyl alcohol, carbon disulfide
3	Flammable and combustible liquids	Nonhalogenated, pourable, organic liquids with flashpoint <140°F (classes I and II)	Acetone, xylene, toluene, methanol, most organic oils, oil-based mud, brine/oil mixtures, oily cuttings solvent wash, invert mud emulsifiers and wetting agents, some lubricants
	Halogenated liquids	Halogenated organic liquids, whether flammable or not	Chloroform, methylene chloride
4	Acids, organic (liq)	Organic liquids with pH <2	Acetic, butyric, formic
	Acids, inorganic mineral (liq) and some concentrated brines	Inorganic liquids with pH <2, generally acids and certain salts	Hydrochloric, hydrobromic, hydrofluoric, dilute sulfuric, phosphoric, conc. brines (low pH, e.g., bromides and iodides)
	Bases, organic (liq)	Organic liquids with pH >12.5	Amines, hydrazines
	Bases, inorganic (liq)	Inorganic liquids with pH >12.5	Ammonia, ammonium hydroxide, sodium hydroxide, potassium hydroxide

5	Unclassified materials (hazardous and nonhazardous)	Calorimetric ampoules (ammonia and phosphate); mercury thermometers; salt gel (attapulgite); chemical spill kits, corrosion inhibitors, field product samples, HTCE residuals, well-cleaning chemicals, some shale inhibitors, tar
6	Nonhazardous materials, salts, clays, etc.	Miscellaneous materials that do not exhibit any of the hazards identified in categories 1–10, including nonoxidizing salts with $2 < \text{pH} < 12.5$ (if in solution) Most clays; nonflammable/noncombustible/ nontoxic polymers (HEC, CMC, PAC); chrome-free lignosulfonates; most empty containers; chrome-free freshwater test fluids, filter cakes, filter media, retort solids residue; solid or aqueous chlorides, formates, carbonates, acetates, dilute bromides, and iodides

*For Additional Information Partial List of Incompatible Chemicals http://www.biosci.ohio-state.edu/~jsmith/safety/IncompatibleChemicals.htm
From "Prudent Practices for Handling Hazardous Chemicals in Laboratories," National Research Council, Washington, D.C., 1995.

2.6.3 Waste Management and Disposal

The drilling-fluid program should address environmental issues concerned with the discharge of drilling fluid, products, and removed solids. Personnel managing the solids-separation equipment must be very familiar with this part of the drilling-fluid program and have a good understanding of governmental regulations and operator requirements. Many drilling operations have strategies in place for drilling-fluid recovery and will have established some general guidelines for the disposal of materials classified as waste. However, situations can arise that present the engineer managing the solids-control equipment with the issue of whether to discard or recycle some types of waste and how to do it. If disposal costs are not a factor, then all waste can be disposed of and treated, if necessary, onsite or sent to a processor offsite. However, if it is possible to recycle some of the products to the mud system, it may prove economical to do so [Hollier *et al*]. Table 2.4 contains some general guidelines approved in the state of Texas for recycling and disposing of waste from a drilling operation. Definitions used in those guidelines for hazardous, class 1, class 2, and class 3 wastes are given below. Solid waste is classified as hazardous by the U.S. Environmental Protection Agency if it meets any of the following four conditions:

- The waste exhibits ignitability, corrosivity, reactivity, or toxicity.
- The waste is specifically listed as being hazardous in one of the four tables of 40 CFR 261: [Code of Federal Regulations]
  1. Hazardous wastes from nonspecific sources (40 CFR 261.31)
  2. Hazardous wastes from specific sources (40 CFR 261.32)
  3. Acute hazardous wastes (40 CFR 261.33(e))
  4. Toxic hazardous wastes (40 CFR 261.33(f)).
- The waste is a mixture of a listed hazardous waste and a nonhazardous waste.
- The waste has been declared to be hazardous by the generator.

Class 1 waste is any material that, because of its concentration or physicochemical characteristics, is considered "toxic, corrosive, flammable, a strong sensitizer or irritant, a generator of sudden pressure by decomposition, heat or other means, or may pose a substantial present or potential danger to human health or the environment when

Table 2.4
Waste Recycle/Disposal Guidelines

M	Recycle / Disposal	Class
Acids (undiluted)	Disposal	Haz
Acids (spent, except hydrofluoric acid)	Dilute and dispose of down sink drain	—
Barite, finished or crude	Recycle	—
Bentonite clays and test fluids	Recycle	—
Biocides	Disposal per MSDS	1, 2, or Haz
Bleach	Dilute and dispose of down sink drain	—
Brines, high-density, new	Recycle	—
Brines, high-density, used	Disposal	2
Brine/oil mixtures (emulsion testing, etc.)	Disposal	1
Broken glass	Disposal	2
Buffer solution	Dilute and dispose of down sink drain	2
Calcium carbonate	Disposal	2
Calcium chloride (solid)	Disposal	2
Calcium chloride (solution)	Dilute and dispose of down sink drain	—
Chemical spill kits	Disposal per kit directions	1, 2, or Haz
Cleaning service rags	Disposal	2
Freshwater test fluids	Recycle	—
Corrosion inhibitors	Disposal per MSDS	1, 2, or Haz
Culture waste (filter media, gravel, etc.)	Disposal	2
Cuttings, neat	Disposal	2
Cuttings, with oil	Disposal	1
Empty containers, hazardous	Disposal	1
Empty containers, nonhazardous	Disposal	2
Enzyme solutions	Dilute and dispose of down sink drain	—
Field product samples	Disposal	1, 2, or Haz
Filter cake, disks. and paper w/chrome-free mud	Recycle	—
Filter cake, disks, and paper w/chrome-containing mud	Disposal	Haz
Freshwater test fluids	Recycle	—
Hydrofluoric acid (handle with extreme care)	Disposal	Haz

(continued)

Table 2.4
Continued

M	Recycle / Disposal	Class
Hydrogen peroxide	Disposal	2
Hydroxy ethyl cellulose	Disposal	2
Lignosulfonate and lignite product test muds	Disposal	???
Mercury thermometers	Disposal	Haz
Well-cleaning chemicals	Disposal per MSDS	1, 2, or Haz
Mud additives		
Emulsifiers		
Fluid loss control		
Lignosulfonates		
Lubricants	Disposal per MSDS	1, 2, or Haz
Shale inhibitors		
Shale stabilizers		
Surfactants		
Wetting agents		
Mud filtrates, oil-based/ synthetic-based mud	Disposal	1
Mud filtrates, water-based mud	Recycle	–
Oil, with non-OBM constituents	Disposal	1
Oil, with OBM constituents required for OBM conditioning	Recycle	–
Oil, mixed with hazardous wastes	Disposal	2
Oil-based/synthetic-based mud and wash chemicals	Disposal	1
Organic peroxides	Disposal	2
Paper towels used to clean up brines and muds	Disposal	2
Persulfates	Disposal	Haz
pH Buffers	Dilute and dispose of down sink drain	–
pH test solution residuals	Return to original container	–
Polymer slurries, mineral oil or other carrier	Disposal	1
Polymers, dry	Disposal	2
Potassium hydroxide (solid)	Disposal	Haz
Potassium hydroxide (solution)	Dilute and dispose of down sink drain	–
Retort cooked solids (chrome-containing mud)	Disposal	Haz

(*continued*)

Table 2.4
Continued

M	Recycle / Disposal	Class
Retort cooked solids (chrome-free mud)	Disposal	2
Salt gel/attapulgite	Disposal	2
Silver nitrate solution	Dilute and dispose of down sink drain	–
Sodium carbonate (solid)	Disposal	Haz
Sodium carbonate (solution)	Dilute and dispose of down sink drain	–
Sodium hydroxide (solid)	Disposal	Haz
Sodium hydroxide (solution)	Dilute and dispose of down sink drain	–
Solvents, chlorinated	Disposal	Haz
Solvents, nonchlorinated	Disposal	Haz
Titration residue	Dilute and dispose of down sink drain	–
Titration solution residue	Dilute and dispose of down sink drain	–
Wash water, laboratory equipment and general	Dilute and dispose of down sink drain	–
WBM spent titrations	Dilute and dispose of down sink drain	–
WBM (Cl < 20,000 ppm and oil < 3%)	Recycle	–
WBM (Cl > 20,000 ppm or oil > 3%)	Disposal	1
WBM, chrome-free saltwater	Disposal	???
WBM, chrome-containing	Disposal	???
WBM, all others	Disposal	???

OBM = oil-based mud; WBM = water-based mud; MSDS = Material Safety Data Sheet.

improperly processed, stored, transported, disposed of, or otherwise managed," as further defined in 30 TAC 335.505 [Texas Administrative Code].

Class 2 waste is any material that cannot be described as hazardous, as class 1, or as class 3.

Class 3 wastes are inert and essentially insoluble materials, usually including, but not limited to, "materials such as rock, brick, glass, dirt and certain plastics and rubber, etc., that are not readily decomposable."

REFERENCES

Ali, A., Kalloo, C. L., and Singh, U. B., "A Practical Approach for Preventing Lost Circulation in Severely Depleted Unconsolidated Sandstone Reservoirs," SPE/IADC 21917, SPE/IADC Conference, Amsterdam, March 11–14, 1991.

Amoco Production Co., *Solids Control Handbook*, March 1994.

API 13A, "Specification for Drilling Fluid Materials," American Petroleum Institute, Washington D.C., 15th ed., May 1993.

API RP 13B1, "Recommended Practice for Field Testing Water-Based Drilling Fluids," American Petroleum Institute, Washington, D.C., 3rd ed., Nov. 2003.

API RP 13B2, "Recommended Practice for Field Testing Oil-Based Drilling Fluids," American Petroleum Institute, Washington, D.C., 3rd ed., 2002.

API RP 13C, "Recommended Practice for Drilling Fluid Processing Systems Evaluation," American Petroleum Institute, Washington, D.C., 2nd ed., March 1996.

API RP 13D, "Recommended Practice on the Rheology and Hydraulics of Oil-Well Drilling Fluids," American Petroleum Institute, Washington, D.C., 4th ed., May 2003.

API RP 13I, "Recommended Practice, Standard Procedure for Laboratory Testing Drilling Fluids," American Petroleum Institute, Washington, D.C., 6th ed., May 2000.

Barnes, H. A., Hutton, J. F., and Walters, K., *Rheology Series 3: An Introduction to Rheology*, Elsevier Science B.V., New York, 1989.

Bourgoyne, A. T., Jr., Millheim, K. K., Chenevert, M. E., and Young, F. S., Jr., *Applied Drilling Engineering*, SPE Textbook Series, Vol. 2, Richardson, TX, 1986, p. 476.

Bowles, J. E., *Engineering Properties of Soil and Their Management*, McGraw-Hill, New York, 1992.

Bradford, D. W., Growcock, F. B., Malachosky, E., and Fleming, C. N., "Evaluation of Centrifugal Drying for Recovery of Synthetic-Based and Oil-Based Drilling Fluids," SPE 56566, SPE Annual Technical Conference and Exhibition, Houston, Oct. 3–6, 1999.

Bush, H. E., "Treatment of Drilling Fluid to Combat Corrosion," SPE 5123, SPE Annual Technical Conference and Exhibition, Houston, Oct. 6–9, 1974.

Cheremisinoff, N. P., Editor, *Encyclopedia of Fluid Mechanics: Slurry Flow Technology*, Vol. 5, Gulf Pub. Co., Houston, TX, 1986.

Code of Federal Regulations (CFR), Washington, D.C., 2002.

Darley, H. C. H., "Designing Fast Drilling Fluids," *J. Petrol. Technol.*, April 1965, pp. 465–470.

Darley, H. C. H., and Gray, G. R., *Composition and Properties of Oil Well Drilling Fluids*, 5th ed., Gulf Pub. Co., Houston, 1986.

Friedheim, J., Toups, B., and van Oort, E., "Drilling Faster with Water-Based Muds," AADE Ann. Tech. Forum, Houston, March 30–31, 1999.

Garnier, A. J., and van Lingen, N. H., "Phenomena Affecting Drilling Rates at Depth," *J. Petrol. Technol.*, Sept. 1959, pp. 232–239.

Gavignet, A. A., and Sobey, I. J., "A Model for the Transport of Cuttings in Directional Wells," SPE 15417, SPE Annual Technical Conference and Exhibition, New Orleans, Oct. 5–8, 1986.

Glenn, E. E., and Slusser, M. L., "Factors Affecting Well Productivity II. Drilling Fluid Particle Invasion into Porous Media," *J. Petrol. Technol.*, May 1957, pp. 132–139.

Growcock, F. B., "New Calcium Chloride Base Drilling Fluid Increases Penetration Rate in Deep Water GOM Well," *The Brief*, Amoco Production Co., March 1997.

Growcock, F. B., Frederick, T. P. and Zhang, J., "How Water-Based Muds Can Be Treated to Increase Drilling Rate in Shale," AADE Ann. Tech. Forum, Houston, March 30–31, 1999.

Hollier, C., Reddoch, J., and Hollier, G., "Successful Optimization of Advances in Disposal and Treatment Technologies, to Cost Effectively Meet New Oil-Based Cuttings Environmental Regulations," AADE 01-NC-HO-13, AADE National Drilling Conference, Houston, March 27–29, 2001.

Lal, M., and Thurber, N. E., "Drilling Waste Management and Closed Loop Systems," presented at the International Conference on Drilling Wastes, Calgary, Alberta, Canada, April 5–8, 1988.

Lundie, P., "Technical Review for UKOOA Alernative Mud Systems for Replacing Conventional Oil/Water Ratio Muds," United Kingdom Offshore Operators Association No. 206647, 1989.

M-I LLC, *Drilling Fluids Engineering Manual*, Houston, TX, 2002.

O'Brien, T. B., and Dobson, M., "Hole Cleaning: Some Field Results," SPE/IADC 13442, 1985 SPE Drilling Conference, New Orleans, March 5–8, 1985.

Robinson, L., private communication.

Svarovsky, L., *Solid–Liquid Separation*, Chemical Engineering Series, Butterworth & Co. Ltd., London, 1981.

Texas Administrative Code (TAC), Austin, TX, 2002.

Walker, R. E., "Drilling Fluid Rheology," *Drilling*, Feb. 1971, pp. 43–58.

Wright, T. R., Jr., "Guide to Drilling, Workover and Completion Fluids," *World Oil*, June 1978, pp. 53–98.

Young, G., and Robinson, L. H., "How to Design a Mud System for Optimum Solids Removal," *World Oil*, Sept.–Nov. 1982.

CHAPTER 3

SOLIDS CALCULATION

Leon Robinson
Exxon, retired

Low-gravity solids in a drilling fluid may be calculated from the equation below:

$$V_{LG} = \frac{100\rho_f}{(\rho_B - \rho_{LG})} + \frac{(\rho_B - \rho_f)}{(\rho_B - \rho_{LG})} V_S - \frac{12}{(\rho_B - \rho_{LG})} \text{MW} - \frac{(\rho_f - \rho_o)}{(\rho_B - \rho_{LG})} V_o$$

where

- ρ_f = density of filtrate, g/cm^3
- ρ_B = density of barite, g/cm^3
- ρ_{LG} = density of low-gravity solids, g/cm^3
- ρ_o = density of oil, g/cm^3
- V_S = volume percentage of suspended solids
- V_o = volume percentage of oil
- MW = mud weight, ppg

The density of the filtrate may be calculated from the equation:

$$\rho_f = 1.0 + 6.45 \times 10^{-7}[\text{NaCl}] + 1.67 \times 10^{-3}[\text{KCl}] \\ + 7.6 \times 10^{-7}[\text{CaCl}_2] + 7.5 \times 10^{-7}[\text{MgCl}_2]$$

where

- [NaCl] = concentration of NaCl in filtrate, mg/L
- [KCl] = concentration of KCl in filtrate, ppb
- [CaCl$_2$] = concentration of CaCl$_2$ in filtrate, mg/L
- [MgCl$_2$] = concentration of MgCl$_2$ in filtrate, mg/L.

The total suspended solids, V_s, may be calculated:

$$V_s = 100 - V_o - \frac{V_W}{(\rho_f - 10^{-6}\{[NaCl]+[CaCl]+[MgCl]\} - 0.00286[KCl])}$$

where V_W is the volume percentage of water.

Low-gravity solids calculated by the equation include bentonite. The quantity of drilled solids in the drilling fluid would be the difference between the volume percentage of (%vol) low-gravity solids and the %vol bentonite.

For freshwater drilling fluids, assuming that the density of low-gravity solids is 2.6 and the density of barite is 4.2, the equation becomes

$$V_{LG} = 62.5 + 2.0\,V_S - 7.5\,MW.$$

A 12.0-ppg drilling fluid with 20% volume of solids would contain 12.5% volume of low-gravity solids:

$$V_{LG} = 62.5 + 2.0(20) - 7.5(12.0\,\text{ppg}) = 12.5\%$$

The retort is normally used to determine the volume percent solids in a drilling fluid.

3.1 PROCEDURE FOR A MORE ACCURATE LOW-GRAVITY SOLIDS DETERMINATION

This procedure requires an oven, a pycnometer, and an electronic balance to weigh samples. A pycnometer can be made by removing the beam from a pressurized mud balance. Any type of balance may be used to determine weight; however, electronic balances are more convenient.

Determine the volume of the pycnometer:

1. Weigh the pycnometer (assembled).
2. Fill with distilled water.
3. Determine the water temperature.
4. Reassemble the pycnometer and pressurize it.
5. Dry the outside of the pycnometer completely.[1]
6. Weigh the pycnometer filled with pressurized water.

---

[1] Extreme care should be taken to ensure that the outside is dry. Even a small amount of water will destroy the accuracy of the test.

7. Determine the density of water using a table of density/temperature of water. (See Appendix.)
8. Subtract the pycnometer weight from the weight of the pycnometer filled with water, to determine the weight of water in the pycnometer.
9. Divide the weight of water in the pycnometer by the density of water to determine the volume of the pycnometer.

Determine the density of drilled solids:

1. Select large pieces of drilled solids from the shale shaker and wash them with the liquid phase of the drilling fluid (water for water-base drilling fluid, oil for oil-base drilling fluid, and synthetics for synthetic drilling fluid.)
2. Grind the drilled solids and dry them in the oven or in a retort.[2]
3. Weigh the assembled, dry pycnometer.
4. Add dry drilled solids to the pycnometer and weigh.
5. Add water to the solids in the pycnometer, pressurize, and weigh.[3]
6. Determine the density of the NAFs using the procedure used to calibrate the pycnometer with water.
7. Determine the density of the water.
8. Subtract the weight of the dry pycnometer from the weight of the dry pycnometer containing the dry drilled solids. This is the weight of drilled solids.
9. Subtract the weight of the dry pycnometer containing the drilled solids from the weight of the water, drilled solids, and pycnometer. This is the weight of water added to the pycnometer.
10. From the temperature/density chart for water, determine the density of the water.
11. Divide the weight of the water (determined in step 9) by the density of the water. This is the volume of water added to the pycnometer.
12. Subtract the volume of the water added to the pycnometer (step 10) from the volume of the pycnometer. This is the volume of drilled solids contained in the pycnometer.
13. Divide the weight of the drilled solids (step 8) by the volume of the drilled solids (step 11). This is the density of the drilled solids.

---

[2]Nonaqueous fluids (NAFs) should be dried in a retort.
[3]Use NAFs for barite and water, or NAFs may be used for drilled solids.

This may be used in the equations previously described to determine the %vol drilled solids in the drilling fluid.

Suggestion: Verify the accuracy of the procedure by using barite, whose density has been determined by the procedure in API RP 13C.

The pycnometer may also be used to make a more accurate determination of drilled solids in a water-base drilling fluid:

1. Determine the density of the drilling fluid with the pycnometer.
   - Weigh the pycnometer filled and pressurized with drilling fluid.
   - Subtract the weight of the dry pycnometer from the weight of the filled, pressurized pycnometer.
   - Divide the weight of drilling fluid in the pycnometer by the volume of the pycnometer. This is the density of the drilling fluid.
2. Weigh a metal or heat-resistant glass dish.
3. Add a quantity of drilling fluid to the dish and weigh.
4. Determine the weight of drilling fluid in the dish. (Subtract the weight of the dish from the weight of the dish and drilling fluid.)
5. Divide the weight of the drilling fluid in the dish by the density of the drilling fluid (step 1). This is the volume of drilling fluid.
6. Dry the material in the dish in the oven at 250°F for at least 4 hours.
7. Weigh the dish with the dry solids.
8. Subtract the dish weight from the weight of the dish and dry solids. This is the weight of the solids.
9. Divide the weight of the dry solids (step 8) by the weight of drilling fluid in the dish (step 4). This is the weight fraction of solids in the drilling fluid.
10. Subtract the weight of the dry solids (step 8) from the weight of drilling fluid in the dish (step 4). This is the weight of liquid in the drilling fluid.
11. Divide the weight of liquid in the drilling fluid (step 10) by the density of the liquid phase of the drilling fluid. This is the volume of the liquid phase.
12. Subtract the volume of the liquid phase (step 11) from the volume of the drilling fluid (step 4). This is the volume of solids in the drilling fluid.
13. Divide the volume of solids in the drilling fluid (step 12) by the volume of drilling fluid. This is the volume fraction of solids in the drilling fluid.

14. Multiply the volume fraction of solids in the drilling fluid by 100 to obtain the %vol solids in the drilling fluid.

3.1.1 Sample Calculation

During the drilling of a relatively uniform 2000-foot shale section, an API 200 continuous screen cloth was mounted on a linear shale shaker. An 11.2-ppg, freshwater, gel/lignosulfonate drilling fluid was circulated at 750 gpm while drilling. A typical set of samples will be described here.

Large pieces of shale were removed from the shaker screen and excess drilling fluid washed from the surface with distilled water. The shale pieces were ground and dried in an oven at 250°F overnight. The shale was placed in a 173.91-cc pycnometer and weighed. Water was added to the pycnometer and pressurized to about 350 psi. The increase in weight of the pycnometer indicated the volume of water added to fill the pycnometer. (Room and water temperature was 68°F, so the density of water was about 1.0 g/cc.) Subtracting this volume of water from the known volume of the pycnometer calculates the volume of shale sample. Once the volume of the shale sample and the weight were known, the density could be calculated. The shale drilled in this well had a density of 2.47 g/cc.

After movement of solids across the shale shaker screen appeared to be relatively uniform for more than 10 minutes, all the shaker discard was collected in a bucket. In 16.21 seconds, 3720.7 g of discard was captured. The discard rate was 13,772 g/min. The discard had a density of 1.774 g/cc or 14.8 ppg.

Calculation Procedure

A sample of the discard was placed in the pycnometer and weighed:

$$\text{pycnometer} + \text{sample weight} = 869.68 \text{ g}.$$

Since the pycnometer weighed 660.61 g dry and empty, the sample weight was 209.07 g.

The pycnometer with shaker discard sample was filled with distilled water, pressurized, and weighed:

$$\text{pycnometer} + \text{sample} + \text{water} = 948.32 \text{ g}.$$

The weight of water added was 948.32 g − 869.68 g = 78.64 g. Volume of 70°F. water added = 78.64 g/0.998 g/cc.
Since the pycnometer volume was 173.91 cc, the sample volume was

$$173.91 \text{ cc} - 78.80 \text{ cc} = 95.11 \text{ cc}$$

The density of the sample was 209.07 g/95.11 cc = 2.2 g/cc.

The objective of the shale shaker is to remove drilled solids, preferably without excessive quantities of drilling fluid. The fraction of the discard stream that is water, barite, and low-gravity solids can be determined by the preceding equations. These calculations indicate that the discard stream had 5.06 %vol barite, 38.38 %vol low-gravity solids, and 56.56 %vol water.

Calculation Procedure to Determine Low-Gravity Solids Discarded

The discard from the screen weighs 14.8 ppg and contains 43.44 %vol solids. We use the equation presented previously:

$$(\rho_B - \rho_{lg})V_{lg} = 100\rho_f + (\rho_B - \rho_f)V_s - 12 \text{ MW}$$

where MW = mud weight.

Assuming a barite density of 4.2 g/cc and a drilled-solids density of 2.47 g/cc, the equation becomes:

$$(4.2 - 2.47)V_{LG} = 100 + (4.2 - 1.0)(43.44) - 12(14.8)$$

$$V_{LG} = 38.38 \text{ %vol.}$$

To determine the quantity of drilled solids discarded by the shale shaker, a sample of the discarded material was placed in a metal dish and dried in an oven overnight. The weight percentage of (wt%) dry solids was 68.11 and had a density of 2.78 g/cc.

The rate of dry solids discarded (RDSD) is calculated from the product of the wet discharge flow rate and the weight fraction of dry solids in the discharge (with the appropriate unit conversion factors):

$$\text{RDSD} = (13{,}772 \text{ gpm})(0.6811)\{1.0 \text{ lb}/453.59 \text{ g}\}\{60 \text{ min/hr}\}$$

$$\text{RDSD} = 1239 \text{ lb/hr}$$

Experimental and Calculation Procedure

A sample of discard was placed in a 40.10-g crucible and weighed:

$$\text{crucible} + \text{sample weight} = 114.94 \text{ g.}$$

The wet sample weight was 74.84 g. Since the wet discard density was 1.77 g/cc, the wet sample had a volume of 74.84 g/1.77 g/cc = 42.19 g/cc.

After heating overnight at 250°F, the crucible and sample weight were 91.08 g. The dry solids weight in the sample was 91.08 g − 40.10 g = 50.98 g.

The wt% dry solids in the discard was the weight of dry solids divided by the wet-sample weight times 100, or

$$[50.98 \text{ g}/74.84 \text{ g}] \times 100 = 68.12 \text{ wt\%.}$$

The volume of the dry sample was calculated by subtracting the volume of water lost from the volume of the wet sample:

The 42.19-cc wet sample lost 114.94 cc − 91.08 cc = 23.86 cc of water.
The volume of the dry sample was 42.19 cc − 23.86 cc = 18.33 cc.
The density of the dry solids was the weight of dry solids divided by the volume of dry solids, or 50.98 g/18.33 cc = 2.78 g/cc.

Calculation of Barite Discarded by Shale Shaker

Assuming that all of the drilled and other low-gravity solids in the drilling fluid have a dried density of 2.47 g/cc and the barite has a density of 4.2 g/cc, the wt% barite in the dry sample may be calculated from the mass-balance equation:

$$\text{Density of Dry Solids} = \frac{\text{Weight of Solids}}{\text{Volume of Solids}}$$

or

$$\begin{bmatrix} \text{Density of} \\ \text{Dry Solids} \end{bmatrix} = \frac{\text{Weight of Barite} + \text{Weight of Low Gravity Solids}}{\text{Volume of Barite} + \text{Volume of Low Gravity Solids}}$$

To determine the terms on the right side of the equation:

1. The volume of barite is the density (4.2 g/cc) divided by the weight of barite.
2. The volume of low-gravity solids is the total volume of dry solids minus the volume of barite.

3. The volume of low-gravity solids in 1 cc of solids equals 1 cc minus the volume of barite in 1 cc of solids.

$$\begin{bmatrix} \text{Volume of low gravity} \\ \text{solids in 1 cc of solids} \end{bmatrix} = 1 \text{ cc} - \frac{W_B}{4.2 \text{ g/cc}}$$

$$\begin{bmatrix} \text{Weight of Low Gravity} \\ \text{Solids in 1 cc of dry solids} \end{bmatrix} = \left[1 - \frac{W_B}{4.2 \text{ g/cc}}\right](2.47 \text{ g/cc})$$

$$\text{Density of Solids } (D) = \frac{W_B + 2.47 \text{ g/cc}\left[1 - \frac{W_B}{4.2 \text{ g/cc}}\right] 1 \text{ cc}}{\frac{W_B}{4.2 \text{ g/cc}} + \left[1 - \frac{W_B}{4.2 \text{ g/cc}}\right]}$$

This equation may be reduced to the expression:

$$D = 0.4119 \, W_B + 2.47$$

or

$$\text{Weight percent barite} = \frac{D - 2.47}{0.4119}$$

The discard density is 2.78 g/cc, so the wt% barite is 27.07. The weight of dry discard from the shaker screen is 1239 lb/hr. The quantity of barite discarded is (0.2707)(1239 lb/hr), or 377 lb/hr. The low-gravity-solids discard rate is 1239 lb/hr − 377 lb/hr, or 862 lb/hr.

Calculation of Solids Discarded as Whole Drilling Fluid

A water-base drilling fluid contains 13% volume of solids in the liquid phase of the shale shaker discard, which could be associated with the whole drilling fluid.

The wt% dry solids discarded from the shaker screen is calculated to be 68.12; so 31.89% of the discard must be liquid. Assume that this liquid is composed of drilling fluid with the solids distribution of the drilling fluid in the pits. The liquid discard rate is (13,772 g/min)(0.3189), or 4391.9 g/min. This liquid should contain 13% volume of solids.

Since the drilling fluid contains 13% volume of solids, a 100 cc sample contains 87 cc of liquid. In this 100 cc sample, the water fraction would weigh 87 g. With an 11.2-ppg (1.343 g/cc) density drilling fluid, the 100 cc sample should weigh 134.3 g. Since the liquid weighs 87 g, the solids must weigh 47.3 g. Or, stated another way, the drilling fluid contains 47.3 g of solids for every 87 g of water. The total liquid discard rate is 4391.9 g/min. The solids discarded by the screen that are associated with the drilling fluid would be

(47.3 g solids/87 g water)(4391.9g/min) = 2387.8g/min, or 315.6 lb/hr.

The wt% barite in the drilling fluid is 77.4 and the wt% low-gravity solids in the drilling fluid is 22.4. From the solids discarded from the screen associated with the whole drilling fluid, 244 lb/hr are barite and 71.2 lb/hr are low-gravity solids.

Previously, the dry solids discarded by the shaker screen were calculated to be 377 lb/hr barite and 861 lb/hr low-gravity solids. Subtracting the solids associated with the drilling fluid from the solids removed by the screen indicates the discarded solids in excess of those associated with the drilling fluid:

Barite:

$$377 \text{ lb/hr} - 244 \text{ lb/hr} = 133 \text{ lb/hr}$$

Low-gravity solids:

$$861.0 \text{ lb/hr} - 71.2 \text{ lb/hr} = 789.8 \text{ lb/hr}$$

This indicates that the API 200 screen is removing 133 lb/hr of barite and almost 800 lb/hr of drilled solids in addition to the quantity contained in the associated drilling fluid.

Note that the technique of using the concentration of barite in the discard does not allow an accurate measurement of the quantity of drilling fluid in the shaker discard. Some measurements even indicate that less barite is in the discard than is in the whole drilling fluid. Shaker screens can pass much of the small-size barite and remove it from the liquid before it is discarded by the shaker screen.

3.2 DETERMINATION OF VOLUME PERCENTAGE OF LOW-GRAVITY SOLIDS IN WATER-BASED DRILLING FLUID

The %vol low-gravity solids can also be calculated from the average specific gravity (ASG) of the solids in the sample:

$$\text{ASG} = \frac{12\text{MW} - (\%\text{vol water})(\rho_w) - (\%\text{vol oil})(\rho_o)}{\%\text{vol solids}}$$

$$\frac{\%\text{vol Low}}{\text{Gravity Solids}} = \frac{(\%\text{vol solids})(\text{Barite SG} - \text{ASG})}{(\text{Barite SG} - \text{Low-Gravity SG})}$$

The ASG of the solids in the drilling fluid can be determined independently with the pressurized mud balance in the same manner as the low-gravity solids and the barite. (See procedure below.)

3.3 RIG-SITE DETERMINATION OF SPECIFIC GRAVITY OF DRILLED SOLIDS

If a laboratory balance is not available at the rig site, have a laboratory determine the volume of a pressurized mud balance and stamp the volume, in cubic centimeters, on the cup. For example, assume 240 cc:

- Collect some representative cuttings (about 100 g) from the shaker screen.
- Wash the cuttings in distilled water (or oil for an oil mud) to remove the drilling fluid.
- Place the shale cuttings in several retorts and remove the liquid from the cuttings.
- Add the dry cuttings to the cup of the pressurized mud balance and determine a mud weight.

Assume that the dry-solids weight is 9.1 ppg. To determine the weight of cuttings:

- Multiply the dry-solids weight by the cup volume and divide by 8.34 ppg/g/cc.

$$[9.1 \text{ ppg} \times 240 \text{ cc}]/(8.34 \text{ ppg/g/cc}) = 262 \text{ g}.$$

To determine the volume of cuttings:

- Add water to the dry solids (stir to remove as much air as possible), pressurize the cup, and read the mud balance.[4]
- Calculate the volume of cuttings by the equation:

$$\text{Volume of Cup} - \{[(\text{Mud Weight} \times \text{Cup Volume})/8.34] - \text{Weight of Cuttings}\}.$$

---

[4] Make certain that the outside of the mud balance is completely dry.

Solids Calculation

Assume that the pressurized balance reads 13.7 ppg. The weight of cuttings and water in the cup can be found by dividing the weight by 8.34 ppg/g/cc and multiplying by the volume of the cup. In the example:

$$(13.7 \text{ ppg} \times 240 \text{ cc})/(8.34 \text{ ppg/g/cc}) = 394.2 \text{ g}.$$

Since the cup contains 262 g of solids, it must contain 132 g of water: [394 g − 262 g = 132 g], or 132 cc of water. The cup holds 240 cc of slurry with 132 cc of water, so it contains 108 cc of cuttings:

$$[240 \text{ cc} - 132 \text{ cc} = 108 \text{ cc}].$$

The density of the cuttings may be determined by dividing the weight of the cuttings by their volume. In this example: Density = 262 g/108 cc, or 2.43 g/cc.

CHAPTER 4

CUT POINTS

Bob Barrett
M-I SWACO

Brian Carr
M-I SWACO

Cut points are used to indicate the separation characteristics of solids-control equipment at a given moment in time. The performance of the equipment, in addition to the condition of the drilling fluid, should be taken into consideration in the assessment of cut point data. Cut point curves are derived from the collected data and indicate, at the actual moment of data collection, the percentage of chance that a particle of a particular size can flow through or be discarded by the solids-control equipment. Therefore, the cut point curve is a function of the physical properties of the solids (i.e., density), particle size distribution of the solids, physical condition of the solids-control equipment (i.e., sealing capabilities), and the drilling-fluid properties.

Cut points may be determined for all drilled-solids removal equipment. The mass flow rate of various-size particles discarded from the equipment is compared with the mass flow rate of the same-size particles presented to the equipment. When testing a particular unit, knowledge of the feed flow rate to the unit and the two discharge flow rates are required. The density of the feed flow multiplied by the volume flow rate provides the mass flow rate into the unit. Discharge mass flow rates are also calculated by multiplying the density of the stream by the volume flow rate. Obviously, the sum of the discharge mass flow rates must be equal to the feed mass flow rate. Usually one of the discharge flow streams is discarded and the other is retained in the drilling fluid.

The material balance—both the volume flow rate balance and the mass flow rate balance—should be verified before measuring the particle sizes of the various streams.

Solids-removal equipment removes only a very small fraction of the total flow into the equipment. For example, a 4-inch desilter processing about 50 gpm of drilling fluid will discard only about 1 gpm of material. Since the discarded material is such a small proportion of the total material processed, the difference between the retained stream and the feed stream is difficult to measure. For this reason, more accurate data are acquired by mathematically adding the value of the discarded solid concentrations to that of the retained solids concentration to determine the feed solids concentration.

To determine the mass flow of a particular-size particle in the feed (or retained) stream and the mass flow of the same-size particle in the discard, flow rate measurements and solids concentrations are needed. The discard volume flow rates are normally relatively low, but the feed rates require using a flow meter or a positive displacement pump.

For shale shakers, the feed to the shaker will be the circulating rate coming from the well. Mud pumps must be calibrated to provide an accurate feed rate. While drilling, move the suction from the suction tank to the slug tank and measure the rate of drop of the fluid leaving the slug pit. The fluid in the slug tank will contain liquid and gas (or air), so the volume percentage of (%vol) gas must be subtracted from the volume of fluid leaving the slug tank. The %vol gas is calculated by dividing the difference between the pressurized mud weight and the unpressurized mud weight by the pressurized mud weight and multiplying by 100. If the desilters or centrifuges are fed by centrifugal pumps, some type of flow meter will be required to accurately determine the feed rate. The flow meter could be a large container whose volume is calibrated and a stopwatch. A centrifuge underflow volume flow rate is difficult to measure because of the high concentration of solids. A barrel or other large container can be split vertically and support beams or pipes welded to provide a support when the container is placed across the top of a mud tank. Calibrated lines are painted inside of the container to provide volume measurements. A quantity of water is placed in the container and the container is positioned adjacent to a decanting centrifuge mounted on top of a mud tank. The stopwatch is started when the container is pushed under the centrifuge, and the rate of water level is observed. The known volume between lines and the time permit calculation of the volume discard rate. Representative samples of the underflow or heavy slurry provide the density measurements of the underflow. After confirming that there is a mass and volume flow balance with the measured values, the particle sizes in the discharge streams are determined.

All of the discard stream may be captured for analysis during a period of several minutes. The contents of the feed stream during that period must be known so that the ratio of discard to feed particle mass can be determined for various particle sizes. The feed stream and retained stream for shakers and desilters, however, would require much larger containers, and it is impractical to try to weigh or measure their volumes directly. Representative samples of the retained stream must be used to determine the mass of various-size particles.

With the centrifuge and the desilters, the particle sizes must be measured with an instrument that discerns particle sizes as small as 1 micron. With the shaker measurements, sieves may be used because the cut point range will be within the range of screens standardized by the American Society for Testing and Materials (ASTM). A variety of different laboratory devices are available that measure small-diameter particles. Instruments using lasers are popular in many laboratories.

The discard sample will contain the solids and the liquid phase of the drilling fluid. With the shale shaker discard, the mass of solids retained on each ASTM test screen may be measured directly by weighing the solids after they are dried. With the desilter underflow and the centrifuge underflow (or heavy slurry) discharge, the density of the solids must be used to determine the mass percentage of solids.

Cut points for shale shakers are measured by determining the particle size distribution of the feed and discard streams with the use of a stack of U.S. Standard Sieves. The flow rate of each stream is determined, and the mass flow rate for each sieve size in each stream is calculated. The mass flow rate of the discard stream for each sieve size is divided by the mass flow rate for the same size introduced into the equipment in the feed stream.

Using this method, the feed-stream sample represents a small fraction of the total overall flow. This can create a problem with material balances. A better method is to sample the discard and underflow streams. Combining these two solids distributions will yield a more accurate cut point curve. This method can be used on solids-control equipment in which the feed-stream flow rate is greater than the discard stream.

Samples of the discard and underflow streams are taken from the solids-control equipment for analysis. The density of all streams is measured. The volume flow rate of the discard stream is measured by capturing all of the discard stream in a container—a section of gutter works well at the discard end of a shaker screen. The volume flow rate of the discard stream is determined by multiplying the mass of fluid captured by the density, or mud weight, of the discard. The volume flow rate of the feed is determined by accurately measuring the flow rate from the rig pump. The mass flow rate of the feed is calculated by multiplying the density of the drilling fluid by the circulating flow rate. Each sample is wet sieved over a stack of U.S. Standard Sieves with a broad distribution of sizes. The excess drilling fluid is washed through the screen with the liquid phase of the drilling fluid. The samples at each individual sieve size are thoroughly dried. Weights of the solids retained at each individual sieve size are measured, and the flow rate for each stream at each individual sieve size is calculated. To determine the screen cut point curve, the quantity of a particular-size particle in the discard must be compared with the quantity of the same-size particles presented to the screen. All of the discard can be captured, and all of the mass of the discard solids of a particular size can be determined. However, it is impractical to try to capture and sieve all of the fluid passing through the screen during the period that the discard is being captured. For example, if the rig flow is 500 gpm and the discard sample is captured during a 3.50-min period, the underflow through the shaker screen would be 1750 gal. If the mud weight is 9.2 ppg, this means that 16,100 lb of drilling fluid has passed through the screen. A 9.2-ppg drilling fluid with no barite and 2.6 specific-gravity low-gravity solids would have 6.5% volume of solids. The total solids that would be presented to the screen during the 3.5-min period would be 113.75 gal [6.5% of 1750 gal] or 2467 lb of solids [(114 gal)(2.6)(8.34 ppg)]. Since it is not practical to capture and sieve this quantity of solids, a representative sample of the underflow through a screen can be used to determine the solids concentration and sizes that did pass through the screen. The flow rate of the underflow sample and the dry weight of the individual sieve sizes must be measured. This is the reason that flow rates of the dry solids are used in the calculations instead of using all of the solids captured in a specific time interval.

The corresponding feed mass flow rate (sum of discard and underflow rates) for each individual sieve size is also determined. The ratio of the discard and the feed flow rates at each sieve size determines the percentage of solids discarded over the solids-control equipment. The size of the sieves (expressed in microns) versus the percentage of solids removed produces a cut point curve.

A cut point curve graphically displays the fraction of various-size particles removed by the solids-control equipment compared with the quantity of that size of particle presented to the equipment. For example, a D_{50} cut point is the intersection of the 50% data point on the Y axis and the corresponding micron size on the X axis on the cut point graph. This cut point indicates the size of the particle in the feed to the solids-control equipment that will have a 50% chance of passing through the equipment and a 50% chance of discharging off of the equipment. Frequently, solids-distribution curves are *erroneously* displayed as cut point curves. Cut point curves indicate the fraction of solids of various sizes that are separated. They also are greatly dependent on many drilling-fluid factors and indicate the performance of the complete solids-control device only at the exact moment in time of the data collection. The cut points of the solids-control equipment will be determined by the physical condition of the equipment and the properties of the drilling fluid.

Following is a procedure detailing the required steps to perform this method of particle-size analysis and the calculations used to create a cut point curve. An example of data collected and analyzed using a shale shaker is included after the detailed procedure. The example demonstrates useful information that can be obtained by following the procedure. This procedure is most applicable to performing cut point analysis with a shale shaker. Therefore, the example data measure solids to only 37 microns (No. 400 sieve).

Calculating cut point curves for hydrocyclones and centrifuges should use methods other than sieving. Measurements with a No. 635 sieve (20 microns) is about the limit of sieve analysis, but information is required about particles much smaller. Particle size analysis equipment, such as laser diffraction, is required for measurements of smaller sizes of solids. However, the assumption that the solids being analyzed have a constant density would have to be made.

4.1 HOW TO DETERMINE CUT POINT CURVES

Feed:

1. If a flow meter is unavailable, determine the flow rate to the solids-control equipment. To calculate the flow rate, one must know the fluid pump's gallons per stroke, strokes per minute, and efficiency. The flow rate can then be calculated by:

    flowrate = (cylinder volume · N)(spm) (pump efficiency)

where
- cylinder volume = (((pump sleeve inner diameter in inches)$^2 \cdot \pi$)/4) · pump stroke length in inches (0.00433 in^3/gal)
- N = number of pump cylinders
- (spm) = strokes per minute

2. Take a representative sample from the feed stream and measure the density.

Underflow:

1. Weigh the sampling container. A minimum container size of 5 gal is recommended in order to capture a large sample of solids.
2. Take a representative sample from the underflow (effluent) stream of the solids-control equipment system (Figure 4.1; note that using a smaller container to fill the larger sampling container will not adversely affect the solids sample).
3. Weigh the sampling container and effluent sample.
4. Calculate the weight of the effluent sample:

   weight of effluent sample = effluent sample and container
   − weight of container

5. Wet sieve and dry the sieved solids thoroughly. Slowly pour the collected sample through a stack of U.S. Standard Sieve screens with a broad distribution of micron opening sizes (see Section 4.2 for a representative distribution of sieve sizes). A gentle stream of water is used to wash the solids and to assist the sieving process (Figures 4.2 and 4.3). Once the sample has completely passed through the stack of sieves, each sample of solids on each individual sieve must be dried. Drying can be accomplished by placing the sample in a static oven[1] and heating at a maximum temperature of 250°F until all of the water has evaporated. If an oven is unavailable, the samples may also be allowed to slowly air dry.
6. Measure the weight of dry solids captured on each size of sieve screen. These will be the *weights of individual dry effluent solids*.

---

[1]This method applies to water-base fluids only. For oil-base fluids, proper cleansing and drying of the sample should be administered in order to extract all residual fluids from the solids.

Figure 4.1. Sample is taken directly from the effluent stream.

Figure 4.2. Effluent sample is wet sieved by pouring over a stack of U.S. Standard Sieves.

Figure 4.3. Water is used to assist the wet-sieving process.

Discard:

1. Weigh the trough that will be used to collect the discard sample.
2. Collect the discard sample off the end of the solids-control equipment (Figure 4.4).
3. Measure the time (in minutes) for which all the discard is collected from the solids-control equipment. This will be the *time of discard sample*.
4. Weigh the discard sample and trough.
5. Calculate the weight of the discard sample in the trough:

    wet discard sample weight = discard sample and trough
    − weight of trough

6. Wet sieve and dry the sieved solids thoroughly.[2]

---

[2]This method applies to water-base fluids only. For oil-base fluids, proper cleansing and drying of the sample should be administered in order to extract all residual fluids from the solids.

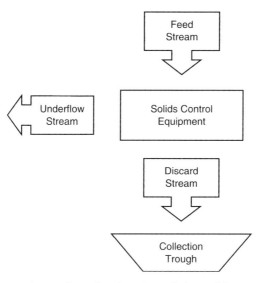

Figure 4.4. Trough is used to collect discard sample from solids-control equipment.

Take a representative sample from the discarded solids and slowly pour through a stack of U.S. Standard Sieve screens. Use the same sizes of sieves used for the underflow sample, and follow the same procedure: Wash the solids with a gentle stream of water, which also assists the sieving process (Figures 4.2 and 4.3). Once the sample has completely passed through the stack of sieves, dry each sample of solids on each individual sieve. Drying can be accomplished by placing the sample in a static oven[3] and heating at a maximum temperature of 250°F until all of the water has evaporated. If an oven is unavailable, the samples may also be allowed to slowly air dry.

7. Measure the weight of dry solids captured on each size of sieve screen. These will be the *weights of individual dry discard solids*.

Plotting the Cut Point Curve

1. Determine the wet discard flow rate:

    wet discard flow rate = wet discard sample weight/
    time of discard sample.

---

[3]This method applies to water-based fluids only. For oil-based fluids, paper cleansing and drying of the sample should be administered in order to extract all residual fluids from the solids.

2. Determine the effluent flow rate:

    effluent flow rate = well flow rate − wet discard flow rate.

3. Calculate the time taken for the effluent sample:

    effluent sample time = weight of effluent sample/effluent flow rate.

4. For each U.S. Standard Sieve screen size, determine the rate of solids collected for the discard sample:

    discard flow rate = weight of individual dry discard solids/ time of discard sample.

5. For each U.S. Standard Sieve screen size, determine the rate of solids collected for the effluent sample:

    effluent flow rate = weight of individual dry effluent solids/ effluent sample time.

6. Determine the feed flow rate for each sieve size:

    feed flow rate = dry discard flow rate + effluent flow rate.

7. Calculate the percentage of discarded solids for each sieve size:

    percentage of discard = (discard flow rate/feed flow rate) · 100.

8. Plot the percentage of discard on the Y axis with the corresponding sieve size (expressed in microns) along the X axis of a graph to produce the cut point curve for the analyzed system. The cut point curve would actually be a series of horizontal lines between sieve sizes. The curve is usually drawn through the center of each segment to produce a smooth curve.

4.2 CUT POINT DATA: SHALE SHAKER EXAMPLE

Create a shale shaker cut point curve using the following known data:

- Well flow rate = 560 gpm
- Density of feed = 8.90 lb/gal
- Container used to collect effluent sample = 1.80 lb
- Total effluent sample and container weight = 41.5 lb
- Trough used to collect discard sample = 38.1 lb
- Time to collect the discard sample = 1.00 minute
- Total discard sample and trough weight = 56.5 lb

1. Calculate the mass flow rate of the system = 560 gpm × 8.90 lb/gal = 4984 lb/min.
2. Determine the weight of the effluent sample = 41.5 lb – 1.80 lb = 39.7 lb.
3. After sieving, drying, and weighing the effluent solids, document the individual weights of the solids on each size sieve.
4. Calculate the weight of the discard sample = 56.5 lb – 38.1 lb = 18.4 lb.
5. After sieving, drying, and weighing the discard solids, record the individual weights of the solids on each size sieve.
6. Calculate the wet discard flow rate = 18.4 lb ÷ 1.00 minute = 18.4 lb/min.
7. Calculate the effluent flow rate = 4984 lb/min – 18.4 lb/min = 4965.6 lb/min.
8. Calculate the effluent sample time = 39.7 lb ÷ 4965.6 lb/min = 0.008 minutes
9. Determine the rate of solids collected on each individual sieve size for the discard sample. Example for 37 micron sieve = 8.80 grams ÷ 1.00 minute = 8.80 grams/minute
10. Determine the rate of solids collected on each individual sieve size for the effluent sample. Example for 37 micron sieve = 17.7 grams ÷ 0.008 minutes = 2214 grams/minute
11. Determine the feed flow rate for each sieve size. Example for 37 micron sieve = 9 grams/minute + 2214 grams/minute = 2223 grams/minute
12. Calculate the percent of discard solids for each sieve size. Example for 37 micron sieve = (8.80 grams/minute ÷ 2223 grams/minute) × 100 = 0.40%

Individual Dry-Solids Weights and Cut Point Curve Calculations

Mesh Size	Opening (microns)	Dry Euent Solids Wt. (grams)	Dry Discard Solids Wt. (grams)	Discard Flow Rate (gm/min)	Euent Flow Rate (gm/min)	Feed Flow Rate (gm/min)	Percent of Discard
400	37	17.7	8.8	9	2214	2223	0
325	44	7.1	8.8	9	888	897	1
270	63	14.2	11.7	12	1776	1788	1
200	74	10.7	38.1	38	1338	1376	3
140	105	2.3	85.2	85	288	373	23
120	118	0.7	339.2	339	88	427	79
100	140	0.2	550.7	551	25	576	96
80	177	0.3	1468.0	1468	38	1505	98
60	234	0	23450.7	23451	0	23451	100

13. Plot the sieve sizes versus the percent discard

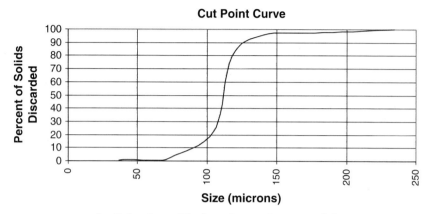

Cut Point Curve Displays Seperation Potential

CHAPTER 5

TANK ARRANGEMENT

Mark C. Morgan
Derrick Equipment Co.

The purpose of a drilling rig surface fluid processing system is to provide a sufficient volume of properly treated drilling fluid for drilling operations. The active system should have enough volume of properly conditioned drilling fluid above the suction and equalization lines to keep the well bore full during wet trips.

The surface system needs to have the capability to keep up with the volume-building needs while drilling; otherwise, advanced planning and premixing of reserve mud should be considered. This should be planned for the worst case, which would be a bigger-diameter hole in which high penetration rates are common. For example for a 14¾-inch hole section drilling at an average rate of 200 ft/hr and with a solids-removal efficiency of 80%, the solids-removal system will be removing approximately 34 barrels of drilled solids per hour plus the associated drilling fluid coating these solids. More than likely, 2 barrels of drilling fluid would be discarded per barrel of solids. If this is the case, the volume of drilling fluid in the active system will decrease by 102 barrels per hour. If the rig cannot mix drilling fluid fast enough to keep up with these losses, reserve mud and or premixed drilling fluid should be available to blend into the active system to maintain the proper volume.

The surface system should consist of three clearly identifiable sections (Figure 5.1):

- Suction and testing section
- Additions section
- Removal section

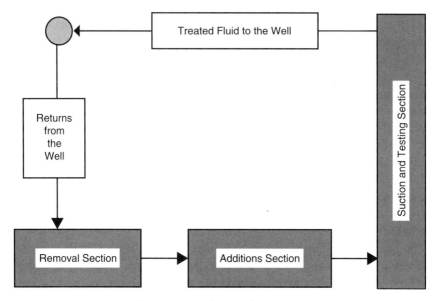

Figure 5.1. Surface circulation system.

5.1 ACTIVE SYSTEM

5.1.1 Suction and Testing Section

The suction and testing section is the last part of the surface system. Most of the usable surface volume should be available in this section. Processed and treated fluid is available for various evaluation and analysis procedures just prior to the fluid recirculating downhole. This section should be mixed, blended, and well stirred. Sufficient residence time should be allowed so that changes in drilling-fluid properties may be made before the fluid is pumped downhole. Vortex patterns from agitators should be inhibited to prevent entraining air in the drilling fluid.

In order to prevent the mud pumps from sucking air, vertical baffles can be added in the tank to break up the possible vortex patterns caused by the agitators. If the suction tank is ever operated at low volume levels, additional measures should be taken to prevent vortexing, such as adding a flat plate above the suction line to break up the vortex pattern.

Proper agitation is very important, so the drilling fluid is a homogeneous mixture in both the tank and the well bore. This is important because if a "kick" (entrance of formation fluid into the well bore due to a drop in hydrostatic pressure) occurs, an accurate bottom-hole pressure can be calculated. The well-control procedures are based on the required bottom-hole pressure needed to control the formation pressures. If this value is not calculated correctly, the well bore will see higher than necessary pressures during the well-control operation. With higher than required pressure, there is always the risk of fracturing the formation. This would bring about additional problems that would be best avoided whenever possible. For agitator sizing, see Chapter 10 on Agitation.

5.1.2 Additions Section

All commercial solids and chemicals are added to a well-agitated tank upstream from the suction and testing section. New drilling fluid mixed on location should be added to the system through this tank. Drilling fluid arriving on location from other sources should be added to the system through the shale shaker to remove unwanted solids.

To assist homogeneous blending, mud guns may be used in the additions section and the suction and testing section.

5.1.3 Removal Section

Undesirable drilled solids and gas are removed in this section before new additions are made to the fluid system. Drilled solids create poor fluid properties and cause many of the costly problems associated with drilling wells. Excessive drilled solids can cause stuck drill pipe, bad primary cement jobs, or high surge and swab pressures, which can result in lost circulation and/or well-control problems. Each well and each type of drilling fluid has a different tolerance for drilled solids.

Each piece of solids-control equipment is designed to remove solids within a certain size range. Solids-control equipment should be arranged to remove sequentially smaller and smaller solids. A general range of sizes is presented in Table 5.1 and in Figure 5.2.

The tanks should have adequate agitation to minimize settling of solids and to provide a uniform solids/liquid distribution to the hydrocyclones and centrifuges. Concerning the importance of proper agitation in the operation of hydrocyclones, efficiency can be cut in half when the suction tank is not agitated, versus one that is agitated. Unagitated suction tanks usually result in overloading of the hydrocyclone or plugged apexes. When a hydrocyclone is overloaded, its removal efficiency is reduced. If the apex becomes plugged, no solids removal

Table 5.1
Size of Solids Removed by Various Solids-control Equipment

Equipment	Size	Median Size of Removed Microns
Shale Shakers	API 80 screen	177
	API 120 screen	105
	API 200 screen	74
Hydrocyclones (diameter)	8-inch	70
	4-inch	25
	3-inch	20
Centrifuge		
Weighted mud		> 5
Unweighted mud		< 5

occurs and its efficiency then becomes zero. Agitation will also help in the removal of gas, if any is present, by moving the gaseous drilling fluid to the surface of the tank, providing an opportunity for the gas to break out.

Mud guns can be used to stir the tanks in the additions section provided careful attention is paid to the design and installation of the mud gun system. If mud guns are used in the removal section, each mud gun should have its own suction and stir only that particular pit. If manifolding is added to connect all the guns together, there is a high potential for incorrect use, which can result in defeating proper sequential separation of the drilled solids in an otherwise well-designed solids-removal setup. Manifolding should be avoided.

5.1.4 Piping and Equipment Arrangement

Drilling fluid should be processed through the solids-removal equipment in a sequential manner. The most common problem on drilling rigs is improper fluid routing, which causes some drilling fluid to bypass the sequential arrangement of solids-removal equipment. When a substantial amount of drilling fluid bypasses a piece or pieces of solids-removal equipment, many of the drilled solids cannot be removed. Factors that contribute to inadequate fluid routing include ill-advised manifolding of centrifugal pumps for hydrocyclone or mud cleaner operations, leaking valves, improper setup and use of mud guns in the removal section, and routing of drilling fluid incorrectly through mud ditches.

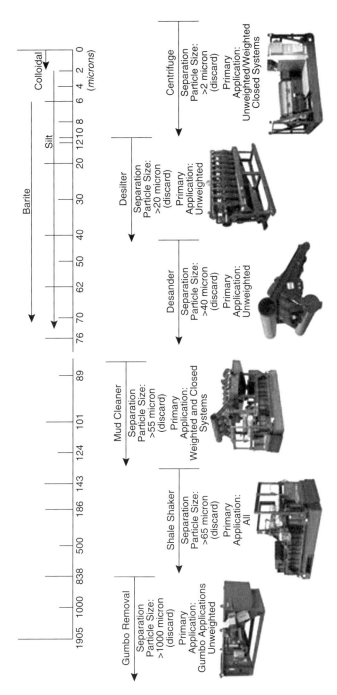

Figure 5.2. General solids control equipment removal capabilities.

Each piece of solids-control equipment should be fed with a dedicated, single-purpose pump, with no routing options. Hydrocyclones and mud cleaners have only one correct location in tank arrangements and therefore should have only one suction location. Routing errors should be corrected and equipment color-coded to eliminate alignment errors. If worry about an inoperable pump suggests manifolding, it would be cost saving to allow easy access to the pumps and have a standby pump in storage. A common and oft-heard justification for manifolding the pumps is, "I want to manifold my pumps so that when my pump goes down, I can use the desander pump to run the desilter." What many drilling professionals do not realize is that improper manifolding and centrifugal-pump operation is what fails the pumps by inducing cavitation. Having a dedicated pump properly sized and set up with no opportunity for improper operation will give surprisingly long pump life as well as process the drilling fluid properly.

Suction and discharge lines on drilling rigs should be as short and straight as possible. Sizes should be such that the flow velocity within the pipe is between 5 and 10 ft/sec. Lower velocities will cause settling problems, and higher velocities may induce cavitation on the suction side or cause erosion on the discharge side where the pipe changes direction. The flow velocity may be calculated with the equation:

$$\text{Velocity, ft/sec} = \frac{\text{Flow rate, gpm}}{2.48(\text{insided diameter in})^2} \quad (1)$$

Pump cavitation may result from improper suction line design, such as inadequate suction line diameter, lines that are too long, or too many bends in the pipe. The suction line should have no elbows or tees within three pipe diameters of the pump section flange, and their total number should be kept to a minimum. It is important to realize that an 8-inch, 90° welded ell has the same frictional pressure loss as 55 feet of straight 8-inch pipe. So, keep the plumbing fixtures to a minimum.

5.1.5 Equalization

Most compartments should have an equalizing line, or opening, at the bottom. Only the first compartment, if it is used as a settling pit (sand trap), and the degasser suction tank (typically the second compartment) should have a high overflow (weir) to the compartment downstream.

The size of the equalizing pipes can be determined by the following formula:

$$\text{Pipe diameter} = \sqrt{\frac{\text{Max. Circulation Rate, gpm}}{15}} \qquad (2)$$

A pipe of larger diameter can be used, since solids will settle and fill the pipe until the flow velocity in the pipe is adequate to prevent additional settling (5 ft/sec).

An adjustable equalizer is preferred between the solids-removal and additions sections. The lower end of an L-shaped, adjustable equalizer, usually field fabricated from 13-inch casing, is connected to the bottom of the last compartment in the removal section. The upper end discharges fluid into the additions section and can be moved up or down. This controls the liquid level in the removal section and still permits most of the fluid in the suction section to be used.

5.1.6 Surface Tanks

Most steel pits for drilling fluid are square or rectangular with flat bottoms. Each tank should have adequate agitation except for settling tanks. Additionally, each tank should have enough surface area to allow entrained air to leave the drilling fluid. A rule of thumb for calculating the minimum active surface pit area is:

$$\text{Area, ft}^2 = \frac{\text{Flow rate (gpm)}}{40} \qquad (3)$$

For example, if the active circulating rate is 650 gpm, the surface area of each active compartment should be about 16 square feet. The depth of a tank is a function of the volume needed and ease of stirring. Tanks that are roughly cubical are most efficient for stirring. If this is not convenient, the depth should be greater than the length or width. If circular tanks are used, a conical bottom is recommended and centrifugal pump suction and/or a dump valve should be located there. Another consideration is that the tanks need to be deep enough to eliminate the possibility of vortexing at the centrifugal pump suction. The depth required is a function of the velocity of the drilling fluid entering the suction lines (Figures 5.3, 5.4, 5.5, and 5.6).

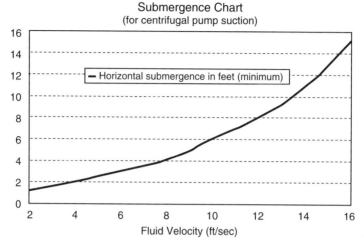

Figure 5.3. Submergence chart for centrifugal pump section.

Tank Design and Equipment Arrangement

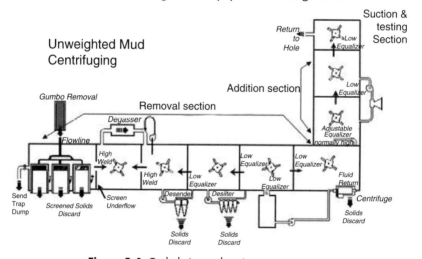

Figure 5.4. Tank design and equipment arrangement.

5.1.7 Sand Traps

After the drilling fluid passes through the main shaker, it enters the mud pit system. When screens 80-mesh and coarser were routinely used, the sand trap performed a very useful function. Large, sand-size particles would settle and could be dumped overboard. The bottom of a sand trap

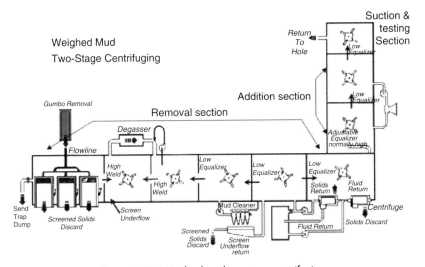

Figure 5.5. Weighted mud two-stage centrifuging.

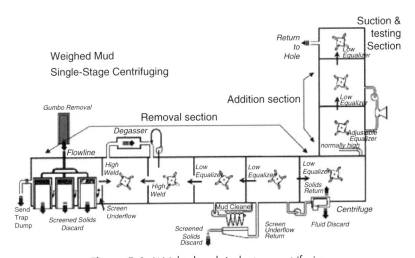

Figure 5.6. Weighted mud single-stage centrifuging.

should be sloped at about 45° to facilitate quick dumping. A sloped bottom 45° or greater will self-clean when dumped. The sand trap should not be agitated and should overflow into the next compartment. Linear and balanced elliptical motion shale shakers have all but eliminated this technique. Small drilled solids generally do not have sufficient residence time to settle. When inexpensive drilling fluid was used, sand traps were dumped once or twice per hour. Today, in the era of fine-mesh screens, expensive waste disposal, and environmental concerns, such dumping is either not allowed or is cost prohibitive.

The preceding illustrations show the solids-removal system with a sand trap. Rigs currently operating may or may not have sand traps. If a rig does not have a sand trap, then the shakers would have their underflow directed to the degasser suction pit and all other functions would remain as illustrated.

5.1.8 Degasser Suction and Discharge Pit

For proper operation of a vacuum-type degasser, the suction pit should be the first pit after the sand trap, or if no sand trap is present, then the first pit. This pit should typically be agitated in order to help roll the drilling fluid and break out as much gas, if present, as possible. The processed fluid flows into the next pit downstream. There needs to be a high equalizer or weir between these two tanks.

The degasser discharge pit is also the suction pit for the centrifugal pump used to pump drilling fluid through the eductor on the degasser. This is commonly called *power mud*. Pumping power mud through the eductor actually pulls the fluid out of the degasser vessel from the degasser suction pit and out to the discharge line due to the Bernoulli effect, causing a low-pressure zone in the eductor. The discharge from the eductor goes back into the same tank used for the suction for the power mud.

The reason that mud is sucked into the vacuum degasser and through the degasser vessel is that a centrifugal pump will not pump gaseous mud; therefore it cannot be pumped through the vessel and has to be sucked into it. (For complete information on operation of degassers, refer to Chapter 9 (Gas Busters, Separators, and Degassers) in this book.)

5.1.9 Desander Suction and Discharge Pits

The degasser discharge pit is also the suction pit for the desander. The desander, as well as the desilter, needs to be downstream of the degasser operation. If the hydrocyclone suction is upstream of the degasser operation and gas is present in the mud, the efficiency of the centrifugal pump will be reduced, or the pump will become gas locked and simply not pump any mud. Additionally, induced cavitation can occur and cause premature wear to the centrifugal pump. This wear can be rapid and severe.

The desander discharge (cone overs) should flow into the next pit downstream, and a low equalizer between these tanks should be opened. This allows backflow through the equalizer when the cone manifold is processing a greater volume than is entering the tank (recommended). This ensures that all of the drilling fluid is processed through the desander manifold.

Desander operation is typically recommended only for unweighted drilling fluids. If operated with weighted drilling fluid, the desander will discard a lot of drilling fluid away, including a lot of weight material.

5.1.10 Desilter Suction and Discharge Pits (Mud Cleaner/Conditioner)

The desilter suction pit is the desander discharge pit. The desilter will remove smaller particles than the desander, so its operation is downstream of the desander. Setup and operation of desilters are the same as with desanders. The manifold discharge is downstream of the suction, with a low equalizer between the two tanks. It is recommended that the desilter process more volume than the rig is pumping so that there is a backflow through the equalizer, ensuring that all of the drilling fluid is processed.

If drilling fluid is being pumped through mud guns from the suction compartment downstream, this fluid must also be processed through the hydrocyclones. For weighted drilling fluids, the underflow of the desilter cones is processed by a shaker. Ideally this shaker will have screens installed that allow the weight material to pass through while rejecting any drilled solids larger than the weight material.

5.1.11 Centrifuge Suction and Discharge Pits

Centrifuge suction is taken from the pit that the desilter manifold discharges into (for unweighted drilling fluids). The drilled solids removed by the centrifuge are discarded, and the cleaned drilling fluid is returned to the active system in the next pit downstream.

For a weighted aqueous drilling fluid, the solids separated by a centrifuge are composed largely of weight material (assuming upsteam processing has been performed correctly) used to increase the density of the drilling. This solids discharge (centrate or cake) is returned to the active system and the effluent or liquid discharge is discarded. The effluent contains the fine particles (colloidal or clay size) that will cause rheological problems with the drilling fluid if allowed to accumulate to a high enough concentration.

For a weighted nonaqueous drilling fluid, it is not feasible to discharge the effluent from a centrifuge, due to environmental and/or economic concerns. In this situation, a dual centrifuge setup is utilized in which the first centrifuge operates at a lower g setting (usually 600–900 g) and the weight material (which is easy to separate due to its higher specific gravity) is returned to the active system. The effluent from the first centrifuge typically flows to a holding tank, and this fluid is not processed by a second centrifuge operating at a higher g force in order to separate finer solids, which are discarded. The solids from the second centrifuge typically are not in the size range that would cause rheological problems, but given time they will degrade into smaller particles that could start causing problems. Therefore, they need to be removed while the equipment can still remove them. The effluent from the second centrifuge is then returned to the active system.

5.2 AUXILIARY TANK SYSTEM

5.2.1 Trip Tank

A trip tank should also be a component of the tank system. This tank should have a well-calibrated, liquid-level gauge to measure the volume of drilling fluid entering or leaving the tank. The volume of fluid that replaces the volume of drill string is normally monitored on trips to make certain that formation fluids are not entering the well bore. When one barrel of steel (drill string) is removed from the borehole, one barrel of drilling fluid should replace it to maintain a constant liquid level in the well bore. If the drill string volume is not replaced, the liquid level may drop low enough to permit formation fluid to enter the well bore due to the drop in hydrostatic pressure. This is known as a kick. Fluid may be returned to the trip tank during the trip into the well. The excess fluid from the trip tank should be returned to the active system across the shale shakers. Large solids can come out of the well and plug the hydrocyclones if this drilling fluid bypasses the shakers.

The addition of trip tanks to drilling rigs significantly reduces the number of induced well kicks. The obsolete or old-system drillers filled the hole with drilling fluid with the rig pumps by counting the mud pump strokes (the volume was calculated for the displacement of the drill pipe pulled). The problem here was that a certain pump efficiency was estimated in these calculations. If the mud pump was not as efficient as estimated, slowly but surely the height of the column of drilling fluid filling the hole decreased. This caused a decrease in hydrostatic head, and if formation pressures were greater than the hydrostatic head of the drilling fluid, a kick would occur.

Another common way to induce a kick was to continue filling the hole with the same number of strokes used for the drill pipe even when reaching the heavy-weight drill pipe or drill collars were pulled. Both the heavy-weight drill pipe and the drill collars have more displacement per stand than the drill pipe; therefore a reduction in the height of the column of drilling fluid in the well bore would occur and problems would result.

5.3 SLUG TANK

A slug tank or pit is typically a small 20- to 50-barrel compartment within the suction section. This compartment is isolated from the active system and is available for small volumes of specialized fluid. Some drilling-fluid systems may have more than one of these small compartments. They are manifolded to a mixing hopper so that solids and chemicals may be added and are used to create heavier slurry to be displaced partway down the drill pipe before trips. This prevents drilling fluid inside the pipe from splashing on the rig floor during trips. These compartments are also used to mix and spot various pills, or slurries, in a well bore. The main pump suction must be manifolded to the slug pit(s).

Proper agitation is needed for this tank because there will be many different types of slurries mixed during drilling operations. Some will be easy to mix, while others will take a lot of energy to mix properly. The addition of a mud gun or guns would be beneficial in mixing various pills as well as keeping solids from settling in the bottom or corners of this tank.

5.4 RESERVE TANK(S)

The reserve tank(s) are for storage of excess drilling fluid, base fluids, or premixed drilling fluid for future mixing/additions. It could even be a completely different type of mud system for displacing the existing drilling fluid.

Land drilling rigs do not have reserve tanks in their systems. Extra tanks are rented as needed for their operation. These tanks are typically called fractionalization (frac) tanks.

Marine drilling rigs incorporate reserve or storage tanks in their design. The volume and number of these tanks depend on the space available and the available deck load capabilities of the rig. If more storage volume is required for marine drilling rigs, extra storage tanks can sometimes be installed on deck depending on space and deck load availability.

The type of drilling fluid stored in the reserve tanks will dictate whether it needs to be agitated. Since the type of fluids stored will vary, adequate agitation should be available if required.

CHAPTER 6

SCALPING SHAKERS AND GUMBO REMOVAL

Mark C. Morgan

Derrick Equipment Co.

Gumbo is formed in the annulus from the adherence of sticky particles to each other. It is usually a wet, sticky mass of clay, but finely ground limestone can also act as gumbo. The most common occurrence of gumbo is during drilling of recent sediments in the ocean. Enough gumbo can arrive at the surface to lift a rotary bushing from a rotary table. This sticky mass is difficult to screen. In areas where gumbo is prevalent, it should be removed before it reaches the main shale shakers.

Many gumbo removal devices are fabricated at the rig site, frequently in emergency response to a "gumbo attack." These devices have many different shapes but are usually in the form of a slide at the upper end of a flowline. One of the most common designs involves a slide formed from steeply sloped rods spaced 1 to 3 inches apart and about 6 to 8 feet long. The angle of repose of cuttings is around 42°, so the slides have a slope of around 45°. Gumbo, or clay, does not stick to stainless steel very well; consequently, some of the devices are made with stainless steel rods. Drilling fluid easily passes through the relatively wide spacing in the rods, and the sticky gumbo mass slides down to disposal (Figure 6.1).

Several manufacturers have now built gumbo removal devices for rig installation. One of these units consists of a series of steel bars formed into an endless belt. The bars are spaced 1 to 2 inches apart and disposed perpendicular to the drilling-fluid flow. The bars move parallel to the

Figure 6.1. Gumbo removal device.

flow. Gumbo is transported to the discharge end of the belt, and the drilling fluid easily flows through the spacing between the bars. Another machine uses a synthetic mesh belt with large openings, like an API 5 to an API 10 screen. The belt runs uphill and conveys gumbo from a pool of drilling fluid. A counterrotating brush is used to clean gumbo from the underside of the belt. The belt speed is variable so it can be adjusted for the solids-loading and fluid properties.

When linear motion shale shakers were introduced into oil well drilling operations, drilling fluid could routinely be sieved through API 200 screens for the first time. This goal was desirable because it allowed the removal of drilled-solids sizes down to the top of the size range for barite that met American Petroleum Institute (API) specifications. Circular motion and unbalanced elliptical motion shale shakers were usually limited to screens of about API 80. Drillers soon found, however, that gumbo could not be conveyed uphill on a linear motion shale shaker. The material adhered to the screen. To prevent this, the circular and unbalanced elliptical motion shakers were used as scalping shakers.

The "rig" shakers remained attached to the flowline to remove very large cuttings and gumbo. These were called scalping shakers. Even in places where gumbo might not be present, scalpers were used to prevent very large cuttings or large chunks of shale from damaging the API 200+ screens. These screens have finer wires, which are much more fragile than wire used on an API 20 or API 40 screens. Scalping shakers also had the advantage of removing some of the larger solids that would enter the mud tanks when a hole appeared in the fine screen.

Tests indicated that fitting the scalper with the finest screen possible, an API 80 or API 100, did not result in the removal of more solids when combined with the API 200 on the main shaker. Apparently, shale in that size range would break apart on the scalper into pieces smaller than 74 microns, or it would damage the cuttings enough so that the linear motion shaker screen broke the cuttings. This action resulted in fewer total solids rejected from the system. Scalping shakers should be used as an insurance package to prevent very large cuttings from hitting the fine screens. Scalping shakers, even with an API 20 screen, will still convey gumbo. Frequently, relatively fine screens on a scalping shaker experience near-size blinding. Coarse screens are preferred.

Scalping shakers should be used with either linear motion or balanced elliptical motion shale shakers. Gumbo must be removed before any screen can convey drilled solids uphill out of a pool of liquid. These motions may be used on shakers to remove gumbo if the screen slopes downward from the back tank (possum belly) to the discharge end of the shaker.

Combination shakers are available that mount a downward-sloping screen on a linear motion shaker above an upward-sloping screen with a linear motion. Gumbo will move down the top screen and be removed before the fluid arrives at the lower screen. Again, however, the scalping screen should be a very coarse screen. Some screens reject gumbo significantly better than others. Some screens also convey gumbo more efficiently than others. Manufacturers have a great assortment of screens that have been tested in various regions of the world that experience gumbo attacks. Hook-strip screens with large rectangular openings—in the range of API 12—have been used very effectively in many regions. Experience indicates that the rectangular openings should be oriented so that the long opening is parallel to the flow.

Several important factors control whether to use a shale shaker/scalping shaker to effectively remove gumbo. If the screen deck can be tilted downward, with an articulated deck, gravity will assist gumbo removal. If the older-style unbalanced elliptical motion shaker is used, the fixed downward angle will usually satisfactorily convey gumbo. If the deck angle is flat, like most circular motion machines, the shaker has to generate a sufficient negative, or downward, force vector, normal to screen, to overcome the adhesion factor (or stickiness) of the gumbo so that the screen separates from the solids. If it does not separate, the gumbo is effectively glued to that spot on the screen. If the drilling fluid is changed to decrease the quantity of gumbo reaching the surface, gumbo removal may not be a problem for main shakers with a high g factor. Scalping shakers, however, will still provide insurance for removal of large cuttings and in case the break in a shale shaker screen goes unnoticed.

CHAPTER 7

SHALE SHAKERS

James Merrill
Fluids Systems

Leon Robinson
Exxon, retired

Shale shaker is a general term for a vibrating device used to screen solids from a circulating drilling fluid
   Many configurations have been used. These include:

- A square or rectangular screening area with drilling-fluid flow down the length
- Revolving, nonvibrating, cylindrical screens with longitudinal flow down the center axis
- Circular screens with flow from the center to the outside

Other configurations have been tried but have not become commercial. The majority of shale shakers flow the drilling fluid over a rectangular screening surface. Larger solids are removed at the discharge end, with the smaller solids and drilling fluid passing through the screen(s) into the active system. All drilled solids above 74 microns are undesirable in any drilling fluid. API 200 (74-micron) screens are so desirable on shale shakers for this reason. Weighting materials that meet American Petroleum Institute (API) specifications still have 3% by weight larger than 74 microns. Screens this size may remove large quantities of barite and may significantly affect the drilling-fluid and well cost.
   Shale shakers are the most important and easiest-to-use solids-removal equipment. They are the first line of defense once drilling fluid is returned

from the well bore. In most cases, they are highly cost-effective. If shale shakers are used with torn screens, fluid bypassing screens, incorrectly sized screen panels, or worn parts, the remaining solids-removal equipment will not perform properly.

A shale shaker can be used in all drilling applications in which liquid is used as the drilling fluid. Screen selection is controlled by circulation rate, shaker design, well-bore properties, and drilling-fluid properties. The large amount of variation in drilling-fluid properties dictates screen throughput to such an extent that shaker capacities are not listed in book.

Most operations involved in drilling a well can be planned in advance because of experience and engineering designs for well construction. Well planners expect to be able to look at a chart or graph and determine the size and number of shale shakers required to drill a particular well anywhere in the world. They expect to be able to determine the opening sizes of the shaker screens used for any portion of any well. But there are too many variables involved to allow these charts to exist. Many shale shaker manufacturers, because of customer demand, publish approximate flow charts indicating that their shakers can process a certain flow rate of drilling fluid through certain-size screens. These charts are usually based on general field experience with a lightly treated water-based drilling fluid and should be treated as approximations at best. These charts should be used to provide only very inaccurate guesses about screens that will handle flow rates for a particular situation.

Rheological factors, fluid type, solids type and quantity, temperature, drilling rates, solids/liquid interaction, well-bore diameters, well-bore erosion, and other variables dictate actual flow rates that can be processed by a particular screen. Drilling fluid without any drilled solids can pose screening problems. Polymers that are not completely sheared tend to blind screens and/or appear in the screen discard. Polymers that increase the low-shear-rate viscosity or gel strength of the drilling fluid also pose screening problems. Polymers, like starch, that are used for fluid-loss control are also difficult to screen through a fine mesh (such as an API 200 screen). Oil-based drilling fluids, or nonaqueous fluids (NAFs), without adequate shear and adequate mixing are difficult to screen. NAFs without sufficient oil-wetting additives are very difficult to screen through mesh finer than API 100.

Screen selection for shale shakers is dependent on geographical and geological location. Screen combinations that will handle specific flow rates in the Middle East or Far East will not necessarily handle the same flow rates in Norway or the Rocky Mountains. The best method to select shale shaker screens and/or number of shale shakers for a particular drilling site is to first use the recommendations of a qualified solids-control advisor from the area. Screen use records should be established for further guidance.

7.1 HOW A SHALE SHAKER SCREENS FLUID

The primary purpose of a shale shaker is to remove as many drilled solids as possible without removing excessive amounts of drilling fluid. These dual objectives require that cuttings (or drilled solids) convey off the screen while simultaneously most of the drilling fluid is separated and removed from the cuttings. Frequently, the only stated objective of a shale shaker is to remove the maximum quantity of drilled solids. Stopping a shale shaker is the simplest way to remove the largest quantity of drilled solids. Of course, this will also remove most of the drilling fluid. When disregarding the need to conserve as much drilling fluid as possible, the ultimate objective of reducing drilling costs is defeated.

The size of drilled cuttings greatly influences the quantity of drilling fluid that tends to cling to the solids. As an extreme example, consider a golf-ball–size drilled solid coated with drilling fluid. Even with a viscous fluid, the volume of fluid would be very small compared with the volume of the solid. As the solids become smaller, the fluid film volume increases as the solids surface area increases. For silt-size or ultra-fine solids, the volume of liquid coating the solids may even be larger than the solids volume. More drilling fluid is returned to the system when very coarse screens are used than when screens as fine as API 200 are used.

Drilling fluid is a rheologically complex system. At the bottom of the hole, faster drilling is possible if the fluid has a low viscosity. In the annulus, drilled solids are transported better if the fluid has a high viscosity. When the flow stops, a gel structure builds slowly to prevent cuttings or weighting agents from settling. Drilling fluid is usually constructed to perform these functions. This means that the fluid viscosity depends on the history and the shear within the fluid. Typically, the low-shear-rate viscosities of drilling fluids range from 300–400 centipoise (cP) to 1000–1500 cP. As the shear rate (or, usually, the velocity) increases, drilling fluid viscosity decreases. Even with a low-shear-rate viscosity of 1500 cP, the plastic viscosity (or high-shear-rate viscosity) could be as low as 10 cP.

Drilling fluid flows downward, onto, and through shaker screens. If the shaker screen is stationary, a significant head would need to be applied to the drilling fluid to force it through the screen. Imagine pouring honey onto a 200-mesh screen (Figure 7.1). Honey at room temperature has a viscosity of around 100 to 200 cP. Flow through the screen would be very slow if the screen were moved rapidly upward through the honey (Figure 7.2), causing the honey to pass through the screen surface and into a collection device. These forces of vibration affect drilling fluid in the same manner. The introduction of vibration into this process applies upward and downward forces to the honey. The upward stroke moves the screen rapidly through the honey. These forces of vibration affect drilling fluid in the same manner. The upward stroke moves drilling fluid through the screen. Large solids do not follow the screen on the downward stroke, so they can be propelled from the screen surface.

When the screen moves on the downward stroke, the large solids are suspended above the screen and come in contact with the screen at a farther point toward the discharge end of the shaker. This is the reason that the elliptical, circular, and linear motion screens transport solids.

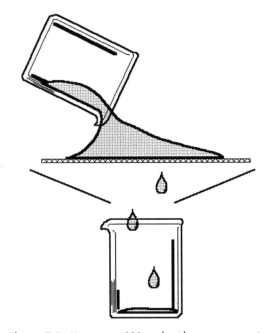

Figure 7.1. Honey onto 200 mesh with screen not moving.

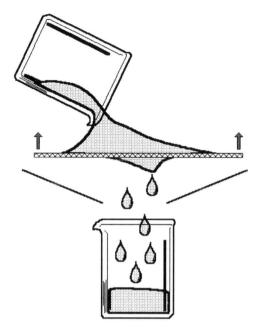

Figure 7.2. Honey onto 200 mesh with screen moving.

Screens are moved upward through the fluid with the elliptical, circular, and linear motion shale shakers. The linear motion shaker has an advantage because solids can be transported out of a pool of liquid and discharged from the system. The pool of liquid creates two advantages: Not only does it provide an additional head to the fluid, but it also provides inertia or resistance to the fluid as the screen moves upward. This significantly increases the flow capacity of the shaker. The movement of the shaker screen through the drilling fluid causes the screen to shear the fluid. This decreases the viscosity and is an effective component to allow the shaker to process drilling fluid.

The upward movement of the shaker screen through the fluid is similar to pumping the drilling fluid through the screen openings. If the fluid gels on the screen wires, the effective opening size is decreased. This would be the same as pumping drilling fluid through a smaller-diameter pipe. With the same head applied, less fluid flows through a smaller pipe in a given period of time than a larger pipe. If a shaker screen becomes water wet while processing NAF, the water ring around the screen opening effectively decreases the opening size available to pass the fluid. This, too, greatly reduces the flow capacity of the shaker.

7.2 SHAKER DESCRIPTION

The majority of shale shakers use a back tank (commonly known as a possum belly or a mud box) to receive drilling fluid from the flowline (Figure 7.3). Drilling fluid flows over a weir and is evenly distributed to the screening surface, or deck. The screen(s) are mounted in a basket that vibrates to assist the throughput of drilling fluid and the movement of separated solids. The basket rests on vibration isolator members, such as helical springs, air springs, or rubber float mounts. The vibration isolation members are supported by the skid. Below the basket, a collection pan (or bed) is used to channel the screen underflow to the active system.

Shale shaker performance is affected by the type of motion, stroke length of the deck, and the rotary speed of the motor. The shape and axial direction of the vibration motion along the deck is controlled by the position of the vibrator(s) in relation to the deck and rotation direction of the vibrator(s). There are many commercially available basket and deck configurations. The deck may be mounted at a slope (Figures 7.4A, B, E, and F) or horizontally (Figures 7.4C and 7.4D). Deck surfaces may be tilted up or down in the basket. The basket may

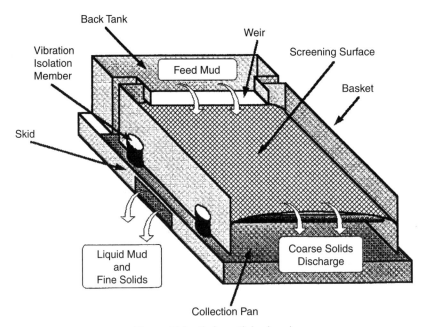

Figure 7.3. Shaker with back tank.

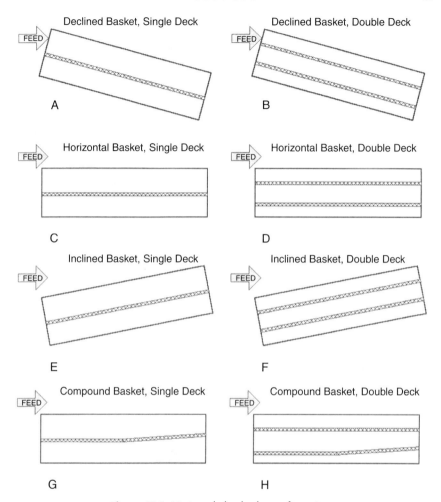

Figure 7.4. Various shaker basket configurations.

be horizontal or at a fixed angle, or have an adjustable angle. An adjustable basket angle allows the deck to be tilted up or down.

On sloped deck units (cascade or parallel flow), the screens may be continuous, with one screen covering the entire deck length (Figures 7.4A and E), may have a divided deck that has more than one screen used to cover the screening surface (Figures 7.4B and F), or may have individual screens mounted at different slopes (Figures 7.4G and H). On multiple deck units, fluid passes through the upper deck before flowing to the next deck (Figures 7.4B, F, and H).

7.3 SHALE SHAKER LIMITS

A shale shaker's capacity has been reached when excessive amounts of drilling fluid (or drilling-fluid liquid phase) first begins discharging over the end of the shaker. The capacity is determined by the combination of two factors:

1. The *fluid limit* is the maximum fluid flow rate that can be processed through the shaker screen.
2. The *solids limit* is the maximum amount of solids that can be conveyed off of the end of the shaker.

The two limits are interrelated in that the amount of fluid that can be processed will decrease as the amount of solids increases.

Any shale shaker/screen combination has a fluids-only capacity (i.e., no solids are present that can be separated by the screen) that is dependent on the characteristics of the shaker (g factor, vibration frequency, type of motion, and angle of the screen deck), of the screen (area and conductance), and of the fluid properties (viscosity characteristics, density, additives, and fluid type). The mechanical features of the shaker are discussed later in this chapter. The fluid-only capacity is the fluid limit with zero removable solids. For the sake of the current discussion, the drilling fluid is assumed to be a fluid with no solids larger than the openings in the shaker screen, although this is not true in many real instances.

The screen cloth can be considered to be a permeable medium with a permeability and thickness (conductance) and an effective filtration area. The fluid capacity will decrease as the fluid viscosity increases (plastic viscosity is important but yield and gel strengths can have a significant impact as well). Capacity will also increase as the fluid density increases due to increased pressure on the screen surface acting as a force to drive fluid through the screen.

The fluid-only capacity will generally be reduced when certain polymers are present in the fluid. Partially hydrolyzed polyacrylamide (PHPA) is most notable in this respect, as it can exhibit an effective solution viscosity in a permeable medium higher than that measured in a standard viscometer. At one time, the effective viscosity of PHPA solutions was determined by flowing the solution through a set of API 100 screens mounted in a standard capillary viscometer. PHPA drilling fluids typically have a lower fluid-only capacity for a given shaker/screen combination than do similar drilling fluids with PHPA because of this higher effective viscosity. This decrease in fluids-only capacity can be as much as 50% compared with a bentonite/water slurry. Adsorption of PHPA polymer may decrease effective opening sizes (as it does in porous media), thereby increasing the pressure drop required to maintain constant flow. This makes the PHPA appear to be much more viscous than it really is. This effect also happens with high concentrations of XC (xanthan gum, a polysaccharide secreted by bacteria of the genus *Xanthomonas campestris*) in water-based fluids, in drilling fluids with high concentrations of starch, in newly prepared NAFs, and in polymer-treated viscosifiers in NAFs.

The solids limit can be encountered at any time but occurs most often during the drilling of large-diameter holes and soft, sticky formations and during periods of high penetration rates. A relationship exists between the fluid limit and the solids limit. As the fluid flow rate increases, the solids limit decreases. As the solids loading increases, the fluid limit decreases. Internal factors that affect the fluid and solids limits are discussed in section 7.5, Shale Shaker Design.

The following are some of the major external factors that affect the solids and fluid limits.

7.3.1 Fluid Rheological Properties

Literature indicates that the liquid capacity of a shale shaker screen decreases as the plastic viscosity (PV) of a drilling fluid increases. PV is the viscosity that the fluid possesses at an infinite shear rate.[1] Drilling-fluid viscosity is usually dependent on the shear rate applied to the fluid. The shear rate through a shale shaker screen depends on the opening size and how fast the fluid is moving relative to the shaker

---

[1] The Bingham Plastic rheological model may be represented by the equation

$$\text{shear stress} = (PV)\text{shear rate} + YP.$$

By definition, viscosity is the ratio of shear stress to shear rate. Using the Bingham Plastic expression for shear stress,

$$\text{viscosity} = [(PV)\text{shear rate} + YP]/\text{shear rate}.$$

Performing the division indicated, the term for viscosity becomes

$$(PV) + [YP/\text{shear rate}].$$

As shear rate approaches infinity, viscosity becomes PV.

screen wires. For example, if 400 gpm is flowing through a 4 × 5-ft API 100 market grade (MG) screen (30% open area), the average fluid velocity is only 1.8 inches per second. Generally the shear rates through the shaker screen vary significantly. The exact capacity limit, therefore, will depend on the actual viscosity of the fluid. This will certainly change with PV and yield point (YP).

7.3.2 Fluid Surface Tension

Although drilling-fluid surface tensions are seldom measured, high surface tensions decrease the ability of the drilling fluid to pass through a shale shaker screen, particularly fine screens, with their small openings.

7.3.3 Wire Wettability

Shale shaker wire screens must be oil wet during drilling with oil-based drilling fluids. Water adhering to a screen wire decreases the effective opening size for oil to pass through. Frequently, this results in the shaker screens not being capable of handling the flow of an oil-based drilling fluid. This is called "sheeting" across the shaker screen, which frequently results in discharge of large quantities of drilling fluid.

7.3.4 Fluid Density

Drilling-fluid density is usually increased by adding a weighting agent to the drilling fluid. This increases the number of solids in the fluid and makes it more difficult to screen the drilling fluid.

7.3.5 Solids: Type, Size, and Shape

The shape of solids frequently makes screening difficult. In single-layer screens, particles that are only slightly larger than the opening size can become wedged into openings. This effectively plugs the screen openings and decreases the open area available to pass fluid. Solids that tend to cling together, such as gumbo, are also difficult to screen. Particle size has a significant effect on both solids and liquid capacity. A very small increase in near-size particles usually results in a very large decrease in fluid capacity for any screen, single or multilayer.

7.3.6 Quantity of Solids

Solids compete with the liquid for openings in the shaker screen. Fast drilling can produce large quantities of solids. This usually requires coarser screens to allow most of the drilling fluid to be recovered by the shale shaker. Fast drilling is usually associated with shallow drilling. The usual procedure is to start with coarser-mesh screens in the fast drilling, larger holes near the top of the well and to "screen down" to finer screens as the well gets deeper. Finer screens can be used when the drilling rate decreases.

Boreholes that are not stable can also produce large quantities of solids. Most of the very large solids that arrive at the surface come from the side of the borehole and not from the bottom to the borehole. Drill bits usually create very small cuttings.

7.3.7 Hole Cleaning

One factor frequently overlooked in the performance of shale shakers is the carrying capacity of the drilling fluid. If cuttings are not brought to the surface in a timely manner, they tend to disintegrate into small solids in the borehole. If they stay in the borehole for a long period before arriving at the surface, the PV and solids content of the drilling fluid increases. This makes it appear that the shale shaker is not performing adequately, when actually the solids are disintegrating into those that cannot be removed by the shale shaker.

7.4 SHAKER DEVELOPMENT SUMMARY

Shale shakers have undergone many improvements since the *Shale Shaker Handbook* was written in the early 1970s. The current design, linear motion shakers, was introduced in the 1980s and has become widely used because of its improved solids conveyance and fluid throughput. The various types of motions are discussed in the next sections. Linear motion has made it possible to move solids toward the discharge end of the deck while it is tilted uphill. The uphill tilt of the deck creates a pool of fluid at the feed end of the deck, which, in combination with the linear motion, exerts greater pressure on the fluid flowing through the screen openings. This allows a finer screen than with all previous shaker designs. The acceleration perpendicular to the screen surface controls the liquid throughput. Orbital (circular or unbalanced elliptical) and linear motion shakers can have the same acceleration (or *g* factor), but the linear motion shaker can process a greater flow rate. The linear motion conveys solids uphill, whereas orbital motion will not. The uphill solids conveyance allows the linear motion or balanced elliptical motion to process a greater flow rate.

The use of linear motion shakers has become feasible with the development of improved screen designs. The life of shaker screens has been extended with the introduction of repairable bonded and pretensioned screen panels. Other design improvements are available in wire cloth, rectangular weaves, nonmetallic screens, and three-dimensional screen surfaces, which have improved the solids-separation capabilities of all shakers.

Although linear motion shale shakers have made a significant impact on solids-removal concepts, the other shale shakers have many advantageous features. *Circular motion* is easier on the shale shaker structure and shaker screens and conveys gumbo better than does linear motion. Linear motion shakers require bonded screens of which 30–50% of the area is forfeited. The liquid pool at the back of the linear motion screens can cause solids to be ground up into many smaller particles and forced through the shaker screens. This liquid pool also gives solids slightly finer than the screen openings more of a chance to go through the screen.

7.5 SHALE SHAKER DESIGN

The purpose of a shale shaker is to induce drilling fluid to flow through a screen, transport solids across a screen surface, and discharge solids off the end of the screen. Its primary function is to remove drilled solids, which is the primary function of all solids-removal equipment. Screening is the result of using the energy developed by a rotating eccentric mass and applying that force to a porous surface. The energy causes the screen to vibrate in a fixed orbit or path.

The elements of shale shaker design focus on several aspects of the machine:

- Shape of motion (orbit or path)
- Vibrating systems
- Screen deck design
- *g* Factor
- Power systems

All of these parameters contribute to the results achieved. We will now examine them in turn.

7.5.1 Shape of Motion

Historically, the progression of the design of shale shakers has been toward allowing the use of finer screens. Shale shakers have developed through the years from relatively simple, uncomplicated designs to today's complex models. In fact, this evolutionary process has seen several distinct eras of shale shaker technology and performance. These developmental time frames can be divided into four main categories:

1. Unbalanced elliptical motion
2. Circular motion
3. Linear motion
4. Balanced elliptical motion

The eras of oilfield shaker (and screening) development may be defined by the types of motion(s) produced by the vibrators and their associated machines.

If a single rotating vibrator is located away from the center of gravity of the basket, the motion is elliptical at the ends of the deck and circular below the vibrator (Figure 7.5). This is an *unbalanced* elliptical motion. If a single rotating vibrator is located at the center of gravity of the basket, the motion is *circular* (Figure 7.6). Two counterrotating vibrators attached to the basket are used to produce *linear motion* (Figure 7.7).

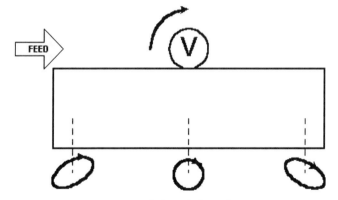

Figure 7.5. Unbalanced elliptical motion.

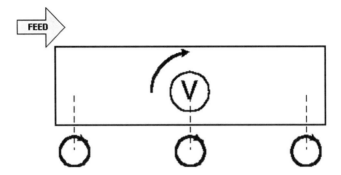

Figure 7.6. Circular motion.

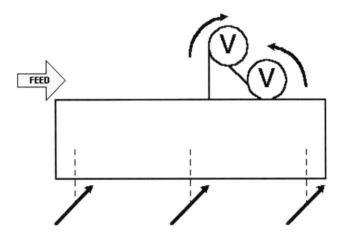

Figure 7.7. Linear motion.

When placed at an angle to the basket, two counterrotating vibrators can produce a *balanced elliptical motion* (Figure 7.8).

Unbalanced Elliptical Motion Shale Shakers

In the 1930s, unbalanced elliptical shale shakers were adapted by the oilfield. These first shakers came from the mineral ore dressing industries (e.g., coal, copper) with little or no modifications. They were basic, rugged, and mechanically reliable but were generally limited to API 20 and coarser screens.

In an unbalanced elliptical motion shaker (Figure 7.5), the movement of the shaker deck/basket is accomplished by placing a single vibrator

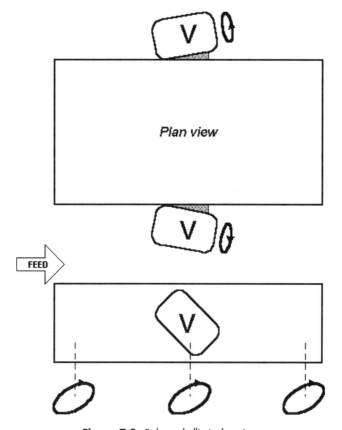

Figure 7.8. Balanced elliptical motion.

system above the shaker deck. That is, the mechanical system of a spinning counterweight (or an elliptically shaped driveshaft) is installed above the center of gravity of the deck. The resulting motion imparted to the bed is a combination of elliptical and circular. Directly below the vibrator, the motion of the basket is circular, while at either end of the deck the motion is elliptical.

The orientation of the major axes of the ellipses formed at the feed end and the solids-discharge end of the basket has a major impact on solids conveyance. Specifically, it is desirable for the major axis of the ellipsoidal trace to be directed toward the solids-discharge end. However, the orientation of the major axis of the ellipse formed at the solids-discharge end is just the opposite; it is directed backward toward the feed end. This discharge-end thrust orientation is undesirable, since it makes discharging solids from the shaker more difficult (Figure 7.9). To assist in solids conveyance, the deck or last screen is tilted downward (Figure 7.10) or the vibrator is moved to the discharge end. Moving the vibrator toward the discharge end reduces the fluid capacity and reduces the screen life of the end screen significantly. This also reduces the residence time of the feed slurry on the screening surface. Advertisers of this style of motion touted the fact that the reverse-tilted ellipse allowed solids to remain on the screen longer, thereby removing more liquid.

Early elliptical motion shale shakers used hook-strip screens that were manually tensioned. A series of tension rails and tension bolt spring assemblies were used to pull the screens tightly over the support bars to ensure proper tightening. Pretensioned screens and pretensioned screen

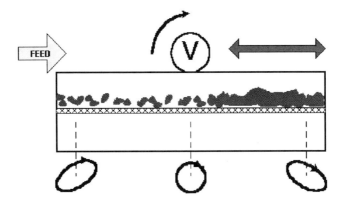

Figure 7.9. Undesirable discharge end thrust orientation.

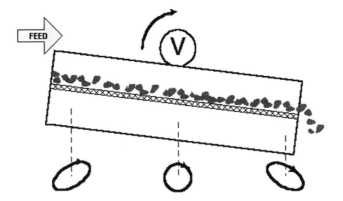

Figure 7.10. Tilted discharge orientation.

panels were not introduced until the 1970s and even then were not commonly used on elliptical motion units.

As with most engineered products, compromises have been made. Achievement of an acceptable balance is sought between the amount of feed slurry the shaker can handle and its ability to effectively move solids along the screen deck. The early elliptical motion shakers typically had one screen surface driven by a motor sheaved to the vibrator with a belt drive. Later models of this design employed additional screen area and/or integral vibrators to increase flow capacity. These shakers were capable of processing drilling fluid through API 60 to API 80 screens.

Unbalanced elliptical motion shale shakers are compact, easy to maintain, and inexpensive to build and operate. They use relatively coarse screens (API 60 to API 80), and for this reason are frequently used as scalping shakers. Scalping shakers remove large solids or gumbo and reduce solids loading on downstream shakers.

Circular Motion Shale Shakers

Circular motion shakers were introduced in 1963. These shakers have a single vibrator shaft located at the center of mass of a horizontal basket. A motor drives a concentric shaft fitted with counterweights, which provides pure circular motion along the entire length of the vibrating deck. This feature improves solids conveyance off the end of the deck compared with unbalanced elliptical designs. The circular motion transports solids along a horizontal screen, thus reducing the loss of liquid without sacrificing solids conveyance.

Circular motion units often incorporate multiple, vertically stacked decks. Coarse screens mounted on the top deck separate and discharge the larger cuttings, thus reducing solids loading on the bottom screens. These multiple deck units allowed the first practical use of API 80 to API 100 screens.

Flowback trays (Figure 7.11) introduced in the late 1970s direct the slurry onto the feed end of the finer screen on the lower deck. The tray allows full use of the bottom screen area to achieve greater cuttings removal with less liquid loss. Even with these units, screens are limited to about API 100 by the available screening area, vibratory motion, and screen panel design. If bonded screens are used, screens as fine as API 150 have been used with flowback trays.

Screens on the circular motion units are installed either overslung or underslung. The open hook strip screen is tensioned across longitudinal

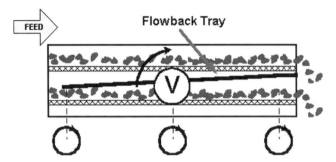

Figure 7.11. Flowback tray.

support members. Both designs have advantages and disadvantages. Overslung screens have reasonable screen life, but the drilling fluid tends to channel to the sides. On underslung screens, drilling fluid tends to congregate around and beneath the longitudinal support members. Grinding of this accumulation of drilled solids between the rubber support and the screen tends to reduce screen life. To overcome this screen life reduction, rubber supports with flatter cross sections are used and strips are installed between the rubber support and the screen.

In the 1980s some circular motion machines began to be fitted with repairable bonded underslung screens that increased screen life and fluid throughput. Even though the use of repairable bonded screens reduced the available unblanked area, the detrimental effect on fluid capacity was more than offset by the use of higher-conductance screen cloths and larger bonded openings.

Linear Motion Shale Shakers

The introduction of linear motion shale shakers in 1983, combined with improved screen technology, resulted in the practical use of API 200 and finer screens. Linear motion is produced by a pair of eccentrically weighted, counterrotating parallel vibrators. This motion provides cuttings conveyance when the screen deck is tilted upward.

Linear motion shakers have overcome most of the limitations of elliptical and circular motion designs. Straight-line motion provides superior cuttings conveyance (except with gumbo) and superior liquid throughput capabilities with finer screens. Linear motion shale shakers generally do not convey gumbo uphill. They can effectively remove gumbo if they are sloped downward toward the discharge end. The increased physical size of these units (and an accompanying increase in deck screen surface area) allows the use of even finer screens than those used on circular or elliptical motion shakers.

Screening ability is the result of applying the energy developed by a rotating eccentric mass to a porous surface or screen. The energy causes the screen to vibrate in a fixed orbit. This transports solids across the screen surface and off the discharge end and induces liquid to flow through the screen.

In conventional unbalanced elliptical and circular motion designs, only a portion of the energy transports the cuttings in the proper direction, toward the discharge end. The remainder is wasted due to the peculiar shape of the screen-bed orbit, manifested by solids becoming nondirectional or traveling in the wrong direction on the screen surface. Linear motion designs provide positive conveyance of solids throughout the vibratory cycle because the motion is in a straight line rather than elliptical or circular. The heart of a linear motion machine is the ability to generate this straight-line or linear motion and transmit this energy in an efficient and effective manner to the vibrating bed.

As shown in Figure 7.12, a linear motion system consists of two eccentrically weighted counterrotating shafts. The net effect of each equal eccentric mass being rotated in opposite directions is that resultant forces are additive at all positions along the vibratory trace, except at the very top and bottom of each stroke, resulting in a thrust (vibration) along a straight line—hence, the term *linear,* or straight-line, motion.

To achieve the proper relationship between the rate of solids conveyance and liquid throughput, the drive system must be mounted at an angle to the horizontal bed. A thrust angle of 90° relative to the screen surface would simply bounce solids straight up and down. Taken to the other extreme, a thrust angle of zero degrees would rapidly move solids but yield inadequate liquid throughput and discharge very wet solids. On most units this angle is approximately 45° to the horizontal (Figure 7.12).

Some machines have adjustable angle drive systems that can be changed to account for various process conditions (Figure 7.13). If a thrust angle were decreased (for example to 30° to the horizontal), the X component of the resultant vibratory thrust (force) would increase and the Y component decrease. Conversely, building a greater angle would cause the X component to decrease and the Y component to increase.

A larger X vector component of thrust will move solids along the deck faster. A larger Y component vector increases liquid throughput and the

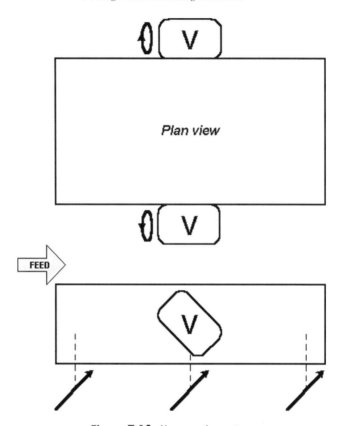

Figure 7.12. Linear motion system.

residence time of material on the screen. Most manufacturers choose a fixed angle near 45°, which gives near-equal values for each vector. This is a logical approach, since the shaker must simultaneously transmit liquid through the screen and convey solids off the screen.

The ability to create linear motion vibration allows the slope of the bed to vary up to a +6° incline (which affects residence time and therefore shaker performance) and to create a liquid pool at the flowline end of the machine. This allows a positive liquid pressure head to develop and help drive liquid and solids through the finer wire cloths. The deck on most linear motion shale shaker designs can be adjusted up to a maximum of +6°. In some cases, the beds can be tilted down to help in cases in which gumbo is encountered. These movements of bed on skid can be accomplished with mechanical, hydraulic, or

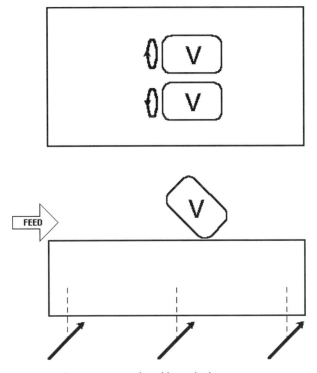

Figure 7.13. Adjustable angle drive system.

combination mechanical/hydraulic systems. On some units these adjustments can be made while the unit is running.

The ability of linear motion to convey uphill allows the use of finer shaker screens. Finer screens allow for smaller particles to be removed from the drilling fluid. Hence, a solids-control system that utilizes fine-screen linear motion shakers will better maintain the drilling fluid and improve efficiency of downstream equipment such as hydrocyclones and centrifuges. When screens are tilted too much uphill, many solids are ground to finer sizes as they are pounded by the screen. This tends to increase—not decrease—the solids content of the drilling fluid.

When linear motion shale shakers were introduced, other solids-removal equipment (like the mud cleaner) was sometimes erroneously eliminated. For a short time, this appeared to be a solution, but solids analysis, discards from other equipment, and particle size analyses proved the need for downstream equipment. Linear motion shale shakers should not be expected to replace the entire solids-removal system.

Balanced Elliptical Motion

Balanced elliptical motion was introduced in 1992 and provides the fourth type of shale shaker motion. With this type of motion, all of the ellipse axes are sloped toward the discharge end of the shale shaker. Balanced elliptical motion can be produced by a pair of eccentrically weighted counterrotating parallel vibrators of different masses. This motion can also be produced by a pair of eccentrically weighted, counter-rotating vibrators that are angled away from each other (Figure 7.14). The ellipse aspect ratio (major axis to minor axis) is controlled by the angle between vibrators or by different masses of the parallel vibrators.

Larger minor axis angles, or angle of vibrators relative to each other, will produce a broader ellipse and slow the solids conveyance. A thin ellipse with a ratio of 3.5 will convey solids faster than a fat ellipse

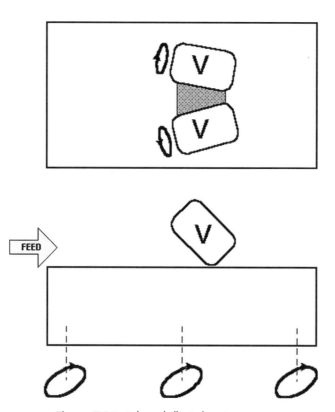

Figure 7.14. Balanced elliptical motion system.

with a ratio of 1.7. The typical operating range is 1.5 to 3.0, with the lower numbers generating slower conveyance and longer screen life. Balanced elliptical motion shale shakers can effectively remove gumbo if they are sloped downward toward the discharge end in the same manner as the linear motion shakers.

The increased physical size of these units and an accompanying increase in deck screen surface area allows the use of even finer screens than are used on other orbital motion shakers.

In conventional unbalanced elliptical and circular motion designs, only a portion of the energy transports the cuttings in the proper direction toward the discharge end. Balanced elliptical motion transports cuttings toward the discharge end of the screen in the same manner as linear motion. Balanced elliptical motion provides positive conveyance of solids throughout the vibratory cycle.

7.5.2 Vibrating Systems

The type of motion imparted to the shaker depends on the location, orientation, and number of vibrators used. In all cases, the correct direction of rotation must be verified.

Unbalanced elliptical motion shakers use a single vibrator mounted above the shaker's center of gravity. Integral vibrators, enclosed vibrators, and belt-driven vibrators are used for this shaker design.

Circular motion shakers use a single vibrator mounted at the shale shaker's center of gravity. Belt-driven vibrators and hydraulic-drive vibrators are used for this shaker design.

Most linear motion shakers use two vibrators rotating in opposite directions and mounted in parallel, but in such a manner that the direction and angle of motion is achieved. Integral vibrators, enclosed vibrators, belt-driven vibrators, and gear-driven vibrators are used for this shaker design.

Balanced elliptical motion shakers use two vibrators rotating in opposite directions but at a slight angle to each other so that they are not parallel. These vibrators must be oriented correctly to achieve the direction and angle of motion desired. The elliptical motion traces must all lean toward the discharge end and not backward toward the possum belly. If two vibrators of different masses are mounted in the same manner as the linear motion vibrators (i.e., parallel), a balanced elliptical motion is also achieved.

Various vibrating systems are used on shale shakers. These systems include:

1. *Integral vibrator:* The eccentrically weighted shaft is an integral part of the rotor assembly in that it is entirely enclosed within the electric motor housing.
2. *Enclosed vibrator:* This is a double-shafted electric motor that has eccentric weights attached to the shaft ends. These weights are enclosed by a housing cover attached to the electric motor case.
3. *Belt-driven vibrator:* The eccentrically weighted shaft is enclosed in a housing and a shaft is attached to one end. A sheaved electric motor is used to rotate the shaft with a belt drive. The electric motor may be mounted alongside, above, or behind the shaker, depending on the model. It may also be mounted on the shaker bed along with the vibrator assembly.
4. *Dual-shafted, belt-driven vibrator:* This system is similar to that of the belt-driven vibrator except that it has two vibrator shafts rotating in opposite directions and is driven by one electric motor with a drive belt.
5. *Gear drive:* A double-shafted electric motor drives a sealed gearbox, which in turn rotates two vibrator shafts in opposite directions.
6. *Hydraulic drive:* A hydraulic drive motor is attached directly to a vibrator shaft, which is enclosed in a housing. The hydraulic motor must have a hydraulic power unit that includes an electric motor and a hydraulic pump. The hydraulic-drive motor powers the vibrator shaft.

7.5.3 Screen Deck Design

Hook-strip screens have been mounted with both underslung and overslung supports. Some previous generations of oilfield shaker designs used screens that were underslung, or *pulled up* from the bottom of a group of support, or "bucker," bars (Figure 7.15). These support bars would divide the flow of material down the screen. Some problem is experienced occasionally when solids are trapped under the rubber bar supports.

Some linear motion shale shakers utilize overslung screens (Figure 7.16). With this approach, screens are attached to the bed of the shaker by being *pulled down* onto the bed from the top. This results in a screening area completely free of obstacles. Modern shale shaker bed

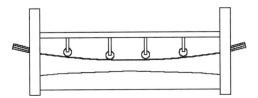

Figure 7.15. Underslung deck support system.

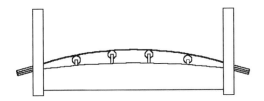

Figure 7.16. Overslung deck support system.

design has also increased the number of support ribs located beneath the screen to aid in fine-screen support and to reduce the amount of "crown," or "bow," necessary to properly tension screen panels. Some problem is experienced occasionally when the fluid leaves the high center of the screen and flows down the sides of the screen.

Most circular motion shale shakers were built with a double deck, meaning that fluid flowed over and through the top screen onto a finer screen immediately below. This arrangement led to some problems in operation, because the bottom screen was not easily visible. (Generally a flashlight was needed to inspect it.) A torn screen could remain in operation for a long time before it was noticed and changed. This created problems with solids removal because the bottom screen would not provide the intended finer screening. Some manufacturers installed backflow pans under the top screen to direct the flow through the entire screen area of the bottom screen, but these just made it even more difficult to see the bottom screen.

Most manufacturers of linear motion shakers have adopted a single-deck design. The units have clear visibility for ease of care and maintenance. This unobstructed approach also makes screen changing much easier. The fluid pool tends to obscure any torn screens until drill pipe connections are made. Therefore, a torn screen on a single-deck shaker reduces solids-removal efficiency until a new screen is installed.

Crews need to be alert to torn screens no matter what shaker is used. This is especially true during slow drilling, when drill pipe connections are infrequently made. When riser-assist pumps are used, flow should be periodically directed to different shakers during connections. This allows screens to be properly inspected and replaced, if needed.

7.5.4 g Factor

The *g* factor refers to a ratio of an acceleration to Earth's gravitational acceleration. Jupiter has a mass of 418.6×10^{25} lb and Earth has a mass of 1.317×10^{25} lb. A person on Earth who weighs 200lb would weigh 320 times as much on Jupiter, or 64,000 lb. A person's mass remains the same on Earth or Jupiter, but weight is a force and depends on the acceleration of gravity. The gravitational acceleration on Jupiter is 320 times the gravitational acceleration on Earth. The *g* factor would be 320. (As a point of interest, Mars has a mass of 0.1415×10^{25} lb, so the *g* factor would be 0.107; a 200-lb person would weigh only 21.4 lb on Mars.)

The term "*g* force" is sometimes used incorrectly to describe a *g* factor. In the preceding example, the *g* force on Earth would be 200 lb and the *g* force on Jupiter would be 64,000 lb. This is because the acceleration of gravity on Jupiter would be 320 times the acceleration of gravity on Earth.

Calculation of g Factor

Accelerations are experienced by an object or mass rotating horizontally at the end of a string. A mass rotating around a point with a constant speed has a centripetal acceleration (C_a) that can be calculated from the equation

$$C_a = r\omega^2$$

where r is the radius of rotation and ω is the angular velocity in radians per second.

This equation can be applied to the motion of a rotating weight on a shale shaker to calculate an acceleration. The centripetal acceleration of a rotating weight in a circular motion with a diameter (or stroke) of $2r$, in inches, rotating at a certain rpm (or ω) can be calculated from the preceding equation,

$$(C_a) = (1/2)(\text{diameter})\{\omega\}^2.$$

$$C_a = \frac{1}{2}(\text{stroke, in inches})\left(\frac{1 \text{ ft}}{12 \text{ inches}}\right)$$
$$\times \left[\text{RPM}\left(\frac{2\pi \text{ radians}}{\text{revolution}}\right)\left(\frac{1 \text{ minute}}{60 \text{ seconds}}\right)\right]^2$$

Combining all of the conversion factors to change the units to ft/sec²:

$$C_a(\text{in ft/sec}^2)\left(\frac{\text{stroke, in inches}}{(\text{RPM})^2}\right)$$

Normally this centripetal acceleration is expressed as a ratio of the value to the acceleration of gravity:

$$\text{No. of g's} = \frac{C_a}{32.2 \text{ ft/sec}^2} = \frac{(\text{stroke, in inches}) \times (\text{rpm})^2}{70490}.$$

Shale shakers are vibrated by rotating eccentric masses. A tennis ball rotating at the end of a 3-ft string and a 20 lb weight rotated at the same rpm at the end of a 3-ft string will have the same centripetal acceleration and the same g factor. Obviously, the centripetal force, or the tension in the string, will be significantly higher for the 20-lb weight.

The rotating eccentric weight on a shale shaker is used to vibrate the screen surface. The vibrating screen surface must transport solids across the surface to discard and allow fluid and solids smaller than the screen openings to pass through to the mud tanks. If the weights rotated at a speed or vibration frequency that matched the natural frequency of the basket holding the screen surface, the amplitude of the basket's vibration would continue to increase and the shaker would be destroyed. This will happen even with a very small rotating eccentric weight. Consider a child in a swing on a playground: Application of a small force every time the swing returns to full height (amplitude) soon results in a very large amplitude. This is a case in which the "forcing function" (the push every time the swing returns) is applied at the natural frequency of the swing.

When the forcing function is applied at a frequency much larger than the natural frequency, the vibration amplitude depends on the ratio of the product of the unbalanced weight (w) and the eccentricity (e) to the weight of the shaker's vibrating member (W); or

$$\text{vibration amplitude} = \frac{we}{W}.$$

The vibration amplitude is one half of the total stroke length.

The peak force, or maximum force, on a shaker screen can be calculated from Newton's second law of motion:

$$\text{force} = \frac{W}{g} a$$

where a is the acceleration of the screen. For a circular motion, the displacement is described by the equation:

$$x = X \sin \frac{Nt\pi}{30}$$

The velocity is the first derivative of the displacement, dx/dt, and the acceleration is the second derivative of the displacement. This means that the acceleration would d^2x/dt^2 be

$$a = -\left(\frac{N}{30}\right)^2 (\pi)^2 X \sin\left[(\pi)^2 \frac{Nt}{30}\right].$$

The maximum value of this acceleration occurs when the sine function is equal to 1. Since the displacement (X) is proportional to the ratio of we/W for high-vibration speeds, the peak force, in lb (from the peak acceleration), can be calculated from the equation

$$F = \frac{weN^2}{35,200}.$$

So the force available on the screen surface is a function of the unbalanced weight (w), the eccentricity (e), and the rotation speed (N).

Stroke length for a given design depends on the amount of eccentric weight and its distance from the center of rotation. Increasing the weight eccentricity and/or the rpm increases the g factor. The g factor is an indication of only the acceleration of the vibrating basket and not necessarily of performance. Every shaker design has a practical g-factor limit. Most shaker baskets are vibrated with a 5-hp or smaller motor and produce 2–7 g's of thrust to the vibrating basket.

Conventional shale shakers usually produce a g factor of less than 3; fine-screen shale shakers usually provide a g factor of between 4 and 6. Some shale shakers can provide as much as 8 g's. Greater solids separation is possible with higher g factors, but they also generally shorten screen life.

Linear Motion Shale Shakers subsection), y transports the cuttings in the proper cal and circular vibration motion designs.

is lost due to the peculiar shape of the :ed by solids becoming nondirectional or ion on the screen surface. Linear motion ;ns provide positive conveyance of solids ycle because the motion is straight-line lar.

ı forces perpendicular to the screen surface l and solids passing through the screen, or leration forces parallel to the screen surface ; transport, or the solids capacity.

:, the motion is generally at an angle to the ary weights are aligned so that the accele- urface. The higher liquid capacity of linear motion shale shakers for the same size openings in the screens on unbalanced elliptical or circular motion shakers seems related primarily to the fact that a pool of drilling fluid is created at the entry end of the shale shaker. The pool provides a liquid head to cause a higher flow rate through the screen. The linear motion moves the solids out of the pool, across the screen, and off the end of the screen.

On a linear motion shaker with a 0.13-inch stroke at 1500 rpm, the maximum acceleration is at an angle of 45° to the shale shaker deck. The g factor would be 4.15. The acceleration is measured in the direction of the stroke. If the shale shaker deck is tilted at an upward angle from the horizontal, the stroke remains the same. The component of the stroke parallel to the screen moves the solids up the 5° incline.

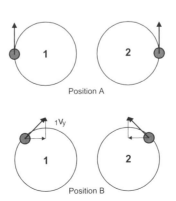

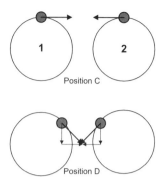

Relationship of g Factor to Stroke and Speed of Rotation

An unbalanced rotating weight vibrates the screen deck. The amount of unbalanced weight combined with the speed of rotation will give the g factor imparted to the screen deck (see preceding paragraphs). The stroke is determined by the amount of unbalanced weight and its distance from the center of rotation and the weight of the shale shaker deck. (This assumes that the vibrator frequency is much larger than the natural frequency of the shaker deck.) Stroke is independent of the rotary speed.

The *g* factor can be increased by increasing the stroke or the rpm, or both, and decreased by decreasing the stroke or rpm, or both. The stroke must be increased by the inverse square of the rpm reduction to hold the *g* factor constant. Examples are given below to hold 5 *g*'s constant while varying stroke length at different values of rpm:

5 *g*'s @ 0.44" stroke at 900 rpm   4 *g*'s @ 0.35" stroke at 900 rpm
5 *g*'s @ 0.24" stroke at 1200 rpm   4 *g*'s @ 0.20" stroke at 1200 rpm
5 *g*'s @ 0.16" stroke at 1500 rpm   4 *g*'s @ 0.13" stroke at 1500 rpm
5 *g*'s @ 0.11" stroke at 1800 rpm   4 *g*'s @ 0.09" stroke at 1800 rpm

7.5.5 Power Systems

The most common power source for shale shakers is the rig electrical power generator system. The rig power supply should provide constant voltage and frequency to all electrical components on the rig. Most drilling rigs generate 460 alternating-current-volt (VAC), 60 Hz, 3-phase power or 380 VAC, 50 Hz, 3-phase power. Other common voltages are 230 VAC, 190 VAC, and 575 VAC. Through transformers and other controls, a single power source can supply a variety of electrical power to match the requirements of different rig components.

Shale shakers should be provided with motors and starters to match the rig generator output. Most motors are dual wound. These may be wired to use either of two voltages and starter configurations. For example, some use 230/460 VAC motors and some use 190/380 VAC motors. Dual-wound motors allow the shaker to be operated properly with either power supply after relatively simple rewiring. Care must be taken, however, to make certain that the proper voltage is used. Electric-motor armatures are designed to rotate at a specific speed. Typically the rotational speed is 1800 rpm for 60-Hz applications and 1500 rpm for 50-Hz applications.

Shale shakers use a vibrating screen surface to conserve the drilling fluid and reject drilled solids. The effects of this vibration are described in terms of the g factor, or the function of the angular displacement of a screen surface and the square of the rotational speed. (For a detailed discussion, see the preceding section on g factor.) Angular displacement is achieved by rotating an eccentric mass. Most shale shakers are designed to be operated at a specific, fixed g factor by matching the stroke to a given machine's rotational speed. It follows that any deviation in speed will affect the g factor and influence the shaker performance.

Deviations in speed may be caused by one or more factors but typically are caused by fluctuations in voltage or the frequency of the alternating current. If the voltage drops, the motor cannot produce the rated horsepower and may not be able to sustain the velocity needed to keep the eccentric mass moving correctly. Low voltage also reduces the life of electrical components. Deviations in frequency result in the motor turning faster (frequencies higher than normal) or slower (frequencies lower than normal). This directly influences rpm and shaker performance.

Slower rpm for a particular motor reduces the g factor and causes poor separation and poor conveyance. Faster rpm increases the g factor and may improve conveyance and separation, but can destroy the machine and increases screen fatigue failures. In extreme cases, higher rpm may cause structural damage to the shale shaker. Thus, it is important to provide proper power to the shale shaker.

For example, a particular shale shaker is designed to operate at 4 g's. The shaker has an angular displacement, or stroke, of 0.09 inches. This shaker must vibrate at 1750 rpm to produce 4.1 g's. At 60 Hz, the motor turns at 1750 rpm, so the *g* factor is 4.1, just as designed. If the frequency drops to 55 Hz, the motor speed reduces to 1650 rpm, which results in a *g* factor of 3.5. Further reduction of frequency to 50 Hz results in 1500 rpm and a *g* factor of 2.9.

Most rigs provide 460 VAC, 60 Hz power, so most shale shakers are designed to operate with this power supply. However, many drilling rigs are designed for 380-VAC/50-Hz electrical systems. To provide proper *g* factors for 50-Hz operations, shale shaker manufacturers rely on one of two methods: increasing stroke length or using voltage/frequency inverters (transformers).

A motor designed for 50-Hz applications rotates at 1500 rpm. At 0.09-inch stroke, a shale shaker will produce 2.9 *g*'s. Increasing the stroke length to 0.13 inches provides 4.1 *g*'s, similar to the original 60-Hz design. However, the longer stroke length and slower speed will produce different solids-separation and conveyance performance. At the longer stroke lengths, shakers will probably convey more solids and have a higher fluid capacity. Conversely, instead of increasing the stroke length, some manufacturers use voltage inverters to provide 460-VAC/60-Hz output power from a 380-VAC/50 Hz supply.

Constant electrical power is necessary for good, constant shale shaker performance. The tables below assist in designing a satisfactory electrical distribution system. Alternating-current motors are common on most shale shakers. The motor rating indicates the amount of electrical current required to operate the motor. The values in Table 7.1 provide some guidelines for various motors. Be wary of all electrical hazards; follow *all* applicable regulatory codes, nationally, internationally, regionally, and locally, as well as manufacturer's safety and installation instructions. *The manufacturer's recommendations should always take precedence over the generalized values in these tables.* The values in the tables are to be used as general guidelines only. Many factors, including insulating material and temperature, control the values.

The amount of electric current that a conductor (or wire) can carry increases as the diameter of the wire increases. Common approximate values for currents are presented with the corresponding wire size designation in Table 7.2. Conductors, even relatively large-diameter wire, still have some resistance to the flow of electric current. This resistance to flow results in a line voltage drop. When an electric motor is located in an area remote from the generator, the line voltage drop may decrease the motor voltage to unacceptably low values. Some guidelines of wire

Table 7.1
Electric Current Required by Motors Running at Full Load

Motor Rating hp	Single Phase		Three Phase		Three Phase	
	115 v	230 v	190 v	230 v	460 v	575 v
1	16	8	8	3.6	1.8	1.4
1½	20	10	10	5.2	2.6	2.1
2	24	12	12	5.8	3.4	2.7
3	34	17	17	9.6	4.8	3.9
5	56	28	28	15.2	7.6	6.1
7½	80	40	40	22	11	9.0
10	100	50	50	28	14	11
15				42	21	17
20				54	27	22

hp = horsepower; v = volts.
WARNING: Electrical Hazard—follow ALL national electric codes, local electric codes, and manufacturer's safety and installation instructions. Always conform to regulatory codes, as apply regionally and internationally.

Table 7.2
Maximum Allowable Electric Current in Various Wire Sizes

Current, amps	35	50	70	80	90	100	125	150	200	225	275
Wire Size, AWG	10	8	6	5	4	3	2	1	0	00	000

AWG = American Wire Gauge.
WARNING: Electrical Hazard—follow ALL national electric codes, local electric codes, and manufacturer's safety and installation instructions. Always conform to regulatory codes, as apply regionally and internationally.

diameter necessary to keep the voltage drop to 3% are presented in Table 7.3.

7.6 SELECTION OF SHALE SHAKERS

Most drilling rigs are equipped with at least one shale shaker. The purpose of a shale shaker, as with all drilled-solids removal equipment, is to reduce drilling cost. Most drilling conditions require limiting the quantity and size of drilled solids in the drilling fluid. Shale shakers remove the largest drilled solids that reach the surface. These solids are the ones that can create many well-bore problems if they remain in the drilling fluid.

Table 7.3
Copper Wire Size Required to Limit Line Voltage Drop to 3%

120-Volt Single Phase

Current, amps	Wire Length, feet			
	50	100	150	200
10	12	12	10	8
20	12	8	6	6
40	8	6	4	3
60	6	4	2	1
80	6	3	1	0
100	4	2	0	00

190-Volt Three Phase

Current, amps	Wire Length, feet			
	50	100	150	200
10	12	12	12	12
20	10	10	8	8
40	8	6	6	4
60	6	4	4	3
80	4	4	3	2
100	4	3	2	1

230-Volt Three Phase

Current, amps	Wire Length, feet			
	150	200	250	300
10	12	12	12	12
20	10	10	8	8
40	8	6	6	4
60	6	4	4	3
80	4	4	3	2
100	4	3	2	1

WARNING: Electrical Hazard—follow ALL national electric codes, local electric codes, and manufacturer's safety and installation instructions. Always conform to regulatory codes, as apply regionally and internationally.

The first consideration is whether a gumbo slide or gumbo-removal device will be needed. This is often necessary when drilling recent sediments. The second consideration is whether a scalping shaker will be needed. Scalping shakers are usually needed when large quantities of drilled solids or gumbo reach the surface. Usually, long intervals of 17½-inch-diameter holes with flow rates above 1000 gpm require scalping shakers in front of fine screens. The final consideration is to decide on the type and quantity of main shakers necessary to process all of the drilling fluid. The goal should be to sieve an unweighted drilling fluid through the finest screen possible. With weighted drilling fluids, the goal should be to screen all of the drilling fluid through an API 200 screen (finer screens may remove too much weighting material).

Many factors affect the liquid capacity of specific shale-shaker and screen combinations. No information has been published that accounts for all variables. Some manufacturers publish curves relating the fluid flow capacity to screen sizes as a function of one or two parameters. These curves are usually generated without a comprehensive testing program. Most knowledgeable authorities fully agree that such curves cannot include consideration of all of the significant parameters. Many manufacturers use generalizations to gauge the number of shakers needed based on the maximum flow rate anticipated. For example, flow rates between 300 and 500 gpm can probably be processed through an API 200 screen on most linear motion shale shakers.

7.6.1 Selection of Shaker Screens

Some proprietary computer programs are available that reportedly allow predictions of screen sizes used on some shale shakers. Most of these computer programs have been verified with data taken from laboratory-prepared drilling fluid with limited property variation. Different drilling-fluid ingredients can reduce the capacity of a shaker system. For example, a drilling fluid containing starch is difficult to screen because starch, acting as a good filtration control additive, tends to plug small openings in screens. Drilling fluids with high gel strengths are also difficult to screen through fine screens. No charts will be presented here that purport to predict screen sizes that will handle certain flow rates. Screen selection for various shale shakers is primarily a trial-and-error evaluation. The best advice is to contact the manufacturer for recommendations for various geographical areas.

7.6.2 Cost of Removing Drilled Solids

Few wells can be drilled without removing drilled solids. However, even for 3000- to 4000-ft wells, one problem created by drilled solids, such as lost circulation, stuck pipe, or a well-control problem, will more than nullify the modest savings resulting from the decision not to properly process the drilling fluid. In expensive operations, the proper use of solids-removal equipment will significantly reduce drilling costs.

Although drilled solids can be maintained by simply diluting the drilling fluid to control the acceptable level or concentration of drilled solids, the expense and impracticality of this approach are evident using the following example. A 12¼-inch-diameter hole 1000 feet deep will contain about 144 barrels of solids. If these solids are to be reduced to a 6% volume target concentration, they must be blended into a 2400-barrel slurry. To create the 2400 barrels, the 144 barrels of drilled solids must be added to 2256 barrels of clean drilling fluid:

$$(144 \text{ bbl}/2256 + 144 \text{ bbl}) = 6\% \text{ volume}$$

Not only would the cost of the clean drilling fluid be prohibitive, but most drilling rigs do not have the surface volume to build 2256 barrels of clean drilling fluid for every 1000 feet of hole drilled. (See Chapter 15 for a complete discussion of dilution calculations.)

Remove as many drilled solids as possible with the shale shaker. Shakers are a very important component of this process, but they are still only one component of a complete drilled-solids removal system. All of the system must be operated with careful attention to details to develop the most efficient drilled-solids removal. Complete processing will decrease the cost of building excess drilling fluid. Proper drilled-solids control is directed primarily at reducing the cost of drilling.

7.6.3 Specific Factors

Some specific factors that should be considered when designing the shale shaker system are flow rate, fluid type, rig space, configuration/power, elevation available, discharge dryness (restrictions).

Most programs extrapolate laboratory-generated performance curves to predict field performance. Unfortunately, laboratory-manufactured drilling fluid does not duplicate properties of a drilling fluid that has been used in a well. High shear rates through drill-bit nozzles at elevated temperatures produce colloidal-size particles that are not duplicated in surface-processed drilling fluid.

Flow Rate

The flow rate that a particular shaker/screen combination can handle depends greatly on the flow properties of the drilling fluid. The lower the values of PV, YP, gel strength, and mud weight, the finer the screen opening sizes that can be used on a shale shaker. The conductance of the shaker screen provides a guide for the fluid capacity but does not reveal how the screen will actually perform. Screens with the same conductance may not be able to handle the same flow rate if used on different shale shakers.

Shaker screen selection programs have been developed by several companies to predict the quantity of solids that can be removed from a drilling fluid by various shaker screens on specific commercial shakers. Most programs start by assuming that the flow rate of drilled solids reaching the surface is identical to the generation rate of the drilled solids. Unfortunately, many drilled solids are stored in the well bore and do not reach the surface in the order in which they are drilled. Frequently, in long intervals of open hole, as many drilled solids enter the drilling fluid from the sides of the well bore as are generated by a drill bit.

One proposed relationship shows that the maximum flow rate (Q) that can be handled by a shaker is inversely proportional to the product of the PV and mud weight and proportional to the screen conductance (K). This relationship would answer the question, If a linear motion shale shaker is handling 1250 gpm of a 10.3-ppg drilling fluid with a PV of 10 cP on a 120 square MG mesh screen, what flow rate could be handled on a 200 square MG mesh screen if the mud weight were increased to 14.0 ppg and the PV becomes 26 cP?

$$Q_2 = \frac{(K_2)(PV_1)(MW_1)}{(K_1)(PV)(MW_2)} Q_1$$

$$Q_2 = \frac{(1.24 \text{ kd/mm})(10 \text{ cP})(10.3 \text{ ppg})}{(0.68 \text{ kd/mm})(26 \text{ cp})(14.0 \text{ ppg})} \times 1250 \text{ gpm}$$

or

$$Q_2 = 645 \text{ gpm}.$$

The problem with this equation is that it fails to account for other rheological variables. For example, if the gel strength of the 10.3-ppg drilling fluid were increased significantly, the shaker could no longer handle the fluid. Some shakers might handle 750 gpm of an 11.0-ppg drilling fluid with a certain PV. If PHPA or a high concentration of starch is added to this fluid, the shaker capacity might be only 350 gpm. In both cases, the PV would change very little but would have a significant effect on the screening capability. If there are no other property changes in a drilling fluid except mud weight and PV, the preceding equation can help predict what flow rate can be handled.

Rig Configuration

On some drilling rigs, the derrick rig floor is not high enough to allow some shale shakers to be used because the flowline is not high enough. Small land rigs frequently have difficulty positioning larger shale shakers so that the flowline has sufficient slope to prevent fluid from overflowing the bell nipple. Whichever shaker or shakers are used, consideration must be given to providing sufficient safe power to the shaker motors. Check with the manufacturer about electrical service needed for the shaker used.

Discharge Dryness

In some areas, drilled solids and drilling fluid cannot be discarded at the rig location. This applies to both land and offshore rigs. In some areas, the cost of handling discarded material may require drying the discard. The fine screens discharge much wetter solids than do very coarse screens. Hence, fine-screen discharge may require additional processing with dryers or dewatering techniques.

7.7 CASCADE SYSTEMS

Cascade systems use one set of shakers to scalp large solids and/or gumbo from the drilling fluid and another set of shakers to remove fine solids. The first cascade system was introduced in the mid-1970s. A scalper shaker received fluid from the flowline and removed gumbo or large drilled solids before the fluid passed through the main shaker with a fine screen. The first unit combined a single-deck, elliptical motion shaker mounted directly over a double-deck, circular motion shaker (Figure 7.17). This combination was especially successful offshore, where space is at a premium. It was, however, subject to the technology limitations of that time period, which made API 80 to API 120 screens the practical limit.

One advantage of multiple-deck shale shakers is their ability to reduce solids loading on the lower, fine-screen deck. This increases both shaker capacity and screen life. However, capacity may still be exceeded under

Figure 7.17. First cascade shaker system.

many drilling conditions. The screen opening size, and thus the size that solids returned to the active system, is often increased to prevent loss of whole drilling fluid over the end of the shaker screens.

Processing drilling fluid through shale shaker screens, centrifugal pumps, hydrocyclones, and drill-bit nozzles can cause degradation of solids and aggravate problems associated with fine solids in the drilling fluid. To remove drilled solids as soon as possible, additional shakers are installed at the flowline so that the finest screen may be used. Sometimes as many as 6 to 10 parallel shakers are used. Downstream equipment is often erroneously eliminated. The improved shale shaker still remains only one component (though a very important one) of the drilled-solids removal system.

A system of cascading shale shakers—using one set of screens (or shakers) to scalp large solids and gumbo from the drilling fluid and another set of screens (or shakers) to receive the fluid for removal of fine solids—increases the solids-removal efficiency of high-performance shakers, especially during fast, top-hole drilling or in gumbo-producing formations, which is its primary application. The cascade system is used where solids loading exceeds the capacity of the fine screens, that is, it has been designed to handle *high solids loading*. High solids loading occurs during rapid drilling of a large-diameter hole or when gumbo arrives at the surface.

The advantages of the cascade arrangement are:

1. Higher overall solids loading on the system
2. Reduced solids loading on fine mesh screens
3. Finer screen separations
4. Longer screen life
5. Lower fluid well costs

There are three basic designs of cascade shaker systems:

- Separate unit concept
- Integral unit with multiple vibratory motions
- Integral unit with a single vibratory motion

The choice of which design to use depends on many factors, including space and height limitations, performance objectives, and overall cost.

7.7.1 Separate Unit

The separate unit system mounts usable rig shakers (elliptical or circular motion) on stands above newly installed linear motion shakers (Figure 7.18). Fluid from the rig shakers (or scalping shakers) is routed to the back tank of a linear motion shaker. Line size and potential head losses must be considered with this arrangement to avoid overflow and loss of drilling fluid. This design may reduce overall cost by utilizing existing equipment and, where space is available, has the advantages of highly visible screening surfaces and ease of access for repairs.

7.7.2 Integral Unit with Multiple Vibratory Motions

This design type combines the two units of the separate system into a single, integral unit mounted on a single skid. Commonly, a circular, elliptical, or linear motion shaker is mounted above a linear motion shaker on a common skid (Figure 7.19). The main advantages of this design are reduced installation costs and space requirements. The internal flowline eliminates the manifold and piping needed for the two separate units. This design reduces screen visibility and accessibility to the drive components.

Figure 7.18. Separate unit cascade system.

Figure 7.19. Integral cascade unit with multiple vibratory motions.

7.7.3 Integral Unit with a Single Vibratory Motion

This design is shown in Figure 7.20. Typically, this device uses a linear motion shaker and incorporates a scalping screen in the upper part of the basket. The lower bed consists of a fine-screen, flowline shaker unit, and the upper scalper section is designed with a smaller-width bed using a coarser screen. Compared with the other cascade shaker units, this design significantly lowers the weir height of the drilling-fluid inlet to the upper screening area. Visibility of and access to the fine-screen deck can be limited by the slope of the upper scalping deck.

7.7.4 Cascade Systems Summary

Cascade systems use two sets of shakers: one to scalp large solids gumbo and another to remove fine solids. Their application is primarily during fast, top-hole drilling or in gumbo formations. This system was designed to handle *high solids loading*. High solids loading occurs during rapid drilling of a large-diameter hole or when gumbo arrives at the surface.

The introduction of high-performance linear motion and balanced elliptical shale shakers has allowed development of fine-screen cascade systems capable of API 200 separations at the flowline. This is particularly important in areas where high circulating rates and large amounts

Figure 7.20. Integral cascade unit with single vibratory motions.

of drilled solids are encountered. After either the flow rate or solids loading is reduced in deeper parts of the borehole, the scalping shaker should be used only as an insurance device. Screens as coarse as API 10 may be used to avoid dispersing solids before they arrive at the linear motion shaker. When the linear motion shaker, with the finest screen available, can handle all of the flow and the solids arriving at the surface, the need for the cascade system disappears, and the inclination may be to discontinue the use of the scalping screen unit. Even when the fine screen can process all of the fluid, screens should be maintained on the scalper shaker. These screens can be a relatively coarse mesh (API 10 to API 12), but they will protect the finer-screen mesh on the main shaker. The use of finer screens on the scalping shaker will result in fewer drilled solids being removed by the scalping and main shakers.

7.8 DRYER SHAKERS

The dryer shaker, or dryer, is a linear motion shaker used to minimize the volume of liquid associated with drilled cuttings discharged from the main rig shakers and hydrocyclones. The liquid removed by the dryers is returned to the active system. Dryers were introduced with the closed loop mud systems and environmental efforts to reduce liquid-waste haul-off. Two methods, chemical and mechanical, are available to minimize liquid discharge. The chemical method uses a system called a dewatering unit, while the mechanical method takes place through linear motion shakers. These systems may be used separately or together.

The dryer deliquifies drilled cuttings initially separated by another piece of solids-separation equipment. These drilled solids can be the discharge from a main shaker or a bank of hydrocyclones. Dryers recover liquid discharged with solids in normal liquid/solids separation that would have been previously discarded from the mud system. This liquid contains colloidal solids, and the effect on drilling-fluid properties must be considered, since dewatering systems are frequently needed to flocculate, coagulate, and remove these solids.

The dryer family incorporates pieces of equipment long used as independent units: the main linear motion shaker, the desander, and the desilter, which are combined in several configurations to discharge their discard across the fine screens (e.g., API 200) of a linear motion shaker to capture the associated liquid. These units, formerly used as mud cleaners, are mounted on the mud tanks, usually in line with the main linear motion shaker. They can be tied into the flowline to assist with fine screening when not being used as dryers. Their pumps take suction from the same compartments as desanders and desilters and discharge their overflow (effluent) into the proper downstream compartments.

A linear motion dryer may be used to remove the excess liquid from the main shaker discharge. The flow rate across a linear motion dryer is substantially smaller than the flow rate across the main shaker. The lower flow rate permits removal of the excess fluid by the linear motion dryer by using a finer screen. The dryer is usually mounted at a lower level than the other solids-separation equipment to use gravity to transport solids to it. Whether by slide or by conveyor, the cuttings dump into a large hopper, located above the screen, that replaces the back tank, or possum belly. As the cuttings convey along the screen, they are again deliquified. This excess fluid, with the fine solids that passed through the screens, is collected in a shallow tank that takes the place of a normal sump. The liquid is pumped to a catch tank that acts as the feed for a centrifuge or back to the active system.

A dryer unit can be used to remove the excess fluid from the underflow of a bank of hydrocyclones (desanders or desilters). This arrangement resembles a mud cleaner system. In this configuration, the dryer unit may be used on either a weighted or an unweighted mud system. The liquid recovered by the linear motion shaker under the hydrocyclones can be processed by a centrifuge, as previously described.

The perfection of the linear motion shaker for drilling-fluid use, coupled with advanced fine-screen manufacturing technology, has made these dryers very efficient. In most configurations, the dryers use the same style of screens, motors, and/or motor/vibration combinations as do other linear motion shakers by the same manufacturer.

Depending on the fluid, saving previously discarded liquid may be financially advantageous. The dryer discard is relatively dry and can be handled by backhoe and dump truck rather than by vacuum truck.

Drilling-fluid properties must be monitored properly when the recovered liquid is returned to the active system. Large quantities of colloidal solids may be recovered with the liquid. This could affect the PV, YP, and gel strengths of a drilling fluid.

7.9 SHAKER USER'S GUIDE

Every solids-removal system should have enough shale shakers to process 100% of the drilling-fluid circulating rate. In all cases, consult the owner's manual for correct installation, operation, and maintenance procedures. If an owner's manual is not available, the following general guidelines may be helpful in observing proper procedures.

7.9.1 Installation

1. Low places in the flowline will trap cuttings. Flowline angle should be such that settling of solids does not occur, that is, a 1-inch drop for every 10 feet of flowline seems to be a good rule of thumb.
2. When a back tank (possum belly), is used, the flowline should enter at the bottom to prevent solids from settling and building up. If the flowline enters over the top, it should be extended to within one pipe diameter of the flowline from the bottom.
3. Rig up with sufficient space and approved walkways around the shaker(s) to permit easy service and maintenance.
4. Branched tees (Figure 7.21) should be avoided. Solids preferentially travel in a straight path, resulting in uneven solids distribution to the shale shakers.
5. Ensure equal fluid and solids distribution when more than one shaker is used, as shown in Figure 7.22.
6. The options shown in Figures 7.22 and 7.23 are better than the distribution system shown in Figure 7.21.
7. An optional top delivery (Figure 7.23) prevents cuttings settling in the back tank.
8. A cement bypass that discharges outside the active system is desirable.
9. Mount and operate the shale shaker where it is level. Both the solids and fluid limits will be reduced if this rule is not followed.

Figure 7.21. Branched flowline.

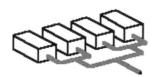

Figure 7.22. Equal distribution flowline.

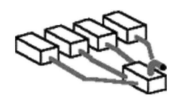

Figure 7.23. Top feed tank distribution flowline.

10. Motors and starters should be explosion-proof. Local electrical codes must be met. Be sure that starter heaters are the proper size.
11. Provide the proper electrical voltage and frequency. Low line voltages reduce the life of the electrical system. Low frequency reduces the motion and lowers the capacity of the shale shaker.
12. Check for correct motor rotation.
13. Check for correct motion of the shale shaker deck.
14. Check drive belts for proper tension according to manufacturer's instructions.
15. Screens should be installed according to manufacturer's instructions.
16. Provide a wash-down system for cleaning.
17. Water-spray bars, if installed, should provide only a mist of water—not a stream.

7.9.2 Operation

1. For double-deck shale shakers, run a coarser-mesh screen on the top deck and a finer-mesh screen on the bottom. The coarser screen should be at least two API sizes coarser than the finer-mesh screen. Watch for a torn bottom screen. During normal drilling operations, cover at least 75–80% of the bottom screen with drilling fluid to maximize utilization of available screen area. Properly designed flowback pans may improve shaker performance. (Gumbo shakers mounted above as an integral part of linear shale shakers are not called double-deck shale shakers, although the operation guidelines above still apply.)
2. For single-deck shale shakers with multiple screens on the deck, try to run screens all of the same mesh. If coarser screens are necessary to prevent drilling fluid loss, run the finer screens closest to the possum belly. All screens should have approximately the same-size openings. For example, use a combination of MG API 100 (140 microns) + MG API 80 (177 microns), but not MG API 100 (140 microns) + MG API 50 (279 microns). Under normal drilling conditions, cover at least

75–80% of the screen area with drilling fluid to properly utilize the screen surface area.
3. Water-spray bars (mist only) may be used for sticky clay to aid conveyance that reduces whole drilling fluid loss. High-pressure washers should not be used on the screen(s) while they are circulating, as solids will be dispersed and forced through the screen openings. Water-spray bars are not recommended for weighted fluids or oil-based NAFs.
4. Do not bypass the shale shaker screens or operate with torn screens; these are the main causes of plugged hydrocyclones. This results in a build-up of drilled solids in the drilling fluid. Dumping the back tank into the pits (to clean the screen or for whatever reason) is a form of bypassing the shale shaker and should not be done.
5. All drilling fluids that have not been processed by solids-removal equipment and are intended to be added to the active system should be screened by the shale shakers to remove undesirable solids. This specifically includes drilling fluid delivered to a location from remote sources.
6. Do not dump the back tank, or possum belly into the system before trips. These solids do not settle and will plug hydrocyclones downstream.

7.9.3 Maintenance

1. For improved screen life with nontensioned screens, make certain that the components of the screen tensioning system, including any rubber supports, nuts, bolts, springs, etc., are in place and in good shape. Install screens according to the manufacturer's recommended installation procedure.
2. For improved life pretensioned screens, ensure the deck rubber support seals are not worn or missing.
3. Lubricate and maintain the unit according to the manufacturer's instructions. (Some units are self-lubricating and should not be "relubricated").
4. With screens that are not pretensioned, check the tension of screens at 1, 3, and 8 hours after installation and hourly thereafter.
5. Check the tension of and adjust drive belts according to the manufacturer's instructions.
6. If only one deck of a multiple-deck shaker is used, be sure that other tension rails are secured.

7. Wash screens at the beginning of a trip so as not to allow fluid to dry on them. We repeat: Do *not* dump the possum belly into the active system or the sand trap below the shaker. The result will be plugging of hydrocyclones downstream and/or an increase in drilled-solids concentration in the drilling fluid.
8. Check the condition of vibration isolator members and screen support rubbers and replace them if they show signs of deterioration or excessive wear.
9. Check the fluid bypass valve and other places for leaks around the shaker screens.
10. Remove drilling-fluid buildup from the vibrating bed, vibrators, and motors. CAUTION: Do *not* spray electrical equipment or motors with oil or water.
11. Make certain that no hose, cables, etc., are in contact with the vibrating bed.

7.9.4 Operating Guidelines

Shale shakers should run continuously while circulating. Cuttings cannot be removed if the shaker is not in motion

1. Drilling fluid should cover most of the screen. If the drilling fluid covers only one fourth or one third of the screen, the screen is too coarse.
2. A screen with a hole in it should be repaired or replaced at once. Holes in panel screens can be plugged. Install screens according to manufacturer's recommended installation procedures. Cuttings are not removed from the drilling fluid flowing through the hole.
3. Shaker screen replacements should be made as quickly as possible. This will decrease the amount of cuttings remaining in the drilling fluid because the shale shaker is not running.
4. Locate and arrange tools and screens before starting to make the replacement. If possible, get help.
5. If possible, change the screen during a connection. In critical situations, the driller may want to stop (or slow) the pumps and stop drilling while the screen is being replaced.
6. For improved screen life with nonpretensioned screens, make certain the components of the screen tensioning system, including any rubber supports, nuts, bolts, springs, etc. are in place and in good shape.
7. Check condition of vibration isolators members and screen support rubbers and replace if they show signs of deterioration or wear.

8. Water should not be added in the possum belly (or back tank) or onto the shale shaker screen. Water should be added downstream.
9. Except in cases of lost circulation, the shale shaker should not be bypassed, even for a short time.
10. Wash screen(s) at the beginning of a trip so fluid will not dry on the screen(s).

The possum belly (or back tank), should not be dumped into the sand trap or mud tank system just before making a trip. If this is done, cuttings will move down the tank system and plug desilters as the next drill bit starts drilling.

7.10 SCREEN CLOTHS

Shale shakers remove solids by processing solids-laden drilling fluid over the surface of a vibrating screen. Particles smaller than the screen openings pass through the screen along with the liquid phase of the drilling fluid. Larger particles are separated into the shaker overflow for discard. The shaker screen acts as a 'go no-go' gauge. That is, particles larger than the screen openings remain on the screen and are discarded. Particles finer than the screen openings go through the screen with the drilling fluid. The criterion for early shale shaker screens was a long screen life. This demand for screen life was consistent with the shaker designs and solids-removal philosophies of the time period. Early shale shakers could remove only large solids from the drilling fluid. The sand trap, reserve and settling pits, and downstream hydrocyclones (if utilized) removed the bulk of drilled solids. Today's shale shakers are capable of utilizing finer screens that remove more solids. Desirable characteristics for a shaker screen are:

1. Economical drilled-solids removal
2. Large liquid flow rate capacity
3. Plugging and blinding resistance
4. Acceptable service life
5. Easy identification

For any particular shale shaker, the size and shape of the screen openings have a great effect on solids removal. This means that the performance of any shaker is largely controlled by the screen cloth used.

The first four items in the preceding list are largely controlled by choice of screen cloth and by the screen panel technology. Large gains in shale shaker performance are a direct result of improved screen cloth and panel fabrication. Screens used on shale shakers have evolved into complex opening patterns.

7.10.1 Common Screen Cloth Weaves

Some of the common cloth weaves available in the petroleum drilling industry are the plain square weave, the plain rectangular weave, and the modified rectangular weave. These are simple over/under weaves in both directions, which can be made from the same or different wire diameters. By making the spacing between the wires the same in both directions, a square weave is created. By making the spacing in one direction longer than the spacing in the other direction, a rectangular weave is made.

Plain square and rectangular weaves are often referred to by the number of wires (the same as openings) in each direction per linear inch. This is the *mesh count*. Mesh is determined by starting at one wire center and counting the number of openings along the screen grid to another wire center 1 linear inch away. For example, an 8 mesh screen has 8 openings per inch in two directions at right angles to each other. When counting mesh, a magnifying glass scale designed for the purpose is helpful.

Use of a single number for screen description implies square mesh. For example, "20 mesh" is usually understood to describe a screen having 20 openings per inch in either direction along the screen grid. Oblong mesh screens are generally labeled with two numbers. A "60 × 20 mesh," for example, is usually understood to have 60 openings per inch in one direction and 20 openings per inch in the perpendicular direction. Referring to a 60 × 20 mesh screen as an "oblong 80 mesh" is confusing and inaccurate.

The actual separation that a screen makes is largely determined by the size of the openings in the screen. The opening size of a square mesh screen is the distance between wires measured along the screen grid, expressed in either fractions of an inch or microns. Screen opening size is most often stated in microns. One inch equals 25,400 microns. Specifying the mesh count does not specify the opening size! This is because both the number of wires per inch and the size of the wires determine the opening size. If the mesh count and wire diameter are known, the opening size can be calculated as follows:

$$D = 25{,}400 \, \{(1/n) - d\}$$

where

- D = opening size, in microns
- n = mesh count, in number of wires per inch (1 per inch)
- d = wire diameter, in inches

The preceding equation indicates that screens of the same mesh count may have different-size openings depending on the diameter of the wire used to weave the screen cloth. Smaller-diameter wire results in larger screen openings, and larger particles will pass through the screen. Such a screen will pass more drilling fluid than an equivalent mesh screen made of larger-diameter wires.

A specialty weave screen is available that consists of large-diameter wires in the long direction and multiple bunches of fine wires in the narrow direction. The long, narrow openings provide low flow resistance and remove spherical and chunky solids.

Layered screens were introduced into the industry in the late seventies. Layered screens are often chosen because they provide a high liquid throughput and a resistance to blinding by drilled solids lodging in the openings. A layered screen is the result of two or more wire cloths overlaying each other. Square and rectangular cloths can be layered. Reducing the diameter of the wires increases liquid throughput. A large assortment of opening sizes and shapes are produced by the multiple screen layers and the particular screen wire diameter. Layered screens have a wide variety of opening shapes and sizes, and therefore a wide variety of sizes of particles pass through the screen.

In 1993 a three-dimensional surface screen was introduced into the industry. This screen surface is corrugated and supported by a rigid frame for use primarily on linear motion shale shakers. As drilling fluid flows down these screens, the solids are moved along in the valleys, and the vertical surfaces provide additional area for drilling fluid to pass. This increases the fluid capacity of a particular mesh size when compared with a flat surface screen.

In summary, specifying the mesh count of a screen does not indicate screen separation performance, since screen opening size, not mesh count, determines the sizes of particles separated by the screen. Because there are almost an infinite number of mesh counts and wire diameters, screen manufacturers have simplified the selection by offering several standard types of cloth series, such as MG, tensile bolting cloth (TBC), and extra-fine wire cloth (XF), as shown in Table 7.4. Notice in this table that an MG 80 cloth has an opening size of 181 microns, whereas a TBC 80 has an opening size of 222 microns. The MG 80 cloth has a smaller opening size than the TBC 80 because the MG cloth's heavier wires take up some of its opening space. As a result, an MG 80 cloth can remove smaller solids than a TBC 80. Furthermore, as a result of the larger wires, the MG 80 cloth will be more resistant to abrasion and will last longer. The major drawback of the MG 80 compared with the TBC 80 is that it will allow less than half the flow rate. That is (see Table 7.5), the screen conductance (ability to transmit fluid) for the TBC 80 is 7.04 kilodarcy/mm, whereas for the MG 80 it is 2.91 kilodarcy/mm. Similar comparisons can be made between the separation/fluid conductance of the TBCs relative to the XF cloths. For instance, a single layer of XF 180 screen cloth has almost the same opening size as a single layer of TBC 165. The XF 180 screen could pass 72% more flow! The screen life of the XF 180 will most likely be shorter than the TBC 165, since the wire diameter of the XF 180 is 30.5 microns and that of the TBC 165 is 48.3 microns. Also, the larger openings would remove fewer drilled solids even though they would pass a larger quantity of fluid.

The National Bureau of Standards has a sieve series that is often used to describe screen opening sizes (Table 7.4). The opening size of this test series plots on uniform increments on semilog paper, making it ideal for use in plotting particle size distributions. Shaker screens used in the industry may be assigned an equivalent National Bureau of Standards sieve mesh count according to their opening sizes as shown in Table 7.4.

From the discussion above, it should be abundantly clear that mesh count alone does not specify the screen opening size. As a result, if mesh count is used, it must be accompanied by a designation of wire diameter, such as MG (mesh count) + mesh count, TBC + mesh count, or National Bureau of Standards Test Sieves equivalent mesh count. One other complicating factor enters with shale shaker screens: Layered screens do not have uniform opening sizes in either direction of the screen. This is the reason that the API has developed a procedure to identify screens.

Just as opening size has been used to measure separation performance, the percentage of open area of a single-layered screen has been used to indicate liquid throughput. The percentage of open area, or the portion

Table 7.4
U.S. STANDARD SIEVE SERIES FOR WIRE CLOTH

Sieve Designation		Nominal Sieve Opening, in.	Permissible Variation of Average Opening from the Standard Sieve Designation	Maximum Opening Size for Not More Than 5% of Openings	Maximum Individual Opening	Nominal Wire Diameter, mm^2
Standard	Alternative					
(1)	(2)	(3)	(4)	(5)	(6)	(7)
125 mm	5 in.	5	±3.7 mm	130.0 mm	130.9 mm	8.0
106 mm	4.24 in.	4.24	±3.2 mm	110.2 mm	111.1 mm	6.40
100 mm	4 in.	4	±3.0 mm	104.0 mm	104.8 mm	6.30
90 mm	3½ in.	3.5	±3.5 mm	93.6 mm	94.4 mm	6.08
75 mm	3 in.	3	±2.2 mm	78.1 mm	78.7 mm	5.80
63 mm	2½ in.	2.5	±1.9 mm	65.6 mm	66.2 mm	5.50
53 mm	2.12 in.	2.12	±1.6 mm	55.2 mm	55.7 mm	5.15
50 mm	2 in.	2	±1.5 mm	52.1 mm	52.6 mm	5.05
45 mm	1½ in.	1.75	±1.4 mm	46.9 mm	47.4 mm	4.85
37.5 mm	1¼ in.	1.5	±1.1 mm	39.1 mm	39.5 mm	4.59
31.5 mm	1⅛ in.	1.25	±1.0 mm	32.9 mm	33.2 mm	4.23
26.5 mm	1.06 in.	1.06	±0.8 mm	27.7 mm	28.0 mm	3.90
25.0 mm	1 in.	1	±0.8 mm	26.1 mm	26.4 mm	3.80
22.4 mm	⅝ in.	0.875	±0.7 mm	23.4 mm	23.7 mm	3.50
19.0 mm	¾ in.	0.750	±0.6 mm	19.9 mm	20.1 mm	3.30
16.0 mm	⅝ in.	0.625	±0.5 mm	16.7 mm	17.0 mm	3.00

(continued)

Table 7.4
Continued

Sieve Designation		Nominal Sieve Opening, in.	Permissible Variation of Average Opening from the Standard Sieve Designation	Maximum Opening Size for Not More Than 5% of Openings	Maximum Individual Opening	Nominal Wire Diameter, mm
Standard	Alternative					
(1)	(2)	(3)	(4)	(5)	(6)	(7)
13.2 mm	0.530 in.	0.530	±0.41 mm	13.83 mm	14.05 mm	2.75
12.5 mm	¼ in.	0.500	±0.39 mm	13.10 mm	13.31 mm	2.67
11.2 mm	⁷⁄₁₆ in.	0.438	±0.35 mm	11.75 mm	11.94 mm	2.45
9.5 mm	⅜ in.	0.375	±0.30 mm	9.97 mm	10.16 mm	2.27
8.0 mm	⁵⁄₁₆ in.	0.312	±0.25 mm	8.41 mm	8.58 mm	2.07
6.7 mm	0.265 in.	0.265	±0.21 mm	7.05 mm	7.20 mm	1.87
6.3 mm	¼ in.	0.250	±0.20 mm	6.64 mm	6.78 mm	1.82
5.6 mm	No. 3½	0.223	±0.18 mm	3.90 mm	6.04 mm	1.68
4.75 mm	No. 4	0.187	±0.15 mm	5.02 mm	5.14 mm	1.54
4.00 mm	No. 5	0.157	±0.13 mm	4.23 mm	4.35 mm	1.37
3.35 mm	No. 6	0.132	±0.11 mm	3.55 mm	3.66 mm	1.23
2.80 mm	No. 7	0.111	±0.095 mm	2.975 mm	3.070 mm	1.10
2.36 mm	No. 8	0.0937	±0.080 mm	2.515 mm	2.600 mm	1.00
2.00 mm	No. 10	0.0787	±0.070 mm	2.135 mm	2.215 mm	0.900
1.70 mm	No. 12	0.0661	±0.060 mm	1.820 mm	1.890 mm	0.810
1.40 mm	No 14	0.0555	±0.050 mm	1.505 mm	1.565 mm	0.725

1.18 mm	No 16	0.0469	±0.0445 mm	1.270 mm	1.330 mm	0.650
1.00 mm	No. 18	0.0394	±0.040 mm	1.080 mm	1.135 mm	0.580
850 μm	No. 20	0.0331	±35 μm	925 μm	970 μm	0.510
710 μm	No. 25	0.0278	±30 μm	775 μm	815 μm	0.450
600 μm	No. 30	0.0234	±25 μm	660 μm	695 μm	0.390
500 μm	No. 35	0.0197	±20 μm	550 μm	585 μm	0.340
425 μm	No. 40	0.0165	±19 μm	471 μm	502 μm	0.290
355 μm	No. 45	0.0139	±16 μm	396 μm	425 μm	0.247
300 μm	No. 50	0.0117	±14 μm	337 μm	363 μm	0.215
250 μm	No. 60	0.0098	±12 μm	283 μm	306 μm	0.180
212 μm	No. 70	0.0083	±10 μm	242 μm	263 μm	0.152
180 μm	No. 80	0.0070	±9 μm	207 μm	227 μm	0.131
150 μm	No. 100	0.0059	±8 μm	174 μm	192 μm	0.110
125 μm	No. 120	0.0049	±7 μm	147 μm	163 μm	0.091
106 μm	No. 140	0.0041	±6 μm	126 μm	141 μm	0.076
90 μm	No. 170	0.0035	±5 μm	108 μm	122 μm	0.064
75 μm	No. 200	0.0029	±5 μm	91 μm	103 μm	0.053
63 μm	No. 230	0.0025	±4 μm	77 μm	89 μm	0.044
53 μm	No. 270	0.0021	±4 μm	66 μm	76 μm	0.037
45 μm	No. 325	0.0017	±3 μm	57 μm	66 μm	0.030
38 μm	No. 400	0.0015	±3 μm	48 μm	57 μm	0.025

The average diameter of the warp and of the shoot wires, taken separately, of the cloth of any sieve shall not deviate from the nominal values by more than the following:

Sieves coarser than 600 μm	5%	Sieves 600 to 125 μm	7½%
Sieves finer than 125 μm	10%		

Table 7.5
Typical Market Grade and Tensile-Bolting-Cloth Shaker Screen Characteristics

	Screen Designation mesh	Separation Potential, in microns			Conductance Kd/mm
		d-16	d-50	d-84	
Market grade	10 × 10	1678	1727	1777	49.68
	20 × 20	839	864	889	15.93
	30 × 30	501	516	531	8.32
	40 × 40	370	381	392	4.89
	50 × 50	271	279	287	2.88
	60 × 60	227	234	241	2.40
	80 × 80	172	177	182	1.91
	100 × 100	136	140	144	1.44
	120 × 120	114	117	120	1.24
	150 × 150	102	105	108	1.39
	200 × 200	72	74	76	0.68
	250 × 250	59	62	63	0.58
	325 × 325	43	44	45	0.44
Tensile bolting cloth	20 × 20	1011	1041	1071	40.93
	30 × 30	662	681	700	24.33
	40 × 40	457	470	483	11.63
	50 × 50	357	368	379	7.94
	60 × 60	301	310	319	5.60
	70 × 70	261	269	277	5.25
	80 × 80	218	224	230	3.88
	94 × 94	175	180	185	2.84
	105 × 105	160	165	170	2.77
	120 × 120	143	147	151	2.51
	145 × 145	116	119	122	2.03
	165 × 165	104	107	110	1.86
	200 × 200	84	86	88	1.49
	230 × 230	72	74	76	1.30

d = 16:84% of particles this size will pass through the screen;
d = 50:50% of particles this size will pass through the screen;
d = 84:16% of particles this size will pass through the screen.

of screen surface not blocked by wire, is calculated as follows:

$$P = \frac{(O)(o)(100)}{(O+D)(o+d)}$$

where

- P = percentage of open area
- O = length of opening in one direction along the screen grid (inches)
- o = length of opening along screen grid perpendicular to the O direction (inches)

- $D =$ diameter of wire perpendicular to the O direction (inches)
- $d =$ diameter of wire perpendicular to the o direction (inches)

Although open area can be used to indicate the ability of a screen to transmit fluid, a better measure of the ability of a screen to pass fluid is the *conductance* (or equivalent permeability of the screen cloth). Conductance takes into account both the openings and the drag of the fluid on the wires. (Conductance is discussed later in this text.)

For years there was confusion in screen designations. Mesh count and percentage of open area simply did not adequately quantify screen cloth performance. Deceptive marketing practices were common. Furthermore, with the advent of the layered cloths, which have a range of hole sizes, there simply were no standards against which to compare screens.

7.10.2 Revised API Designation System

Shale shaker screens made of two or three layers of screen cloth of different mesh sizes present openings that cannot be easily characterized. A technique to describe these openings has been adopted by the API as the "Recommended Practice for Designations of Shale Shaker Screens," API RP 13C, to be issued soon. This recommended practice supersedes the second edition (1985) of API RP 13E, which was valid for only single-layer screens.

The new designation system was chosen to convey information on screen opening size distribution and the ability of nonvibrating screens to pass fluid. Information for each of the following is legibly stamped on a tag attached to the screen panel in such a way as to be visible after the screen is installed on the shale shaker:

- Manufacturer's designation
- API number
- Flow capacity
- Screen conductance
- Conductance
- Total nonblanked area

Manufacturer's Designation

The screen manufacturer may name a particular screen in any manner it desires. This designation is used when ordering a shaker screen with particular characteristics.

API Number

Shaker screen designation has been complicated by the advent of multi-layered screens. When two or three screens are layered together, the opening sizes are not uniform. Experience has shown that the flow rate through these layered screens is much higher than anticipated, and the solids-removal rate is maintained. Since the screens have different irregular shapes, the standard mesh equivalent cannot be used to describe the screens. API RP 13C was recently rewritten by a task group composed of most of the authors of this book. The task group selected a mechanical method of designating shale shaker screens and comparing them to equivalent square mesh opening sizes. This section describes this method for determining the API U.S. sieve number equivalent of a shaker screen using a laboratory sieve shaker, U.S. Standard Test Sieves, and sized grit samples. Screens are rated on the U.S. sieve number scale by the separations they achieve in dry sieving standard grit samples and comparing these separations to the separations of the same standard grit samples with standard U.S. sieves.

For example, a shaker screen that separates the grit sample similar to a U.S. 100 mesh test sieve is designated an API 100. Standard U.S. Test Sieves applicable to this procedure are as follows:

Screen Designation		Permissible opening		Maximum Opening Size Microns
Standard Microns	Alternative Microns	Microns	Microns	
25	500	22	28	34
32	450	29	35	42
38	400	35	41	48
45	325	42	48	57
53	270	49	57	66
63	230	59	67	77
75	200	70	80	91
90	170	85	95	108
106	140	100	112	126
125	120	118	132	147
150	100	142	158	174
180	80	171	189	207
212	70	202	222	242
250	60	238	262	283
300	50	286	314	337
355	45	339	371	395
425	40	406	444	471

500	35	480	520	550
600	30	575	625	660
710	25	680	740	775
850	20	810	890	925
1000	18	955	1045	1080
1180	16	1130	1230	1270
1400	14	1340	1460	1505
1700	12	1630	1770	1820
2000	10	1920	2080	2135

- Hold the test screen securely between the top and bottom parts, which are designed to bolt together and to nest with regular 8-inch U.S. test sieves.
- Arrange the sieves in consecutive order with the coarsest on top and the finest on the bottom. Nest the sieve stack with the sieve pan on the bottom.
- Place the grit test sample on the top sieve, cover it, and shake it for about 5 minutes with the RoTap test sieve shaker. Determine the weight of the grit remaining on the test screen. The fraction of the weight sample retained on the test screen determines the API screen number. Calculate the cumulative weight percentage retained for each individual sieve (beginning with the coarsest) by summing up the results.
- Prepare a plot of cumulative weight percentage retained versus the U.S. sieve opening (in microns) using a linear plot from point to point.
- Sieve and size the test grit through square mesh ASTM (American Society for Testing and Materials) screens. Place equal quantities of five different sizes of the test grit on a test screen on a RoTap for 5 minutes. The quantity and sizes of solids presented to the test screen would be: (1) no solids from an ASTM 80 mesh (180 microns); (2) 10 g from an ASTM 100 mesh (150 microns); (3) 10 g from an ASTM 120 mesh (125 microns); (4) 10 g from an ASTM 140 mesh (106 microns); (5) 10 g from an ASTM 170 mesh (90 microns); and (6) 10 g from an ASTM 200 mesh (75 microns). Present the total sample of 50 g of solids to the test screen, shaken for 5 minutes on a RoTap, and weight the residue on the screen.
- Graphically determine the D100 separation, in microns, from the plot. The value of the D100 separation usually falls between two U.S. sieve openings.
- When the D100 separation falls at a point that is 0.5 or less of the difference between the openings of a finer and the next coarser consecutive U.S. sieve, rate the test screen as the finer U.S. test sieve.

When the D100 separation falls at a point that is more than 0.5 of the difference between the openings of a finer and the next coarser consecutive U.S. sieve, rate the test screen as the coarser U.S. test sieve. For example, if the D100 separation is between a U.S. 170 (90 microns) and a U.S. 200 (75 microns), the test screen is rated as an API 170 if the D100 separation is greater than 82.5 microns, and as an API 200 if the D100 separation is 82.5 microns or less. API numbers are assigned with the following D100 separations, in microns.

D100 Separation (microns)	API No.
> 3075.0 to 3675.0	API 6
> 2580.0 to 3075.0	API 7
> 2180.0 to 2580.0	API 8
> 1850.0 to 2180.0	API 10
> 1550.0 to 1850.0	API 12
> 1290.0 to 1550.0	API 14
> 1090.0 to 1290.0	API 16
> 925.0 to 1090.0	API 18
> 780.0 to 925.0	API 20
> 655.0 to 780.0	API 25
> 550.0 to 655.0	API 30
> 462.5 to 550.0	API 35
> 390.0 to 425.0	API 40
> 327.5 to 390.0	API 45
> 275.0 to 327.5	API 50
> 231.0 to 275.0	API 60
> 196.0 to 231.0	API 70
> 165.0 to 196.0	API 80
> 137.5 to 165.0	API 100
> 116.5 to 137.5	API 120
> 98.0 to 116.5	API 140
> 82.5 to 98.0	API 170
> 69.0 to 82.5	API 200
> 58.0 to 69.0	API 230
> 49.0 to 58.0	API 270
> 41.5 to 49.0	API 325
> 35.0 to 41.5	API 400
> 28.5 to 35.0	API 450
> 22.5 to 28.5	API 500

In the graph that follows, 33 weight percentage of the grit sample was captured on the test screen. This screen would have an API number of 140 and an opening size of 102 microns.

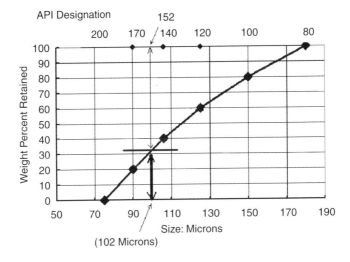

The designation would be an API number of 140, with the actual separation point of 102 microns in parentheses:

API 140 (102 microns)

The "mesh" designation is now called an API number so that the new designation will be more rig-user friendly. Rig crews recognize "mesh size" even though they may not actually know the definition. The change is necessary, since the API Recommended Practices are being converted to International Standards Organization (ISO) documents. ISO uses the metric system, consequently the number of openings per inch would need to be converted to the number of openings per centimeter or meter. This number would be meaningless to most rig crews.

The API designation number is specified to appear at least three times larger in physical appearance than any other letters or numbers on the screen tag. The ASTM 140 screen has openings of 106 microns, and an ASTM 170 screen has openings of 90 microns. The number in parentheses will indicate that the screen designation was actually measured and provide an indication of how close the openings are to standard screens.

Flow Capacity/Screen Conductance

A screen that makes an extra-fine separation is not useful in the drilling industry if it will not pass a high-volume flow rate. The amount of fluid that a screen will process is dependent on the screen construction as well as solids conveyance, solids loading, pool depth, deck motion and acceleration, drilling-fluid properties, and screen blinding. Although it is difficult to calculate the expected fluid processing capacity of a shaker, screens can be ranked according to their ability to transmit fluid.

Conductance is a measure of the ease with which fluid flows through a screen cloth. It is analogous to permeability per unit thickness of the screen, $C = k(\text{darcy})/l(\text{mm})$. To calculate the flow through a porous medium, Darcy's law is used as follows:

$$V = K \times \Delta p / (\mu \times l).$$

Now conductance, C, can be calculated where $Q = V \times A$ as follows:

$$C = K/l = V \times \mu/\Delta p = Q \times \mu/(A \times \Delta p)$$

where

- C = conductance (darcy/cm)
- K = permeability (darcy)
- l = screen thickness (cm)
- V = velocity (cm/sec)
- μ = fluid viscosity (cP)
- Δp = pressure drop across screen (atm)
- Q = volume flow rate (cm^3/sec)
- A = screen area (cm^2)

Higher conductances mean that for a given pressure drop across the screen, more fluid is able to pass through the screen.

To measure the conductance, a 50-gal container of motor oil is mounted above the test screen. A flow valve is adjusted so that some of the oil overflows the screen into catch pans outside the apparatus. The oil that flows through the screen is captured in a container on a balance. The weight of the container and oil is observed and recorded. When the flow becomes steady and uniform, the weight of the oil flowing through the screen is measured as a function of time. The temperature of the oil is measured and is kept constant. The density and viscosity of the oil as a function of temperature is determined prior to the test.

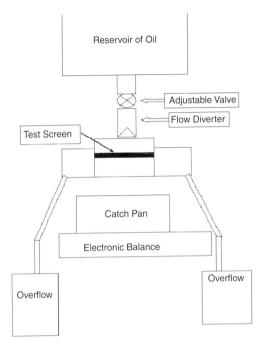

Conductance Measuring Device

From the height of overflow fluid above the test screen, the head can be measured. From the density/temperature charts, the pressure applied to the screen can be calculated. From the density/temperature charts and the weight measurements, the volume of motor oil flowing per unit of time can be calculated. Care is taken to ensure a low flow rate to prevent turbulence in the oil flowing through the screen. From these measurements and the equations described above, the permeability per unit thickness of the screen can be calculated. This is the conductance.

Total Nonblanked Area

Continuous cloth screens present all available screen area to the drilling fluid to remove solids. Panels are popular because screen tears are minimized and limited to only one small area of the screening surface. The screen panels, however, remove some of the screening area that would be available with continuous cloth screens. The *nonblanked area* allows an evaluation of the surface area available for liquid transmission through the screen.

7.10.3 Screen Identification

The screen tag would appear as:

API Designation (micron opening size) Manufacturer's Designation Country of Manufacture	Manufacturer's Name NONBLANKED AREA: 7.23 sq.ft. CONDUCTANCE 1.4 Kd/mm Conforms to API RP13C

API RP 13 C SCREEN DESIGNATION **API 170** **(92 Microns)** Catalog Number XX-13	**Unblanked Area** **0.67 sq. meters.** **Conductance** **9.0 Kd/mm**	We - R - Shakers 311 Fantasy Lane Utopia, Texas 00000

7.11 FACTORS AFFECTING PERCENTAGE-SEPARATED CURVES

The relationship between the size and shape of the particles being separated and the size and shape of the screen openings will influence how fine a separation is made. This is reflected in the percentage-separated curve. If all of the solids being drilled are spherical, then the distribution of the narrowest dimension of the screen openings will establish the percentage-separated curve. For wells with poor drilling practices, cuttings are tumbled in the annulus and arrive well rounded at the surface. For wells that have good cuttings transport in the annulus, the cuttings may be long, thin slivers of rock.

Solids have mobility in a pool of fluid to seek a screen opening large enough to go through. As a result, the conveyance velocity, contact time with the screen, and presence of other solids all affect the ability of the solids to go through the holes in the screen. These variables therefore affect the percentage-separated curve.

Surface tension of the fluid causes solids to agglomerate together as they exit a pool of fluid. If solids finer than the screen openings make it out of the pool of fluid, then they are held by the surface tension and have very little chance to go through the screen. Adding a spray wash to the last screen panel disperses these patties, which will allow finer solids to be washed through the screen.

Blinding or plugging of screen cloth, as shown in Figure 7.24, dramatically affects not only the amount of fluid that will pass through the screen, but also the separation the screen makes. Many of the screen openings effectively become smaller, and fewer solids will pass through. The screen then makes a much finer separation than originally intended, and the screen capacity decreases significantly.

Reported values for percentage-separated curves are also affected by the way the measurement is made in the laboratory. The greatest error is often the measurement of particle size distribution. Particle sizing by sieve analysis is the best way to characterize solids being screened, since the sieving process is similar to screening. Unfortunately, sieving is a tedious and slow process. Forward laser-light-scattering particle size analyzers such as the Malvern and Cilas granulometers tend to report size distributions somewhat larger than sieve analysis. These instruments report particle sizes in terms of equivalent spherical diameters. Some drilled solids may be more rectangular in shape, so the equivalent spherical diameter may not exactly agree with the sieve analysis. Clay particles in the 1-micron size are broad, flat surfaces, similar to a tabletop. These are difficult to describe in terms of a diameter.

In summary, the percentage-separated curve represents the fraction of solids rejected by the screen as a function of size. From the preceding discussion, it may be noted that the percentage-separated curve is dependent on the conditions that existed when the data were taken. As a result, in actual drilling conditions, the percentage-separated

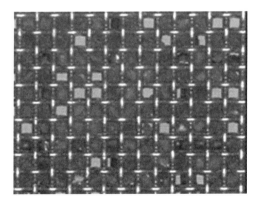

Figure 7.24. Particles plugging wire mesh.

curve probably varies as drilling-fluid properties and the shapes of the solids change and as the screen blinds.

7.11.1 Screen Blinding

Screen blinding occurs when grains of solids being screened find a hole in the screen just large enough to get stuck in. This often occurs during the drilling of fine sands in the Gulf of Mexico. The following sequence is often observed during screen blinding:

- When a new screen is installed, the circulating drilling fluid falls through the screen a short distance.
- After anywhere from a few minutes to even several hours, the fluid endpoint slowly or even quickly travels to the end of the shaker.
- Once this occurs, the screens must be changed to eliminate the rapid discharge of drilling mud off the end of the shaker.
- After the screens have been washed, fine grains of sand are observed stuck in the screen.
- The surface of the screen will feel like fine sandpaper because of the sand particles stuck in the openings.

Most every screen used in the oilfield is blinded to some extent by the time it has worn out. This is the reason that when the same screen size is reinstalled, the fluid falls through the screen closer to the feed.

A common solution to screen blinding is to change to a finer or coarser screen than the one being blinded. This tactic is successful if the sand that is being drilled has a narrow size distribution. Another solution is to change to a rectangular screen, although rectangular screens can also blind, with multiple grains of sand. Unfortunately, the process of finding a screen that will not blind is expensive.

In the late seventies the layered screen was introduced to avoid screen blinding. This hook-strip type of screen was mounted on a downhill sloping unbalanced elliptical motion shale shaker vibrating at 3600 rpm. The two fine layers of screening cloth, supported at 4-inch intervals, tended to dislodge fine grains of sand and would blind only about 25% of the screen in severe laboratory tests, leaving 75% of the screen nonblinded! The nonblinding feature is assumed to be the result of the deceleration of the two screens. The wire diameter is in the range of 0.002 inch and the opening sizes are in the range of 0.004 inch. In the upward thrust of a layered screen, the screens must come to a stop at the upward end of the motion. They would tend to each have an inertia that would prevent them from stopping at exactly the same time. This would create an opening size slightly larger than the original opening size of the layered screen during the upward part of the thrust. Solids would be expelled from the screen. On the downward thrust of the motion, the two layers remain together until the screen starts deceleration. At the bottom of the stroke, again the inertial forces could cause the screens to slightly separate, allowing larger solids to pass through the screen. This probably also explains why the separation cut point curve shows poorer separation characteristics for a layered screen than for a single square mesh screen. Many particles larger and smaller than the median opening size are found in the discard from a layered screen.

Unfortunately, the downhill sloping basket and high frequency limits the amount of liquid that can pass through the screen. Furthermore, lost circulation material has a high propensity to get stuck in the screen due to the high-frequency, short-stroke vibration. These problems have been ameliorated by reducing the vibration to 1800 rpm and flattening the basket slope. In the early 1980s, linear motion was introduced so that solids could march up an incline out of a pool of liquid. This fluid pool provided additional pressure to force fluid through the screen. Unfortunately, linear motion, combined with marginal support, tore layered screens apart. The only way to obtain satisfactory screen life on a linear motion machine was to support the layered screen in 1-inch squares.

7.11.2 Materials of Construction

The materials used to weave the cloth screens are quite varied. Screens are made from metal wires, plastic wires, and molded plastic cloths.

Metals

Alloys that are most weavable and resistant to corrosion are nickel/chrome steels; 304, 304L, 316, and 316L. These alloy wires are available in sizes down to 20 microns. The finest wire available is 304L, which is available to 16 microns. Other materials, including phosphor bronze, brass, copper, monel, nickel, aluminum alloys, plain steel, and plated steel, are also available. Within the drilling industry, 304 stainless cloth is the most common.

Plastics

Two types of synthetic screens are available: woven synthetic polymer and molded one-piece cloth, called a platform.

Conventional looms can be used to weave synthetic polymer screens. Polymers, such as polyesters, polypropylene, and nylon, are drawn into strings having diameters comparable to those of wire gauges and woven into screen cloth. Synthetic screens exhibit substantial stretch when mounted and used on shale shakers. Because of this, plastic screen openings are not as precise, although this variability is not nearly as great as in layered metal steel screens.

One-piece injection molded synthetic cloths are typically made from urethane compounds. These synthetic cloths have limited chemical and heat resistance but display excellent abrasion resistance. The designs range from simply supported molded parts having very few open areas to complex structures with up to 55% open area. Molded cloths are very popular in the mining industry, where abrasion resistance is important. These screens make a coarser separation than screens used in the oilfield. Development of molded cloth screens capable of making a fine separation that have heat and chemical resistance necessary for oilfield application is under way.

Cloth selection for shale shaker screens involves compromises among separation, throughput, and screen life.

7.11.3 Screen Panels

Shale shaker screens have changed as demands on the shale shaker have increased. Shaker screens have three primary requirements:

- High liquid and solids handling capacity
- Acceptable life
- Ability to be easily identified and compared

Early shale shaker screens had to last a long time. This demand was consistent with the shaker designs and solids-removal philosophies of the period. Shakers could remove only the large, coarse solids from the drilling fluid, sand trap, and reserve pit, while downstream hydrocyclones (if utilized) removed the bulk of the drilled solids.

Drilling-fluid changes, environmental constraints, and a better understanding of solids/liquid separation have modified the role of the shale shaker. Generally, the effectiveness of the downstream equipment is greater when more solids are removed at the flowline. Reserve pits can be smaller or in most cases eliminated. Cleanup costs are lower than not removing the solids at the flowline and overall drilling efficiency is increased.

As important as the mechanical aspects of new-design shale shakers may be, improvements in screen panels and screen cloths have also significantly increased shaker performance—shakers of older design have benefited from these improvements as well. Two design changes have been made to extend the economic limit of fine-screen operation: (1) a coarse backing screen, which protects the fine screen from being damaged, extends life, and provides additional support for heavy solids loading, and (2) tensioned cloth bonded to a screen panel.

Pretensioned Panels

The most important advance in screen panel technology has been the development of pretensioned screen panels. Similar panels have been used on mud cleaners since their introduction. Earlier shakers did not possess the engineering design to allow their use. With the advent of linear motion machines, the pretensioned panels extended screen life and permitted more routine use of API 200 screens.

Pretensioned panels consist of a fine-screen layer (or layers) and a coarse backing cloth bonded to a support grid. The screen cloths are pulled tight, or tensioned, in both directions during the fabrication process. This ensures the beginning of proper tension of every screen. Correct installation procedures and post-run retightening of screen panels can add significantly to shaker performance and screen life.

Manufacturers employ different geometric apertures in screen panel design. Some of the more common panel shapes are square, rectangular, hexagonal, and oval. The apertures in the panels can vary from 1-inch to 3-inch squares to 7×33-inch rectangles to 1.94-inch hexagons to 2×6-inch ovals.

The panels can be flexible (of thin-gauge metal or plastic) to be stretched over crowned shakers, or they can be flat (of heavy-gauge mechanical tubing) for installation, as on flat-decked (noncrowned) shakers.

Regardless of configuration, the function of the pretensioned panel is to provide mechanical support for the fine-screen cloth bonded to it, and at the same time occlude as little potential flow area as possible with the supporting grid structure.

Some screened panels are made with no support grid at all, but simply by bonding of the finer-mesh cloth directly to a coarser backing wire using a heat-sensitive adhesive. Essentially this becomes a hook-strip design, with certain support refinements.

Pretensioned screen panels address two of the three original design goals: capacity and screen life. The remaining goal of easy identification is a function of better labeling techniques to display important screen characteristics.

7.11.4 Hook-Strip Screens

Hook-strip screens (named for the method of hooked edging that provides the tension along the screen) are also available. Because of the superior life characteristics of panel mount units, this type of screen has been relegated to a minor role on linear motion machines. They are used extensively on circular and unbalanced elliptical motion machines. Proper tensioning (and frequent retensioning) of all types of screens is good screen management and adds significantly to screen life. Individual manufacturer's operation manuals should be consulted to obtain the proper installation methods and torque requirements, where applicable, for specific screens/panels.

7.11.5 Bonded Screens

Several types of bonded screens are available. The repairable perforated plate screen has one or more layers of fine mesh cloth bonded to a sheet of metal or plastic with punched, patterned holes. Perforated plate designs are available in various opening sizes and patterns. Additional designs include a special application in which backing and fine-screen materials are bonded together, eliminating the need for perforated plates. Flat-surfaced pretensioned screen panels are becoming more even tensioned, easy to install, and capable of even distribution of liquids and solids across the screen deck.

7.11.6 Three-Dimensional Screening Surfaces

Three-dimensional screen panels were introduced in the mid-1990s. These typically offer between 75 and 125% more screening area than flat-panel repairable plate screens, while retaining the ability to be repaired. Compared with nonrepairable hook-strip screens, most three-dimensional screen panels offer up to 45% more screening area.

This type of screen panel adds a third dimension to the previous, two-dimensional screens. The screen surface is rippled and supported by a rigid frame. Most three-dimensional screen panels resemble the metal used in a corrugated tin roof. Construction consists of a screen cloth that is in fact corrugated, pretensioned, and bonded to a rigid frame.

Like bonded flat screens, the three-dimensional screen panel needs only to be held firmly in place with a hook strip or other means to prevent separation between the shaker bed and the screen panel during vibration. Three-dimensional screen panels can be used to support any type or style of wire cloth and with any type of motion. They improve any shaker performance over comparable flat-screen surfaces under most drilling conditions. Three-dimensional screens may not improve shaker performance when drilling gumbo or large, pliable, sticky cuttings.

Three-dimensional screen panels allow solids to be conveyed down into the trough sections of the screen panel. When submerged in a liquid pool, this preferential solids distribution allows for higher fluid throughput than is possible with flat-screen panels by keeping the peaked areas clear of solids. A three-dimensional screen panel improves distribution of fluid and solids across the screen panel.

7.12 NON-OILFIELD DRILLING USES OF SHALE SHAKERS

Trenchless drilling is one of the fastest growing areas for shale shaker use other than in drilling oil and gas wells. Many of these shakers are used in conjunction with hydrocyclones, creating a mud cleaner.

7.12.1 Microtunneling

Microtunneling has become very popular in Europe and is being used more and more in the United States. Microtunneling is horizontal boring of a large-diameter hole (from 27 inches up to 10 feet) while simultaneously laying pipe. This is typically done in cities for laying or replacing water and sewer pipe under buildings and heavily traveled roads.

To prepare for these operations, large-diameter vertical holes, or caissons, are excavated so that the drilling equipment and hydraulic rams are set up at the desired depth. The caisson is excavated slightly below the equipment level, creating a sump for the returned drilling fluid and associated drilled solids. The returns are pumped to the surface by a submersible pump to a compact solids-removal system, which typically consists of a shale shaker and mud cleaner mounted over a small tank.

7.12.2 River Crossing

To run a pipeline under a river, a small-diameter hole is directionally drilled under the riverbed. The pipe for the pipeline is attached to the end of the drill string and pulled back under the river while a larger hole to accommodate the pipe is back reamed.

During the laying of large-diameter pipelines, a substantial solids-control system must be set up with multiple shakers, desanders, desilters, and centrifuges. Mud cleaners reduce drilling-fluid disposal volumes.

7.12.3 Road Crossing

Pipelines or cables often need to cross under roads. Drilling beneath a road does not disrupt traffic or destroy the road surface. Frequently, the hole volume is small enough that no solids-removal equipment is used. If drilling-fluid accumulation could cause a problem, or for large-diameter holes and wide roadbeds, a shaker or mud cleaner is used.

7.12.4 Fiber-Optic Cables

The laying of fiber-optic cables does not require large-diameter holes, and it is often done in residential or business areas where drilling fluid and drilled solids must be contained. Solids-control systems for this application usually consist of only a small tank, pump package, and small shaker.

CHAPTER 8

SETTLING PITS

Leon Robinson
Exxon, retired

Drilling fluid enters the removal-tank section after it passes through the main shale shaker. Immediately below the main shaker is the first pit, called a settling pit or sand trap. Fluid passing through the shaker screen flows directly into this small compartment. The fluid in this compartment is not agitated. This allows solids to settle. The fluid overflows from the sand trap into the next compartment, which should be the degasser suction pit. The sand trap is the only compartment not agitated in the mud tank system.

The sides of a sand trap should slope at 45° or more to a small area in front of a quick opening discharge valve. When the solids are dumped, the valve can be closed quickly when drilling fluid begins to flow from the trap. The purpose of the quick-opening valve is to allow only settled solids to leave the compartment, with minimal loss of drilling fluid. In many cases during periods of fast drilling, with coarse or damaged shaker screens in use, the sand trap will fill several times per day.

An effective sand trap requires an overflow weir of maximum length to create a liquid column as deep as possible. A common, and recommended, practice is to utilize the full length of the partition between the sand trap and the degasser suction pit.

8.1 SETTLING RATES

The rate at which solids settle depends on the force causing the settling, the dimensions of the solid, and the fluid viscosity in which the solid is settling. Analysis of forces acting on irregularly shaped objects is extremely complicated. Analysis of forces acting on spheres is not as complicated and is addressed here: For simplicity, the solid will be assumed to be spherical and settling in a quiescent fluid. The forces acting on the sphere would be the gravitational force causing it to fall and the buoyant force tending to prevent settling. The force causing settling could also be centrifugal, from a device such as a hydrocyclone or a centrifuge. This section will develop the equation relating to solids settling through a drilling fluid in the sand trap.

Settling rates of spherical particles in liquid can be calculated from Stokes' law:

$$F = 6\pi\mu v R \qquad (8.1)$$

where

- F is the force applied to the sphere by the liquid, in dynes
- μ is the fluid viscosity, in Poise
- v is the particle velocity, in cm/sec
- R is the radius of the sphere, in cm.

Stokes' law was developed when the centimeter/grams/second (cgs) unit system was popular with scientists. Viscosity is defined as the ratio of shear stress in a liquid to the shear rate. One Poise has the units of dynes-sec/cm^2 in absolute units, or [g/cm-sec] in cgs units. The unit of dyne also has the units of gcm/sec^2.

A sphere falling through a liquid experiences a downward force of gravity and an upward force of the buoyancy effect of the liquid. The buoyancy force is equal to the weight of the displaced fluid:

$$\text{buoyant force} = \frac{4\pi}{3}(R^3)\rho_l. \qquad (8.2)$$

The downward force is mass times acceleration, or the weight for gravity settling. The mass of the sphere is the volume of the sphere times the density of the sphere (ρ_s):

$$\text{mass of a sphere} = \frac{4\pi}{3}[R^3]\rho_s. \qquad (8.3)$$

Equation 1 now becomes

$$\frac{4\pi}{3}[R^3]\rho_s - \frac{4\pi}{3}(R^3)\rho_l = 6\pi\mu v R \tag{8.4}$$

$$\frac{4}{18}[R^2](\rho_s - \rho_l) = \mu v. \tag{8.5}$$

Solving this equation for velocity and changing the radius R to diameter d, in cm:

$$v = \frac{d^2}{18\mu}(\rho_s - \rho_l)g \tag{8.6}$$

$$v: \text{cm/sec} = \frac{(d: \text{cm})^2}{18(\mu: (\text{poise}))}[(\rho_s - \rho_l): \text{g}_m/\text{cm}^3](g: \text{cm}/\text{sec}^2) \tag{8.7}$$

$$v: \text{cm/sec} = \frac{(d: \text{microns} \times 10^{-4})^2}{18(\mu: (\text{cP}/100))}[(\rho_s - \rho_l): \text{g}_m/\text{cm}^3](980 \text{ cm}/\text{sec}^2) \tag{8.8}$$

$$v: \text{ft/sec} = \frac{(d: \text{microns} \times 10^{-4})^2}{18(\mu: (\text{cP}/100))}[(\rho_s - \rho_l): \text{g}_m/\text{cm}^3]$$
$$\times (980 \text{ cm}/\text{sec}^2)(\text{ft}/30.48 \text{ cm})(60 \text{ sec}/\text{min}) \tag{8.9}$$

$$v: \text{ft/min} = \frac{1.07 \times 10^{-4}(d: \text{microns})^2}{\mu: (\text{cP})}[(\rho_s - \rho_l): \text{g}_m/\text{cm}^3] \tag{8.10}$$

where

- v = settling or terminal velocity, in ft/min
- D = particle equivalent diameter, in microns
- ρ_s = solid density, in g/cm^3
- ρ_l = liquid density, in g/cm^3
- μ = viscosity of liquid, centipoises (cP)

A 2.6-g/cm^3 drilled solid passing through an API 20 screen (850-micron diameter) would fall through a 9.0-ppg, 100-cP drilling fluid with a terminal velocity of 1.6 ft/min. This could be calculated from equation 10:

$$v: \text{ft/min} = \frac{1.07 \times 10^{-4}(d)^2}{\mu}(\rho_s - \rho_l)$$

$$v: \text{ft/min} = \frac{1.07 \times 10^{-4}(100 \text{ microns})^2}{100 \text{ cp}}(2.6 - [9.0 \text{ ppg}/8.34 \text{ ppg}])$$

$$= 1.62 \times 10^{-2} \text{ ft/min}$$

If the rig is circulating 500 gpm through a 50-bbl settling tank or sand trap, the fluid remains in this tank for a maximum of 4.2 min. If the sand trap holds 100 bbl of drilling fluid, the retention time is 8.4 min. Solids can settle about 6 inches during the 4.2-min retention time or 1 foot during the 8.4-min retention time.

The selection of a viscosity to use in the equation is complicated. On drilling rigs, normally the lowest viscosity measurement made is with the 3-rpm viscometer reading. Some drilling rigs using polymer drilling fluids use Brookfield viscometers, which measure very low shear rate viscosities. Drilling-fluid viscosity is a function of shear rate, as discussed in Chapter 2 on Drilling Fluids. As particles settle, the fluid viscosity impeding the settling depends on the settling rate. As the velocity decreases, the viscosity of the fluid increases. The K viscosity is the viscosity of a fluid at one reciprocal second, which is within the shear rate range of a small solid falling through a drilling fluid and can be determined on most drilling rigs. Some drilling fluids are constructed to have very large low-shear-rate viscosities, to facilitate carrying capacity as the solids are moved up the borehole. Many drilling-fluid systems have K viscosities in the range of 1000 effective cP instead of the 100 cP used in the example above. Solids settling will be greatly hindered in these fluids because they are designed to prevent settling.

8.2 COMPARISON OF SETTLING RATES OF BARITE AND LOW-GRAVITY DRILLED SOLIDS

Stokes' law can be used to describe the anticipated settling rate for spheres of barite or low-gravity drilled solids:

$$v_B = \frac{d_B^2}{18\mu}(\rho_B - \rho_l)g \qquad (8.11)$$

$$v_{LG} = \frac{d_{LG}^2}{18\mu}(\rho_{LG} - \rho_l)g. \tag{8.12}$$

Equations 11 and 12 can be used to solve for the ratio of diameters that will cause the settling velocity of barite to be equal to the settling velocity of low-gravity solids:

$$D_B = 0.65 D_{LG}. \tag{8.13}$$

Equation 13 indicates that a 20-micron barite sample settles at the same rate as a 30-micron low-gravity solid; or a 48-micron barite sample settles at the same rate as a 74-micron low-gravity solid. Note that this is true regardless of the viscosity of the fluid in which these particles are settling.

8.3 COMMENTS

Linear motion and balanced elliptical motion shale shakers permit the use of finer screens than were used in the past. Consequently, sand traps are frequently ignored in a system using them. Considering the inescapable fact that screens regularly tear and wear out, sand traps offer the ability to capture some of the solids that would normally be left in the drilling fluid.

When API 80 screens were used on shale shakers and represented the smallest openings possible for processing drilling fluid, sand traps were a very important component of the surface drilling-fluid system. Normally, screens as coarse as API 20 to API 40 (850 microns to 425 microns) were used in the upper part of a borehole. The solids that passed through these screens settled quite rapidly. When API 200 screens are installed on the main shakers, the largest solid presented to the fluid in the tank is 74 microns. These solids settle much more slowly than the larger 850-micron (API 20) solids that were separated earlier.

The sand trap is still used in a system to provide backup for failures in the main shaker screen. These screens sometimes break, and the failure may go unnoticed for a long period of time. The sand trap offers the possibility of capturing some of the solids that pass through the torn screen.

Although not intended to be used as an insurance shield, scalping shakers also provide the opportunity to remove solids larger than API 20 to API 40 before the fluid reaches the main shaker. This provides some relief from large solids reaching the sand trap if the finer wires of the main shaker break.

8.4 BYPASSING THE SHALE SHAKER

One rule cited frequently is "Do not bypass the shale shaker." Cracking the bypass valve at the bottom of the shaker back tank allows a rig hand to mount fine screens on the shaker. However, the solids that are not presented to the shaker screen are not removed and cause great damage to the drilling fluid. Sand traps provide some insurance against this activity; but they do not capture all of the larger solids that bypass the screen—so it is still a very bad practice. This is frequently the reason hydrocyclones are plugged.

Another activity common on drilling rigs bypasses the shaker screen more subtly. Before making a trip, the "possum belly," or back tank, is dumped into the sand trap to clean the shaker. Drilling fluid left on a shaker screen dries during a trip and causes the screen to flood. In an effort to prevent any screen plugging, the possum belly is also cleaned and all of the settled solids are dumped into the sand trap. All of the dumped solids, however, do not settle. When circulation is restored, these suspended solids migrate down the removal system until they reach the apex of a hydrocyclone. These solids plug many cones on drilling rigs. Possum bellies should be dumped into a waste pit, NOT into the drilling-fluid system.

CHAPTER 9

GAS BUSTERS, SEPARATORS, AND DEGASSERS

Bill Rehm

Drilling Consultant

9.1 INTRODUCTION: GENERAL COMMENTS ON GAS CUTTING

Solids-control equipment can be severely affected by gas in the drilling mud. This condition is misinterpreted and misunderstood in most field applications. The primary problems caused by gas cutting in solids control are blinding of the shaker screens and degradation of pump output to hydrocyclones and centrifuges. Gas cutting always occurs during drilling of a gas-bearing formation.

If there is enough gas to displace drilling fluid to the surface (and increase pit volume), bottom-hole pressure is reduced. This occurs when the pressure exerted by the drilling fluid is less than the formation pressure and there is some significant permeability. This condition requires surface control, gas busters or separators, and a degasser.

If there is no pit volume increase but the drilling fluid is gas cut and the flowline mud density reduced, bottom-hole pressure is not significantly reduced and this condition in general calls for only a degasser (see Box 9.1).

Box 9.1 Bottom-Hole Pressure Reduction Due to Gas Cutting [Goins & O'Brien]

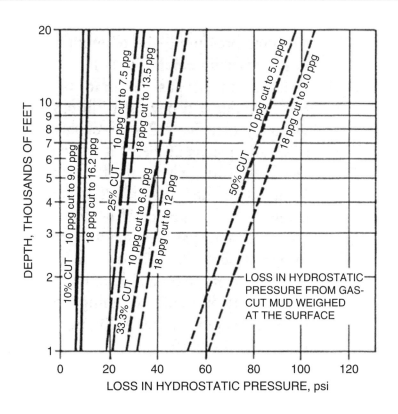

Calculation of bottom-hole pressure reduction in a *static* mode is expressed by the iterating Strong-White equation [Strong], a simplistic model of which is given as:

$$\Delta P = n \cdot 2.3 \text{Log}_{10} \cdot p$$

where

- ΔP = reduction in bottom-hole pressure, atm
- n = ratio of gas to mud
- p = hydrostatic pressure, atm
- $n = (1 - x)/x$
- x = weight of cut mud/weight of uncut mud

Or, the equation can be rearranged as:

$$\Delta P = ((w_1 - w_2)/w_2) \cdot 33.81 \cdot \text{Log}(p/14.7)$$

where

- ΔP = bottom-hole pressure reduction, psi
- w_1 = weight of uncut mud, lb/gal
- w_2 = weight of cut mud, lb/gal
- p = hydrostatic pressure of mud, atm

The addition of flow to the static mode is an iterating set of equations that generally are accepted to introduce errors of $\pm >100\%$. These equations are most fully developed in the underbalance models of the various service and engineering companies.

The simplest mathematical solution for a drilling operation is to calculate height in the annulus of the mud displaced by the gas cutting and reduce that to a pressure loss:

$$(-)\Delta P = (\Delta V/Av) \times p \times 0.052$$

where

- $(-)\Delta P$ = bottom-hole pressure decrease, psi
- ΔV = pit volume increase due to gas cutting, bbl
- Av = annular volume behind drill pipe, bbl/ft
- P = density of the (uncut) drilling fluid, ppg
- 0.052 = units constant

Trip gas and connection gas are indications of the swabbing effect of upward pipe movement. The swabbing force reduces the pressure exerted by the drilling fluid to below the formation pressure. Generally this problem is temporary and gas cutting can be handled by a gas buster and a degasser.

Increasing mud density only because of gas-cut mud is not generally a good solution. If the gas cut is the result of drilling a gas-bearing formation, increasing mud density may lead to lost circulation. However, increasing connection gas may be an indication of increasing formation pressure.

9.2 SHALE SHAKERS AND GAS CUTTING

Shale shaker screening is dependent on a constant flow of drilling fluid with cuttings. The fluid must pass through the screen, and the cuttings must either pass through or be rejected by the screen. Gas cutting in the drilling mud can have up to three different effects that upset the screening process.

1. Gas heading can cause volume surges in the mud flow that exceed the ability of the screen to handle fluid flow. This is usually from gas, intermixed in the mud, rapidly expanding at the surface and pushing large surges of drilling fluid out the flowline. Gas busters and gas separators are the solutions to this problem.
2. Gas cutting from tiny gas bubbles entrained in the drilling fluid can cause screen blinding when the bubbles expand to fill the area between the screen wires. This problem is usually handled by a degasser that removes the entrained gas from the drilling fluid.
3. Foaming associated with gas cutting leaves a film of very light, wet foam on the shaker screen. The foam is too light to be gravity-pulled through the screen and carries extra liquid off the end of the shaker. Often the fluid loss is not significant and can be ignored. In other cases shaker sprays and defoamers are needed to break the foam. However, shaker sprays tend to wash extra cuttings through the screen.

A fourth problem, not related to gas, is an extreme rise in the viscosity of the drilling fluid from saltwater or salt. This may cause the drilling fluid to flow over and not pass through the shaker screen. This problem often requires the bypassing of the contaminated mud to a reserve tank.

9.3 DESANDERS, DESILTERS, AND GAS CUTTING

The operation of any desilter or desander is dependent on the head pressure, which is in part a function of the volume pumped by the centrifugal pump. Centrifugal pumps are very sensitive to gas cutting. The gas collects in the reduced-pressure area at the center of the impeller and quickly reduces the pump output. In cases in which there is very low fluid head on the suction of the centrifugal pump, a small amount of gas in the drilling fluid may gas-lock the pump and stop or limit the flow of drilling fluid.

Reduced flow from the centrifugal pumps tends to dump whole mud from the bottom of the desilter or desander hydrocyclone to waste. Gas locking of the centrifugal pump stops desilter or desander action. Any time there is gas in the flowline, a degasser should be used ahead of the solids-control feed pumps.

9.4 CENTRIFUGES AND GAS CUTTING

The centrifuge feed is typically a Moyno pump, also called a progressing cavity pump. The Moyno pump is a positive displacement pump and does not gas-lock, as does a centrifugal pump, but the input feed is reduced. The reduction is in direct proportion to the gas in the mud and the feed pressure to the centrifuge. The gas at atmospheric pressure in the mud is compressed by the Moyno pump to the feed pressure. The compression of the gas reduces the output of the pump (Box 9.2). Reduction in feed to the centrifuge reduces the output and may change the cut point.

9.5 BASIC EQUIPMENT FOR HANDLING GAS-CUT MUD

Gas busters are a simple cylinder or baffle box at the flowline where mixed drilling fluid and gas are roughly separated while flowing. The drilling fluid goes to the shale shaker, and the gas is allowed to flow away or is sent to a flare line.

Separators are holding tanks where mixed water, oil, and gas are allowed to separate by gravity. They have evolved in the last 50 years from simple open tanks to complex closed and pressurized tanks. Separators can be informally divided into two groups: (1) atmospheric, or unpressurized, and (2) pressurized, or closed.

Degassers are somewhat different devices from the preceding two. The degasser is a tank in which a vacuum and/or spray removes entrained gas from the mud system. Degassers handle much smaller gas volumes than do gas busters or separators but do a more complete job of removing the gas.

The distinction between gas busters, separators, and degassers is not precise. There are unpressurized (atmospheric) degassers, and there are separators that use centrifugal force or an involute spiral to help bring

Box 9.2 Reduction in Positive Displacement Pump Output Due to Gas Cutting

Note: The following estimates are minimum values. As gas cutting increases, the flow properties of the fluid increase and the head above the suction decreases, all of which will further degrade the performance of a positive displacement pump.

The following would apply to Moyno feed pumps as well as duplex and triplex mud pumps.

What would be the reduction in output of a positive displacement pump if the gas cut at the suction of the pump reduced the fluid density from 12 ppg (1440 kg/m^3) to 10 ppg (1198 kg/m^3)?

1. Given: The displacement of the pump is 1^3 volume. The reduction in liquid volume due to gas at the suction is

    $$12 \text{ ppg} \times Y^3 \text{ vol} = 10 \text{ ppg} \times 1^3 \text{ vol}$$

    where $Y^3 = 0.83$, so 83% of 12-ppg mud and 17% gas by volume will give a 10-ppg drilling fluid.

2. Given: The suction pressure is 1 atm absolute (14.7 psi gage or 101 kPa), and the discharge pressure is 5 atm absolute (73.5 psi gage, 518 kPa). The reduction in volume as the gas is compressed to the discharge pressure would follow Boyle's law:

    $$P_1 V_1 = P_2 V_2$$

    $$1 \text{ atm abs} \times 0.17 V_1 = 5 \text{ atm abs} \times V_2$$

    $$V_2 = 0.03 V_1$$

3. Then the final discharged volume of fluid would be only 86% of the uncut volume. If the pump under consideration were supercharged (as a triplex mud pump would be) part 2 of this example would have to be calculated first for compression of the gas as a result of the charging pump, then for the new volume used in part 1, and then recalculated for part 2 and summed to get a solution.

about separation. However, the purpose of a gas buster or separator is to separate *mixed* gas and water, or water and oil, while the purpose of a degasser is to remove *entrained* gas from the drilling fluid.

In all of these devices, the method of separation involves one or several of the following processes.

9.5.1 Gravity Separation

Gravity separation depends on the difference in density of the materials, the depth of the liquid column, the size of the gas bubbles, and internal resistance to flow in the liquid. A simple mud, or fracturing (frac), tank is a gravity separator that holds liquid, solids, and gas until they separate naturally. Gas rises and exits from the top of the tank. Oil separates from the other fluids and floats to the top. After the oil overflows a weir or plate inside the tank, it is pumped away. Drilling fluid or other liquids such as saltwater are removed from near the bottom of the tank. Solids settle to the bottom and can be left there or they can be stirred into the liquid and removed with conventional solids-control equipment. Gravity separation is the dominant method used in oil/water separator tanks and in closed system pressurized separators.

9.5.2 Centrifugal Separation

The fluid is spun by sending it tangent to the inside of a round vessel, or is spun in a rotating cylinder. The gas, oil, water, and solids are separated by the artificial gravity caused by the centrifugal force in the spinning liquid. This method is used in a number of the "atmospheric" or West Texas separators. An involute spiral is used in some of the closed pressurized systems to cause centrifugal separation.

9.5.3 Impact, Baffle, or Spray Separation

In impact, baffle, or spray separation, the fluid is directed onto a baffle at high velocity. The impact of the fluid containing gas starts separation. The fluid velocity may be the result of flow from the flowline or blooie line or it may be picked up by a pump and sent as a fluid stream or spray.

9.5.4 Parallel-Plate and Thin-Film Separation

Parallel-plate or thin-film separation works with gas in a liquid. The fluid is spread out as a thin film over a plate, which allows the gas to escape more easily.

- In a parallel-plate separator, the fluid containing the gas is forced between parallel plates, which distort the bubbles of gas and help them break. This is common in commercial demisting or defoaming operations.
- Thin-film separation is the process whereby the liquid is flowed in a thin film over a plate that allows the gas bubbles to expand and break. The thin-film process is part of most vacuum degasser operations.

9.5.5 Vacuum Separation

The vacuum degasser, which separates gas entrained in a liquid, uses reduced pressure (a partial vacuum) that causes the gas bubbles to expand and break. This method is used primarily in degassers to remove entrained gas.

9.6 GAS BUSTERS

The purpose of a gas buster is to remove gas mixed with the drilling fluid before the drilling fluid goes over the shale shaker. A gas buster works well in fluid with large bubbles of free gas. (Often the gas is starting to break free in the flowline.) A problem with the basic gas buster is that the heavier gases will not rise and be dissipated in the air but settle around the rig.

An old-fashioned but effective gas buster is made from a piece of 9- or 11-inch (228.6- or 279.4-mm) casing (Figure 9.1). An inlet, tangent to the side but tilted up about 5°, is welded into a 6-ft (2-m) length of casing about one third of the distance from the bottom. The mud entering the inlet spins, and the centrifugal force allows the gas to go to the center and out while the mud goes to the sides and down. A pipe on the top carries the gas away, and the bottom of the casing is open to the shale tank on the shaker.

The tangential intake is used on many land rigs where the gas buster is installed in the possum belly (back tank) of the shaker. The tangential

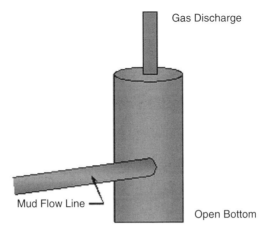

Figure 9.1. Simple Gas Buster.

intake balances the force of the drilling fluid and expanding gas so that the gas-buster tank does not need heavy bracing.

Another version of the pipe gas buster uses the same casing size as the standard gas buster but directs the mud and gas mixture onto a blast plate or "baffle," which breaks up the flow pattern and separates gas and drilling fluid. This system is unbalanced, and the pipe needs to be restrained. In heavy mud and with higher than normal initial gel strength, the baffle system may cause some entrainment of the gas, which appears as gas cutting.

The offshore version of the gas buster uses an 11-inch casing up to 20 feet tall ahead of the shale shaker. The offshore gas buster is closed and usually has a U tube on the line to the shaker to build backpressure and force gas to the discharge line.

9.7 SEPARATORS

9.7.1 Atmospheric Separators

The purpose of a separator is to separate free gas or mixed gas and oil from the drilling fluid and convey the gas or oil to a flare or holding tank.

Mud Tanks

The simplest atmospheric separator is a mud tank that uses gravity separation. With gravity separation, gas rises and escapes; the oil separates from the drilling fluid and rises to the top of the tank. Oil/water separation can then be carried out with a simple siphon. Several extra mud tanks are commonly used when there is significant amount of free oil to be separated from a water-base drilling fluid. The drilling-fluid mixture may be first passed through a West Texas type of separator (see the section to follow) to remove the free gas, and then it is sent to the first of several mud tanks. In the tank sections, the oil is successively skimmed off with siphons or skimmers, and concentrated oil is pumped to a shale tank. The drilling fluid is taken from the bottom and sent to the solids-control system.

Mud tank separators, often used in the southern end of the Austin Chalk trend in Texas, allow the limited hydrocarbon vapor to escape and settle many of the solids in the bottom of the skimmer tanks. They are easy to set up and inexpensive to operate.

A simplified tank separator system is used with some workover and small drilling rigs. The drilling-fluid/gas/oil mixture is sent directly to an open tank, normally a frac tank, with the flowline in line with the long axis of the tank. The open tank then acts as a separator, with the gas escaping to the atmosphere, the oil and drilling fluid separating, and the solids settling out on the bottom. Drilling fluid is taken from the bottom of the far end of the tank, and during connections oil is pumped to a shale tank.

9.7.2 West Texas Separator

The basic West Texas separator was developed in the 1950s to drill the Permian Red Beds in West Texas and is still the "gold standard" atmospheric separator. It is simple, inexpensive, and efficient. It works best with heavy gas cutting, or with gas/liquid (gaseated) mud. It is classified as an atmospheric, nonpressurized tank, but some of the tanks develop from 2 to 5 psi of internal pressure to force the gas to the flare stack (Figure 9.2).

There are many variations of the West Texas separator. One system uses centrifugal separation with a collection tank. Another basic version uses an impact baffle and a collection tank. One method of discharge is via a U tube to hold a liquid seal in the tank. Another version uses a float and valve. Pressure can be held in the tank as a result of the height of the flare stack, or with a balancing valve on the flare line.

The basic design is so old that little math was ever applied to it, but engineering simulations/solutions have been applied to confirm some of

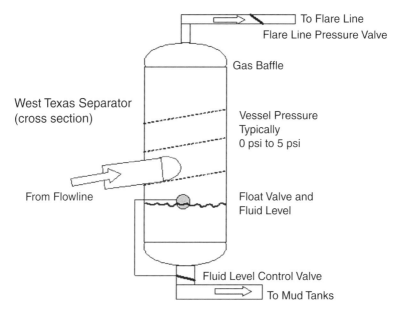

Figure 9.2. "West Texas" Separator.

its actions. Tank sizes vary from 4 to 7 feet in diameter and are generally about 12 feet high, mounted in a frame so that the discharge can be gravity-fed to the shaker. Tank or throughput capacity is a function more of tank and discharge line size than anything else. The original design seems to be have been developed for drilling with 350 gpm of drilling fluid in an $8\tfrac{3}{4}$ inch hole while the hole produced 1 to 2 MMscf/day (600 to 1200 scf/m) of gas. Present manufacturers appear to feel confident with values twice as high.

The Super Mud Gas Separator (SWACO SMGS^{TM}), 22 feet high × 6 feet wide, was first used for underbalance drilling in the northeastern section of the Austin Chalk trend, where there are very high flow rates. In size, the SMGS may be the largest separator of this type. Capacity is quoted as 65 MMscf/day of gas or 38,000 bpd of liquid.

9.8 PRESSURIZED SEPARATORS

9.8.1 Commercial Separator/Flare Systems

There are a number of portable rental flare systems that use a closed separator or free water knockout (FWK) ahead of the flare stack.

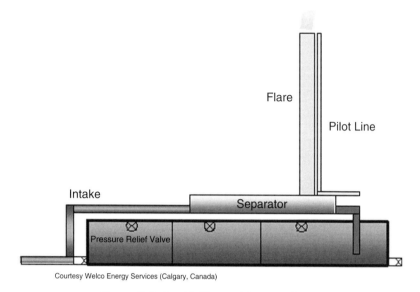

Figure 9.3. Portable Flare and Separator System.

The systems range from a trailer-mounted flare stack with a minimal FWK to skids with a 6 × 10-feet, 1.5-atmosphere (7 psig) cylindrical separator (Figure 9.3). The key to using the portable flare is not to overload the liquid end of the separation system, since the primary purpose is flaring.

The separator/flare systems are commonly used with recompletion and workover operations, or where there are strict regulations about safety and flare stack heights.

9.8.2 Pressurized, or Closed, Separators: Modified Production Separators

The major advantages of a closed pressurized system are that it (1) controls gas from the well and sends it to a flare line under pressure and (2) is serviced by a special crew.

The separator is usually operated under 3 to 5 atmospheres of pressure (45 to 75 psig). Horizontal units are typically about 9 feet in diameter and 50 feet long, with a throughput of 5 MMcf or 500 bbl fluid. These are typical numbers, and sizes and pressure vary according to special jobs.

In areas where H_2S is common, where there are strict flare regulations, or where gas could accumulate in a closed area on an offshore platform, keeping a separator operating under pressure is a critical factor in drilling safety. Further, the separator system is generally instrumented to give a history of annular pressure and measurements of gas, water, and oil volumes.

Canada began using modified production test separators as closed drilling separators in the 1990s. Production separators already had the capability to separate and measure the amount of gas, oil, water, and BS&W (bottom sediment and water), so they were easily adapted for use in underbalance drilling projects where H_2S was to be expected or where the drilling fluid was an oil/nitrogen mixture.

Closed separators are gravity separators, often with an involute spiral feed (a type of centrifugal input). Baffles separate areas inside the tank and start to isolate cuttings, oil, and water. Gas rises and under separator pressure is forced to the flare line. The oil rises out of the drilling fluid and is pumped to an oil tank. Cuttings that start to settle out in the separator are recycled through the tank, and the drilling fluid is returned with the cuttings to the shale shaker (Figure 9.4).

Closed separator systems typically have measurement systems for water, oil, and gas; fluid height; and tank pressure. There is a separate upstream trapping system to collect cutting samples. Horizontal closed separators are the most common, but vertical separators are available and are typically used on offshore rigs.

Cuttings removal has always been a problem in closed separators. Most of the cuttings are trapped in the first compartment. In some early

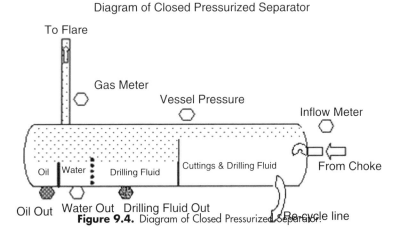

Figure 9.4. Diagram of Closed Pressurized Separator.

designs, a screw, or Moyno, pump automatically pumped the cuttings out. However, generally cuttings are so sticky that they have a high angle of repose and will not fall into the pump suction, and so build up in the front of the tank. Later systems recirculated a part of the mud from the separator back through the cuttings compartment and circulated solids through the system to the rig shale shaker.

The closed separator is operated by a special crew. As a result of the large equipment and extra crew, the closed separator system is expensive to mobilize and operate but provides a degree of safety not found with other types of separator systems.

9.8.3 Combination System: Separator and Degasser

The West Texas separator has been combined with the vacuum tank degasser as the TOGA^{TM} system (the SWACO Total Gas Containment System, or Tri-Flow Separator System). The mud from the flowline enters the separator, where the vacuum degasser controls the fluid flow, so that the degasser is always in balance and can never be overloaded. A "T" in the separator/degasser line allows mud from the pit to enter the degasser if there is surging and periods of no flow from the flowline. Gas from the degasser is pumped to the flare line by the discharge of the vacuum pump.

The extra volume in the base of the separator tank acts as a buffer against surges in the system. The separator tank is pressured to 15–50 psi by a backpressure valve at the base of the exit to the flare line. The vacuum breaker line on the degasser is connected to the output of the separator. No free or entrained gas is released except to the flare line. This system has seen extensive use in urban areas and with H_2S gas and tends to be used in the United States in place of the closed (pressurized) modified production separator used in Canada. International operations vary with the vendor or local rules (Figure 9.5).

9.9 DEGASSERS

The purpose of a degasser is to remove entrained gas from the drilling fluid. By this definition, then, the degasser has a limited capability for handling large quantities of gas—typically anything more than about 50–100 scfm (20 scfm of gas at surface pressure will gas-cut 400 gpm of 16-ppg drilling fluid to 10 ppg). Large volumes of gas need to be removed first by a separator or gas buster.

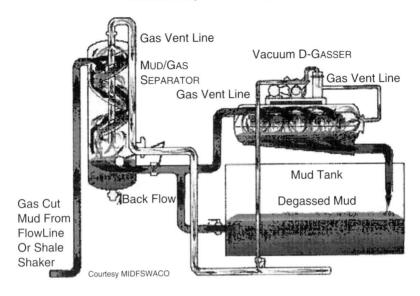

Figure 9.5. Swaco "Total Gas Containment System"™.

Entrained gas is composed of such small bubbles that the gravity displacement of the bubbles is near the viscous drag of the fluid. The bubbles do not rise to the surface, or rise very slowly. Initial gel strength is also a powerful force in retarding the bubbles' rise.

The bubbles must rise to the surface and expand enough to overcome the liquid film enveloping them before they can break and release the gas. Fine solids and long-chain polymers in the drilling fluid often make the film around the bubbles stronger. As an example, foaming in drilling fluid occurs when the liquid film around the bubbles becomes strong enough to contain the gas at atmospheric pressure.

The importance of a vacuum to help remove entrained gas from the drilling fluid can be shown by some simple calculations (see Box 9.3).

9.9.1 Degasser Operations

The effective throughput of a degasser depends on a number of variables:

1. The vacuum level is limited in part by how high the drilling fluid has to be lifted to enter the vacuum chamber. Lifts of more than 10 feet (3 m) are probably counterproductive.

Box 9.3 The Effect of a Vacuum on Entrained Gas

Bubble-Volume Increase

Bubble-volume increase is expressed by a simplified form of the general gas law, when ignoring the effect of temperature:

$$P_1 V_1 = P_2 V_2$$

where

- P = pressure in absolute terms, psia
- V = volume of the bubble

Using a 2-mm-diameter entrained gas bubble, with 1-mm radius, and 14.7 as absolute pressure, from ($P_1 V_1 = P_2 V_2$) it can be seen that reducing the atmospheric pressure by one-half doubles the volume of the bubble.

$$\text{volume of a sphere} = 4/3 \Pi r^3 \text{ or } 4.189 r^3$$

$$P_1 \times V_1 = P_2 \times V_2$$

So the radius of the bubble with twice the volume becomes 2.099 mm.

Bubble Surface Area Increase

To go a bit further and relate volume to radius to final surface area:

$$\text{surface area of the bubble, } S = 4 \Pi r^2 \text{ or } 12.566 r^2.$$

But the surface area increase will be the ratio of the radii squared or 4.406/1.

So, decreasing the pressure from 1 atm (14.7 psia or 29.9 in. Hg) to ½ atm (7.35 psia or 14.95 in. Hg) doubles the volume of the bubble and increases the surface area of the original 2-mm-diameter bubble by 4.4 times.

Two things are happening simultaneously: The bubble volume is getting larger, so it will rise to the surface faster, where there is further decreased pressure; and the skin, or surface area, of the bubble is becoming larger and weaker.

2. The denser the drilling fluid, the more the displacement force of the bubble upward, but this is not generally as important as the properties of the fluid fraction of the drilling fluid. A higher drag coefficient reduces the ability of the bubble to rise. This is related to viscosity effects and initial gel strength and is generally greater in higher-density drilling fluids.
3. Polymers and fine solids in the drilling fluid tend to build a tougher film around the bubble.
4. The more fluid is pumped or ejected, the less the residence time in the vacuum; or contrary-wise, with more gas, there will be less fluid throughput.

With any particular drilling fluid, volume of entrained gas, and height of the degasser suction, there is a limit to the ability of the degasser to remove all the gas from the drilling fluid. Since it is not possible to predict all the drilling fluid/gas conditions, degasser planning is based on experience in the area. Some manufacturers have test curves that will show the real output of the degasser under fixed conditions.

9.9.2 Degasser Types

Vacuum-Tank Degassers

The original degasser, and the most common form of degasser, is the vacuum tank (Figure 9.6). The tank may be a horizontal or a vertical cylinder. The drilling fluid is pulled into the tank by vacuum action. The primary vacuum force for filling the tank is created by the jet that is discharging the drilling fluid, or in some cases by the *pump* that is discharging the drilling fluid. The fluid level in the tank is controlled by a float that opens or closes a vacuum breaker valve.

The separation of gas and liquid starts as the drilling fluid is pulled up the suction. When the liquid enters the tank, it is distributed over a plate or series of plates where it flows as a thin film. As entrained bubbles increase in size, come to the surface, and break, the vacuum pump discharge pumps the released gas to a disposal line. The size of tank degassers varies widely, but the standard horizontal tank degasser on a skid is generally about 12 feet long × 4 feet wide and weighs about 3000 pounds. Some units are quoted at about 1000 gpm of fluid and a maximum vacuum of about 13 inches Hg.

206 *Drilling Fluids Processing Handbook*

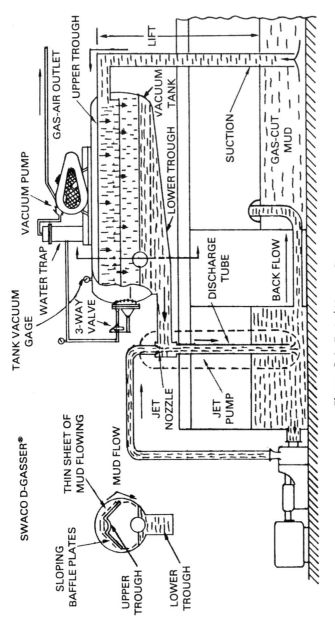

Figure 9.6. Typical Vacuum Degasser.

The throughput of the vacuum tank degasser is controlled by the discharge jet or pump. The higher the tank is above the surface of the drilling fluid, the more of the energy from the jet or pump is used to lift the fluid. The throughput volume of the tank decreases with height. Most problems with the vacuum tank degasser are because the tank lift is too high or the jet discharge is not strong enough (Box 9.4).

9.9.3 Pump Degassers or Atmospheric Degassers

The size and weight of the tank degasser lead to the development of smaller and lighter degassing units. There are several configurations, but the pump degassers are typically about 3½ feet in diameter at the top and 8 feet long. An impeller in the head pulls up the drilling fluid and discharges it against the inside of the degassing chamber. Degassing is accomplished by the reduction in pressure as the drilling fluid is pulled up to the impeller and then by the impact of the spray discharge.

9.9.4 Magna-Vac™ Degasser

The Burgess Magna-Vac [Burgess Manufacturing Ltd.] is the most sophisticated and complex design among drilling fluid degassers. It combines the more efficient vacuum removal of gas with the lighter weight and smaller size of the pump systems (Figure 9.7).

The drilling fluid is drawn up from the pits through a rotating pipe by a vacuum provided by the regenerative vacuum blower on the top of the unit. The drilling fluid enters the vacuum chamber of the unit through holes in the top of the rotating pipe, and at that point is further accelerated and sprayed outward against the walls of the vacuum chamber. The gas is pulled to the vacuum pump through a narrow gap at the upper edge of the vacuum chamber that excludes liquids. Pressurized gas is then sent to the flare or discharge line.

The drilling fluid flows to the bottom of the vacuum chamber, where it is picked up by an evacuation (centrifugal) pump and discharged.

The system is controlled by a buoyant scheduling ring in the vacuum chamber that controls the height of the liquid in the vacuum chamber by restricting or opening the entrance to the vacuum chamber.

Depending upon the model type, the unit is about 3½ feet in diameter, about 6 feet long, and weighs 900 to 1500 pounds. Flow volume is quoted at 1000 gpm and maximum vacuum at 10–15 inches Hg.

Box 9.4 Basic Treatment Calculations

Mathematical calculations of separation ability for the various separation methods have a number of variables that make it impractical to mathematically predict separation of gas or oil from the drilling fluid.

Basic variables:

- Liquid flow rate
- Gas volume
- Fluid properties, plastic viscosity, gel strength, density, etc.
- Residence time
- Centrifugal force
- Vacuum
- Gas bubble size
- Gas bubble film strength
- Emulsifiers (for drilling fluid and oil)

Some basic ideas and numbers are:

- Retention time in the vessel
  - $Tr = Ve / Q \times k$, where
    - Tr = retention time, sec
    - Ve = effective volume of the vessel, ft^3
    - Q = flow rate, gpm
    - k = units constant, 7.48 gal/ft^3
- Vacuum measurements:
  - Atmospheric pressure = 14.7 psia
  - = 29.9 Hg
  - = 33.94 water
  - = 1 atm absolute
  - = 101.3 kPa
- Hydraulic horsepower (Hhp):
  - $Hhp = P \times Q/1714$, where
  - P = gage pressure, psi
  - 1714 = units constant, psi/gpm.

Electrical or mechanical equivalent hp is at least twice the theoretical Hhp (see Box 9.1).

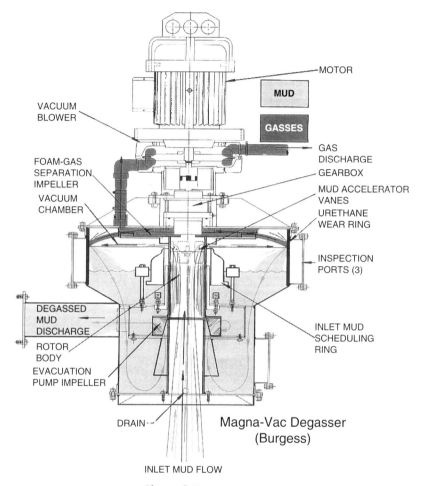

Figure 9.7. Pump Degasser.

9.10 POINTS ABOUT SEPARATORS AND SEPARATION

Following are points about liquid/gas separation.

- Gas and water in a mixture with no solids or emulsifiers separate naturally and quickly by gravity because gas is lighter, floats up in big bubbles, and breaks out.
- Water and oil in a mixture with no solids or emulsifiers separate naturally by gravity because oil is normally lighter than water. Bear in

mind, however, that some oil is heavier than water and that some oil is soluble in water (and all gas is soluble in oil).
- Gas and oil may require extra energy and time to separate. For example, air and nitrogen separate from diesel oil very quickly. On the other hand, methane and ethane are soluble in oil and may require a long time to separate without the addition of heat or chemicals.
- The time required for drilling fluid and gas to separate depends on the fluid properties. If the drilling fluid approaches water, separation is quick. If, however, the drilling fluid has high apparent viscosity (thick), separation can be difficult and time-consuming.
- If emulsifiers are present in the mud system, it is difficult to separate oil from drillling fluid. In some cases it is simply not practical to work on line because it requires time, heat, and breaking chemicals.
- Foaming makes it difficult to separate gas from liquid. The best practical defoamers are aluminum stearate and alcohol.
- In foam, bentonite and some polymers, such as CMCs (ceramic matrix composites), make such a stable foam that it is time-consuming and difficult to break them out.
- It is vital to know how the separator and/or degasser in use works, both mechanically and chemically. Moreover, there must be enough separators and degassers to handle the anticipated volumes.
- In general, the old rule of simpler is better also holds true for separators. However, safety considerations offshore or when dealing with H_2S gas are of paramount importance.

REFERENCES

Burgess Manufacturing Ltd. Web site: www.burgess-mfg.com. Brochure "Burgess Magna-Vac™ Degassers." Personal conversation with Harry Burgess, 8/11/03.

Derrick Equipment. Web site: www.derrickequipment.com/products/vacu_flo.htm

Goins & O'Brien, *Blow Out Control and What You Need to Know About It.* Series from *Oil & Gas Journal*, 1960.

Lawson, G. & Liljestrand, W. *Mud Equipment Manual Handbook 5, Degassers.* Gulf Publishing, 1976.

Rehm, WA. *Practical Underbalance Drilling and Workover.* University of Texas, PETEX, 2002

Strong, *Journal of the SPE*, 1956, "Gas Cut Mud."

SWACO

  CD, *Complete Pressure Control*

  CD, *SWACO Product Catalog*, ver. 1.0

Tri Flo

  General Catalog, *Fluids Processing Equipment*, 6/01

  Service and operating manuals:

Flow-Line Degasser

  Advanced Vacuum Degasser

  800 Compact Degasser

  Mud Gas Separator

Weatherford, personal communication, Scott Nolan, Calgary, Canada, 8/19/03.

Welco Energy Services, Inc., Calgary, Canada. Compact disc, *Portable Flare Tank Specifications*.

CHAPTER 10

SUSPENSION, AGITATION, AND MIXING OF DRILLING FLUIDS

Mike Richards

Brandt, A Varco Co.

10.1 BASIC PRINCIPLES OF AGITATION EQUIPMENT

Drilling fluids are used for a variety of purposes: to control well pressures, deliver power to downhole motors, remove drilled cuttings from the bit and transport them to the surface, and stabilize the well bore by chemical or mechanical means.

As drilling geometry becomes more challenging in regard to directional steering, or in over- or underpressurized formations or formations that are susceptible to sloughing and swelling, special precautions must be taken to ensure that mud properties are adequate for the situation. Viscosity-enhancing agents, thinners, weighting material, and special additives are used to produce a drilling fluid that will meet site-specific requirements. At the same time, drilled solids must also be removed from the fluid to allow reuse without excessive dilution. This chapter will demonstrate why agitation is important to the mud circulation system. It will also define the pitfalls of improper agitation, suspension, and mixing of drilling fluid. Mixing, shearing, blending, and addition of chemicals are equally important. Proper sizing and installation will allow the mixing of fluid additives and chemicals at maximum speed with minimal problems.

The purpose of a surface mud system is to allow maintenance of the mud before it is pumped down the hole. This is accomplished by the effective use of solids-control equipment to remove undesirable solids, while simultaneously recovering as much drilling fluid as is feasible. Secondary to the solids-removal process is the addition of chemicals and the rapid and thorough mixing of mud materials. Thorough agitation is necessary to effectively accomplish both of these tasks. In the majority of drilling fluids, agitation equipment must be used to suspend solids in the surface tanks and maintain a homogeneous drilling fluid. There are two types of equipment to do this:

- mechanical agitators, and
- fluid motive devices, also called mud guns.

Both are widely used, and will be discussed by showing basic principles, installations, and sizing guidelines.

Drilling-fluid components such as bentonite (gel), barium sulfate (barite), lost circulation material, polymers, and chlorides (to name a few) must be wetted before they are dispersed throughout the system. Proper introduction of these materials into the drilling fluid is equally important and can enhance drilling efficiency as well as reduce the amount of additives needed. To gain the most advantage, correct addition at the surface will enhance their effectiveness in the well. Several technologies are described to blend, add, hydrate, shear, or mix drilling-fluid additives and enhance their effectiveness.

As stated previously, the purpose of agitation equipment is to suspend solids, completely mix mud materials, and maintain a homogeneous mixture throughout the surface system. For all of these requirements to be met, agitation equipment must create an upward velocity within the mud tanks that is greater than the *settling velocity* (see chapters 8, 11, and 13 for a complete explanation) of the suspended solids. There must be adequate shear and stirring to dissolve, wet, and disperse mud additives. The ability to blend drilling fluid helps maintain consistency. The following discussion illustrates how mechanical agitators and mud guns operate and accomplish these goals.

10.2 MECHANICAL AGITATORS

Mechanical agitators are used extensively for drilling-fluid surface tanks. Regardless of manufacturer, mechanical agitators have similar components, namely, a drive motor, a geared reducer (also called a gearbox), a gearbox output shaft, and impeller(s). The objective of a properly designed mechanical agitation system is uniform suspension of all solids, appropriate application of shear, homogeneous fluid properties throughout the system, and economical application of applied power.

Most mechanical agitators are driven by electric motors. These motors must be rated for explosion-proof duty (to ensure that motor, starters, and wiring meet specifications for local codes and operating criteria) and may be mounted horizontally or vertically (Figure 10.1 shows a horizontally mounted unit, while Figure 10.2 shows a vertically mounted one). Motors may be coupled to or direct-face mounted to a gear reducer that in turn drives the impeller shaft. Impellers are mounted on the shaft at a specified distance off the tank bottom to achieve desired results.

10.2.1 Impellers

Impellers (sometimes called turbines) convert mechanical power into fluid movement, much like the impeller of a centrifugal pump. Considerable study has been devoted to proper impeller design and placement. Every impeller transmits power to the fluid in two ways:

1. pumping ability
2. shearing ability.

Impeller design will promote one of these components by sacrificing effectiveness of the other. The amount of fluid that is moved by an impeller is its pumping capacity, or displacement capacity, and most manufacturers have undergone extensive testing to determine flow characteristics and capacities for type and size of impeller. Less predictable is shear rate. Shear rate should be thought of as the velocity gradient of fluid with distance. Shear rate can be measured at a point some distance from the impeller blade tips, and maximum and average values calculated. Shear stress is the product of shear rate multiplied by the viscous properties of the fluid. With Newtonian fluids in laminar flow, shear stress and shear rate are nearly synonymous. However, most drilling fluids are non-Newtonian, and therefore predictability of laboratory or paper models is less relevant to real-world applications. When discussing agitation needs, shear rate depends on many variables, including impeller design, tip speed, distance to compartment walls, baffling, particulate concentration, particle size distribution, fluid density,

Figure 10.1. Horizontally mounted agitator.

plastic viscosity, gel strength, and yield point, among others. Laboratory modeling suggests that proper mixing of drilling-fluid components is achieved by the eddy currents present in the turbulent areas created by the fluid flow and associated fluid boundaries within a compartment. Therefore, most manufacturers have a track record of knowing which impeller is suitable for specific conditions.

Impeller configuration will depend on the type of duty and tank geometry. The resultant flow of an impeller design may be categorized as predominantly *radial* or *axial*. This describes the type of flow produced

Figure 10.2. Vertically mounted agitator.

within the compartment. Impellers may have as few as two blades, but in oilfield drilling-fluid applications are usually supplied with four or more blades. The blades are usually carbon steel but can be stainless steel when merited by economics and fluid properties. The blades may be flat (Figure 10.3); canted (Figure 10.4); or swept-face, also known as contoured (Figure 10.5). Blades may be welded to a central disk (Figure 10.6) or bolted to a patterned plate that in turn is mounted to a disk or coupling (Figure 10.7).

Radial Flow Impellers

Radial flow results when the impeller blades are vertically mounted, that is, are in line with the agitator shaft (Figure 10.8). In radial flow, the

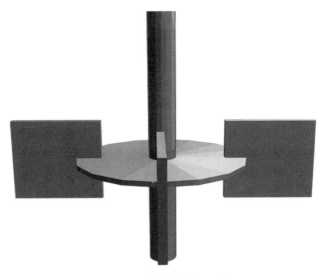

Figure 10.3. Flat blade impeller.

Figure 10.4. Canted blade impeller.

impellers move fluid in a predominantly horizontal, circular pattern within the compartment. Ideally, the fluid will then travel upward once it contacts the tank wall and maintain uniform suspension throughout the compartment. When used alone, radial-type impellers should be

Figure 10.5. Contour blade impeller.

Figure 10.6. Canted blades welded to hub, mounted to solid shaft.

placed near the bottom of the tank, typically less than 12 inches (about 30 cm). For uniform agitation at both the top and the bottom of the compartment, tank depth must be limited to about 6 feet (1.83 m). When mounted higher in the tank, radial flow impellers can generate two zones of fluid movement; one above and one below the impeller.

Figure 10.7. Contour blades bolted to hub, mounted on hollow shaft.

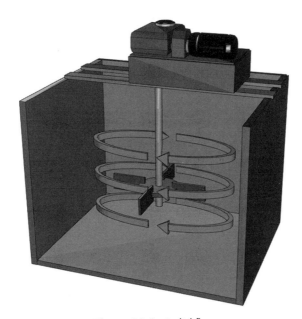

Figure 10.8. Radial flow.

The upper and lower zones share a boundary with one another and therefore exhibit varying degrees of effectiveness. This scenario should be avoided and highlights the need for proper impeller placement on the shaft.

Axial Flow Impellers

Impeller blades that are pitched at some angle toward the tank bottom, typically 45°–60° from vertical (see Figure 10.4), induce a predominant axial fluid movement. The spinning motion of the blade also promotes some degree of radial flow as well. Axial flow impellers draw fluid from the compartment top along the axis of the impeller shaft and push, or "pump," the fluid downward to the bottom (Figure 10.9), then along the bottom to the side wall, which forces the fluid upward and to the surface, where it completes the journey and begins again. When used alone, these impellers should be placed within two thirds to three quarters of the impeller diameter off bottom. Fluid also travels in a radial pattern within the compartment, due to spinning of the impeller. The combination of both radial and axial motion induces more thorough mixing in most instances. Tanks deeper than 6 feet (1.83 m) will require some type of axial flow impeller and may require more than one impeller per shaft.

Most axial flow impellers have a constant blade angle (Figure 10.4). This produces more flow at the blade tip and less toward the hub. These type of impellers pump less, but induce more shearing force.

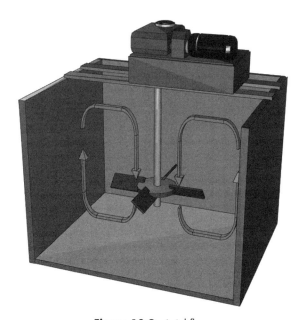

Figure 10.9. Axial flow.

Contour Impellers

Impellers manufactured with variable pitch (see Figure 10.5), called contour or swept-face impellers, promote both radial and axial flow patterns to a greater or lesser degree. The inclination and pitch of the impeller determines whether it will induce more or less of one component. These impellers typically impart less shear force to the fluid than traditional single-plane blades, therefore compartment usage must be known in order to ensure the correct degree of shear. These impellers are designed more closely comparable to airplane propellers or ships' screws.

10.2.2 Gearbox

There are many types of gear reduction suitable for use with mechanical agitators. Most units fall into two categories: worm/roller type or helical/bevel type. There is no direct evidence that one type is superior to the other, although most experts agree that helical/bevel requires less horsepower, and it is therefore offered for larger applications over worm/roller. Worm/roller types are usually offered in right-angle drive with a single reduction in speed. Helical/bevel gearboxes may be either right-angle or parallel drive, that is, the motor shaft is parallel to or in line with the impeller shaft. They may have single, double, or triple gearing reductions. Both types work well and have advantages when it comes to maintenance and economy of operation (the particulars of a discussion of which are too lengthy for the body of this document). However, as with all mechanical equipment, routine scheduled maintenance and adherence to manufacturers' recommendations will prolong equipment life. Keeping a thorough record of maintenance is also advisable. The highest-wear parts on most gearboxes are the bearings and seals. Routine inspection for leaks will indicate the need to replace seals. Excess noises and elevated temperatures are indications that bearings need replacement. Prompt repair or replacement of these components will reduce more costly repairs to the gearbox.

10.2.3 Shafts

Two types of shafts are commonly used: solid shafts (see Figure 10.6) and hollow shafts (see Figure 10.7). Either may be supplied in a variety of materials, with carbon steel being the most popular.

Solid shafts of mild carbon steel are generally cut to desired length and joined to the gearbox output shaft, usually with a rigid coupling.

A machined key slot at the bottom provides a range of adjustment for proper impeller height installation. A bottom end stabilizer should be installed when tank depths exceed 6 feet (1.83 m).

Hollow steel shafts are especially suited for deep tanks. They may be supplied in flanged sections and bolted together, making installations possible where overhead height is limited. Most hollow shafts use contour (swept-face) impellers that are bolted in place (see Figure 10.7). Hollow shafts deflect less than solid steel shafts of the same cross-section modulus; that is, for the same mass of material, they resist bending due to their larger overall diameter. Therefore, they are stiffer than solid shafts of equal or less weight. This equates to longer allowable shaft lengths. When shaft lengths are equal, the reduced weight reduces loads transmitted to the output bearing. Critical speed (vibration caused by shaft flexing under loads at startup) is also less of a concern with hollow shafts.

10.3 EQUIPMENT SIZING AND INSTALLATION

As with all equipment, agitation and mixing equipment must be sized and installed properly. Poor performance will result from improper sizing of equipment and improper installation.

10.3.1 Design Parameters

The following information must be known to properly size an agitator system:

- Tank and compartment dimensions
- Compartment shape
- Compartment duty (solids removal, testing, suction, storage, or pill/slug)
- Maximum mud density expected

Once this information is collected, the design process focuses on the size and type of impeller and the amount of energy required. As mentioned earlier, there are two basic types of flow patterns for mechanical agitators: radial and axial.

Choosing the Right Impeller

Radial flow impellers, as used in the drilling industry, are typically fabricated from mild carbon steel (stainless steel is less often used and generally not required, but is available for certain situations) and have rectangular blades (typically three or four per impeller) mounted in a vertical position on some type of hub. In square and rectangular compartments, properly sized radial flow will produce some axial flow when the fluid impacts the compartment side walls. As with all square or rectangular compartments, dead spaces or spots will occur. Complete elimination of dead spots is not practical; however, when properly sized, they are negligible. The impeller should be mounted about 3 to 6 inches ($\sim$ 7.5 to 15 cm) from the tank bottom. If the impeller is positioned too far off bottom, the flow will not sweep the tank bottom properly and dead spots can occur. Dead spots not only reduce usable tank volume (thereby decreasing the effective circulating volume) of drilling fluid, but can also increase drilling-fluid costs by allowing barite or other commercial solids to settle. Additionally, the settled solids will increase the time and expense of cleaning tank compartments before rig demobilization or when displacing or converting a drilling fluid from one type to another. It cannot be emphasized enough that good agitation results from proper agitator sizing and impeller placement.

In deep tanks, or where two or more impellers are on the same shaft, the impeller shaft should be stabilized at the tank bottom. Stabilizers are typically short lengths of pipe with an internal diameter large enough to accept an agitator shaft without hindering rotation. The stabilizer pipe will usually have drainage holes cut in it and be welded perpendicular to a small portion of flat steel plate that is affixed to the tank bottom, or the stabilizer may be directly welded to the tank bottom. Stabilizers will limit excess side loads on the bearings, can extend gearbox output bearing and oil seal life, and help prevent the shaft from bending. This is especially useful in situations where rig components are stored in the tanks during rig moves.

With axial flow impellers, there is more flexibility in the choice of shaft length than with radial flow impellers. A well-designed installation will usually have the axial flow impeller mounted two thirds to three quarters of the impeller diameter from the tank bottom. Proper placement of the impeller will make best use of the tank geometry to deflect the fluid flow from the impeller, along the bottom and against the compartment walls. The impeller shaft should be stabilized in deep tanks.

Since axial flow impellers are positioned higher in the tank, higher liquid levels must be maintained to prevent vortex formation (Figure 10.10) and entrapment of air in the mud. If the axial flow impeller is mounted too low, bottom scouring may occur, which could lead to excess

Figure 10.10. Vortex formation in fluid.

erosion of the tank bottom. Another consequence of an improperly mounted impeller is ineffective agitation of the fluid at the surface, which can lead to poor homogenization of the fluid.

Variable pitch impellers (contour, or swept-face, impellers) depart from flat and axial flow impellers by incorporating more than one contact angle on the impeller face. They require less horsepower to move the same amount of fluid as axial or radial impellers; therefore, they can have larger blades or smaller gearbox/motor combinations, or higher torque-capacity gearboxes with smaller motors than radial or axial impeller agitators. Often, they are used in extremely large-volume compartments. Because they impart a more efficient movement of fluid as it contacts the blade surface, less shear force is imparted. Shear is desirable in some compartments of a mud system and not needed in others. When dealing with "freshly made" fluid or in a slug or pill compartment, shear aids in speeding the process of blending and homogenization. Compartments predominantly designated for long-term storage of drilling fluid, such as large-volume holding tanks, will not need strong shear forces. Contour-type impellers have the added benefit of requiring less horsepower per unit of fluid displaced, so they are ideally suited for this purpose.

The placement and sizing of impellers, whether radial, axial, or contour, are extremely important. If all other phases of design and installation are correct and impellers are improperly placed, the result can negate the efforts of an otherwise efficient design. Be sure to consult the manufacturer for proper sizing and placement of all components.

10.3.2 Compartment Shape

Agitators work best when they are placed in symetrically sized round or square compartments (as viewed from the top). Round compartments are ideal for many reasons, including their having less dead space in which solids can settle.

When fitted with a center drain or clean-out, "round compartments" are easier to clean and require less wash fluid than rectangular and square tanks. They are symmetrical; therefore, mixing is usually thorough and optimal; however, there are some drawbacks to round compartments. They use space less effectively and will therefore occupy more room for the same capacity compared with rectangular or square tanks. Round tanks require baffles to prevent solid-body swirl and promote good suspension patterns, which raises fabrication cost.

10.3.3 Tank and Compartment Dimensions

Proper agitator sizing is based upon the amount of fluid to be stirred. Therefore, knowledge of tank dimensions is required. Under most circumstances, all compartments other than the sand trap require agitation. Some systems convert the sand trap to an active compartment; in this case, agitation is required. This can be problematic, considering that many systems have shakers mounted above that compartment, with little or no space allotted for mechanical agitators. If such a contingency is anticipated prior to tank fabrication, arrangements can be made to place the agitators where they will not interfere with shaker placement. Alternatively, mud guns may be used. They will provide fluid movement to stir the compartment. This topic is discussed later in the chapter.

10.3.4 Tank Internals

An important consideration when constructing mud tanks is the placement of internal piping. If these are positioned wrong, effective agitation may be impossible. The best advice when installing pipes in any type of system is to use common sense. Think about what effect the piping will have on the flow patterns within the compartment. The flow path of the agitated fluid should not be obstructed by pipe or structural support members.

10.3.5 Baffles

Round Tank Baffling

For a round or cylindrical tank, baffles are essential. Baffles convert swirling motion into a flow pattern to help suspend solids and maintain homogeneity. Baffling can also help prevent vortex formation. In both cases, baffles promote efficient application of power. Baffle width should range from one tenth to one twelfth of the tank diameter and be positioned at 90° increments around the tank. Baffles are generally more effective when placed a short distance from the vessel wall. A gap of $1/60$ to $1/72$ of the vessel diameter is recommended between the baffle and the vessel wall.

Square Tank Baffling

Good fluid suspension in a properly sized square tank is similar to a fully baffled circular tank. The sharp corners of square and rectangular tanks induce nearly the same motion as baffles in round tanks. However, as the length to width ratio of a rectangular compartment increases, the chance for dead zones increases in the far ends of the tank. Strategically placed baffles at the midpoint of the long section of the compartment will counteract this negative effect. When the ratio exceeds 1.5 : 1, more than one agitator is recommended.

Some manufacturers recommend that baffles be installed around each impeller to enhance agitation and prevent air vortices. A typical steel plate baffle is $1/2$ to $3/4$ inch thick by 12 inches wide and extends from the tank bottom to at least 6 inches above the top agitator blade (about 1 to 2 cm thick by 30 cm wide and extends 15 cm above). Four baffles are installed around each agitator at 90°-spacing along lines connecting the agitator shaft center with the four corners of a pit (Figure 10.11). For a long rectangular pit with two or more agitators, the tank is divided into imaginary square compartments and a baffle is pointed at each corner (either actual or imaginary).

10.3.6 Sizing Agitators

Regardless of what style of agitator or impeller is used, proper sizing of components is critical. Once compartment size has been determined, the impeller diameter and corresponding horsepower requirements must

Figure 10.11. Baffles: pointed at corners; note poor placement of pipe.

be calculated. If the maximum mud weight to be used with the rig is not known, it is best to base all calculations on 20 lb/gal fluid (2.4 specific gravity [SG]). This will give a sufficient safety factor to allow agitation of most any fluid without fear of overloading the motor. Most oilfield agitators range in shaft speed from 50 to 90 rpm.

10.3.7 Turnover Rate (TOR)

Impeller sizes are determined by calculating the TOR (sometimes called time of rollover) for each compartment. This is the time, in seconds, required to completely move the fluid in a compartment (Table 10.1) and can be calculated by knowing the tank volume and impeller displacement:

$$\text{TOR} = \frac{V_t}{D} \times 60$$

where

- V_t = tank volume, in gallons or liters
- D = impeller displacement, in gpm or lpm (as displayed in Table 10.2).

For flat and canted impeller applications, TOR should range between 40 and 85 seconds. As the TOR approaches 40 seconds, the chance for vortex formation and possible air entrainment increases. At values

Table 10.1
Typical Turnover Rate Values, in seconds

Impeller Type	Removal	Addition	Suction	Reserve	Pill
Canted/flat	50–75	50–75	65–85	50–80	40–65

Table 10.2
60-Hz Impeller Displacement D Values

Diameter		Flat		Canted		Contour	
in	Mm	gpm	lpm	gpm	lpm	gpm	lpm
20	508	1051	3978	909	3441	N/A	N/A
24	610	1941	7347	1645	6226	N/A	N/A
28	711	2839	10746	2468	9341	5861	22185
32	813	4635	17543	3764	14247	N/A	N/A
34	864	N/A	N/A	N/A	N/A	8790	33270
36	914	6273	23743	5402	20447	9180	34746
38	965	7342	27789	6343	24008	10604	40136
40	1016	8411	31836	7284	27570	N/A	N/A
42	1067	N/A	N/A	N/A	N/A	13940	52762
44	1118	11300	42771	9928	37577	N/A	N/A
45	1143	N/A	N/A	N/A	N/A	16812	63633
48	1219	14401	54508	12512	47358	20020	75776
52	1321	18630	70515	16100	60939	24852	94063
54	1372	N/A	N/A	N/A	N/A	27602	104475
56	1422	NA	NA	NA	NA	30353	114887
60	1524	NA	NA	NA	NA	36567	138404
64	1626	NA	NA	NA	NA	43533	164771

greater than 85 seconds, proper suspension may be jeopardized and solids will begin to settle.

For contour impeller applications, values must be significantly faster (i.e., smaller numbers) to achieve the same results, but because of the impeller design, air entrainment is less probable. In symmetrical compartments, the fluid has a nearly equal distance to travel from the center of the impeller shaft or from the impeller blade tip before it contacts the vessel wall. Agitators should be placed where the shaft is centered in the tank or compartment.

When defining the area in which to mix, it is best to work with symmetrical shapes like squares or circles (as viewed in a plan drawing or overhead view of the tank layout). Rectangular tanks should be converted to nearly square compartments if possible. Maximum fluid working volumes in compartments should not be higher than 1 foot (about 3/10 m) from the top of the tank. This will allow for a little extra capacity in emergencies, slightly out of level installations, and/or fluid movement on floating rigs.

Working volume for square or rectangular tanks is calculated by knowing dimensional values for length (L), width (W), and height (H; in feet for gallons, in meters for liters):

For gallons:

$$V_t = L \times W(H - 1) \times 7.481$$

For liters:

$$V_t = L \times W(H - 0.3) \times 1000.$$

The working volume for round tanks with flat bottoms is:
For gallons:

$$V_t = \pi r^2 (H - 1) \times 7.481$$

For liters:

$$V_t = \pi r^2 (H - 0.3) \times 1000.$$

For round tanks with dish or cone bottoms, calculations for working fluid volume are based on straight wall height (i.e., this height is measured from the tank top to where the tank joins the cone or dish at the bottom). This leaves adequate free space above the maximum fluid operating level. In all cases, if $H < 5$ feet ($\sim$1.5 m), a radial flow impeller should be specified.

Example

A compartment is 30 feet long, 10 feet wide, and 10 feet high (Figure 10.12). Maximum mud weight is anticipated to be 16 lb/gal (1.92 SG). If the maximum mud weight is not known, use 20 lb/gal fluid density (2.4 SG).

All compartments will be solids-removal sections.

Convert the compartment to symmetrical shapes. In this case, three compartments 10 × 10 feet square ($\sim$3 × 3 m). Determine the volume for one compartment (V_t).

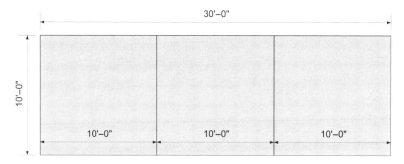

Figure 10.12. Example for sizing exercise.

$$V_t = 10 \times 10 \times (10 - 1) \times 7.48$$
$$V_t = 900 \times 7.48$$
$$V_t = 6732 \text{ gal}.$$

Since the tank is deeper than 6 feet, flat (turbine) impellers cannot be used; therefore, canted impellers are chosen. Locate the appropriate impeller diameter from the impeller displacement table D (see Table 10.2) so that TOR is within the recommended range. As a rule, choose an initial impeller displacement value so that D is close to V_t to increase the accuracy of the selection. In this case, the gpm value for 60-Hz service is close to the 38-inch impeller (6343 gal).

$$\text{TOR} = [V_t/D] \times 60$$
$$\text{TOR} = (6732/6343) \times 60$$
$$\text{TOR} = 1.06 \times 60, \text{ or about 64 seconds}.$$

Compare this TOR to values in Table 10.1 and determine suitability for compartment purpose. In this example, the TOR is sufficient for solids-removal compartments. If a lower numeric value to cause faster fluid movement is desired, then choose a larger impeller. If less movement is desired, choose a smaller impeller and recalculate until the appropriate values from Table 10.1 are achieved.

After determining which impeller will produce the effective TOR, locate the size in the Impeller Diameter columns in Table 10.3. In this case we anticipate using a 38-inch impeller and 16-ppg fluid. Since there is no 38-inch value in the 16-ppg column in Table 10.3, we must round up to the next highest value, that is, 40 inches. This allows for a safety factor should the mud density increase slightly. Follow the 40-inch impeller value horizontally to the left in Table 10.3 and determine

Table 10.3
Power Requirements for Canted Impellers per Fluid Density

Power		Impeller Diameter							
		20 ppg (2.40 SG)		16 ppg (1.92 SG)		12 ppg (1.44 SG)		8.3 ppg (1.00 SG)	
hp	KW	in	cm	in	cm	in	cm	in	cm
1.0	0.7	22	56	24	61	26	66	29	74
2.0	1.5	26	66	28	71	30	76	32	81
3.0	2.2	29	74	31	77	33	83	34	86
5.0	3.7	32	81	34	86	36	91	38	97
7.5	5.6	35	89	37	94	39	100	42	107
10.0	7.5	37	94	40	102	43	109	46	117
15.0	11.2	42	107	45	114	49	123	51	130
20.0	14.9	46	117	50	126				
25.0	18.6	49	124						
30.0	2.4								

the horsepower needed for that application. In this case, a 10-hp motor is sufficient. Therefore, three 10-hp agitators with 38-inch canted impellers are suitable for the tank in this application.

It should be noted that this system is for one brand of agitator, and the values in the tables may not apply to all brands. Deep tank designs require that multiple blades be mounted on a single shaft. This involves more thorough calculations than those shown. Because of this, it is highly recommended that the manufacturer or supplier be consulted before final design or placing of an order for any application.

10.4 MUD GUNS

Both high-pressure and low-pressure mud guns agitate mud by means of rapid fluid movement through a nozzle. Pressure and flow are delivered via either high-pressure main mud pumps or, most often, centrifugal pumps. There are two points of view as to where pumps for mud guns should draw suction from: (1) the compartment that they stir or (2) a compartment downstream, ideally immediately after the last solids-control device.

Consider option 1 (drawing suction from the compartment to be stirred): When mud guns are placed in compartments that are designated for solids-removal equipment (e.g., hydrocyclones, centrifuges), the possibility of pumping solids-laden fluid exists. Continuous circulation of solids-laden fluid (especially sand) will result in faster wear on pump parts and mud gun nozzles. Manufacturers provide nozzle inserts made of wear-resistant plastics to increase life and reduce replacement costs. If parts are not replaced when needed, agitation efficiency decreases. Another argument against pumping into the same compartment from which suction is drawn is the possibility for particle size degradation, which will also reduce solids-removal efficiency and force use of more costly solids-control mechanisms or dilution. Option 1 is feasible if used in compartments downstream of the last solids-control device, provided the upstream solids-control system is properly designed and operated.

Option 2 (drawing suction from a compartment downstream of the solids-control system) is more desirable for two reasons. First, it prevents continuous circulation of fluid that may be laden with drilled solids, assuming efficient upstream solids removal. This maintains mud gun efficiency and reduces wear to fluid-end pump parts. Secondly, it places cleaned fluid into the compartment to be stirred. This is especially effective when solids concentration is high or fluctuates (e.g., high penetration rates and/or sweep returns to the surface). However, option 2 necessitates more process capacity from the solids-control equipment (including the degasser and hydrocyclone compartments). Rig circulating rate and flow through mud guns must be added and then accounted for. Particular examples are discussed later in this chapter.

10.4.1 High-Pressure Mud Guns

High-pressure guns typically come in 3000 and 6000 psi ratings and require heavy-walled piping. Gun nozzle sizes range from ¼ to ¾ inch (6.4 to 18.4 mm). The rig's main mud pumps (positive displacement piston types) pressurize the guns. The high-pressure system requires heavy piping and connections but relatively small nozzles. It is a high-pressure, low-volume system. The agitation is a result of high-velocity fluid coming from the jet nozzle.

10.4.2 Low-Pressure Mud Guns

Low-pressure mud guns usually require about 75 feet of head for effective operation (see chapter 18 for head and pump sizing). Nozzle sizes range from ½ to 1 inch (12.7 to 25.4 mm). Centrifugal pumps pressurize the nozzles through standard wall piping (typically schedule-40 pipe is used). The low-pressure system does not require heavy-walled piping. Because of higher flow rates, larger-diameter pipe is used to prevent excess friction loss. The jet nozzles are larger than in the high-pressure system. Effective agitation occurs from the large volume of fluid entering the mud tank through the nozzle. Fluid shear is applied by the velocity of the fluid exiting the nozzle. This is called a high-volume, low-pressure system.

With either type of system, it is economically as well as functionally desirable to keep the jet nozzle feed lines as short and straight as possible. With a low-pressure system, this is not merely desirable, but critical for efficient operation.

Eductors

Eductors (also called jets) are used on some systems and can achieve up to four times the fluid movement over conventional nozzles. Eductors create a low-pressure area around the discharge of the nozzle that draws in and entrains fluid immediately around the eductor. This is similar to the action in a mixing hopper. The two fluids are mixed in the venturi section and discharged into the tank at a much higher rate than the jet nozzle alone could achieve. The high-velocity fluid from the jet can either be the same fluid in the tank or be from elsewhere (e.g., mixing and blending compartments).

10.4.3 Mud Gun Placement

Mud guns are usually placed about 6 inches (~15 cm) from the tank bottom and typically come with a 360° swivel that allows directional positioning to stir dead spots. Dead spots can occur in right-angle compartments that have inadequate mechanical agitation or can be caused by piping or other mechanical obstructions.

It is generally accepted that low-pressure nozzles are effective within a 5- to 9-feet diameter depending on the mud weight, viscosity, and nozzle velocity. Nozzle size and feed pressure determine how much fluid will pass through the nozzle and at what velocity. Greater volumes of fluid through the nozzle produce more circulation of fluid, while higher velocity creates more shear. Bernoulli's theorem describes the relationship of pressure and velocity. High-pressure fluid exits a jet nozzle at a high velocity, where head is converted to velocity. One manufacturer

Figure 10.13. Multiple eductors mounted on static pipe.

uses nozzles that are housed within an eductor body that also includes multiple induction ports. The eductor body acts as a mixing chamber and diffuser to induce a swirling flow out of the body. Fluid is pumped into the central nozzle, which then draws fluid through the side induction ports. The manufacturer states that the use of this type of nozzle will increase the amount of fluid moved by up to four times what is delivered to the orifice. An example of how this would be applied is shown in Figure 10.13.

10.4.4 Sizing Mud Gun Systems

Since most systems employ low-pressure mud guns, it is critical to ensure that the proper centrifugal pump and piping are chosen. A great deal of the literature has been devoted to sizing pumps and will not be duplicated here (see chapter xx for full information). It must be emphasized that for a low-pressure system to work properly, several items must be considered, including:

- Jet nozzle size and number
- Number of turns, tees, valves, and reducers
- Pipe length
- Pipe sizes used

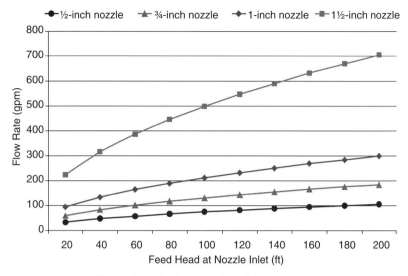

Figure 10.14. Flow rates through ideal nozzles.

All are important, and the effect of each must be determined before a pump can be selected. In most systems, the pressure drop across the nozzles will consume most of the total head delivered by the pump, but discharge line length, pipe diameter, and fittings must be considered to avoid oversizing or undersizing a pump and motor.

To achieve a given flow rate through a jet nozzle, a specific total head will be required to overcome friction through the system. Figure 10.14 shows the discharge capacity (flow rates) for various sizes of *ideal* nozzles at several heads. If reducers or swages are used, the capacity will be lower. The longer the piping, the greater the number of fittings; and/or the smaller the diameter of the piping, the greater the total head required. General rules are:

- Fluid velocity should be maintained between 5 (to avoid solids settling) and 10 ft/sec (to avoid excessive pipe erosion) (1.5 to 3 m/sec).
- Fittings and connections should be minimized.
- Piping should be kept as short as possible.

It is important to consider total system efficiency, by accounting for the added flow into the compartment, when using mud guns in the degasser or hydrocyclone compartments if the mud gun supply comes from elsewhere in the surface system. The volume fraction treated can be calculated by determining the equipment capacity, for example, for a desilter, and the circulating rate of the fluid entering the desilter suction compartment. As a rule, hydrocyclone and degasser systems should be sized to process between 110 and 125% of the maximum flow volume through their compartments. For example, if the rig's circulating rate is 600 gpm, the desilter capacity is 1000 gpm, and the mud gun rate (with suction from a downstream compartment) is 200 gpm, then $1000/[200+600] = (1000/800)$, or $\frac{10}{8}$ (125%) process efficiency. If the desilter capacity is 700 gpm, then $700/[200+600] = (700/800)$, or only $\frac{7}{8}$ process efficiency. In this situation, an undesirable 87.5% process efficiency is achieved. To counteract this inefficiency and achieve at least 100% process efficiency, the desilter capacity must be increased to greater than 800 gpm or the mud gun rate reduced to less than 100 gpm. A combination of the two technologies yields more desirable process efficiency, that is, greater than 100%.

10.5 PROS AND CONS OF AGITATION EQUIPMENT

Many rigs use a combination of agitators and mud guns. To summarize the preceding discussions on mechanical agitators and mud guns, a list of pros and cons are presented to aid in the proper selection and application of agitation equipment. The list is not necessarily complete but will help both the designer and rig personnel consider both immediate and long-term consequences of decisions.

Mechanical agitators may be supplemented with mud guns strategically located to stir dead zones. Round tanks nearly eliminate these dead spaces and thus the need for mud guns and their supporting hardware (pumps, piping, valves, etc.), providing a simpler tank system. Additionally, round tanks demonstrate advantages over square or rectangular tank designs, such as:

- Ease of cleaning out
- Extra space in which to place piping and pumps
- No need for compartmentalization, since each tank is its own compartment
- Symmetrical design, ensuring mixing that is usually very good

With square or rectangular tanks, limitations on mechanical agitators may make the inclusion of mud guns a necessity. For most instances, there should be properly sized mechanical agitators to adequately stir the tanks, and strategically placed mud guns to eliminate dead zones.

10.5.1 Pros of Mechanical Agitators

- Variety of gearbox and impeller combinations to suit most needs
- Effective on deep and large tanks
- Can be designed to induce more or less shear as needed
- Help cool mud by exposing more fluid to the atmosphere

10.5.2 Cons of Mechanical Agitators

- Cannot blend different tanks of mud or transfer mud between compartments
- Higher initial cost
- More surface space required than mud guns
- Heavier than mud guns
- Electricity is required (in most cases)
- Possible dead zones
- May require installation of baffles

10.5.3 Pros of Mud Guns

- Lower capital investment than mechanical agitators
- May use existing rig pumps (if there is sufficient pump capacity available)
- Flow may be concentrated to a given area to reduce or eliminate dead spots
- Lower weight than mechanical agitators (excluding pump and piping)
- Can accelerate shear rates
- May be used to transfer and blend mud between tanks and/or compartments

10.5.4 Cons of Mud Guns

- If mud guns alone are used, there will be a need for many of them, with significant pump and piping costs as well
- High-pressure surface nozzles may aerate mud
- Nozzle wear will cause higher flow, which requires more horsepower and can lead to pump motor overload if nozzles are not replaced in time
- If solids have settled and a gun is directed toward the buildup, a slug of solids can plug pumps, cones, and or centrifuges

- May increase requirements for solids-control equipment capacity and associated hardware

10.6 BERNOULLI'S PRINCIPLE

The Bernoulli principle, first formulated by Daniel Bernoulli in 1738, is one means of expressing Newton's second law of physics, concerning conservation of energy. Roughly stated, this principle demonstrates that the sum of pressure and velocity through or over a device represents is equal, neglecting the effects of losses due to friction and/or increases by adding energy with external devices such as pumps. The basic concept of Bernoulli's principle can be observed in routine daily activities: A ship's sail can push a vessel into the wind; an airplane's wing produces lift; a pitcher induces spin on a baseball and generates high- and low-pressure zones forcing the ball into a curved pattern. Bernoulli's principle can also be demonstrated in the flow of fluids through pipe.

Heavier-than-air flight was not achieved until a wing was developed that engaged the Bernoulli principle. Airplane wings generate lift by creating high- and low-pressure zones (Figure 10.15). Ignoring losses for friction, the total energy at any point along the wing is equal to the sum of the pressure (P) and the velocity (V). Pressure and velocity are equal at points A and C, that is, $P_1 + V_1 = P_2 + V_2$. Because aircraft wings are curved on top, air travels farther and thus moves faster above the wing than underneath it. Therefore, velocity at point B is greater above the wing than below it. The law of conservation of energy indicates that pressure is affected inversely: If V increases, then P decreases. This creates a differential pressure, or ΔP: higher pressure beneath the wing adds lift. As speed increases, so does lift.

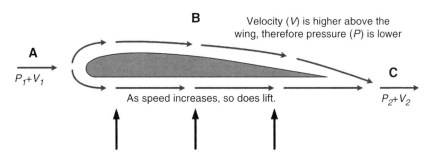

Figure 10.15. Airplane wing cross section.

10.6.1 Relationship of Pressure, Velocity, and Head

Likewise, in piping systems, velocity and pressure are measured as fluid flows internally in the pipe, rather than externally, as over a wing. Again assuming that there is no friction loss and no energy added to the system (e.g., pumps), the sum of pressure h, or head of fluid, and velocity will be a constant at any point in the fluid. Consider an example with an ideal fluid and frictionless pipe. Figure 10.16 shows the relationship between pressure and velocity under steady flow conditions. Remember, ignore losses caused by friction. Attaching manometers to the pipe will indicate h, or head levels, at three points in the pipe. Pressure gauges will also indicate the level of head. Notice that at points A and C, the levels of h are equal, while at point B it is lower. This is because pipe size is reduced; therefore, velocity (v) is higher. Once the pipe expands again, v will decrease, and h will again increase. In actual practice, losses or energy increases or decreases are encountered and must be included in the Bernoulli equation.

Figure 10.16. Relationship of pressure to velocity.

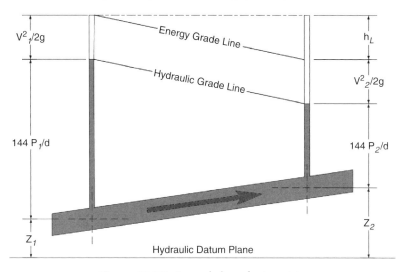

Figure 10.17. Energy balance for twopoints.

Example 1 [Crane Company]

An energy balance may be written for two points in a fluid, as shown in Figure 10.17 (in English measurement units). Note that the energy loss from point 1 to point 2 is h_L foot-pounds per pound of flowing fluid; this is sometimes referred to as the head loss in feet of fluid.

$$Z_1 + \frac{144P_1}{d} + \frac{v_1^2}{2g} = Z_2 + \frac{144P_2}{d} + \frac{v_2^2}{2g} + h_L$$

where

- Z = potential head (feet) or elevation above datum plane
- P = gauge pressure (pounds per square inch)
- d = fluid density (pounds per cubic foot)
- v^2 = mean velocity of flow (feet per second)
- g = gravitational acceleration (32.2 feet per second)
- h_L = foot-pounds per pound of flowing fluid

All practical formulae for the flow of fluids are derived from Bernoulli's theorem, with modifications from empirical studies to account for losses due to friction.

Another means to mathematically define the effects of Bernoulli's theorem can be written where the sum of the individual terms equals a constant.

Example 2 [Weisstein]

Written for use with metric terminology:

- P = static pressure, in Newtons per square meter
- ρ = fluid density, in kilograms per cubic meter
- v = velocity of fluid flow, in meters per second
- h = height above a reference surface, in meters.

In the following equation, the second term is known as the dynamic pressure. Dynamic pressure is the component of fluid pressure that represents fluid kinetic energy (i.e., motion), while static pressure represents hydrostatic effects, so

$$P_{total} = P_{dynamic} + P_{static}$$

The dynamic pressure of a fluid with density ρ and speed u is given by

$$P_{dynamic} \equiv \tfrac{1}{2}\rho u^2,$$

which is precisely the second term in Bernoulli's law. The effect described by this law is called the Bernoulli effect.

Picture a pipe through which an ideal fluid is flowing at a steady rate (Figure 10.18). Let W denote the work done by applying a pressure P over an area A, producing an offset of Δl or volume change of ΔV. Let a subscript 1 denote fluid parcels at an initial point down the pipe, and a subscript 2 denote fluid parcels farther down the pipe. Then the work done by pressure force

$$dW = PdV \qquad (10.1)$$

at points 1 and 2 is

$$\Delta W_1 = P_1 A_1 \Delta l_1 = P_1 \Delta V \qquad (10.2)$$

$$\Delta W_2 = P_2 A_2 \Delta l_2 = P_2 \Delta V \qquad (10.3)$$

and the difference is

$$\Delta W = \Delta W_1 - \Delta W_2 = P_1 \Delta V - P_2 \Delta V. \qquad (10.4)$$

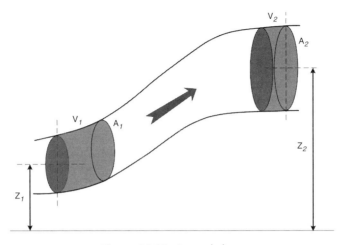

Figure 10.18. Energy balance.

Equating this with the change in total energy (written as the sum of kinetic and potential energies) gives

$$\Delta W = \Delta K + \Delta U$$
$$= \tfrac{1}{2}\Delta m v_2^2 - \tfrac{1}{2}\Delta m v_1^2 - \Delta m g z_2 - \Delta m g z_1. \quad (10.5)$$

Equating (10.5) and (10.4),

$$\tfrac{1}{2}\Delta m v_2^2 - \tfrac{1}{2}\Delta m v_1^2 + \Delta m g z_2 - \Delta m g z_1 = P_1 \Delta V - P_2 \Delta V, \quad (10.6)$$

and this, upon rearranging, gives

$$\frac{\Delta m v_1^2}{2\Delta V} + \frac{\Delta m g z_1}{\Delta V} + P_1 = \frac{\Delta m v_2^2}{2\Delta V} + \frac{\Delta m g z_2}{\Delta V} + P_2, \quad (10.7)$$

so writing the density as $\rho = m/V$ then gives

$$\tfrac{1}{2}\rho v^2 + \rho g z + P = [\text{constant}]. \quad (10.8)$$

This quantity is constant for all points within the pipe, and this is Bernoulli's theorem. Although it is not a new principle, it is an expression of the law of conservation of mechanical energy in a convenient form for fluid mechanics.

10.7 MUD HOPPERS

Prior to the use of centrifugal pumps on drilling rigs, the standby reciprocating mud pumps were customarily used to operate the mud hopper. As with high-pressure mud guns, this required high-pressure pipe and connections. This was costly because the pump required enormous power and expensive piping. A small orifice in the hopper delivered a low flow at a high velocity. The jet velocity was suitable for adequate mixing, but the volume was usually less than 500 gpm. This, of course, would limit the speed of material addition.

Since the 1950s the centrifugal pump has been the predominant tool for charging mud hoppers on drilling rigs. This permits the use of low-pressure equipment and the movement of large volumes of fluid rapidly. Savings were realized through reduced piping cost and higher addition rates that lowered operating cost. It also released the standby main pump from mud mixing duty.

A low-pressure mud hopper is shown in Figure 10.19. Since high-pressure hoppers work in the same manner, this discussion will be limited to low-pressure equipment. The fluid velocity in a low-pressure

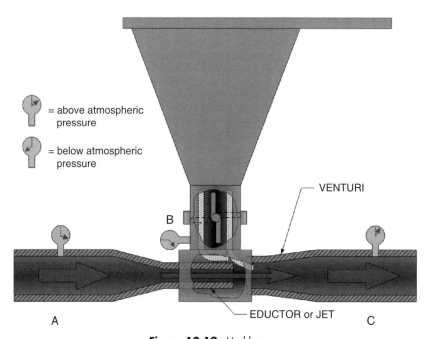

Figure 10.19. Mud hopper.

mud hopper will be around 10 ft/sec on the pressure side of the jet nozzle. The pressure line is reduced in size usually from 6 to 2 inches (152 to 51 mm) and exits the jet nozzle at a much higher velocity but lower pressure. The high-velocity jet stream crosses the gap between the eductor nozzle and the downstream venturi and creates a partial vacuum within the mixing chamber (or tee). This low-pressure area within the tee, along with gravity, actually draws mud materials from the hopper into the tee and fluid stream. The high-velocity fluid wets and disperses the mud additives into the fluid stream. This reduces lumping of material and is an initial shear of the additives.

Most mud hoppers use a venturi for two reasons: (1) to increase the shearing action of the mud and (2) to help regain some of the pressure head to move the mud downstream or upward as it returns to the mud system.

The venturi reduces the time to build viscosity when bentonite is added. This is due to the increased shearing of the fluid that takes place within a well-designed venturi. Similar results have been observed with other drilling fluid additives.

At point A in Figure 10.19, the pressure head is high, while the velocity head would be low, as in the previous examples. Assume that the system is level, so that the elevation head is be zero. As the fluid moves downstream through the venturi, the total head at point C would theoretically remain the same, with the velocity and pressure head being equivalent to that at point A. As in the previous examples, the main problem with this equation is that it ignores friction head losses that in practical applications can be about 50%. In actual practice, friction head must be accounted for, and if the venturi were not present, there would be tremendous turbulence as the flow expanded into the larger-diameter pipe. The venturi simply helps to streamline the flow back to the large pipe with less turbulence. This results in a minimum friction loss and will provide the maximum available head to push the fluid and newly added commercial additives from the hopper into the piping system. Mud hoppers come with valves to isolate the hopper from the mixing chamber. When closed, the space between the venturi chamber and the valve is less than atmospheric, i.e., a partial vacuum is formed. The amount of vacuum at point B influences the resultant addition rates of the device and is determined by several factors, including feed pressure, nozzle diameter and length, venturi design, fluid properties, and downstream piping restrictions.

Figure 10.20. Mud hopper with premix chamber.

One manufacturer modifies this concept by installing a premix wetting chamber between the hopper and the eductor and also by modifying the eductor. A tee installed upstream of the eductor diverts part of the fluid into the premix chamber, where it swirls and forms a vortex that radiates outwardly to the wall of the premix chamber. The lower pressure at the vortex center draws material into the center as well as downward and into the eductor. The eductor is also different in that it features a star-shaped cross-sectional feed area, as opposed to a circular nozzle. The advertised benefits are the swirling action and more thorough mixing (Figure 10.20).

10.7.1 Mud Hopper Installation and Operation

As with any piece of equipment, the mud hopper and related equipment must be sized, adjusted, and installed properly to achieve optimum performance.

Once a mud hopper has been selected, the pump, motor, feed line, and discharge line must be designed to allow the proper flow rate at the recommended head. For example, assume that a mud hopper has been selected that requires 550 gpm (2082 lpm) at 75 feet of head (23 m). With this established, feed- and discharge-line sizes would be determined by recommendations from a recognized authoritative source. The friction head would be determined for all lines and connections at 550 gpm and this would be added to the 75 feet of head for the hopper. The next step is to select a pump and impeller that will provide 550 gpm at the total head required. With the pump selected, the motor size can be determined and adjusted for mud weight. If the maximum mud weight to be used is unknown, it is standard practice in most companies to base all calculations on 20 lb/gal mud (2.4 SG).

Feed and discharge lines of the mud hopper should be kept as short as possible. This is dictated by economics (i.e., less power, piping, and smaller pump) as well as operations. The backpressure from the downstream piping is crucial to effective operations. Figure 10.21 shows the effect of system backpressure on mixing chamber pressure. When using a 2-inch (51-mm) jet nozzle with a 3⅛-inch (~79-mm) gap, there is a very strong vacuum at normal operating pressures of 70 to 75 feet of head until the backpressure reaches 50% of the inlet pressure, at which point mud will actually try to backflow through the mud hopper. Obviously, the downstream pressure drop must be less than 50% of the inlet pressure, and systems should be sized accordingly. System backpressure also has an effect on the rate at which materials can be

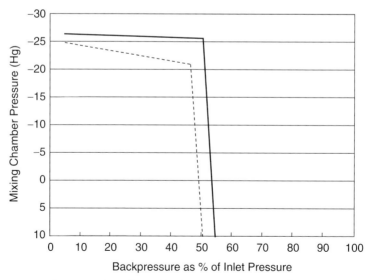

Figure 10.21. Effects of backpressure on hopper performance (2-inch nozzle @ 3⅛-inch venturi gap).

added to the system. It should be noted that the vacuum was about 25 inches of mercury (0.85 bar) when a venturi was utilized. When the venturi was removed, the vacuum was measured at only 9½ inches Hg (0.32 bar). A small change in the backpressure causes a measurable change in the addition rate. An even more noticeable effect on the addition rate is the relationship between the gap between the jet nozzle and the venturi.

When the gap is between 1 and 3⅛ inches (25.4 to 79.4 mm) for this particular mud hopper, there is very little difference in barite addition rate. When the gap is larger than 3⅛ inches, the rate of barite addition is reduced significantly. This further illustrates the need to have equipment and piping sized and adjusted properly.

Another important consideration when installing addition equipment is the amount of lift required. If the discharge goes to a tank on a different deck or there are tall tanks and the hopper is on the ground, mixing rates can be reduced. Sack barite addition is reduced by 17% when the lift is increased from 6 feet to 12 feet (1.83 to 3.66 m). If the lift requirements are severe and the pump is undersized, the mixing capabilities will suffer.

From the preceding discussion, it is obvious that the mud hopper can be highly efficient, while improper mud hopper installations can create many problems. Low addition rates lead to increased rig time and operator hours to treat the mud system. If the discharge from the mud hopper is routed into a tank that is not stirred properly, a large quantity of commercial materials could settle, even if the hopper is properly dispersing and wetting the materials. If the addition system is installed, sized, and adjusted properly, there will be a reduction in system maintenance, a decrease in fluid costs, and a decrease in operator hours.

10.7.2 Mud Hopper Recommendations

The following recommendations will promote efficient mud hopper installation and use:

- Select a mud hopper that is properly sized for the mud system. Generally, a single hopper is sufficient for most rigs. If the mud circulating rate is greater than 1200 gpm (~4550 lpm), then consider using a hopper with 1200-gpm capacity. Generally, there is no need to

add chemicals faster than this. For many operations, 600 to 800 gpm (~2270 to 3030 lpm) is adequate.
- Keep the lines to and from the hopper as short and straight as possible. Size the pump and motor based on the system head and flow rate requirements. A venturi is beneficial in all operations, but especially when the system backpressure may reduce the mud hopper efficiency and operation. The venturi will allow fluid to move vertically higher than the hopper height. On many rigs, hoppers are placed at ground level and the downstream pipe is raised to a height equal to or greater than the top of the mud tanks.
- Use new or clean fittings to reduce friction loss. After each operation, flush the entire system with clean fluid to prevent the mud from drying and plugging the system. Clean the throat of the hopper to prevent material from bridging over that will cause poor performance the next time the hopper is used.
- A table should be attached to or located near the hopper to hold sacks of material. The table should be at a convenient height (36 to 42 in., or ~0.9 to 1.1 m) so that personnel can add material easily with minimal strain. Power-assisted pallet and sack handlers will enhance addition rates and minimize personnel fatigue.
- As with all equipment on the rig, develop a regular maintenance and inspection program for the mud hopper. The mud hopper is normally simple and easy to operate, but worn jets and valves will hinder the operation. Inspect the entire system every 30 to 60 days. Maintain an inventory of spare nozzles, valves, and bushings.

If the jet action and dispersion appear to be substandard, check the following:

1. Determine that the pump is providing an adequate volume of mud at the normal operating discharge head. A gauge upstream of the hopper will verify the pressure or head. A reduced pump discharge volume is normally caused by air entering the air pump packing, an object lodged in the piping, a worn impeller, gas or air causing the pump to airlock, a connection or piping leak, or dry mud packed around the jet and restricting the mixing area within the throat.
2. With the pump shut down, remove the hopper and valve from the tee. Unions installed upstream from the hopper allow disassembly as well

as inspection to determine whether the inside passage is eroded or an object is lodged in the eductor.

10.7.3 Other Shearing Devices

Like the mud hopper, a number of devices are available to increase shear, speed hydration, and enhance curing and saturation. Prehydrating has several advantages that it:

- improves the efficiency of addition, thus reducing over- or undertreatment. Overtreating may force dilution and the possible removal of excess fluid to make room for the increased fluid volume. Undertreating will hinder fluid performance;
- reduces water loss to the formation by enhancing filter cake properties;
- prevents formation of unblended globular masses, or "fish-eyes," in polymers;
- prevents chaining or stringing of polymers;
- stabilizes properties faster for more accurate checks; and builds saturated salt solutions, pill, or slug volumes quickly.

Other shearing devices can be as simple as a small tank (less than 50 bbl, $\sim$8 m^3) with a centrifugal pump and nozzles to circulate the fluid, thus exposing it to high shear in a short period. Many systems enhance centrifugal pump performance with a specialized fluid-end design that features a plate, porting, and multiple internal nozzles (Figure 10.22). These have proven effective by producing up to 83 ft/sec (25.4 m/sec) of mechanical shearing velocity within the impeller housing and 100 ft/sec (30.5 m/sec) of liquid velocity. The close particulate proximity and fluid channeling prevent large particles from exiting the pump without first breaking into much smaller particles. These devices are suitable for enhanced shearing and reducing curing times, and they are ideal for prehydrating additives prior to blending with the active system.

10.8 BULK ADDITION SYSTEMS

A variety of systems are available to transport and contain large volumes of dry bulk chemicals. Commonly known as P-tanks (pressure tanks), these vessels are usually mounted upright and connected via piping to a mud mixing hopper. They may also be mounted remote from the

Figure 10.22. Specialized shearing pump.

hopper and deliver material through a piping system by pneumatic force. Other designs mount the tanks above the hopper and gravity-feed the material. These tanks accommodate most dry bulk chemicals, including barite, bentonite, and cement.

Sack lifting and addition chores on a rig can be hazardous to personnel. Machines specifically designed to lift pallets and individual bags are now available to reduce or relieve personnel from repetitive motion or back injuries.

Sack-lifting equipment uses large suction cups to grasp individual bags without harming or deforming them. Suspended overhead, the manually operated arm permits users to move heavy sacks with ease from pallets onto mud mixing tables. Floor-mounted sack lifters are also available. They use either hydraulic or pneumatic pistons to elevate pallets of sacked material to a comfortable height for transferring onto the hopper table (Figure 10.23).

Automated sack handling systems feed individual sacks onto a conveyor belt, where they are slit, emptied, and compressed. These systems can be operated manually or programmed for automatic feed at predetermined levels. Automatic feed regulates addition rates and has proven to improve fluid properties through accurate dosing rates and reduced labor costs (Figure 10.24).

Mud additives are also supplied in big bags (typically 1 m^3 or more). Most large bags are made of durable synthetic fabric with lifting hooks included for use with overhead cranes. The big bags fit into specifically

Figure 10.23. Pallet lifter.

Figure 10.24. Automated mixing equipment.

designed carriages that hold them in place for dispensing out the spout on the bottom (Figure 10.25). Some models come complete with automated hardware/software packages that regulate addition rates. Manual-control overrides allow operators to directly regulate addition rates. The bags reduce dust generated from sack cutting operations and are collapsible

Figure 10.25. Big bag mixing equipment.

once empty, thus reducing waste volumes and transportation space on boats.

10.9 TANK/PIT USE

The surface mud system consists of the flow line, active tanks, reserve tanks, trip tank, agitators, pumps, motors, solids- and gas-removal devices, mixing and shearing devices, and associated piping. The surface mud system can be considered to be composed of the following sections. Each section has unique agitation and suspension requirements. Please refer to Table 10.1 for TOR requirements for the compartments. Remember that TOR is an indication of how vigorously the fluid must move within the compartment.

10.9.1 Removal

All compartments of a system that use solids-control equipment or degassing equipment require proper agitation, except the sand trap, if the sand trap is used as such. Solids-control equipment works best when solids loading remains constant. Slugs of a large quantity of solids tend to plug hydrocyclones and centrifuges. This leads to downtime and decreased solids-removal efficiency, with the resulting consequences.

10.9.2 Addition

The equipment and tanks utilized in the addition and blending of mud additives require proper agitation. As stated in the early part of this chapter, the purpose of the surface mud system is to allow maintenance of the mud before it is pumped back down the hole. Part of the reconditioning process involves the addition of mud materials and chemicals to the mud system. The addition of materials is also required when the system volume is increased or the mud properties (such as weight or viscosity) are changed. In critical situations, such as well-control problems, it is desirable to be able to mix the mud materials rapidly and thoroughly. The purpose of any addition equipment is to mix mud materials and chemicals into a fluid with minimum balling, maximum speed, and safety to personnel.

A properly designed mud system will have adequate storage and mixing capacity. For existing systems whose capacity is taxed, auxiliary premix systems should be used; especially when rapid shearing is required. Premix systems are useful for blending bentonite or hard-to-mix polymers, such as CMCs (ceramic matrix compounds), PHPA (partially hydrolyzed polyacrylamide), XC (xanthan gum, from the bacterial genus *Xanthomonas campestris*), and many other fluid additives. They also provide extra capacity, which is especially useful if the mud type is to be modified and requires isolation of one mud type from another (e.g., when changing from an aqueous fluid to a non-aqueous fluid).

10.9.3 Suction

All compartments in this section require proper agitation. This section contains the tank(s) and/or compartment(s) from which the rig pumps take suction, including any associated pumps (such as charging pumps) used to deliver fluid to the well or trip tank. Usually included is a pill/slug compartment. The pill/slug tank is used to prepare a drilling fluid with a higher density or extra hole-sweeping ability than usual; generally a 20- to 50-bbl volume is sufficient. The denser nature of the fluid requires that pill/slug compartments have more vigorous agitation than any other compartment. An agitator and mud gun combination is ideal for these compartments. Mud gun suction should come from the slug tank. Thus, recirculation within the compartment promotes maximum homogeneity and precludes dilution of the pill or slug with fluids from other compartments.

10.9.4 Reserve

The tank(s) or pit(s) and associated equipment used to isolate mud from the active system all require proper agitation. Compartments predominantly designated for long-term storage of drilling fluid (e.g., large-volume holding tanks) will not need strong shear forces or high TOR. Contour-type impellers have the added benefit of requiring less horsepower per unit of fluid displaced, so they are ideally suited for this purpose.

10.9.5 Discharge

The tank(s) or pit(s) and equipment located at the well site used to store and process mud and cuttings for disposal form the discharge function. Special consideration for discharge tanks may be required. Due to the variety of solid and fluid combinations encountered with discharges, it is outside the scope of this discussion to make recommendations. Please contact a qualified service provider to determine agitation requirements.

10.9.6 Trip Tank

This tank, and associated equipment, is used to isolate mud from the active systems for gauging pipe displacement during tripping operations. No agitation is required under normal conditions.

REFERENCES

Crane Company, *Flow of Fluids Through Valves, Fittings, and Pipe*. Crane Technical Paper No. 410. Copyright 1988.

Weisstein, E., *Eric Weisstein's World of Physics*. At Wolfram Research website, http://scienceworld.wolfram.com/physics. Copyright 2004. This entry contributed by Dana Romero.

CHAPTER 11

HYDROCYCLONES

Mark C. Morgan
Derrick Equipment Co.

Hydrocyclones are essentially simple devices that convert pressure generated by a centrifugal pump into centrifugal force, causing suspended solids in the mud to be separated from the fluid. This separation is actually accelerated settling due to the increased gravitational force cause by the centrifugal action inside the cone. The action inside the hydrocyclone can multiply gravitational force by as much as 200 times. In drilling operations, hydrocyclones use these centrifugal forces to separate solids in the 15- to 80-micron range from the drilling fluid. This solids-laden fluid is discharged from the lower apex of the cone, and the cleaned drilling fluid is discharged from the overflow discharge.

Hydrocyclones consist of an upper cylindrical section fitted with a tangential feed section, and a lower conical section that is open at its lower apex allowing for solids discharge (Figure 11.1). The closed, upper cylindrical section has a downward-protruding vortex finder pipe extending below the tangential feed location.

Fluid from a centrifugal pump enters the hydrocyclone tangentially, at high velocity, through a feed nozzle on the side of the top cylinder. As drilling fluid enters the hydrocyclone, centrifugal force on the swirling slurry accelerates the solids to the cone wall. The drilling fluid, a mixture of liquid and solids, rotates rapidly while spiraling downward toward the apex. The higher-mass solids move toward the cone wall. Movement progresses to the apex opening at the cone bottom. At the apex opening, the solids along the cone wall, together with a small amount of fluid, exit the cone. The discharge is restricted by the size of the apex. Fluid and smaller-mass particles, which have been concentrated away

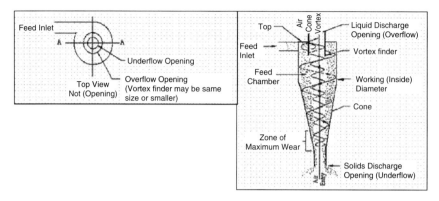

Figure 11.1. Hydrocyclone flow illustration.

from the cone wall, are forced to reverse flow direction into an upward-spiraling path at the center of the cone to exit through the vortex finder.

The vortex finder is a hollow tube that extends into the center of the cone. It diverts drilling fluid from flowing directly to the overflow outlet, causing the drilling fluid to move downward and into the cone. The swirling liquid is forced inward and, still rotating in the same direction, reverses the downward flow and moves upward toward the center of the vortex finder. In a balanced cone, the inner cylinder of swirling fluid surrounds a cylinder of air that is pulled in through the cone apex. Solids and a small amount of liquid are discharged from the lower apex of the cylinder. The apex opening relative to the diameter of the vortex finder will determine the dryness of the discharged solids.

Many balanced cones are designed to provide maximum separation efficiency when the inlet head is 75 feet. To be sure what the recommended inlet head is, check with the manufacturer's technical group. Fluid will always have the same velocity within the cone if the same head is delivered to the hydrocyclone inlet. Pressure can be converted to feet of head with the equation frequently used in well-control calculations but rearranged slightly:

$$\text{head (in feet)} = \frac{\text{PSI}}{(0.052)(\text{mud weight in ppg})}.$$

The relationship between manifold gauge pressure and drilling-fluid weight at constant 75-feet feed head is summarized in Table 11.1.

Hydrocyclones separate solids according to mass, which is a function of both density and particle size. However, in unweighted drilling fluids, the solids density has a comparatively narrow range, and size has

Table 11.1
Pressure for 75 Feet of Head for Various Mud Weights

Pressure (psig)	Feed Head (ft)	Mud Weight (ppg)
32.5	75	8.34
35	75	9.0
37	75	9.5
39	75	10.0
41	75	10.5
43	75	11.0
45	75	11.5
47	75	12.0
49	75	12.5
51	75	13.0

the greatest influence on their settling. Centrifugal forces act on the suspended-solids particles, so those with the largest mass (or largest size) are the first to move outward toward the wall of the hydrocyclone. Consequently, large solids with a small amount of liquid will concentrate at the cone wall, and smaller particles and the majority of liquid will concentrate in the inner portion.

Larger-size (higher-mass) particles, upon reaching the conical section, are exposed to the greatest centrifugal force and remain in their downward spiral path. The solids sliding down the wall of the cone, along with the bound liquid, exit through the apex orifice. This creates the underflow of the hydrocyclone.

Smaller particles are concentrated in the middle of the cone with most of the drilling fluid. As the cone narrows, the reduced cross-sectional area restricts the downward-spiraling path of the innermost layers. A second, upward vortex forms within the hydrocyclone, and the center fluid layers with smaller solids particles turn toward the overflow. At the point of maximum shear, the shear stress within a 4-inch desilter is on the order and magnitude of 1,000 reciprocal seconds.

The upward-moving vortex creates a low-pressure zone in the center of the hydrocyclone. In a balanced cone, air will enter the lower apex in counterflow to the solids and liquid discharged from the hydrocyclone. In an unbalanced cone, a rope discharge will emerge from the cone, resulting in excessive quantities of liquid and a wide range of solids in the discard. An unbalanced cone is little more than a settling pot, similar in operation to a sand trap.

There are two countercurrent spiraling streams in a balanced hydrocyclone, one spiraling downward along the cone surface and the other spiraling upward along the cone center axis. The countercurrent directions, together with turbulent eddy currents, concomitant with extremely high velocities, result in an inefficient separation of particles. The two streams tend to commingle within the contact regions, and particles are incorporated into the wrong streams. Hydrocyclones, therefore, do not make a sharp separation of solids sizes. The efficiency of a hydrocyclone can be improved by extending the vortex finder farther into the cone, which eliminates some of the commingling. The farther the vortex finder is extended, the better the separation.

Hydrocyclone sizes are designed arbitrarily by the inside cone diameter at the inlet. By convention, desanders have a cone diameter of 6 inches and larger; desilters have internal diameters smaller than 6 inches. Normally, discharges from the apex of these cones are discarded when used on unweighted drilling fluids. Prolonged use of these cones on a weighted drilling fluid will result in a significant reduction in drilling-fluid density caused by the discard of weighting material. When these cones are used as part of a mud cleaner configuration, the cone underflow is presented to a shaker screen. The shaker screen returns most of the barite and liquid to the drilling-fluid system, rejecting solids larger than the screen mesh. This is a common application of unbalanced hydrocyclones, since the cut point is determined by the shaker screen and not the cone.

Since most hydrocyclones are designed to operate with 75 feet of head at the input manifold, the flow rate through the cones is constant and predictable from the diameter of the cone for a typical tangential feed cone inlet of given orifice size (Table 11.2). Obviously, manufacturers may select different orifice sizes at the inlet of the cone. The orifice size determines the flow rate through the cone and the internal geometry of the cone.

Other design variations of the feed chamber, although more difficult to manufacture, can minimize backpressure and therefore increase the cone feed rate or capacity while also increasing separation efficiency. Hydrocyclones with feed chambers that reshape the incoming drilling fluid from a round-pipe profile to a rectangular one that conforms better to the feed chamber geometry can increase the amount of drilling fluid the cone can handle by 35–45% for the same pressure while at the same time increasing separation efficiency. The increased performance is due to the fact that the dead spaces in the feed chamber are eliminated. If a ramp feed is included in the feed inlet, the drilling fluid is forced down, which allows incoming drilling fluid to enter without the extreme

Table 11.2
Flow Rates Through Hydrocyclones*

Designation	Cone Diameter (in.)	Flow Rate Through Each Cone (gpm)
Desilter	2	10–30
Desilter	4	50–65
Desilter	5	75–85
Desilter	6	100–120
Desilter	8	200–240
Desilter	10	400–500
Desilter	12	500–600

*These are general values. Some cones will vary.

turbulence caused by impingement of the drilling-fluid streams. This reduces backpressure, which increases throughput even more. It also minimizes turbulence inside the cone, which also increases the separation efficiency.

The D_{50} cut point of a solids-separation device is usually defined as the particle size at which one half of the weight of those particles goes to the underflow and one half of the weight goes to the overflow. The cut point is related to the inside diameter of the hydrocyclone. For example, a 12-inch cone is capable of a D_{50} cut point of around 60 to 80 microns; a 6-inch cone is capable of around 40 to 60 microns; and a 4-inch cone is capable of around 20 to 40 microns. These cut points are representative for a fluid that contains a low solids content. The cut point will vary according to the size and quantity of solids in the feed and the flow properties of the fluid. Cut point determination procedures are explained in Chapter 4.

When hydrocyclones are mounted above the liquid level in the mud tanks, a siphon breaker should be installed in the overflow header or manifold from the cones. Otherwise, a high vacuum will occur and will actually vacuum up a lot of the solids that would otherwise be discarded; instead, these solids are reintroduced back into to the active system. In some extreme cases, no solids will exit the cone apex if the vacuum is high enough. The siphon breaker installed as illustrated should be one quarter of the diameter of the overflow header pipe (Figure 11.2).

11.1 DISCHARGE

Most hydrocyclones are of a balanced design. A properly adjusted, balanced hydrocyclone has a spray discharge at the underflow outlet and

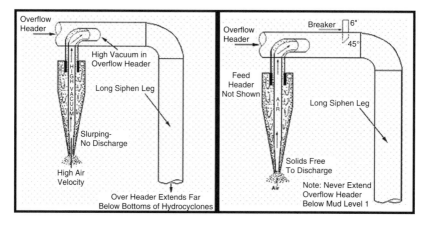

Figure 11.2. Solution to common discharge header problem.

exhibits a central air suction core. Many balanced hydrocyclones can be adjusted so that when water is fed under pressure, nothing discharges at the apex. Conversely, when coarse solids are added to the feed slurry, wet solids are discharged at the apex. Even with this adjustment, there still should be a large opening in the bottom of the cyclone. This will confirm that the cyclone is hydraulically balanced and discharges at the bottom (apex) only when solids, which the cyclone can separate, are in the feed slurry (drilling fluid).

A balanced cyclone should be operated with *spray discharge*. In this process, coarser solids separate to the outside in the downward spiral and pass over the lip of the apex as an annular ring. The apex is actually a weir, or dam, not a choke or valve.

The high-velocity return stream spinning upward near the center of the cone into the vortex finder generates a column of lower pressure, which sucks air inward through the center of the apex opening.

To set a cone to balance, slowly open the apex discharge while circulating water through the cone. When a small amount of water is discharged and the center air core is almost the same diameter as the opening, the cone is said to be balanced (see Figure 11.3).

With spray discharge, the device removes the maximum amount of solids, and discarding of whole mud is minimized. The umbrella-shaped spray discharge indicates that a uniform solids loading is presenting to the cone, with proper separation occurring. The pattern of the apex discharge provides a good indication of cone operation. The discharge should have a hollow center and appear as a cone spray. A wide cone

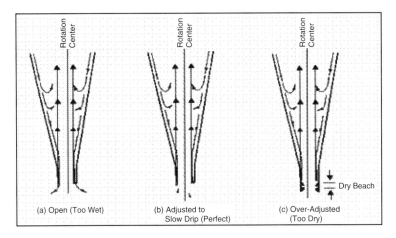

Figure 11.3. Proper balanced cone adjustment.

spray may indicate that the apex orifice is too large. When the apex orifice is larger than required, an excess amount of liquid will exit, carrying with it finer feed solids, thereby reducing sharpness of separation and underflow density. However, if more of the fine solids are discarded through the apex, then the density of the mud returned to the active system will be reduced. This is not necessarily a bad thing. A larger opening minimizes plugging of the apex. It depends on what is most important for your drilling operation. Is it solids removal or liquids conservation that is most important to you? This is a question that should be asked for each and every drilling operation so that the equipment can be chosen and set up accordingly.

It should be mentioned that there are some balanced hydrocyclones that will not balance in the way described above. They are instead balanced by operating with a central air core but always having a spray discharge, even if only pumping water. A cone operation like this is due to its feed-chamber design and/or the ratio of the apex-opening diameter to the vortex finder opening diameter. Which type of cone is best depends on your drilling situation, needs, and parameters.

Several conditions restrict separations and exiting of solids that have spiraled along the cone wall. These include:

- Excessive solids concentration
- Excessive volumetric feed rate per cone (going past the balance point on a balanced cone)
- Excessive fluid viscosity

- Excessive vacuum (caused by a long siphon leg)
- Restricted (too small) apex
- Inadequate feed pressure

A greater number of larger solids are entrained within the central vortex stream to exit with the overflow. The discharge pattern changes from spray to *rope discharge*, which is characterized by a cylindrical, or rope-like, appearance. With the rope discharge, no air core occurs through the center of the cone. In this case, the apex acts as a choke that restricts flow, rather than as a weir.

Rope discharge is a process in which material pours from the cone apex as a slow-moving cylinder (or rope). A hydrocyclone operating like this is performing an inefficient solids/liquid separation. The apex velocity in rope discharge is far less than that in spray discharge; therefore, separation is less efficient because fewer solids are discarded (Figure 11.4).

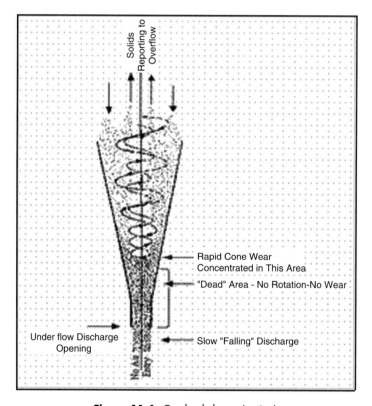

Figure 11.4. Overloaded cone (roping).

A rope discharge can create a false sense of success, as the heavier rope stream appears to contain more solids and does in fact have a higher density than a cone operating in spray discharge. In reality, however, a rope discharge indicates that not all solids that have been separated inside the cone can exit through the apex opening. Solids become crowded at the apex and cannot exit the cone freely. The exit rate is slowed significantly, and some solids that would otherwise be separated become caught in the inner spiral and are carried to the overflow. Dry discharge can also produce cone plugging.

With rope discharge, the exiting solids stream is heavier than that under spray-discharge conditions. All discharged solids will have a surface film of bound liquid. Since finer solids have a greater ratio of surface area to volume (size or mass), finer solids' streams involve greater volumes of bound water. More bound water causes a less dense underflow stream (the finer the particle separation, the wetter the apex stream). This explains why spray discharge stream densities are less than rope discharge stream densities.

The amount of fluid lost in cone underflow is important. A hydrocyclone operating with spray discharge gives solids a free path to flow (to exit the cone). Rope discharge is a dry discharge. Therefore, spray discharge removes significantly more solids than rope discharge. More fluid may be lost in spray discharge, but the greater solids-separation efficiency makes the additional fluid loss insignificant. If fluid loss is a concern, the underflow can be screened (see Chapter 12 on Mud Cleaners) or centrifuged for liquid recovery.

Rope discharge should be immediately corrected to reestablish the higher volumetric flow and greater solids separation of spray discharge. A rope discharge indicates that equipment is overloaded and additional hydrocyclones may be necessary. If the desilter cones are roping and the desander cones are not being run, then turn on the desanders to remove some of the load from the desilters. If possible, install finer screens on the shakers to take some of the load off of the hydrocyclones, which may reestablish spray discharge. If the apex weirs are pinched down, open them up fully to allow more solids to exit the cones.

11.2 HYDROCYCLONE CAPACITY

Since most hydrocyclones are designed to operate at a constant 75 feet of head at the input manifold, flow rate through any cone is constant at constant inlet pressure for a given fluid viscosity.

The smaller desilter hydrocyclones are rated from 40 to 100 gpm of liquid removal, depending on the cone design. The normal 4-inch cones will remove 4 gpm of solids, or 5.7 barrels per hour of solids, per cone. Therefore, a 16-cone desilter manifold will accommodate removal of 510 cubic feet of solids per hour. For a 17-inch hole, this equates to penetration rates averaging 297 feet per hour. Clearly, if design and operational characteristics are adequately maintained, more than ample solids separation can be effected.

For deepwater drilling applications, higher flow rates are encountered when boosting the riser. Additional cones may be needed to handle the additional flow rates to ensure that the cone manifolds process all of the mud.

The accelerated gravitational forces generated in hydrocyclones are inversely proportional to the radius of the hydrocyclone cylinder. Thus, the larger the diameter of the cone, the coarser the separation. In general, the larger the hydrocyclone, the coarser its cut point and greater its throughput. The smaller the cone, the smaller the size of particles the cone will separate. In other words, the median particle size removed decreases with cone diameter. Median particle size also increases with increasing fluid viscosity and density, but decreases as particle-specific gravity increases. Oilfield hydrocyclones range between 4 and 12 inches, based on the inner diameter of the intake cylinder. A small hydrocyclone diameter is used for ultra-fine separations.

11.3 HYDROCYCLONE TANKS AND ARRANGEMENTS

Hydrocyclones are arranged with the unit of larger cone size upstream of the smaller unit. A separate tank is needed for each size unit. Generally, a desander and a desilter manifold are available as part of the rig equipment. Hydrocyclones should process all drilling fluid entering their suction compartments independently of the drilling-fluid circulation rate.

Suction is taken from the tank immediately upstream of the discharge tank. The number of cones in use should process at least 100% but preferably greater than 100% of the flow rate of all fluids entering the suction tank for the hydrocyclone. A backflow between the hydrocyclone discharge and suction compartments of at least 100 gpm usually ensures adequate processing. Estimations based on rig circulation rates are usually inadequate if the plumbing is not arranged properly. For example, if a 500-gpm hydrocyclone overflow is returned to the suction compartment and a 400-gpm rig flow rate enters the suction compartment,

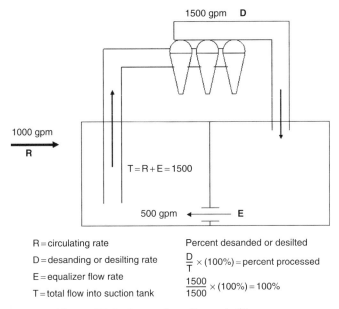

Figure 11.5. Correct desanding or desilting setup.

adequate processing is not achieved even though more fluid is processed than is pumped downhole. In this case, the flow entering the hydrocyclone suction compartment is 900 gpm. The fraction of drilling fluid processed would be 500/900 gpm, or 56%. For proper desanding or desilting setup see Figure 11.5.

11.3.1 Desanders

Desanding units are designed to separate drilled solids in the 50- to 80-micron range and barite in the 30- to 50-micron range. Desanders should be used in unweighted mud when shakers are unable to have API 140 screens (100 microns) or finer installed. They are used primarily to remove high solids volume associated with fast drilling of large-diameter top holes. In water-base drilling fluids, desanders make a median separation cut of 2.6–specific gravity (SG) solids in the 50- to 80-micron size range. The desander removes sand-size and larger particles that pass through the shale shaker screens.

Desanders are installed immediately downstream of the shaker and degasser. Suction is taken from the immediate upstream tank, usually the degasser discharge tank. Discharge from the desander is made into the tank immediately downstream. Suction and discharge tanks are equalized through valves or an opening located on the bottom of each tank.

Desanders are used continuously during drilling surface holes. Plumbing can be arranged to process all total surface pit volume after beginning a trip.

Use of desanders is generally discontinued after barite and/or expensive polymers are added to the drilling mud, because a desander discards such a high proportion of these materials. Use of desanders is generally not cost-effective with an oil-base drilling fluid because the larger cones discharge a significant amount of the liquid phase (see Chapter 12 on Mud Cleaners).

11.3.2 Desilters

Desilter cones are manufactured in a variety of dimensions, ranging from 2 to 6 inches, and make separations of drilled solids in the 12- to 40-micron range. They will also separate barite particles in the 8- to 25-micron range. Desilters are installed downstream from the shale shaker, sand trap, degasser, and desander.

Desilter cones differ from desander cones only in dimensions and operate on exactly the same principles. Common desilter cone sizes are between 2 and 5 inches. A centrifugal pump should be dedicated to provide fluid to the desilter manifold only. Setting up a centrifuge pump to run multiple pieces of equipment is not a good idea, as doing so requires compromises in performance and opportunities for incorrect operation.

These units make the finest particle-size separations of any full-flow solids-control equipment—down to 12 microns of drilled solid. The desilter, therefore, is an important device for reducing average particle size and reducing drilled solids.

Desilter suction is also taken from the immediate upstream tank, usually the desander discharge tank. Desilter suction and discharge tanks are, again, equalized through a valve, or valves, or an opening located on the bottom of each tank. The size of the valve or opening should be

$$^1\text{Diameter (inches)} = \sqrt{\frac{\text{max gpm}}{15}}$$

---

[1]Formula for standard rectangular tanks.

The maximum (max) gpm will be the maximum flow rate expected, not the backflow, as there will be times when the unit will not be operating and the total rig flow rate will have to pass through the valve or opening. Suction should not be taken from the tank into which chemicals and other materials (barite and bentonite) are added because valuable treating materials may be lost.

11.3.3 Comparative Operation of Desanders and Desilters

The role of desanders is to reduce loading downstream on desilters. Installing a desander ahead of the desilter relieves a significant amount of solids loading on the desilter and improves desilter efficiency. High rates of penetration, especially in unconsolidated surface hole, where the largest-diameter bits are used, results in generating larger concentrations of drilled solids. This may place desilters in rope discharge. For this reason, desanders, which have greater volumetric capacity and can make separations of coarser drilled solids, are placed upstream of desilters. Desanders remove a higher mass (i.e., coarser drilled solids) during periods of high solids loading. Desilters can then efficiently process the reduced solids-content overflow of the desanders.

If the drill rate is slow, generating only a few hundred pounds per hour of drilled solids, the desander may be turned off and the desilter used to process the entire circulating system.

Desilters should be used on all unweighted, water-base mud. These units are not used on weighted muds because they discard an appreciable amount of barite. Most barite particles fall within the silt-size range. Desilter operation is important for all unweighted fluids; however, in oil-base muds with high viscosity (as found in deepwater drilling), the apex discharge may be centrifuged for oil-phase salvage.

11.3.4 Hydrocyclone Feed Header Problems

Never make a horizontal bend into an inline hydrocyclone manifold. Centrifugal force will force solids to the outside of the bend, which can overload some cones and cause solids to bypass others. It is a good idea to have straight pipe in front of the first cone that is three times the pipe diameter. This allows time for the solids and liquid to remix evenly (Figure 11.6).

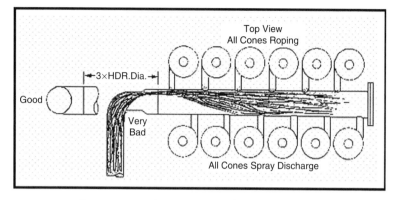

Figure 11.6. Solution to common feed header problem.

11.4 MEDIAN (D_{50}) CUT POINTS

In spray discharge, for any set of cone diameter, feed slurry compositions, flow properties, volumetric flow rates, and pressure conditions, some particles' size (mass) is 100% discarded from the apex.

For every size and design of cone operating at a given pressure with feed slurry of a given viscosity, density, and solids distribution, there is a certain size (mass) of particle that shows no preference for either top or bottom discharge. As a result, 50% of this particular size exits through the vortex and 50% exits through the apex. This particle size is termed the *median cut*, the *median-size particle,* or more frequently in drilling operations, the D_{50} *cut point.*

The median cut, or D_{50} cut point, does not mean that all larger particles exit at the apex and smaller particles exit at the vortex. The D_{50} cut point of a solids-separation device is defined as that particle size at which one half of the weight of specific-size particles go to the underflow (discard) and one half of the weight go to the overflow (returned to the active system). For example, a D_{30} cut point references a particle size that is concentrated 30% in the underflow and 70% in the overflow.

As stated earlier, the cut point is related to the inside diameter of the hydrocyclone. For example, a 12-inch cone has a D_{50} cut point for low-gravity solids in water of approximately 60 to 80 microns; a 6-inch cone, around 30 to 60 microns, and a 4-inch cone, around 15 to 20 microns (Table 11.3). However, the cut point will vary with the size and amount of solids in the feed, as well as fluid viscosity.

Table 11.3
Hydrocyclone Size Versus D_{50} Cut Point (CP)

Cone Diameter (in.)	D_{50} CP in Water	D_{50} CP in Drilling Fluid
2	8–10	15+
4	15–20	35–70
6	30–35	70–100
12	60–70	200+

For comparative purposes, consider a 50-micron-equivalent drilled-solid diameter. Relatively speaking, the percentage of discharge is as follows:

- A 6-inch cone discharges 80% at underflow.
- A 4-inch cone discharges 95% at underflow.
- A 3-inch cone discharges 97% at underflow.

Now consider a 10-micron-equivalent drilled-solid diameter:

- A 6-inch cone discharges 7% at underflow.
- A 4-inch cone discharges 11% at underflow.
- A 3-inch cone discharges 17% at underflow.

If a graph of particle size versus percentage of particles recovered to underflow is plotted, the portion of the curve near the D_{50}, or 50%, recovery point (median cut point) is very steep when separations are efficient.

Particle separations in hydrocyclones vary considerably. In addition to the proper feed head and cone apex setting, drilling-fluid properties such as density, percentage of solids (and solids distribution), and viscosity all affect separations. Any increase in these mud properties will increase the cut point of a separation device.

11.4.1 Stokes' Law

Stokes' law defines the relationship between parameters that control the settling velocity of particles in viscous liquids, not only in settling pits but also in equipment such as hydrocyclones and centrifuges.

The force of gravity and the viscosity of the suspending fluid (drilling mud) control separation in a settling pit. A large, heavy particle settles faster than a small, light particle. The settling process can be increased by reducing the viscosity of the suspending fluid, increasing the gravitational forces on the particles, or increasing the effective particle size with flocculation or coagulation.

Hydrocyclones and centrifuges increase settling rates by application of increased centrifugal force, equivalent to higher gravity force. Stokes' law for the settling of spherical particles in a viscous liquid applies:

$$V_s = \frac{CgD_E^2(\rho_s - \rho_L)}{\mu}$$

where

- V_s = Settling or terminal velocity, in ft/sec
- C = Units constant, 2.15×10^{-7}
- g = Acceleration (gravity or apparatus), in ft/sec^2
- D_E = Particle equivalent diameter, microns
- ρ_s = Specific gravity of solids (cutting, barite, etc.)
- ρ_L = Specific gravity of liquid phase
- μ = Viscosity of media, centipoise

Particles of differing densities and varying sizes can have the same settling rates. That is, there exists an *equivalent diameter* for every 2.65-SG drilled solid, be it limestone, sand, or shale, that cannot be separated by gravimetric methods from barite particles of corresponding equivalent diameter. Presently, it is not possible to separate desirable barite particles from undesirable drilled-solid particles that settle at the same rate.

The recognized rule of thumb is: A barite particle (SG 4.25) will settle at the same rate as a drilled-solids particle (SG 2.65) having 1½ times the barite particle's diameter. This rule of thumb may be verified applying Stokes' law.

Example 1

A viscosified seawater fluid (SG 1.1, plastic viscosity (PV) 2.0 centipoise (cP), and yield point (YP) 12.0 lb/100 sq ft) is circulated to clean out a cased well bore. What size of low-gravity solids will settle out with 5-micron barite particles? What is the settling velocity of 10-micron barite particles in rig tanks? From Stokes' law, settling velocity is:

$$V_s = \frac{Cg(\rho_s - \rho_L)D_E^2}{\mu}$$

Hydrocyclones

For equivalent settling rates, $V_s = V_{s'}$. And for $\mu_1 = \mu_2$ (the same fluid, therefore the same viscosity):

$$D_1^2(\rho_1 - \rho_L) = D_2^2(\rho_2 - \rho_L)$$

$$\rho_1 = 2.65$$
$$\rho_2 = 4.25$$
$$\rho_L = 1.1$$

$$(D_1^2 \times 1.55) = (D_2^2 \times 3.15)$$

$$\frac{D_1^2}{D_2^2} = \frac{3.15}{1.55} = 2.03$$

$$\frac{D_1}{D_2} = 1.42$$

Thus a 5-micron barite particle will settle at the same rate as a 7-micron low-gravity particle, and a 10-micron barite particle will settle at the same rate as a 14-micron low-gravity particle.

Again, Stokes' law applied to settling velocity is:

$$V_s = \frac{Cg(\rho_s - \rho_L)D_E^2}{\mu}$$

Settling velocity, for 5 micron barite (or 7 micron drilled solid) particle is:

$$V_s = \frac{(2.15 \times 10^{-7}) \times 32.2 \text{ ft/sec}^2 \times (4.25 - 1.1) \times 25}{16 \text{ cP}}$$

$$= 3.4 \times 10^{-5} \text{ ft/sec (or 0.24 inches per minute)}$$

and for 10-micron barite (or a 14-micron drilled solid) particle,

$$V_s = \frac{(2.15 \times 10^{-7}) \times 32.2 \text{ ft/sec}^2 \times (4.25 - 1.1) \times 100}{16 \text{ cP}}$$

$$= 13.6 \times 10^{-5} \text{ ft/sec (or 0.98 inches per minute)}$$

Example 2

What are the equivalent diameters of barite (SG 4.25) and drilled solids (SG 2.65) in an 11.5-ppg drilling fluid with PV = 20 cP and YP = 12 lb/100 sq ft?

From Stokes' law, settling velocity is:

$$V_s = \frac{Cg(\rho_s - \rho_L)D_E^2}{\mu}$$

For equivalent settling rates, $V_s = V_{s'}$. And for $\mu_1 = \mu_2$ (the same fluid, therefore the same viscosity):

$$D_1^2(\rho_1 - \rho_L) = D_2^2(\rho_2 - \rho_L)$$

$$\rho_1 = 2.65$$
$$\rho_2 = 4.25$$
$$\rho_L = 1.38$$

$$(D_1^2 \times 1.27) = (D_2^2 \times 2.87)$$

$$\frac{D_1^2}{D_2^2} = \frac{2.87}{1.27} = 2.26$$

$$\frac{D_1}{D_2} = 1.5$$

Thus a 10-micron barite particle will settle at the same rate as a 15-micron low-gravity particle; a 50-micron barite particle will settle at the same rate as a 75-micron low-gravity particle, and so forth.

Stokes' law shows that as fluid viscosity and density increase, separation efficiency decreases. If the drilling-fluid weight is 14.0 ppg (SG 1.68),

$$\frac{D_B^2}{D_{ds}^2} = \frac{4.25 - 1.68}{2.65 - 1.68} = \frac{2.57}{0.97} = 2.65$$

or

$$\frac{D_B}{D_{ds}} = 1.63$$

In drilling fluid weighing 14 ppg, a 10-micron barite particle will settle at the same rate as a 16-micron drilled-solid particle, and a 50-micron barite particle will settle at the same rate as an 80 (or 81.4)-micron drilled-solid particle.

Frequently disregarded is the fact that the efficiency of a separator is viscosity dependent. The D_{50} cut point increases with viscosity, as shown by Stokes' law:

$$V_s = \frac{CgD_E^2(\rho_s - \rho_L)}{\mu}$$

Example 3

A 4-inch cone will separate half of the 12-micron low-gravity (SG 2.6) particles in water, (that is, the D_{50} cut point is 12 microns). What is the D_{50} cut point in a fluid of 50 cP viscosity of the same density?

For constant settling velocity (if fluid density is unchanged) and other constant parameters:

$$\text{For } \rho_1 - \rho_2$$

$$\frac{D_1^2}{\mu_1} = \frac{D_2^2}{\mu_2}$$

$$D_1 = 12 \text{ micron}$$
$$\mu_1 = 1 \text{ cP}$$
$$\mu_2 = 50 \text{ cP}$$

$$D_2^2 = 7200$$
$$D_2 = 84.8 \text{ microns}$$

Thus, if fluid SG remains 1.1, and viscosity is 50 cP, the D_{50} is raised to 85 microns. Similarly, for a 4-inch hydrocyclone:

Fluid Viscosity, cP	D_{50}, microns (SG 2.6)
1.0	12.0
10.0	37.9
20.0	53.7
30.0	65.8
40.0	75.8
50.0	84.8

Cut-point performance can be further projected by dividing by the projected SG at various viscosities. Thus, in terms of the preceding example, for 20-cP viscosity and 1.4 (11.7 ppg)-density fluid, the D_{50} would be:

$$53.7 \text{ microns} \times 1.1/1.4 = 42.1 \text{ microns.}$$

The D_{50} cut points for 6-inch and 8-inch hydrocyclones (common desilter and desander sizes) are usually given as 25 microns and 60 microns.

The variation of D_{50} cut points with viscosity for a 6-inch desilter hydrocyclone (SG 1.1 fluid) can be seen to be:

Fluid Viscosity, cP	D_{50}, microns (SG 2.6)
1.0	25.0
10.0	79.0
20.0	112
30.0	137
40.0	158
50.0	176

Again, cut-point performance can be further projected by dividing by the projected SG at various viscosities. Thus, in the preceding situation, for a 4-inch hydrocyclone, with 20-cP viscosity and 1.4 (11.7 ppg)-density fluid, the D_{50} would be:

$$67.1 \text{ microns} \times 1.1/1.4 = 52.7 \text{ microns.}$$

11.5 HYDROCYCLONE OPERATING TIPS

Other than cone and manifold plugging, improperly sized or operated centrifugal pumps are by far the greatest source of problems encountered with hydrocyclones. Centrifugal pump and piping sizing are critical to efficient hydrocyclone operation. A pressure gauge should be mounted on the hydrocyclone inlet manifold.

Hydrocyclones should always have a pressure gauge installed on the inlet to quickly determine whether proper feed head is supplied by the centrifugal pump.

Hydrocyclones are usually mounted in the vertical position but may be mounted inclined or even horizontally. Cone orientation does not matter, as the separating force is supplied by the centrifugal pump.

The feed slurry must be distributed to a number of hydrocyclones operating in parallel. A *radial manifold* provides each cyclone with the same slurry regarding feed-solids concentration and particle size distribution at the same pressure. Higher-mass (larger-diameter) particles in an *inline manifold* tend to bypass the first cyclones, due to their high energy, and enter the last cones. Thus, the last cones in an inline manifold receive a higher concentration of coarse feed particles. Performance from cones in an inline manifold will not be uniform, as feed concentrations and particle size distributions are different for the various cones. Further, if the last cyclone(s) in an inline manifold are taken offline, the end of the manifold has a tendency to plug.

To minimize loss of head along the feed line and backpressure on the overflow (top) discharge line, keep all lines as short and as straight as possible with the minimum of pipe fittings, turns, and changes in elevations. Required pipe diameters are dependent on the flow rate. Frictional pressure losses need to be minimized while maintaining high enough flow velocity to keep the pipes clear of any settling. To do this, the suction lines should have a minimum velocity of 5 ft/sec (1.5 mps) and a maximum of 9.1 ft/sec (2.73 mps). On the discharge side, the minimum velocity remains the same, but the maximum velocity can be increased to 12.1 ft/sec (3.63 mps). The velocity can be calculated by using the formula:

$$\text{velocity, ft/sec} = \frac{(\text{flow rate, gpm})(0.4087)}{\text{pipe diameter}^2}$$

Operate balanced cones in spray discharge with a central air suction core, and check cones regularly to ensure that apex discharge is not plugged. Operate unbalanced cones according to the manufacturer's instructions.

When balanced cones no longer spray-discharge, either too many solids are being presented for design processing, or large solids have plugged the manifold or apex, or the feed pressure is not correct. If feed pressure is set according to the manufacturer's recommendations, often the inability to maintain spray discharge can be traced to the shale shakers. Check for torn or improperly mounted or tensioned screens, or open bypass. Otherwise ensure that there are sufficient hydrocyclones to process the total fluid being circulated by the mud pumps.

Install a low ("bottom") equalizer to permit backflow from the discharge tank into the suction compartment. Removable centrifugal pump suction screens reduce plugging problems.

Operate hydrocyclones at the recommended feed head. For many (but not all) hydrocyclones, this is around 75 feet. Efficiency will be decreased if the feed head is too low. Too high a feed head can cause plugging problems and will decrease separation efficiency.

Do NOT use the same pump to feed the desilter and the desander. Each unit should have its own centrifugal pump(s).

Run the desilter continuously while drilling with unweighted mud, and process at least 100% of the total surface pit volume after beginning a trip.

Run desanders when shale shakers cannot be run with API 140 (100 microns) or finer screens.

If the cones continually plug due to trash, install a guard screen with approximately ½-inch openings to prevent large trash from entering and plugging the inlets and/or cone apexes. With good crews, this is usually not necessary.

Regularly check cones for bottom plugging or flooding. Desilter cones will plug more often than desander cones. Plugged cones may be cleaned with a welding rod. A flooded cone indicates a partially plugged feed or a worn cone bottom section.

Between wells or in periods when drilling is interrupted, flush manifolds with water and examine the inside surfaces of the cones. If there is significant wear, change those parts.

Keep shale shakers well maintained, and never bypass them.

Hydrocyclones discard absorbed liquid with the drilled solids. Dryness of discharge solids is a function of the apex opening relative to the diameter of the vortex finder.

Mud cleaners and/or centrifuges can be used to process the cones' underflow.

11.6 INSTALLATION

Hydrocyclones should be located so that underflow can be moved away with a minimum of trouble and washdown water and so that they are accessible for maintenance and evaluation.

Discharge overflow should return to the circulating system into a compartment immediately downstream of the suction compartment; however, the discharge overflow compartment must be bottom equalized with the suction compartment, and drilling fluid should backflow from the discharge tank into the suction tank at all times. The hydrocyclones should process all of the drilling fluid entering the suction compartment. Ensure that all drilling fluid entering the discharge compartment has been processed by the hydrocyclones.

Centrifugal feed pumps should be located so that they have flooded suction and minimum suction line length with few elbows and turns, to keep friction losses at a minimum.

A centrifugal pump may stir its suction compartment with mud guns. Mud jet mixers should not be supplied with fluid from other parts of the drilling-fluid system. Preferably, mechanical agitators will be used in the removal section.

Keep the end of the discharge line above the surface of the mud in the receiving tank, to avoid creating a vacuum. Overflow should be introduced into the next compartment downstream at approximately a 45° angle so that lines will be kept full and a siphoning vacuum (which would pull more solids into the overflow discharge) avoided. When hydrocyclones are mounted more than 5 feet above the liquid level of the mud tanks, a siphon breaker should be installed in the overflow manifold from the cones.

For a hydrocyclone troubleshooting guide, see Table 11.4.

11.7 CONCLUSIONS

Hydrocyclones are simple, easily maintained mechanical devices without moving parts. Separation is accomplished by transfer of kinetic input or feed energy into centrifugal force inside the cone. The centrifugal force acts on the drilling fluid slurry to rapidly separate drilled solids and other solid particles in accordance with Stokes' law.

The solids that are generated by drilling in some formations are too fine for shale shakers to remove. Hydrocyclones must be relied on to remove the majority of these solids. Here the shale shaker protects the hydrocyclones from oversized particles that may cause plugging.

Hydrocyclones should be designed to provide maximum removal of solids with minimum loss of liquid. Sufficient hydrocyclones should be arranged in parallel to process all drilling fluid arriving into the additions compartment of the mud tank system.

Hydrocyclones produce a wet discharge compared with shale shakers and centrifuges. Underflow density alone is not a good indicator of cone performance, as finer solids will have more associated liquid and the resultant slurry density will be lower than for coarser solids.

As the solids content increases, separation efficiency decreases and the size of particles that can be separated increases.

Hydrocyclones provide:

- Simple design
- No moving parts

Table 11.4
Hydrocyclone Troubleshooting Guide

Symptom	Probable Cause(s) and Action
Cones continually plug at apex, some cones OK	Partially plugged feed inlet or outlet. Apex(es) too small. Remove cone, clean out lines. Check shaker for torn screens or bypassing.
Some cones losing whole mud in a stream	Plugged cone feed inlet allows backflow from overflow manifold
Low feed head	Check centrifugal pump operation: rpm voltage, etc. Check shaker for torn screens or bypassing. Check for obstructions in line, solids settling, partially closed valve.
Cones discharge dry solids	Apex too small. Need more cones?
Vacuum in manifold discharge (long drop into pits?)	Install siphon breaker
Mud % solids increasing	Insufficient cone capacity, install more cones. Solids may be too small, finer shaker screens?
Heavy discharge stream	Overloaded cones. Increase apex size and/or install more cones
High mud losses	Cone apex too large, reduce. Centrifuge cyclone discharge? Reduce cone sizes.
Unsteady cone discharge, varying feed head	Air or gas in feed
Aerated mud downstream of hydrocyclone	Route overflow into trough to allow air breakout

- Easy maintenance
- Good separation ability

Hydrocyclones have the following disadvantages:

- Limited separation of ultra-fines
- Inability to handle flocculated materials

It is impractical to desand or desilt a mud containing appreciable amounts of barite. Silts and barite have about the same size range. The majority of barite particles are between 2 and 44 microns, some between 44 and 74 microns, and unfortunately some 8–15% are between 0 and

Hydrocyclones

2 microns. A desander median cut (D_{50} cut point) falls between 25 and 30 microns. A desilter median cut falls between 10 and 15 microns. Since much of barite falls above these cuts, it would be discarded along with the silt and sand.

Generally, hydrocyclones are most efficient when solids have a diameter greater than 10 microns and are spherical in shape. If the solids are flat, like mica, movement tends to be random and dependent on whether the flat surface or edge is toward the gravitational force created in the vortex. Since separation efficiency depends somewhat on the freedom and velocity of the solid moving through the liquid phase, it is logical to use a fluid of as low viscosity as possible.

Hydrocyclones have the following ADVANTAGES:

1. Replacement of pump fluid end parts is reduced, and pumps operate more efficiently.
2. Less drill string torque and drag equates to less wear on the string and less key-seating (a major potential for stuck pipe). Casing is run easier.
3. Bit life is extended; again, due to less abrasion.
4. Penetration rates increase.
5. Water dilution to maintain low mud weights is reduced. This is reflected in smaller waste pits and drilling-fluid volumes to clean up at the end of drilling activity.
6. Material additions are decreased.
7. Additions of weight material are made with little or no difficulty.
8. Downhole tools set and release with little or no interference from drill cuttings.

11.7.1 Errata

Sizing hydrocyclones:

9. Separation needed
10. Volume of feed slurry
11. Concentration and distributions of solids in feed slurry

In sizing, the bases from which measurements are made are:

1. Free liquid: water at 20°C, 68°F. Viscosity = 1 cP.
2. Solids: sand (spheres), SG 2.65

3. Feed concentration, extremely dilute: less than 1.0% solids by weight, 0.04% by volume

Because fine solids have more surface area per unit volume (specific area), the amount of liquid discharged per pound of solids is higher with fine solids than with coarse solids. Therefore, the difference between the feed and underflow densities is not a reliable indicator of hydrocyclone performance.

Pressure drop is a measure of the energy being expended in the cone, and thus a higher pressure drop results in a finer separation.

If the D_{50} cut point is increased to 75 microns, 25% of the 100-micron particles are retained and only 25% of the 55-micron particles are discharged.

The purpose of a hydrocyclone is to discharge maximum abrasive solids with minimal fluid loss. Larger particles have a greater probability to discharge through the bottom underflow (apex), while smaller and lighter particles have greater probability to move through the top, or overflow opening.

Cone diameter, cone angle, underflow diameter, feed head, and plastic viscosity have the largest effects on hydrocyclone performance.

Barite particles 3 microns and smaller have a deleterious effect on drilling-fluid viscosity due to surface charge imbalance resulting from unsatisfied broken bonds on the ultra-fine's surface. Therefore, if a centrifuge is set such that its median (D_{50}) cut in 14 ppg of mud is a 3-micron barite particle, its median cut for drilled solids will be 5 (or 4.9) microns.

CHAPTER 12

MUD CLEANERS

Leon Robinson
Exxon, retired

Mud cleaners are a combination of hydrocyclones mounted above shaker screens with small openings. Mud cleaners can be leased, rented, purchased as independent units, or assembled on location. When mud cleaners were invented in 1971, main shale shakers on drilling rigs were either unbalanced elliptical or circular motion machines. The finest screen at that time could separate solids only larger than about 177 microns (API 80) from a normal drilling-rig circulating rate. Barite specifications restricted solids to sizes predominantly smaller than 74 microns (API sand). Hydrocyclones removed large quantities of barite from a weighted drilling fluid and were not generally used. This meant that all of the drilled solids between 74 and 177 microns were available for removal but could not be removed with equipment available at that time. The mud cleaner was invented to remove those drilled solids.

Most 4-inch hydrocyclones discard around 1–3 gpm for every 50-gpm input. With a 1000-gpm flow rate input, the underflow from twenty 4-inch hydrocyclones would be only 20–60 gpm. Most shale shakers in the early 1970s could process this small flow rate through 74-micron (API 200) screens. Barite smaller than the openings, as well as drilled solids smaller than the openings, returned to the drilling fluid. Solids larger than the screen opening, and the solids clinging to those larger solids, were discarded. Solids larger than 74 microns made filter cakes incompressible and also increased the coefficient of friction between the drill string and the well bore. This was not unlike making a coarse-sandpaper filter cake; so, stuck pipe was common.

The D_{50} cut point of a 4-inch hydrocyclone is usually reported to be around 15 to 20 microns for an unweighted drilling fluid. With the same cut point in a drilling fluid weighted with barite, an enormous amount of barite would be contained in the underflow, even in an 11- to 12-ppg drilling fluid. The cut-point curve of a weighted drilling fluid processed through a desilter is not sharp, nor is the D_{50} cut point as low as 20 microns. The barite in an 11-ppg drilling fluid does not have a constant size distribution after several days of circulation. The separation curve in Figure 12.1 indicates that the cut point curve for 4-inch hydrocyclones is not a sharp separation, as might exist for a shale shaker screen or even a 4-inch hydrocylone with an unweighted drilling fluid.

Separation, or cut-point, curves can be misleading. Normally a uniform feed distribution is assumed. In the field, however, this may not actually exist. The cut point curve does not indicate the quantity of material in each size range. Removal of one pound out of two pounds, or 100 pounds out of 200 pounds is the 50% point.

Very little barite has a size above 50 microns in a circulating drilling fluid. In the separation curve, about 50% of the solids between 50 and 120 microns are separated and 50% remain in the drilling fluid. Perhaps this is surprising until the barite distribution in the 11-ppg drilling fluid is considered (Figure 12.2). If no solids larger than 10 microns were in a drilling fluid processed through a desilter cone, the cut point for this cone would obviously be lower than 10 microns.

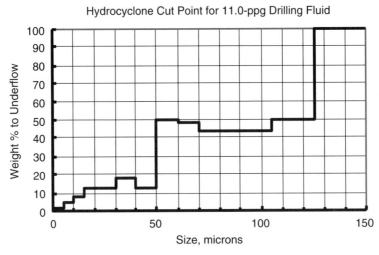

Figure 12.1. Separation curve for 4-inch hydrocyclones processing an 11-ppg drilling fluid.

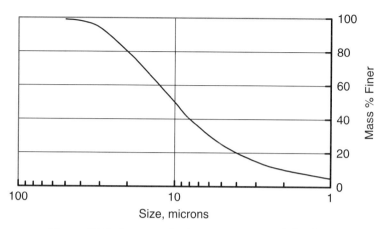

Figure 12.2. Barite size distribution in an 11-ppg drilling fluid.

Not much barite is in the large size range that will report to the underflow of the hydrocyclones. If this were not true and the cut point of a 4-inch hydrocyclone were about 20 microns, as reported for unweighted drilling fluid, the mud cleaner screens would fail from a weight overload during the first few minutes of operation. The 20-micron cut point normally reported for 4-inch hydrocyclones should also contain some statement about the size distribution injected into the cones. If there were no solids larger than 12 microns, the cut point would obviously be lower than the 12 microns.

Since the mud cleaner screen separates solids larger than the opening size, a low and sharp cut point is not needed from the hydrocyclones. The underflow of the hydrocyclones should be very wet. A better separation is made on the shaker screens if ample fluid is available to allow the screen to separate the solids from the slurry.

Some mud cleaners are designed to return some of the cone overflow to the screen to assist solids separation. This is preferable to spraying water or oil onto the screen to enhance the solids separation. Water or oil sprayed onto the screen dilutes the drilling fluid and requires the addition of more barite and chemicals. For example, if only 2 gpm were sprayed onto the mud cleaner screen, 2880 gal would be added to the system in one day. This 68.6 bbl of liquid phase will require the addition of all of the solids and chemicals required to create clean drilling fluid. This would be acceptable if about 18 bbl of drilled solids were discarded from the solids-control equipment in a slurry of about 35% volume of drilled solids in a drilling-fluid slurry. This quantity of drilled solids could be generated if only 180 feet of a 10-inch-diameter hole were drilled during the 24-hour period.

Exercise: Validate the equations and calculations listed in the preceding paragraph.

When linear motion shale shakers were introduced, mud-cleaner utility appeared obsolete. Linear motion shale shakers could separate drilled solids larger than barite while handling all of a drilling rig's circulation. However, many times desilters were plugged with solids too large to pass through a shale shaker screen in unweighted drilling fluids. These drilled solids reached the desilters and plugged the bottom apex. Their method of entry into the drilling-fluid system varied from (1) holes in shaker screens to (2) carryover from the discard end of the shale shakers to (3) improper mounting of shaker screens, including even gaps left between the upper end of the screens and the back tanks. Plugged desilter cones are a common sight around drilling rigs if no one is assigned the task of unplugging. Because this is such a common occurrence, it is not surprising that mud cleaners downstream of API 200 (74-micron) shaker screens are usually loaded with solids. Sometimes solids overload an API 150 screen on a mud cleaner even though API 200 screens are mounted on the main shale shakers.

12.1 HISTORY

The first mud cleaner was a combination of two 12-inch-diameter and twenty 4-inch hydrocyclones mounted above a specially built, 5-foot-diameter, stainless steel, round Sweco shaker. Even though the mud cleaner was invented for use with a weighted drilling fluid, the first application was on a well drilled using one of the first unweighted, potassium chloride drilling fluids, in a shallow, 2200-foot research well near Houston, Texas, in 1971. An API 80 screen was mounted on an unbalanced elliptical motion Linkbelt shale shaker to process the fluid as it left the well bore. The fluid was then pumped to the hydrocyclones. The well was to be used, initially, to evaluate the use of air injection into risers to reduce annular pressure at the seafloor and, subsequently, as a research test well.

The second mud cleaner was a bank of twenty 4-inch Pioneer hydrocyclones mounted above another specially built 4-foot-diameter, double-deck, stainless steel, round Sweco shaker. The screens could be arranged in parallel or in series to process the underflow from the cones. This unit was placed on an exploration well in South Louisiana. A gas-bearing formation at about 11,000 feet contained an 11.0-ppg pore pressure. The gas-bearing formations between 11,300 feet and 16,000 feet had been depleted to pore pressures as low as 2.2 ppg. Plans were to drill through this interval with the 11.0-ppg water-base drilling fluid, set casing, and drill the exploration part of the hole. Through this interval, differential pressures, between the fluid in the well bore and the formation pressure, varied up to 6000 psi. A centrifuge was used along with the mud cleaner and dilution to keep the drilled-solids concentration very low to make a compressible filter cake. While drilling this interval, no lost circulation or stuck pipe was experienced. Torque and drag on the drill string were minimal. After reaching about 80 feet above the predicted casing depth, the mud cleaner was shut down. (Actually, the company man suggested that the research team go home for Christmas because they were having no problems and the new experimental equipment was not really needed.) As the last 80 feet was drilled, considerable drill string drag and torque developed. The casing point was actually 120 feet below the predicted depth. The drilled-solids concentration in the drilling fluid greatly increased. Wiper trips were needed between each logging run prior to setting casing. So much drill string torque and drag were experienced that the research team was asked to return to the rig and turn on the mud cleaner. So many solids were discarded by the desilters and presented to the mud cleaner screen that an API 200 screen could not handle all of the flow. An API 150 screen was mounted on the mud cleaner to reduce the drilled solids during two circulations. Then a final cleanup was made with two circulations using an API 200 screen. The drilled-solids concentration could not be returned to the lower levels achieved during earlier drilling, but a sufficient number of drilled solids were removed to allow casing to be run without incident and cemented in the borehole.

Note that this lucky event of turning off the mud cleaner was probably the reason that mud cleaners became commercial. No drilling program schedules stuck pipe. None was programmed here. The research plan should have included a procedure to validate the mud cleaner performance. Since no problems were encountered and none were expected (although stuck pipe and lost circulation are common with 6000-psi overbalance in a well bore), no comparative data were acquired to prove that the mud cleaner was performing properly—until the machine was luckily shut off. This was also a great lesson in planning research for the drilling processes. The research team concentrated on keeping the drilling fluid in good shape and minimizing the impact of drilled solids; unfortunately, the primary function should have been to prove the machine beneficial. At that time, not all drilling personnel believed that drilled solids were detrimental or evil.

The mud cleaner's U.S. patent—No. 3,766,997 (October 23, 1973), granted to J. K. Heilhecker and L. H. Robinson and assigned to Exxon Production Research Company—was found to be invalid because of prior art later discovered in the British Patent Office. A German inventor had been granted a patent on a similar device in the late 1800s. Although his invention had never been reduced to practice or used in the oil patch, and screens were much coarser in those days, the existence of the information in the public domain prevented collection of royalty for application of this technology. (As an interesting note, all of the companies providing mud cleaner service had offered to pay a nominal $5 a day per unit as a technology transfer fee for help in developing the product. This offer had been rejected and the service companies had been told that a much larger royalty payment would be required when the patent was issued. With an invalid patent, the service companies never had to pay a royalty and certainly did not pay a technology transfer fee.)

The first commercial application, less than a year after the initial patent was submitted, involved two wells, one production and one exploration, in the Pecan Island field in Louisiana. For the first time in that field, the intermediate long string of casing could be reciprocated during the cement job because of the lower drilled-solids concentration in the drilling fluid. The torque and drag in these wells were spectacularly lower than in previous wells.

When the mud cleaner was first introduced, many would try to decide whether to use a mud cleaner or a centrifuge. The problem with this decision is that mud cleaners do not compete with centrifuges in solids removal. In weighted drilling fluids, mud cleaners are designed to remove drilled solids larger than barite (larger than 74 microns). Centrifuges remove solids mostly smaller than most barite (less than 5 to 7 microns).

12.2 USES OF MUD CLEANERS

The principal use of mud cleaners has always been the removal of drilled solids larger than barite. Sufficient drilling fluid bypasses shale shakers so that even with 74-micron (API 200) screens on the shakers, many drilled solids are removed from a weighted drilling fluid. When linear motion shakers permitted API 200 screens to process all of the rig flow, mud cleaner usage declined rapidly. However, whenever a mud cleaner was used downstream, the screens were loaded with solids. Verification of larger solids bypassing the shale shaker has been evident from the prevalence of plugged desilters on a rig. Plugged desilters are very common. Usually, no one is assigned to unplug desilters, so they do not get unplugged.

These larger solids can come from holes in the main shaker screen or a variety of other places. One of the most prevalent is from the back tank (possum belly) of a shale shaker. Derrickmen are told to wash the screens off before a trip to prevent dried drilling fluid from plugging the shale shaker screens. The procedure usually involves opening the shale shaker bypass valve and dumping the accumulated solids into the tank immediately below the shaker. This compartment is usually called a sand trap, or settling tank. Unfortunately, all of these larger solids do not settle in this tank. Within minutes of starting circulation with a new bit on bottom, hydrocyclones plug. Another prevalent method of getting large solids downstream of a shale shaker is dumping the trip tank into the compartment below the shale shaker. Again, solids do not settle but do plug desilters. When solids plug the discharge port of a desilter, that desilter no longer removes solids. The solids-removal efficiency of the system decreases accordingly. If over half of the hydrocyclones are plugged, a properly designed removal system will suffer significantly; costs and trouble will increase accordingly.

A secondary use of mud cleaners is the removal of drilled solids from unweighted drilling fluids with a very expensive liquid phase—such as the initial application with the potassium chloride drilling fluid. In this case, the underflow from desilters is screened to remove solids larger than the screen opening. Solids and liquids pass through the screen and remain in the system. This is beneficial for nonaqueous drilling fluids as well as saline water–base drilling fluids.

In unweighted drilling fluids, the desilter underflow could be directed to a holding tank. A centrifuge could separate the larger solids for discard and return the smaller solids and most of the liquid phase to the drilling-fluid system. This method is easy to apply if a centrifuge is already available on a drilling rig. More drilled solids would be rejected by the centrifuge than by the mud cleaner screen; however, renting a centrifuge for this purpose may be more expensive. Both techniques have been used extensively in the field.

The cost of the liquid phase provides an incentive to recover as much of the fluid as possible: *but* this liquid phase from either a centrifuge or a mud cleaner screen throughput also recovers all of the small solids. In some wells these solids do not have a significant impact on drilling or trouble costs. In general, drilled solids have a much more obvious impact on drilling costs in weighted drilling fluids than in unweighted drilling fluids. In an unweighted drilling fluid, mud weights can be maintained around 8.8 to 8.9 ppg with proper solids control. Poor solids control, in equipment or arrangement, may make it almost impossible to keep the mud weight below 9.5 to 10 ppg. A 1-ppg difference will make a 520-psi difference in bottom-hole pressure at a depth of 10,000 feet. This reduces the drilling rate from chip hold-down and from rock strengthening. Not only does it prevent achieving a good drilling rate, but it also affects hole cleaning. With unweighted drilling fluids, the plastic viscosity can be reduced from 12 cP to 6 cP when the mud weight is decreased by removing solids. For a yield point of 10 lb/100 sq ft, this would increase the K value from 224 to 466 effective cP. This means that bore cleaning would be significantly improved. (See the discussion in Chapter 2 on hole cleaning.) Good solids control requires generating large cuttings and bringing them to the surface without deterioration.

In some areas, mud cleaner discards do contain almost all barite. This is an indication that the formations being drilled are dispersing into the drilling fluid. Formations containing large quantities of smectite drilled with a freshwater drilling fluid will tend to disperse into very small particles. This is an indication that the drilling fluid should be more inhibitive. Generally, the mud cleaner should be shut down and a centrifuge used to remove the very small particles.

Most operations go through serious expensive, agonizing learning experiences when switching from unweighted to weighted drilling fluids. Decisions to retain the liquid phase and entrained solids may save some money with liquid recovery but may also affect the final drilling cost because it more directly affects trouble costs. Trouble costs, or failure to make hole because of problems, are obvious to anyone watching operations. The increase in drilling cost because of poor performance with unweighted drilling fluids may be more subtle and not detected in a review of operations. For this reason, more attention seems to be directed toward solids control while drilling with weighted than with unweighted drilling fluids.

Avoid the common mistake of evaluating systems with immediate temporary cost evaluations in mind instead of total impact on drilling costs.

For example, an analysis of replacing barite with drilled solids appears to be an easy calculation. If drilled solids are allowed to accualate and increase mud weight, the barite additive cost is avoided until the mud weight approaches 11 ppg. The amount of barite "saved" is easily calculated. Even at 5 cents per pound, this barite savings cost is significant; HOWEVER, this will result in a much more expensive well. Review the problems associated with not removing drilled solids discussed earlier and expect to see most of those problems arise from weighting a drilling fluid with too many drilled solids.

12.3 NON-OILFIELD USE OF MUD CLEANERS

One use of mud cleaners that has been very profitable has been in microtunneling. Tunneling under roadways, lakes, or streams for pipelines, fiber-optic conduits, or other installations requires drilling with a circulating fluid. Frequently, the liquid is difficult to acquire and disposal is a problem. So these small drilling systems install a mud cleaner on top of a one-tank circulating system. The solids from the mud cleaner screen are removed, and the excess liquid is returned to the tank. The screens used in this application usually have larger openings and larger wire diameters than the screens used for oil well applications. The purpose of these screens is to dewater the cuttings and retain as much fluid as possible. Drilled solids in this application are not as detrimental as drilled solids in oil well drilling, so the openings in the screens do not need to be as small as with oil well drilling.

12.4 LOCATION OF MUD CLEANERS IN A DRILLING-FLUID SYSTEM

Mud cleaners are normally positioned in the same location as desilters in a drilling-fluid system. Frequently, the desilters, or hydrocyclones, are used in the unweighted portion of a borehole by diverting the underflow away from the mud tanks. When a weighting agent, barite or hematite, is added to the system, screens are placed on the mud cleaner shakers. Solids discarded from the hydrocyclones are sieved to discard solids mostly larger than barite and return solids smaller than the screen size with most of the liquid phase. Practical tip: Barite goes in, screens go on.

Another method has been used to create a mud cleaner using the main shale shaker, mostly on offshore rigs. When several linear, or balanced elliptical, shale shakers are needed to handle flow in the upper part of a well bore, fewer shakers can handle the flow after the hole size decreases and the mud weight is increased. Rigs have been modified so that as many as twenty 4-inch hydrocyclones have been mounted above one of the main shakers. The feed and overflow (light slurry) from the hydrocyclones are plumbed as usual (see Chapter 5 on Tank Arrangements). All of the desilter underflow is discarded in the unweighted part of the hole as usual, while the shale shaker is screening drilling fluid from the flow line. This is normally the largest flow rate expected while drilling the well. As the well gets deeper, a weighted drilling fluid is required, and usually the flow rate is lower. When barite goes in, a valve prevents flowline drilling fluid from going to one of the main shale shakers. The desilter underflow is diverted onto the shale shaker screen so that the shale shaker becomes a mud cleaner.

12.5 OPERATING MUD CLEANERS

When the first mud cleaners were introduced into the field, they had to be shut off during weight-up. A significant amount of barite was discarded during the first circulation. Actually, this revealed that the mud tanks were plumbed improperly. Drilling fluid was frequently pumped through mud guns from the additions or suction section back upstream to the removal tank. Barite can meet American Petroleum Institute (API) specifications and still have a large amount that will be removed with an API 200 (74-micron) screen. If the barite has passed through a drill-bit nozzle, the particles are split so that it will not be removed with such a screen.

One comment frequently heard when a weighted drilling fluid is initially passed through a mud cleaner is: "It's throwing away all my barite!" What creates such a comment? The mud weight is decreasing, and more barite than normal is required to maintain mud weight. When solids are removed from a drilling fluid (barite or drilled solids), mud weight will decrease. Actually, removal of solids larger than 74 microns is beneficial to drilling a trouble-free hole, whether those solids are drilled solids, barite, gold, silver, or diamonds. Those solids make a poor, incompressible filter cake and lead to stuck drill strings. The appearance of the screen discard from a mud cleaner is similar to the underflow, or heavy slurry, from a centrifuge. Although visually it appears to have mostly barite, tests will reveal that this is not true.

API specifications for barite state that 3% by weight may be larger than 74 microns. If 100,000 lb of barite is added to a drilling fluid during a weight-up, 3000 lb of barite could be removed by an API 200 screen. This is one reason that fluid from the additions compartment should not be circulated upstream. The main shale shaker will also discard most of this size of barite from an API 200 screen. The barite is not as noticeable because of the quantity of drilling fluid normally clinging to the shaker discard.

The screen discard from a mud cleaner looks like the underflow from a centrifuge. The solids concentration is usually around 60% volume, and that of liquid about 40%. This initially appears to be an irrationally large quantity of liquid. Researchers frequently pack columns with loose sand to examine various oil recovery procedures. Dry sand is poured into a cylinder while vibrating the pack. If the porosity of the sand pack is 33–35% volume, the packing is about as tight as can be achieved. Loose sand on the beach, immediately after a wave has washed back out to sea, has about 40% volume of water in it. It can be scooped up without water draining from the sand pile. Mud cleaner screen discards and the heavy, or underflow, slurry from a centrifuge have about the same volume percentage of liquid.

The mud cleaner is designed to continuously process drilling fluid just like the main shale shakers. The mud cleaner screen keeps larger particles from entering the system. Operating the equipment for only part of the time allows solids to remain in the drilling fluid system. These solids grind into smaller particles that become more difficult to remove. Centrifuges will be able to remove these solids from a weighted drilling fluid, but they generally do not process all of the rig flow. The mud cleaner can remove these solids before they grind into smaller particles if the mud cleaner is used continuously.

Again, note that the mud cleaner and the centrifuge are complementary to each other—not competitive with each other. The mud cleaner removes solids larger than barite; centrifuges remove solids smaller than most barite.

12.6 ESTIMATING THE RATIO OF LOW-GRAVITY SOLIDS VOLUME AND BARITE VOLUME IN MUD CLEANER SCREEN DISCARD

An estimate of the low-gravity solids content of the mud cleaner screen discard can be made by weighing the discard. Since the solids concentration will be around 60% volume, the mud weight will be a reasonable predictor of the low-gravity solids concentration. For low-gravity solids with a specific gravity (SG) of 2.6 and a barite SG of 4.2, the equation to determine the low-gravity solids concentration, V_{LG}, is:

$$V_{LG} = 62.5 + 2.0\,V_s - 7.5\,\text{MW}$$

where V_s is the volume percentage of total suspended solids, and MW is the mud weight, in ppg.

Assume that the V_s is 60% volume for a mud weight of 19.0 ppg, and the equation will calculate that the volume concentration of low-gravity solids is 40%. This means that 20% of the volume is barite. So, twice the volume of low-gravity solids is being discarded as barite. Even if the actual V_s were 57% instead of 60%, the low-gravity solids concentration would be 34% volume. In most cases the decision to continue running the mud cleaner would not be affected by this inaccuracy. Even if the barite in the discard exceeds the low-gravity concentration, the benefits of removing those larger solids will be evident. Accurate results, of course, can be obtained by retorting the solids; but this is a tedious process because the solids are difficult to handle—care must be taken to obtain a representative sample and pack it into the retort cup without leaving void spaces. A much more accurate method is to use the gravimetric procedure, in which larger quantities can be used and no volume measurements are made.

Generally the discard from a mud cleaner screen is relatively dry and contains around 60% volume of (%vol) solids. The density of this slurry can be measured with a mud balance, but the solids concentration is difficult to measure with a retort. Generally decisions about the performance of a mud cleaner or a centrifuge can be made by weighing the heavy, or underflow, discard from a centrifuge or the screen discard. Very accurate measurements are not really needed. The chart in Figure 12.3 allows an estimate of the concentration of low-gravity solids.

For example, if the mud weight of a mud cleaner shaker screen discard weighs 18.0 ppg, the low-gravity solids concentration would be about 40%vol if the solids concentration were 58%vol. Barite would be 18%vol. The concentration of low-gravity solids would be about 48%vol if the discard total solids concentration is 60%vol. Barite would be only 12%vol for this condition. Note that in either case, the mud cleaner is doing a great job of removing drilled solids or low-gravity solids from the drilling fluid. An accurate measurement is not needed to make the decision to continue running the mud cleaner.

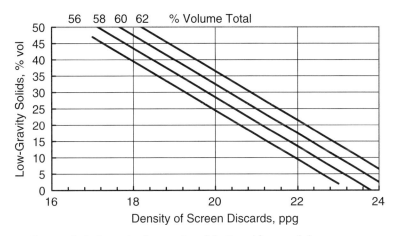

Figure 12.3. Estimating low-gravity solids discard from mud cleaner screens.

12.7 PERFORMANCE

Drilling soft, dispersible shales with a freshwater drilling fluid usually results in drilled solids that cannot be removed with mud cleaners. In these cases, centrifuges should be planned for use in weighted drilling fluids. Usually, if solids are being removed with shale shakers, a mud cleaner will probably be beneficial. Solids removed by mud cleaners will cover a wide range of quantities depending on formations drilled, borehole stability, dispersion of solids as they move up the borehole, type of drill bit, type of drilling fluid, and other variables.

Some data acquired from one well are presented in Table 12.1. The pressure at the entrance to the desilters was varied and the mud cleaner discard examined for drilled solids and barite. With an unweighted drilling fluid, a head of 75 feet was recommended for this brand of desilter. This head creates a balanced hydrocyclone with good separation of low-gravity solids; however, this may not necessarily be true for weighted drilling fluids.

In Table 12.1, during the first circulation after the new drill bit has reached bottom, higher quantities of drilled solids are discarded by the mud cleaner screen as the head is increased on the desilter feed. One method of analysis is to compare the concentration of barite lost with the drilled solids discarded. In Figures 12.4, 12.5, and 12.6, the total solids discarded and the quantity of drilled solids are shown as functions of the desilter manifold pressure. The lowest ratio of barite to drilled

Table 12.1
Drilling with 9⅞" Bit Between 9300 Feet and 9400 Feet with an 11-ppg Drilling Fluid and Six 4-inch Hydrocyclones Above an API 200 Screen

Cyclone Manifold Pressure (psi)	Manifold Head (ft)	Screen Discharge (sec/qt)	Discharge Density (ppg)	Volume % Solids	Drilled Solids Removed (lb/hr)	Barite Discarded (lb/hr)
First circulation: Bottoms up after TIH with new bit						
33	57.7	16	16.8	58	640	108
48	83.9	7	17.7	58	1275	552
60	104.9	7	17.5	58	1317	484
72	125.9	6	17	58	1659	368
Second circulation						
32	55.9	25	16.4	50	308	132
43	75.2	19	16.7	58	547	79
55	96.2	9	17.6	58	1008	403
72	125.9	5	17.5	58	1844	678
Third circulation						
38	66.4	40	16.6	58	263	32
50	87.4	20	17.1	57	471	138
60	104.9	10	17.3	58	951	292
75	131.1	9	17.3	58	1057	324

TIH = tool in hole.

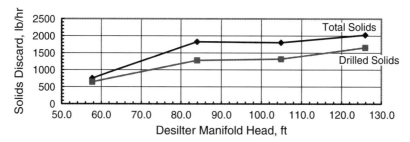

Figure 12.4. Solids discard from mud cleaner screen: First circulation.

solids occurs when the manifold head (or pressure) is low. This might be misleading, however. Larger quantities of drilled-solids discards are much more desirable, even if some additional barite is lost. Higher manifold pressures are preferred to eliminate the largest quantity of drilled solids from the drilling fluid.

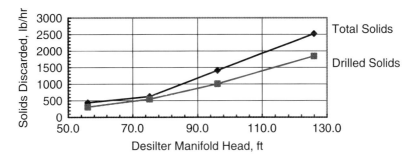

Figure 12.5. Solids discard from mud cleaner: Second circulation.

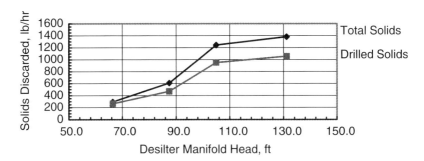

Figure 12.6. Solids discard from mud cleaner screen: Third circulation.

The bottoms-up sample indicates that the smallest discard rate has the highest ratio of drilled solids to barite. However, the largest flow rate of drilled solids, 1659 lb/hr, is still a flow rate 4.5 times as large as the barite flow rate (Table 12.2).

A word of caution is appropriate here. The purpose of solid-control equipment is to remove drilled solids. Economics certainly justify sacrificing a small additional amount of barite for good removal of drilled solids. If solids concentrations are reduced by dilution only, the cost would be many times higher than sacrificing some barite to remove these drilled solids. So, evaluating performance by comparing discard ratios can be very misleading. This is discussed in more detail at the end of this section.

12.8 MUD CLEANER ECONOMICS

The question frequently arises, "Would it be cheaper to simply jet drilling fluid from the system instead of using solids-control equipment?"

Table 12.2
Discard Rates

Drilled Solids Removed (lb/hr)	Barite Discarded (lb/hr)	Ratio of Drilled Solids to Barite
640	108	5.9
1275	552	2.3
1317	484	2.7
1659	368	4.5
308	132	2.3
547	799	0.7
1008	403	2.5
1844	678	2.7
263	32	8.2
471	138	3.4
951	292	3.3
1057	324	3.3

To answer this question, examine the discard rates in Table 12.2 and compare them with the volume of drilling fluid that must be discarded to eliminate the same volume of drilled solids. Cost of drilling-fluid ingredients vary from company to company and from contract to contract. Rather than actually calculate a cost, the comparison will be made between using the mud cleaner and pumping drilling fluid to eliminate the same quantity of low-gravity solids from the system using the measurements in the preceding tables.

An 11-ppg drilling fluid with 13%vol total solids would contain 6.4%vol low-gravity solids and 6.6%vol barite. With the first circulation and 60 psig manifold pressure, 1317 lb/hr of drilled solids and 484 lb/hr of barite are discarded. Since the solids in the drilling fluid are presented in terms of volumes instead of weights, the mass flow rate needs to be translated into volumes.

With a SG of 4.2, the density of barite would be 4.2 times the density of water (8.345 lb/gal), or 35.05 lb/gal. With a SG of 2.6, the density of low-gravity solids is 2.6 times the density of water, or 21.7 lb/gal. The volume flow rate of solids would be the mass flow rate divided by the density. The volume flow rate of barite would be (484 lb/hr)/(35.5 lb/gal), or 13.6 gal/hr, and the volume flow rate of low-gravity solids would be (1317 lb/hr)/(21.7 lb/gal), or 60.7 gal/hr.

In this drilling fluid there is 6.4%vol drilled solids. If the discarded low-gravity solids are 6.4% of the mud volume and 60.7 gal/hr of drilled solids are discarded, the total volume discarded to contain that amount of solids would be:

$$V_{LG} \text{ discarded} = 6.4\%(\text{volume of 11-ppg drilling fluid})$$

or, 60.7 gal/hr = (0.064)(volume of 11-ppg drilling fluid). The volume of 11-ppg drilling fluid discarded is 948 gal/hr.

This quantity of drilling fluid contains 6.6%vol barite, or (0.066) (948 gal/hr) = 62.6 gal/hr of barite. Converting barite volume to barite weight, (62.6 gal/hr)(35.05 lb/gal), means that 2194 lb/hr of barite would be lost to eliminate the 1317 lb/hr of low-gravity solids. Compare this with the measured loss of only 484 lb/hr of barite loss from the mud cleaner screen. In addition to the cost of the excess barite, the pit levels would decrease by 948 gal/hr. This is equivalent to losing (948 gal/hr) (24 hr)/(42 gal/bbl), or 541.7 bbl of drilling fluid daily.

Clearly, concentrating the low-gravity solids with equipment is preferable to dumping drilling fluid to eliminate drilled solids (Table 12.3).

Table 12.3
Eight 4-inch Hydrocyclones Above an API 150 Screen

Cyclone Manifold Pressure (psi)	Manifold Head (ft)	Screen Discharge (sec/qt)	Discharge Density (ppg)	Volume % Solids	Drilled Solids Removed (lb/hr)	Barite Discarded (lb/hr)
First circulation: Bottoms up after TIH with new bit						
38	66.4	16	16.8	58		108
52	90.9	7	17.7	58	1275	552
60	104.9	7	17.5	58	1317	484
78	136.4	6	17	58	1659	368
Second circulation						
40	69.9	25	16.4	50	308	132
50	87.4	19	16.7	58	547	79
60	104.9	9	17.6	58	1008	403
70	122.4	5	17.5	58	1844	678
Third Circulation						
42	73.4	40	16.6	58	263	32
55	96.2	20	17.1	57	471	138
60	104.9	10	17.3	58	951	292
74	129.4	9	17.3	58	1057	324

12.9 ACCURACY REQUIRED FOR SPECIFIC GRAVITY OF SOLIDS

Determination of drilled solids in a drilling fluid depends on an accurate determination of the mud weight, the total solids in the drilling fluid, and the density of the drilling-fluid ingredients. For example, with a freshwater 11-ppg drilling fluid containing 2.6 SG low-gravity solids and 4.2 SG barite, a change in only 1%vol measured solids concentration makes a 2%vol change in calculated low-gravity solids. In Table 12.4, for an 11-ppg drilling fluid, a 13%vol solids concentration would indicate 6%vol, and a 12%vol solids concentration would indicate 4%vol.

12.10 ACCURATE SOLIDS DETERMINATION NEEDED TO PROPERLY IDENTIFY MUD CLEANER PERFORMANCE

A small change in the density of the low-gravity solids and barite affects the low-gravity solids calculation. Table 12.5 indicates that 14%vol total nonsoluble solids in 11-ppg drilling fluid could have between 6.3 and 9.8%vol low-gravity solids, depending on the density of the barite and the low-gravity solids.

Table 12.4
Solids Concentration in an 11-ppg Drilling Fluid

V_{LG} (vol %)	V_s (vol %)	MW (ppg)
8	14	11
6	13	11
4	12	11
2	11	11

V_{LG} = volume of low-gravity solids; V_s = volume of total suspended solids; MW = mud weight.

Table 12.5
Values for an 11-ppg Drilling Fluid with 14% Volume of Total Solids

Density of LGS	Density of Barite	Vol Fraction of LGS
2.4	4.2	7.1
2.9	4.2	9.8
2.4	4.0	6.3
2.9	4.0	9.1
2.6	4.2	8.0

LGS = low-gravity solids.

Obviously the density of the low-gravity solids and the barite must be known for an accurate determination of the concentration of solids in the drilling fluid or the discard from any solids-control equipment. The most common error in viewing the discard from a mud cleaner screen is the conclusion that large quantities of barite are contained in the discard. This is the reason that the new API solids analysis recommends that the density of the barite and solids on the shaker screen be determined on location.

One frequent comment while running a mud cleaner is that the mud cleaner is discarding so much barite that large quantities of it must be added to keep the mud weight constant. Actually, discarding any solids from a drilling fluid decreases density. Replacing low-gravity solids with barite will decrease the total solids concentration, make filter cakes more compressible, decrease the propensity for stuck pipe and lost circulation, and improve the possibility of faster drilling (by increasing the founder point). However, some formations that disperse significantly in a freshwater drilling fluid and very few large solids arrive at the surface. Solids discarded from the mud cleaner screen should be examined to determine whether drilled solids are being removed. If they are not, centrifuges should definitely be considered to remove low-gravity solids.

Actually, mud cleaners frequently do not discard as much barite as do the main shale shakers. Visual observations tend to give an erroneous view. The discard from a mud cleaner screen (and the underflow from a decanting centrifuge) contains around 60%vol solids, while the shale shaker discard contains around 35–40%vol solids. The more liquid (shale shaker) discard does not appear to contain as much barite. However, the shale shaker discard may concentrate barite or deplete barite from the flowline drilling fluid. The way drilling fluid discarded from a shale shaker appears does not reveal the concentration of barite.

12.11 HEAVY DRILLING FLUIDS

Various arbitrary procedures seem to be developed by operating personnel in the field. Often a centrifuge is run for a specific number of hours per day, or a hydrocyclone bank is used for only a short period. Mud cleaners seem to also attract a variety of erroneous rules of thumb.

In heavily weighted drilling fluids, above 13–14 ppg, mud cleaners are frequently shut off because of excessive barite discards. When a mud tank system is plumbed incorrectly, specifically when mud guns transport freshly added barite fluid into the removal system, a significant amount of barite will be discarded. Barite that meets API specifications may have as much as 3% weight larger than 74 microns. If a thousand 100-lb sacks of barite are added to a drilling-fluid system during drilling of an interval (used as clean drilling fluid to dilute remaining drilled solids), 3000 lb of the 100,000 lb added will be larger than 74 microns. Now the question becomes, what damage will these large particles create? Any particles larger than 74 microns, whether they are drilled solids, barite solids, diamonds, or pieces of gold, will make a poor-quality filter cake and enhance the probability of incurring all of the problems created by a large quantity of drilled solids in the drilling fluid.

In a drilling-fluid system that is arranged properly, barite will have traveled down the drill string, passed through the bit nozzles, and then traversed the borehole before reaching the mud cleaner. Any particle larger than 74 microns should be removed from the drilling-fluid system. Barite larger than 74 microns should be removed from the system. In situations in which the system is not arranged properly, the mud cleaner might be shut down for two or three circulations after a significant weight-up. This is a good test for correct mud tank arrangement. If the mud cleaner starts discarding barite immediately upon weight-up, the system has serious flaws. The objective is to remove large particles from the drilling fluid so the filter cake will be thin, slick, impermeable, and compressible. With solids larger than 74 microns in the cake, this objective will not be achieved.

One common problem with sieving hydrocyclone underflow with a mud cleaner is the tendency of the material to dewater, or deliquefy, before reaching the discard end of the screen. Clumps of material traveling down a screen with too little liquid will not separate solids properly. A small reflux of drilling fluid from the hydrocyclone overflow should be sprayed onto the screen to break these clumps into material that can be separated. A spray of water or oil (depending on the liquid phase) could be used but will generally dilute the system too much. The drilling fluid reflux enhances the screen capability to separate solids from the slurry. If the slurry remains liquid until all separation has been completed, the mud cleaner screen should not remove any more barite than would be removed on the main shaker with the same-size screen.

There are situations in which the solids disperse as they travel up the borehole. Usually only slivers and cavings from the borehole wall are removed by the main shaker. When this occurs, a centrifuge is needed to remove the very small particles. A change in drilling-fluid systems might be contemplated before drilling the next well.

CHAPTER 13

CENTRIFUGES

Eugene Bouse

Consulting Engineer (Drilling Fluids, Solids Control, Drilling Waste Minimization)

13.1 DECANTING CENTRIFUGES

Decanting centrifuges are mechanical devices used for the separation of solids from slurries in many industrial processes. In oilwell drilling, centrifuges are used to condition drilling fluids by dividing the fluid into high-density and low-density streams, permitting one to be separated from the other. The division is achieved by accelerated sedimentation. As the drilling fluid is passed through a rapidly rotating bowl, centrifugal force moves the heavier particles to the bowl wall, where they are scraped toward the underflow (heavy slurry) discharge ports by a concentric auger, also called a scroll or conveyor, which rotates at a slightly slower rate than the bowl. The separation of the heavier particles divides the processed fluid into two streams: the heavy phase, also called the underflow or cake; and the lighter phase, which is called the overflow, light slurry, effluent, or centrate (Figure 13.1).

If time were not a factor, sedimentation could be accomplished in any container. To reduce the time required, the geometry could be manipulated to limit the depth of the fluid and, consequently, the distance the settling particles would have to traverse before reaching the bottom of the container (Figures 13.2, 13.3, and 13.4). If this approach were used, a scraping device could remove the settled solids from the bottom of the container, and one end of the container could be sloped to permit the solids to be removed from the liquid by the scraping mechanism.

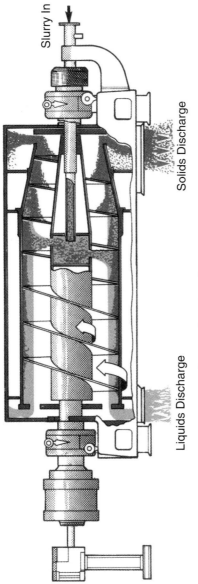

Figure 13.1. Centrifuge cutaway.

Figure 13.2. Deep sedimentation vessel.

Figure 13.3. Shallow sedimentation vessel.

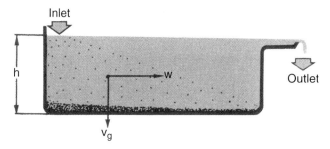

Figure 13.4. Sedimentation vessel with outlet.

They could then be left to dry on this "beach," known as the drainage deck, before being discharged.

Conditions on drilling rigs obviously preclude the use of this approach. The centrifuge, however, utilizes essentially the same process. The inner surface of the rotating bowl receives the settled solids, as the container bottom does in the procedure just described, and the scroll functions as the scraper, conveying the settled solids to, and across, the beach, where they are dried by the removal of free liquid, then to the underflow discharge ports. Essentially, the centrifuge design wraps the surface corresponding to the bottom of a sedimentation container around the scraping device, the conveyor (Figures 13.5, 13.6, and 13.7).

Basically, a decanting centrifuge is a simple machine: a rotating bowl containing a concentric conveying scroll. The rotation of the bowl forces

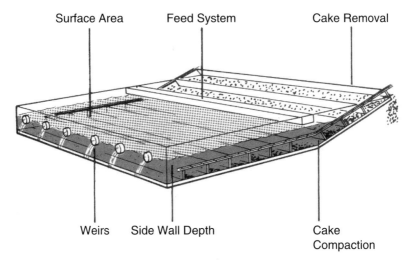

Figure 13.5. Sedimentation tray.

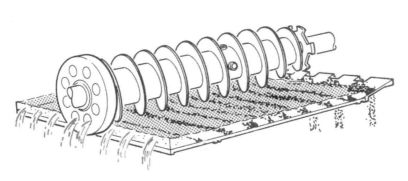

Figure 13.6. Sedimentation tray with auger.

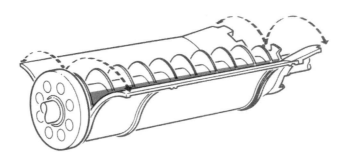

Figure 13.7. Sedimentation tray wrapped around auger.

solids to the wall, where the scroll transports the solids, as previously described. Two different bowl designs are available: conical, in which the entire bowl is cone shaped; and cylindrical/conical, in which the effluent (or light slurry) end of the bowl is cylindrical. This configuration is preferred for drilling-fluids applications because it offers greater capacity.

The elevated centrifugal forces created by the rotation of the bowl accelerate the sedimentation process so that separation that might take hours or days under the normal gravitational force of $1\,g$ in an undisturbed container is achieved in seconds at the 400–$3000\,g$ generated by the centrifuge.

Inasmuch as sedimentation is used to achieve the separation, an understanding of the factors influencing the process is required for the proper use of centrifuges. An Irish mathematician and physicist, Sir George Stokes, who described the basic principals of fluid mechanics in the mid-nineteenth century, defined sedimentation in Stokes' law:

$$v = \left[kgD^2(d_s - d_f)\right]/\mu$$

where

$v =$ terminal velocity
$k =$ a constant that is dependent on the units in use
$g =$ the gravitational constant
$D =$ the diameter of the solid particle
$d_s =$ the density of the particle
$d_f =$ the density of the fluid
$\mu =$ the viscosity of the fluid

The equation permits the calculation of the maximum sedimentation rate achieved by spherical particles. Note that it confirms what is intuitively clear, that particles settle more rapidly in less viscous fluids, and that heavy particles settle more rapidly than light particles. The utility of Stokes' law lies in the fact that when used with proper and consistent units, it permits independent evaluation of each of the variables: particle size, the density of the particle and of the fluid, and the viscosity of the fluid. (For additional discussion of this topic, see Chapter 8 on Settling Pits/Sand Traps.)

The mass of a particle depends on its size and density. While there is a technical difference, mass is essentially equivalent to weight. Stokes' law shows that at any given viscosity and fluid density, the sedimentation rate depends directly on the mass, or weight, of the particle.

13.1.1 Stokes' Law and Drilling Fluids

Drilling fluids normally contain two categories of solids: (1) commercial clays and drilled solids, both low gravity, with specific gravities (SGs) of about 2.6, and (2) weighting agents, usually barite, with an assumed SG of 4.2. If all of the solids particles were of the same size, centrifuges could be used to separate the weighting agent from the low-gravity solids, because the barite particles, due to their higher SG, would be heavier. Drilling fluids, of course, are not slurries of particles of equal size. Weighted drilling fluids always contain solids of both categories, ranging from colloidal particles too fine to settle, even in pure water, to particles 70 microns (μ) in size and larger. Consequently, *the centrifuge cannot separate barite from low-gravity solids*. What it does, when operated properly, is separate larger barite particles from smaller ones and larger low-gravity-solids particles from smaller ones. Failure to recognize this very important fact frequently leads to the misuse of centrifuges.

Stokes' law shows that the terminal velocities of barite and low-gravity-solids particles are equal when they have equal mass. This equality exists when the low-gravity particle is approximately 50% larger than the barite particle. This leads to the conclusion that if a centrifuge is achieving a D_{50} cut point of, for example, $4\,\mu$ on barite, it will be making a D_{50} $6\,\mu$ cut on low-gravity solids. In other words, most of the barite particles larger than $4\,\mu$ and the low-gravity-solids particles larger than $6\,\mu$ are routed to the underflow, while the smaller particles remain with the centrate.

13.1.2 Separation Curves and Cut Points

Solids-separation performance is often described using cut points. The cut point is the size at which a stipulated percentage of the feed solids are separated. If a percentage is not stated, it is usually assumed to be 50%. For example, if a shale shaker is removing 50% of the $100\,\mu$ particles and 50% remain in the mud, the D_{50} cut point is said to be $100\,\mu$. If it is removing 90% of the $120\,\mu$ particles, its D_{90} cut point is $120\,\mu$.

Figures 13.8 and 13.9 indicate the importance of this concept and also the significance of the "sharpness" of cuts. When interpreting these graphs, note that the solids particles of the size range represented by the

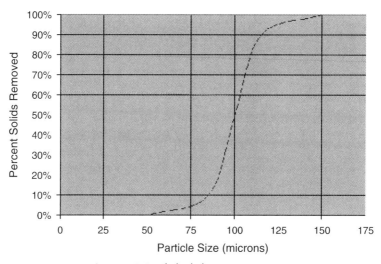

Figure 13.8. Shale shaker separation curve.

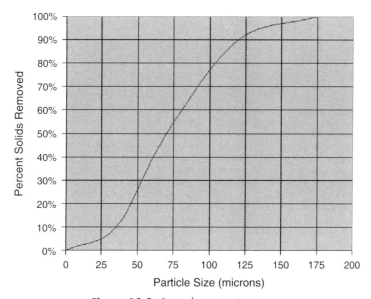

Figure 13.9. Desander separation curve.

area to the right of the curve are separated, while those in the area to the left remain in the fluid. Note also that in the shaker screen example (Figure 13.8), there is little difference between the cut points at 10% and at 90%. The closer the cut point curve is to vertical, the "sharper" the cut is considered to be. With an ideal screen and perfect screening (neither of which exist in the real world), the cut point curve would be a vertical line and all solids smaller than the designed opening size would remain in the fluid, while all of those larger than that size would be separated. Typically, cut point curves for hydrocyclones are far from vertical, which is to say that if the D_{50} cut point is, for example, $25\,\mu$, significant quantities of material larger than $25\,\mu$ will remain in the system, while significant quantities of finer material will be separated. This is also true of centrifuges operated under less than ideal conditions, as they usually are.

Ideal conditions include maximum pool depth, maximum retention time, maximum difference between particle and liquid densities, minimum solids content, and minimum viscosity. Under these ideal conditions, many centrifuges are capable of achieving a D_{50} cut point of $2\,\mu$ on barite. Under the conditions typically encountered in drilling applications, actual cut points can be expected to be significantly higher.

13.1.3 Drilling-Fluids Solids

Commercial solids are used in drilling fluids to provide desired density, viscosity, and filtration control. Additional drilled and sloughed solids become part of the fluid during the drilling process. These formation solids, as well as colloidal barite particles, when present in excessive concentrations, are detrimental to drilling-fluid performance and are considered to be contaminants. The coarser solids, though they can be troublesome, are ordinarily the least injurious to drilling-fluid performance and are the most easily separated.

Barite and bentonite are the most widely used commercial solids in drilling fluids. Current American Petroleum Institute (API) specifications (API 13A, 1996) permit as much as 30% of barite to be finer than $6\,\mu$ in size, and much of this 30% can be assumed to be colloidal when purchased. Pure barium sulfate is a very soft mineral, but the hardness of commercial barite depends on the impurities associated with it. Much of it is rather soft, and particle size can diminish rapidly with use. After several hours, days, or weeks in the mud system, more of the finer material can be expected to become colloidal. Bentonite is ground much finer than barite and can be assumed to be colloidal when purchased.

Drilled solids are unavoidably incorporated into the drilling fluid while drilling. In softer sediments drilled with water-based muds, significant proportions of the drilled material are dispersed and can first reach the surface as colloidal particles, too fine to be separated from the base fluid. Coarser particles, if recirculated, can be expected to break up before returning to the surface. Drilled and sloughed solids that are not removed during their first passage through the surface mud system are unlikely to be separated later.

The solids particles that cause solids problems in oil well drilling are those that create viscosity problems and contribute to poor hole conditions. These are the finest solids, the colloids and ultra-fine solids that, because of their small size and great number at any given solids concentration, have a disproportionate amount of surface area per unit of volume. These solids, which are generally considered to be the most detrimental to drilling-fluid performance, are too fine to be separated by screens or hydrocyclones. Their concentration can be reduced only by dilution or centrifuging.

Note that solids surface area and the concentration of solids particles per unit of liquid volume, rather than the solids volume itself, are the usual sources of solids problems. Consequently, while retort solids can provide clues to the possible causes of drilling-fluid problems, solids problems can arise due to decreasing particle size even though the concentration of solids in the fluid remains unchanged. As particle size decreases, the resultant increase in solids area and number of particles increases plastic viscosity and can create or exacerbate hole problems, even though the solids concentration has not increased. Figures 13.10 and 13.11 illustrate the increase of surface area and number of particles corresponding to a reduction in average particle size of a given volume of solids.

13.2 THE EFFECTS OF DRILLED SOLIDS AND COLLOIDAL BARITE ON DRILLING FLUIDS

Consideration must be given to the effects of the presence of drilled solids and barite on different types of drilling fluids: water-based and nonaqueous fluids (NAFs), unweighted and weighted.

The effects of solids in unweighted fluids, regardless of what their base fluid may be, do not present a problem unless their concentration is allowed to reach excessive levels. With water-based fluids, many operators limit drilled-solids content to 5% by volume. Assuming the presence of approximately 2% bentonite, this is a total of 7% low-gravity solids. Most experts agree that in water-based fluids, concentrations of low-gravity solids in excess of 10% make hole trouble likely. Therefore, in

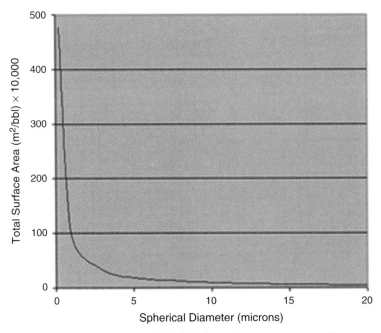

Figure 13.10. One barrel of drilled solids: Surface area versus particle size.

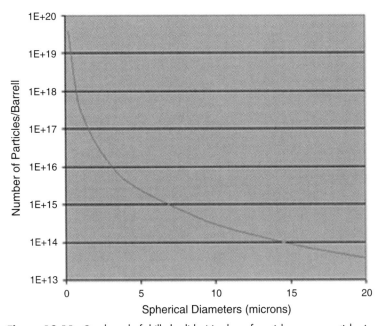

Figure 13.11. One barrel of drilled solids: Number of particles versus particle size.

unweighted water-based fluids, drilled-solids content of 8% or more by volume can be considered to be excessive. Any solids-related problems encountered with these fluids are the result of depending entirely on drilled solids to achieve densities in excess of 9.4–9.6 lb/gal, the density of a fluid containing 9–10% low-gravity solids in freshwater. At lesser concentrations of low-gravity solids, particle size appears to be irrelevant.

NAFs can tolerate higher solids concentrations. Drilled-solids concentrations as high as 12% by volume may be acceptable with these fluids when they are unweighted. This corresponds to an unweighted density of about 9.2 lb/gal.

Solids problems are much more frequent and serious in weighted fluids because of their higher solids content, their reduced tolerance for fines, and the fact that degradation of larger particles leads to a progressive increase in the concentration of fine particles. Excessive concentrations of fine and colloidal solids are known to reduce penetration rates and, by reducing filter cake quality, lead to troublesome hole conditions by increasing torque, drag, and the risk of sticking the drill string. In order to avoid these problems, barite—or an alternative weighting agent—should be used when the desired density is above the 9.6 lb/gal that can be reached with water-based fluids, or 9.2 lb/gal with NAFs.

13.3 CENTRIFUGAL SOLIDS SEPARATION

Before hydrocyclones and centrifuges became available for drilling applications, shale shakers and dilution were the only means of controlling the solids content of drilling fluids. Consequently, solids too fine to be separated by the shaker screens in use could be controlled only by dilution. During drilling with weighted muds, once the solids content reached the maximum acceptable level, the continuing and unavoidable incorporation of drilled solids made it necessary to add a continuous stream of water to control viscosity, while adding barite to control the mud weight. This was, obviously, a costly procedure that generated large quantities of excess drilling fluid.

The centrifuge, in splitting the processed fluid into two streams—the underflow, or "cake," containing the coarser solids; and the overflow, centrate, or effluent, containing most of the liquid and the finer particles—provides a means of selectively removing the finest, most damaging, solids from the drilling fluid. The removal of these solids in order to control rheology and filter cake quality is the primary reason for centrifuging weighted drilling fluids. When the finest solids are not removed, the only alternative means of reducing their concentration is dilution, always an expensive process.

Centrifuging weighted drilling fluids routes the coarser solids (both barite and low gravity) to the underflow, and the finer solids (bentonite, barite, and low gravity) to the overflow. Separation of the overflow, consisting of the finer solids together with most of the processed liquid and the dilution fluid, reduces the concentration of the viscosity-building solids, alleviates solids problems, and reduces the need for dilution.

This application is often described as *barite recovery,* a term that does not accurately describe the process, leads to confusion, and is frequently the reason for improper centrifuge use. The validity of the term depends on the preliminary acceptance of the idea that the fluid entering the centrifuge would otherwise be discarded and that the barite is recovered by the centrifuge. Few, if any, drilling people think of centrifuging in these terms. Centrifuges, like shale shakers and hydrocyclones, are solids-removal devices. Centrifuging weighted muds while drilling is correctly thought of as an alternative to dilution for the reduction of viscosity; not as a means of recovering barite from discarded fluid. Another objection to the phrase is that it lends support to the idea that barite recovery is the reason for centrifuging. It is not. The objective of centrifuging in this manner is the removal of colloids and ultra-fine solids to improve drilling-fluid quality. A third objection is that the use of the term tends to create the totally erroneous impression that the process separates barite from low-gravity solids and that the recovered material is all barite. A natural consequence of this belief is that the underflow from barite-recovery centrifuges is sometimes stored and used to weight up freshly prepared drilling fluids. Inasmuch as the recovered slurry often contains high concentrations of drilled solids, it can be severely contaminated, and is rarely suitable for reuse.

A much better term for the process of discarding the centrate while returning the overflow to the mud system is *traditional centrifuging.*

Refer again to Figure 13.1. The mixture of feed mud and any dilution fluid enters the acceleration chamber, or feed chamber, from which it is ejected through the feed ports by centrifugal force. Centrifugal force then carries the slurry to the pool, or pond, where the increased centrifugal forces produced by the rotation of the bowl cause the larger, heavier particles to settle to the bowl wall. For larger solids, this happens almost immediately, while it takes longer for the smaller solids that are large enough to settle. Solids that reach the wall are scraped toward the beach (drainage deck) and solids-discharge ports by the scroll. The pool is the mud in the bowl at any given time, and the beach is the area between the end of the pool and the solids-discharge ports. During their passage across the beach, most of the free liquid is removed from the solids. The discharged solids will, unavoidably, be wet with adsorbed liquid, but no free liquid should be present. The degree of dryness of the solids in the cake is primarily a function of solids size, the characteristics of the feed fluid, and the operating parameters of the centrifuge. Smaller particles have greater surface area per unit of volume and consequently adsorb more liquid.

The depth of the pool is controlled by the adjustment of the weirs, or effluent discharge ports, at the large liquid-discharge end of the machine. Increasing pool depth increases residence time and separation efficiency while reducing flow capacity. On the other hand, if the other parameters are unchanged, reducing pool depth decreases residence time and the separation of finer particles.

Contour bowls, those with a cylindrical shape for part of their length, as opposed to entirely conical bowls, are able to handle higher feed rates at any given cut point. Adjustment of the level of the solids (or cake)-discharge ports controls the flow capacity. When the pool depth reaches the level of these ports, the floodout point has been reached, and liquid is lost with the separated solids. This is ordinarily undesirable.

The relative motion between the scroll and the bowl, which controls the rate at which cake is removed from the machine, is set by the gearbox. Typically available gearbox ratios include 40:1, 52:1, 80:1, and 125:1. In each case, the scroll makes one less rotation than the bowl at the specified number of bowl rotations. For example, at 80:1, the scroll rotates 79 times each time the bowl rotates 80 times. Solids conveyance is faster at the lower ratios. The relative conveyor rpm can be calculated by dividing the rpm of the bowl by the gearbox ratio. For example, with a 40:1 gearbox, a bowl rotating at 1800 rpm has a differential speed of 45 rpm.

Although the fluid within the bowl is rotated rapidly, it is important to note that there is no shear within the fluid itself once it enters the bowl. Consequently, the low-shear-rate viscosity must be low to allow settling of solids.

The primary variable controlling sedimentation rate is the centrifugal force, which is proportional to bowl diameter and the square of rpm. Centrifuges used in drilling applications usually have diameters of 14–28 inches and bowl lengths of 30–55 inches. Rotational speeds are generally from 1500 to 4000 rpm, with most machines operating toward the lower end of this range. High-*g* centrifuges can produce more than

3000 g: 3000 times the acceleration of gravity. The g force can be calculated from the following equation:

$$g = (\text{rpm}^2)(1.42 \times 10^{-5})(\text{bowl diameter, in.}).$$

For example, for a 14-inch bowl rotated at 2000 rpm:

$$g = (2000)^2(1.42 \times 10^{-5})(14)$$

$$g = 795.$$

A useful rule of thumb concerning rpm and maintenance costs merits consideration when deciding upon the desirable g force. It states that maintenance requirements are proportional to the cube of the rpm. Doubling the rpm can be expected to increase maintenance cost and downtime by a factor of 8, while an increase of 25% would almost double them. Another consideration is that at higher g forces, more solids are separated, and they tend to become more tightly packed, making them more difficult to transport and requiring more torque. Inasmuch as the available torque is limited, the feed rate may have to be limited to avoid stalling the scroll.

13.3.1 Centrifuge Installation

The centrifuge suction should be in the compartment receiving the discharge from the desilters and mud cleaners, and the returned stream should be directed to the next compartment downstream. Installation should permit the discharge of either the underflow (as is usual when processing unweighted fluids) or the overflow, as in traditional centrifuging. In either case, the recovered stream should be returned to a well-stirred area of the receiving compartment. This is particularly important in the case of recovered underflow, which usually contains too little liquid to permit it to flow and can be difficult to remix into the drilling fluid. If the underflow is returned to the mud, or discarded via a chute, the chute must be at an angle of no less than 42° with the horizontal. It is particularly important that the underflow be returned to a compartment that normally has a high fluid level and is well agitated.

13.3.2 Centrifuge Applications

The two primary reasons for the use of centrifuges with drilling fluids are (1) the selective separation of colloidal and ultra-fine solids from weighted fluids to improve their flow properties and (2) the removal of fine solids from unweighted fluids.

13.3.3 The Use of Centrifuges with Unweighted Drilling Fluids

With properly designed and engineered unweighted muds, which have low total solids content, in which particle size is not a primary cause for concern, the objective of centrifuging is the removal of drilled solids. Massive dilution is required to compensate for the incorporation of these solids if their concentration is to be controlled at the low levels that are normally desired. Centrifuging, and separating the underflow, can remove significant quantities of otherwise inseparable solids. This significantly reduces dilution requirements and drilling waste volume.

To control the density of a freshwater-base fluid at 8.8 lb/gal, 27.6 barrels of dilution are required by the incorporation of each barrel of drilled solids. With seawater, and a 9.0 lb/gal drilling fluid, each barrel of incorporated solids requires 37.5 barrels of dilution. Regardless of the type of drilling fluid in use, the preparation and eventual disposal of large quantities of drilling fluid can be very costly. Economics almost always favors the removal of undesirable solids, rather than the reduction of their concentration by dilution.

With unweighted fluids, solids loads are low, and torque rarely presents a problem. Hence, it is best to operate the centrifuge with high g force and a deep pool to maximize solids separation. Since residence time is a factor in solids separation, the maximum feed rate, which reduces residence time, may not produce the best results. The maximum efficient rate rarely exceeds 250 gpm of feed fluid plus dilution. Dilute the feed as necessary to control the funnel viscosity of the effluent at 35–37 sec/qt. This can be expected to control the low-shear-rate viscosity at levels low enough to permit efficient separation. Target processing capacity should be about 25% of the circulation rate. This may require more than one centrifuge. If multiple centrifuges are used, they must be operated in parallel, not in series.

13.3.4 The Use of Centrifuges with Weighted Drilling Fluids

With the unavoidably higher solids content of weighted muds, the increase of gel strengths and viscosities, as well as the degradation of filter cake quality associated with the diminution of particle size with time, can become a serious problem. The problem is caused by the increasing concentration of colloidal and near-colloidal particles that are too fine to be separated from the base fluid. Centrifuges are used to selectively remove these fine solids.

In this application, the underflow is returned to the mud and the overflow; the liquid and finest solids are separated and are either stored for later use as packer fluid, to be reconditioned through dilution and used in another drilling application, or discarded.

Processing 10–15% of the circulating volume is usually sufficient. The capacity of machines designed for this application is much less than that of the high-volume centrifuges designed for use with unweighted fluids. Dilution of the feed mud is almost always necessary with these fluids. As with unweighted muds, the dilution should reduce the effluent viscosity to 35–37 sec/qt.

Due to the high solids content of weighted fluids, torque is often a problem when processing it. Torque can be reduced by increasing pond depth to permit some mud to spill over and become mixed with the separated solids, through reduction in rpm or running a higher conveyor differential, or by processing at a reduced feed rate continuously, rather than at a higher rate intermittently.

At higher mud weights, solids loading in the centrifuge can be reduced by processing the drilling fluid through hydrocyclones and feeding the centrifuge with the overflow (while returning the underflow to the mud system). The temporary removal of the coarser solids, which would remain in the mud in any case, lightens the load on the centrifuge, permitting more efficient isolation and separation of the colloidal and near-colloidal solids.

13.3.5 Running Centrifuges in Series

There are those who tout the practice of series centrifugation as a beneficial procedure. *This is not true* and is the result of misunderstanding the reason for centrifuging weighted fluids. As has been stated, the objective in centrifuging weighted drilling fluids is the removal of colloidal and near-colloidal solids, not the separation of drilled solids.

As the process is described, the first stage recovers the barite, while the second rejects the drilled solids and returns clean fluid to the system. Inasmuch as the centrifuge cannot separate drilled solids from barite, and the second stage cannot separate the finest drilled solids and barite particles from the fluid, the description is clearly inaccurate. At the first stage, the underflow, consisting of the larger solids particles (both barite and low-gravity solids), is returned to the system. The overflow, that is, the liquid and finer solids (both barite and low gravity), is then routed to another centrifuge that is operated at higher g force and makes a finer cut. At this stage, the underflow is discarded and the overflow, containing the finest and most damaging solids, is returned to the circulating system.

Assume for illustrative purposes that the D_{50} cut points on barite are $8\,\mu$ at the first stage and $4\,\mu$ at the second. The net effect is the removal of most of the barite between 4 and $8\,\mu$, and most of the low-gravity solids between 6 and $12\,\mu$. The barite that is removed is in a perfectly acceptable size range, and no benefit is derived from its removal. While the removal of low-gravity solids before they become small enough to become troublesome is beneficial, the benefit cannot be expected to justify the loss of desirable-sized barite.

There is an additional factor to consider in deciding whether or not to utilize this questionable practice. Unless the mud density is being reduced, the desirably sized barite that is disposed of in the first stage has to be replaced. As much as 30% of API-quality fresh barite can be particles smaller than $6\,\mu$ and much of this can be expected to be colloidal. Replacement of the discarded 4–$8\,\mu$ material with fresh barite, which can include as much as 15–20% colloidal particles, increases the concentration of colloids, thereby exacerbating rheological problems and increasing the need for dilution of the active mud system.

Another consideration is that while low-gravity solids tend to be more troublesome than barite, solids problems in weighted fluids, which tend to contain much more barite than low-gravity solids, are frequently caused by excessive concentrations of colloidal and ultra-fine barite. This process returns these problem solids to the drilling fluid.

The validity of this analysis is not dependent on the assumed cut points. Whatever they may be, the result is the same. The solids that are removed are those between the cut points of the two stages, and the finest—most damaging—solids are returned to the mud system.

Respected experts have been counseling against running centrifuges in series for decades. (See George Ormsby's chapter, "Drilling Fluids Solids Removal," in Preston Moore's *Drilling Practices*, for an example.) They point out that the process does not, and cannot, work as described. Some go a step further and point out that the process reduces drilling-fluid quality and is harmful rather than beneficial. There are many situations in which the use of multiple centrifuges is clearly economically attractive; however, they are added for increased capacity and must be operated in parallel, not in series.

13.3.6 Centrifuging Drilling Fluids with Costly Liquid Phases

Weighted NAFs present a special problem because of the cost of the base fluid, which can exceed $200/bbl. The objections to series centrifugation raised in the previous section are equally valid for these fluids. Fortunately, NAFs are more solids tolerant than water-based fluids. However, when NAFs are used long enough, ultra-fines and colloids can accumulate to problem levels and force difficult choices.

In most drilling operations, the most expensive variable costs are those associated with rig time. When solids problems cause losses in penetration rate or hole problems that consume rig time and increase the risk of losing the hole and the drill string, a choice must be made between accepting the cost of conditioning the fluid to relieve the problem and accepting the additional expense, and risk, of continuing with a solids-contaminated fluid. Aside from dilution, the only means of reducing the colloidal and ultra-fine content of NAF is traditional centrifuging, which removes costly liquid from the system together with the finest solids. Economics clearly favors reconditioning the fluid by removing the colloids so that it can be reused. Unfortunately, centrifuges cannot separate colloids from the base liquid, and no other means of removal has proven to be economical. The advantage of centrifuging, thereby selectively separating the problem solids, over diluting and creating excess volume that must also be separated from the system is that centrifuging permits the desired improvement in mud quality to be achieved at lower cost and with the preparation of less new fluid.

For short-term use in which colloid accumulation is not expected to be a problem, dilution with fresh, uncontaminated fluid provides a means of using some of this colloid-laden fluid. It can also be used as packer fluid or stored for emergency use in the event of lost circulation.

13.3.7 Flocculation Units

With unweighted water-based fluids, chemical flocculation can be used to cause colloidal and ultra-fine particles to form aggregates large enough to settle and to be separated by centrifuges. This technique has been used to recover and permit the reuse of water from discarded drilling fluid. Care must be taken to ensure that the residual chemicals from the treatment do not upset the drilling fluid chemistry when returned to the active system.

13.3.8 Centrifuging Hydrocyclone Underflows

Centrifuges can be used to recover fluid from hydrocyclone underflows, thereby reducing drilling waste volume. The process returns the finest solids to the mud system, which can present problems. It is important that the solids content of the recovered fluid be monitored. As long as it is less than twice the desired solids content, the results are beneficial. At higher concentrations, it increases dilution requirements, making the process counterproductive.

13.3.9 Operating Reminders

- Before startup, rotate the bowl by hand to be sure that it rotates freely.
- Start the centrifuge before starting the feed pump or dilution fluid flow.
- Observe the manufacturer's recommendations concerning feed and dilution rates.
- When shutting down, shut the feed off, then the dilution, then the machine.

13.3.10 Miscellaneous

Caution: When working with NAFs, remember that the viscosity of the base fluid is very temperature sensitive and that viscosity is one of the primary factors influencing sedimentation. The effectiveness of centrifuges is very significantly reduced with high-viscosity fluids.

As with other solids-separation devices, it is imperative that centrifuge use be monitored. The volume and composition of the discharge stream should be checked daily to determine the approximate volume of high- and low-gravity solids being separated. Occasional particle size analyses should be run on feed, underflow, and overflow. (See Chapter 14 on Capture Equations for additional information on monitoring separator performance.)

13.4 ROTARY MUD SEPARATOR

The great majority of centrifuges used in drilling are decanting devices, as described previously. The rotary mud separator (RMS), also known as a perforated rotor centrifugal separator, was developed by Mobil in the 1960s. Although it is not, strictly speaking, a centrifuge because the outer barrel is not rotated, it serves the same function that decanting centrifuges do with weighted drilling fluids by discarding ultra-fine and colloidal solids while salvaging silt-size barite, and is often spoken of as another type of centrifuge.

The usual RMS configuration utilizes a perforated rotor 40 inches long and 6 inches in diameter in a nonrotating horizontal housing with an internal diameter of 8 inches. The fluid being processed is diluted and fed into the annular space between the tubes. The centrifugal force created by the rotation of the inner cylinder concentrates the larger particles against the outer wall of the annulus. The larger particles and part of the fluid exit the annulus at the underflow discharge port at the downstream end of the annulus, while the remainder of the fluid and the finer solids, having passed through the perforations, exit via the overflow discharge port at the downstream end of the rotor. The division of the flow between the two exit ports is controlled by a choke in the underflow line.

Both of the discharge streams include enough liquid to permit them to flow or be pumped freely without clogging the lines or hoses carrying the flow. This permits flexibility in choosing a location for the RMS, which does not have to be mounted over a mud tank, as decanting centrifuges do. This has proven to be a very useful feature. Another beneficial characteristic is that the capacity of the RMS is greater than that of the decanting centrifuge.

The factor that has limited the use of the RMS is that it requires 70% dilution of the processed fluid. Each barrel of processed drilling fluid requires the addition 0.7 bbl of diluent. While this left many applications in the 1960s and 1970s with the growing need to reduce the volume of drilling waste, it now severely limits drilling applications, although it may still be a useful tool in mud plants.

13.4.1 Problem 1

Given:
An interval of 7000 feet of $12\frac{1}{4}$-inch hole is to be drilled, through Miocene sediments, below $13\frac{3}{8}$-inch casing set at 2000 feet with water-based mud weighing 9.0 lb/gal. It is expected that the actual average hole diameter at the end of the interval will be 13.5 inches and that the interval will require six bits and 10 days to drill.

Assume the following:

- The only solids present in appreciable concentrations are drilled solids and commercial bentonite.
- Commercial bentonite content is maintained at 2% by volume.
- Mud cost is $10/bbl and liquid disposal costs are $5/bbl.
- Formation porosity can be disregarded.
- Two hundred rotating hours will be required to drill the interval.
- Prior experience in the area indicates that the installed solids-control system (shale shakers, desanders, and desilters) can be expected to remove only 60% of the drilled solids.
- A centrifuge capable of processing 100 gpm of this mud is available at a rental cost of $600/day plus $1500 (total) for mobilization and demobilization.
- The centrifuge, if used, will be operated only while drilling.

Questions:

1. What volume of new drilling fluid will have to be prepared while drilling this interval if no additional solids are removed?
2. What is the cost of this new fluid?
3. What volume of solids would the centrifuge have to remove from the drilling fluid during the drilling of the interval to make its rental economically beneficial?
4. Assuming that no dilution is required at the centrifuge, how much mud will be centrifuged during the drilling of the interval?
5. Assuming that the mud being centrifuged weighs 9.1 lb/gal, what is the weight of drilled solids entering the centrifuge/hr?
6. What percentage of the drilled solids entering the centrifuge has to be removed to offset the cost of its use?
7. Assuming that one third of the drilled solids are removed from the processed drilling fluid, what is the net financial benefit or loss arising from the use of the centrifuge?
8. Making the same assumptions, what would the net financial benefit or loss be if the processing capacity were doubled by providing two centrifuges of the same 100-gpm capacity?
9. Comment on the cost of *not* using a centrifuge for this interval.
10. What effect, if any, will running the single centrifuge for 200 hours have upon bentonite requirements under these conditions?

13.5 SOLUTIONS TO THE QUESTIONS IN PROBLEM 1

Useful relationships:

1. Drilled volume, bbl/1000 ft $= 0.97$ (D, in.)^2
2. Where % = the desired drilled solids content, the barrels of new drilling fluid that must be prepared to compensate for each barrel of incorporated drilled solids is given by the following: bbl/bbl = $(100 - \%)/\%$
3. In slurries consisting of low-gravity solids and freshwater, the solids content in % by volume $= 7.5$ (fluid density, lb/gal–8.33)

13.5.1 Question 1

Drilled solids volume $= 7 \times 0.97 \times (13.5)^2 = 1237$ bbl. Incorporated drilled solids are 40% of this, or 495 bbl. The solids content in this 9.0 lb/gal fluid is 7.5 (9.0 − 8.33) = 5%. Of this, commercial bentonite composes 2%, leaving 3% drilled solids.

The second useful relationship in the preceding list shows that $[(100 - 3)/3] = 32.33$ bbl of new mud are required to compensate for each barrel of incorporated solids. Therefore, the answer to this first question is that $(32.33 \times 495) = 16{,}003$ bbl of new fluid must be prepared to compensate for the incorporated solids and maintain the desired density of 9.0 lb/gal. Note that the total new drilling fluid volume will be 16,498 bbl, the volume of the newly prepared clean drilling fluid plus the volume of the incorporated solids.

13.5.2 Question 2

At $10/bbl preparation cost and $5/bbl disposal cost, the total cost is $242,520: $160,030 for the preparation of 16,003 bbl of new fluid and $82,490 for the ultimate disposal of 16,498 bbl.

13.5.3 Question 3

Rental, mobilization, and demobilization costs for 10 days' use totals $7500. At a total of $15/bbl for each new barrel of fluid mixed, the breakeven point on the use of the centrifuge is reached when the new mud prepared is reduced by 500 bbl ($7500 @ $15/bbl). Inasmuch as each bbl of incorporated solids requires the preparation of 32.33 bbl of new fluid to be added to each bbl of incorporated solids, the removal of each bbl of drilled solids reduces the new mud volume by 33.33 bbl. To reduce the volume of new drilling fluid prepared by 500 bbl, we must separate 15 bbl (500/33.33) of drilled solids during the 10 drilling days.

13.5.4 Question 4

In 200 rotating hours, processing 100 gpm, the centrifuge will process 28,570 bbl of drilling fluid [(100 gal/min × 60 min/hr × 200 hr)/42 gal/bbl].

13.5.5 Question 5

At 9.1 lb/gal, the processed drilling fluid contains 5.8% low-gravity solids and 3.8% drilled solids. At 100 gpm, the centrifuge processes 2.38 bpm, 142.8 bph. The drilling fluid processed in an hour contains 4.28 bbl of drilled solids.

13.5.6 Question 6

At 3.8% drilled solids, the 28,570 bbl of drilling fluid processed in 200 operating hours contain a total of 1086 bbl of drilled solids. In question 3 we calculated that the breakeven removal volume is 15 bbl. This is 1.4% of the drilled solids in the processed fluid.

13.5.7 Question 7

The removal of one third of the drilled solids that were not removed by the upstream solids-removal equipment, 165 bbl, eliminates the need for the preparation of (165 × 32.33) = 5334 bbl of new fluid at a cost of $53,340, and the disposal of (165 × 33.33) = 5499 bbl at a cost of $27,495. Use of the centrifuge reduced net operating cost by ($53,340 + 27,495 − 7500) = $73,335.

13.5.8 Question 8

Doubling the removal of drilled solids that are not separated upstream would double the total cost savings, increasing it to $146,670.

13.5.9 Question 9

The cost of *not* using a centrifuge is very significant: $146,670 when compared with the use of two centrifuges; $73,335 when compared with the use of a single centrifuge.

13.5.10 Question 10

In this application, the underflow—consisting of the larger solids particles and the liquid wetting them—is discarded. The bentonite is expected to remain in the overflow and be returned to the drilling-fluid system. Bentonite consumption should be unaffected.

CHAPTER 14

USE OF THE CAPTURE EQUATION TO EVALUATE THE PERFORMANCE OF MECHANICAL SEPARATION EQUIPMENT USED TO PROCESS DRILLING FLUIDS

Eugene Bouse
Consulting Engineer (Drilling Fluids, Solids Control, Drilling Waste Minimization)

Mike Morgenthaler
Cutpoint, Inc.

When mechanical separation equipment is used to remove suspended solids from liquids, capture determinations provide a simple means of evaluating its performance. *Capture* is defined as the fraction of incoming suspended solids that report to the discarded stream.

Capture is usually expressed as a percentage (%C) and is easily calculated if the concentration, by weight, of suspended solids is known for the process streams entering and leaving the separator. If the samples of the three process streams that are collected are representative of steady-state operation of the separator, then calculated capture is a good measure of the effectiveness of the separation process. The capture equation is written:

$$\%C = \frac{u \cdot (f - o)}{f \cdot (u - o)} \times 100$$

where

f = weight percentage of suspended solids in the feed
u = weight percentage of suspended solids in the underflow
o = weight percentage of suspended solids in the feed

Capture analysis has not been widely used in the drilling-fluids solids-control industry for two reasons: first, the need for representative samples precludes successful use of capture to evaluate shale shaker performance; and second, solids concentration in drilling fluids has traditionally been reported in terms of volume rather than weight; volumetrically rather than gravimetrically. Capture analysis is a useful tool recommended for the evaluation of the performance of solids-control equipment used on drilling-fluid systems. It must be understood that the data generated apply only to the moment in time at which the samples are collected. Capture data can be extrapolated to predict the solids removed by the separator over longer time periods only if (1) the separator is operating under steady-state conditions with consistent and homogeneous feed, and (2) sufficient data are collected to establish average performance for the time period studied.

This discussion will be limited to the application of the capture equation to centrifuges and hydrocyclones. The process stream terminology and abbreviations used are illustrated in Figure 14.1 and defined in the derivation that follows.

The calculation of capture is based on gravimetric analysis of the three process streams common to solid/liquid separators (i.e., feed, underflow, and overflow). The procedure yields good results when homogeneous and representative samples of the process streams are collected, as is usually the case with hydrocyclone units and centrifuges. The procedure cannot be applied to shale shakers because of the difficulty in obtaining representative samples of the three process streams and the inherent inconsistency of shale shaker feed conditions.

The capture calculation, which is derived in the following equation, is based on a material balance of solids entering and leaving the separator. Analysis of small samples of the feed, underflow, and overflow streams

Mechanical Separation Equipment Used to Process Drilling Fluids

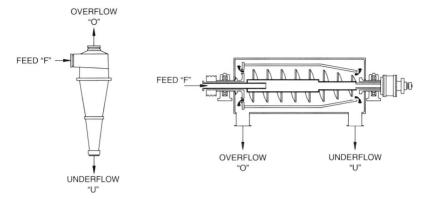

Figure 14.1. Process stream terminology for centrifugal separators.

permits the calculation of the percentage of the feed solids that are separated—the percent capture.

The derivation of the capture equation follows:

eq. (i)	$\%C = \dfrac{u \cdot U}{f \cdot F} \times 100$	Definition of capture expressed as a percentage
eq. (ii)	$F = U + O$	Conservation of total mass in and out
eq. (iii)	$f \cdot F = u \cdot U + o \cdot O$	Conservation of suspended solids
eq. (iv)	$o \cdot F = o \cdot U + o \cdot O$	Multiply eqn. ii by "o"
eq. (v)	$o \cdot O = o \cdot F - o \cdot U$	Rearrange eqn. iv
eq. (vi)	$-o \cdot O = u \cdot U - f \cdot F$	Solve eqn. iii for "oO"
eq. (vii)	$0 = u \cdot U - o \cdot U + o \cdot F - f \cdot F$	Add eqn. v & eqn. vi
eq. (viii)	$\dfrac{U}{F} = \dfrac{(f-o)}{(u-o)}$	Rearrange and simplify eqn. vii
eq. (ix)	$\dfrac{u \cdot U}{f \cdot F} = \dfrac{u \cdot (f-o)}{f \cdot (u-o)}$	Multiply by u/f and then the left side is identical to equation i.
eq. (x)	$\%C = \dfrac{u \cdot (f-o)}{f \cdot (u-o)} \times 100$	Capture expressed in terms of suspended solids concentrations.

where

F = feed mass flow rate
f = weight percentage of suspended solids in the feed
U = underflow mass flow rate
u = weight percentage of suspended solids in the underflow
O = overflow mass flow rate
o = weight percentage of suspended solids in the feed

14.1 PROCEDURE

14.1.1 Collecting Data for the Capture Analysis

A sample set of each of the three process streams should be obtained, sealed, and labeled for identification. The size of each sample should be 50–100 ml. For each set, the time between catching each of the samples should be as brief as possible.

14.1.2 Laboratory Analysis

The required laboratory work consists of determining the weight percentage of suspended solids in the samples. With water-based samples, the simplest method is to determine the weight of the sample using a precise analytical balance, remove the water by dehydration at 200 °F in an oven, and weigh the remaining solids. Correction factors should be determined and applied in cases in which the base liquid contains more than 10,000 ppm salt, or emulsified oil.

Determining Percentage of Suspended Solids in Water-Base Samples

For water-base samples, dehydration ovens are the most convenient heating devices for removing the liquid from the sample. Retorts are required for oil-based samples. Regardless of the dehydration method used, the data must be evaluated on the basis of weight, not volume.

Determining Percentage of Suspended Solids in Oil-Based Samples: Unweighted Fluids

For fluids that do not contain barite or another high specific-gravity (SG) weighting agent, the procedure outlined in the preceding section completes the laboratory work. The quantitative determination of the effects of the solids-removal process are then obtained using the capture calculation and by simply multiplying the volume fraction of the solids in the discharge stream by the rate at which solids enter in the feed stream. The mass flow rate of solids in the discharged stream can be expressed in dry tons per hour or in other units.

14.2 APPLYING THE CAPTURE CALCULATION

On drilling mud applications for both centrifuges and hydrocyclones, the discarded stream can be either the heavy phase or the light phase. Care must be taken to apply the equation correctly for the two different cases. It must be understood that the precision of the calculation of the rate of solids separation is dependent on the accuracy of the feed rate determination and that this is difficult to measure. Even though circumstances may require imprecise measurement, or estimation, of the feed rate, useful and meaningful results are obtained with this procedure.

14.2.1 Case 1: Discarded Solids Report to Underflow

In applications in which the underflow is discarded, consideration of the feed rate, together with the percentage of capture as determined with the capture equation (see Section 14.1), permits the calculation of the rate at which solids are being removed.

14.2.2 Case 2: Discarded Solids Report to Overflow

If the discarded solids exit with the light phase, then the percentage of capture can be determined by subtracting the capture calculated with *equation i* from the solids in the feed stream. In applications in which the overflow is discarded, this calculation is used, together with the feed rate, to determine the rate of solids separation.

14.2.3 Characterizing Removed Solids

Separated solids are characterized on two bases: their SG and their particle size.

Weight Material and Low-Gravity Solids

A dried and weighed sample of the separated solids can be added to a measured volume of water, and the average SG of the solids can be determined from the increase in volume and weight. The percentages of weight material and low-gravity solids can then be determined using the following equations.

$$\text{\%weight material} = (\text{ASG} - 2.6)/(\text{SGWM} - 2.6)$$

$$\text{\%low-gravity solids} = \text{total solids} - \text{weight material}$$

where

ASG = average solids gravity
SGWM = specific gravity of weight material

In order to obtain meaningful and repeatable results, a high degree of precision is required in obtaining these data. The use of a pressurized pycnometer is recommended.

Particle Size

The primary function of centrifugal processing of oil well drilling fluids is the removal of viscosity-building fines. Removal of these particles limits the need for dilution. Given the undeniable influence of average particle size on drilling-fluid quality, it is recommended that occasional particle size analyses be used to monitor the concentration of colloids and near-colloids and ensure that their concentration does not become excessive.

14.3 USE OF TEST RESULTS

14.3.1 Specific Gravity

This procedure provides a means of monitoring the concentration of low-gravity solids in the feed and the discarded material. Inasmuch as we can easily remain aware of the concentration of desirable low-gravity solids by simple record keeping, it also provides a means of monitoring the volume of drilled solids in the mud and in the discarded material.

14.3.2 Particle Size

Experience in an area, coupled with the knowledge that penetration rates tend to decrease while torque, drag, and the likelihood of sticking the drill string increase with increasing concentrations of fines, can provide area-specific guidelines concerning tolerable concentrations of these colloidal and near-colloidal particles. The monitoring of particle size helps determine when the removal of fines by centrifuging is desirable.

14.3.3 Economics

Unweighted Fluids

The economics of discarding the underflow of centrifuges used for solids reduction with unweighted muds can be evaluated by comparing the cost of the solids removal with the cost of the dilution required by the incorporation—rather than removal—of the separated solids, and the differences in waste disposal costs.

The effect of centrifuging upon the mud cost can be determined by calculating the volume of dilution that would have been required to compensate for the incorporation of the separated solids, and multiplying it by the unit cost of the fluid. Inasmuch as all dilution adds directly to waste volume, the cost of disposing of the dilution volume must be added to the cost of preparing it.

Weighted Fluids

Hydrocyclone use is not ordinarily recommended with weighted fluids because high solids content interferes with their operation, and weight material is concentrated in the underflow and discarded.

Traditionally, centrifuging has been used with weighted fluids to reduce dilution requirements and barite consumption. Comparison of the cost of the centrifuging with the value of the barite recovered from the discarded fluid is often used as a measure of its economic effectiveness. While this is a valid, and important, basis for evaluation, the fact that drilling-fluid quality tends to be much better when centrifuges are used can be of much greater economic importance.

The effect of centrifuging on waste volume must be considered also. Dilution volume at the centrifuge feed and the disposal of the liquid in the overflow are obvious factors. Less obvious, but of greater importance, is the fact that the disposal of the colloids and near-colloids discarded with the liquid reduces dilution requirements and therefore mud cost and the volume of waste generated.

A rough approximation of the waste volume reduction achieved by discarding centrifuge overflow can be arrived at by calculating the volume of solids that are being discarded, assuming that they are all colloids or smaller and troublesome ultra-fines, and calculating the dilution that would be required to maintain their concentration at 5% by volume if they were not separated. This dilution, 19 times the volume of the separated solids, can be taken as an approximation of the reduction in dilution.

An evaluation of the economics of centrifuging must include the cost of the preparation of the mud used for dilution, the reduction in waste volume, and the value of the barite returned to the system via the underflow.

Nontraditional use of centrifuges with weighted fluids should be evaluated by comparing the benefits with the value of desirable materials discarded in the process.

14.4 COLLECTION AND USE OF SUPPLEMENTARY INFORMATION

In addition to the samples and feed rate data, a fully completed copy of the most recent Mud Report and the following information should be obtained at the well site:

- Drilling fluid composition and unit cost
- Density of feed, overflow, and underflow
- Funnel viscosity of feed and overflow

This information can be helpful in economic analyses and in evaluation of the operation of the centrifuge(s).

CHAPTER 15

DILUTION

Leon Robinson
Exxon, retired

Dilution refers to the process of adding a liquid phase to a drilling fluid to decrease the drilled-solids concentration. Dilution is used in several ways. If no solids-control equipment is used or if the equipment is used ineffectively, dilution may be the principal method of keeping drilled solids to a reasonably low level. This is an expensive solution to the problem. For example, to decrease drilled solids by 50% requires that 50% of the system be discarded and replaced with clean drilling fluid. Usually dilution is used after processing by solids-removal equipment to dilute drilled solids remaining in the drilling fluid. Dilution may be added as a clean drilling fluid or as the liquid phase of a drilling fluid with the other necessary drilling fluid ingredients, usually through a chemical barrel and a mud hopper. In this discussion, *dilution* will refer specifically to the clean drilling fluid necessary to decrease drilled-solids concentration. Clean drilling fluid is the liquid phase with all necessary additives such as barite, polymers, clay, etc.

As an example of dilution, consider a well in which the drilling-fluid specifications suggest that the volume percentage of (%vol) drilled-solids concentration should be, and is, 6%vol. Assume that 10 bbl of formation solids are brought to the surface and that no solids-removal equipment is used. All 10 bbl would be retained in the drilling-fluid system. These solids would require dilution to maintain the 6% volume concentration in the new drilling fluid.

The new drilling-fluid volume would consist of drilled solids (10 bbl) and clean drilling fluid (dilution). If the drilled solids (10 bbl) are 6% volume of the new drilling fluid built, the volume of new drilling fluid built would be 167 bbl (or 10 bbl/0.06). The volume of clean drilling fluid (dilution) required may be calculated from the statement that the new drilling-fluid volume, 167 bbl, would consist of 10 bbl of drilled solids plus the clean drilling fluid. Obviously the dilution, or the clean drilling fluid, needed would be 157 bbl. This dilution would increase the pit volume by 157 bbl. When nothing is removed from the system, the pits would overflow if they were originally full. The only volume available to accept dilution would be the volume (liquid and solids) removed from the pit system. The volume removed would include all fluid and solids removed by the solids-removal equipment and any drilling fluid removed from the system to be stored or discarded. This is an important concept and is not trivial. Dilution calculations are based on simple material balances—addition and subtraction of volumes added and removed.

Before the dilution (clean drilling fluid) is added, the pit levels would remain the same as they were before the formation was drilled. The solids that were drilled occupy the same volume as they did before they were pulverized by the drill bit. The new hole volume is exactly the volume of material added to the drilling fluid system; consequently, drilling new hole does not change the pit levels. This, too, is an important concept and is the basis of determining the liquid levels in a drilling fluid system before and after using the solids-removal equipment.

Material added to the drilling-fluid system during drilling consists of solids and fluid contained within the formation. (Actually the pit levels would increase slightly by the volume of drill pipe steel added during the drilling of that interval and would decrease by the volume of gas released at the surface. The volume of gas released would be calculated on the basis of the bottom-hole volume that entered the drilling-fluid system. This factor will be ignored in the calculations; however, for purposes of this calculation, the drill pipe will be returned to its original position.)

For example, consider a 13.9-ppg freshwater-based drilling fluid with 2%vol bentonite and 25%vol total solids. Given the specific gravity (SG) of low-gravity solids as 2.6 and of barite as 4.2, the drilled solids would be 8%vol. This may be calculated from the equations presented in Chapter 3 on solids calculations. In the preceding problem, the new drilling fluid built would include 10 bbl of drilled solids consisting by volume of 17% barite, 2% bentonite, and 81% water. The dilution liquid would thus be 81% of the volume of the 167 bbl, or 135 bbl of water.

15.1 EFFECT OF POROSITY

Thus, 10 bbl of formation solids arrive at the surface. However, consider drilling 10 bbl of hole. Ground rock occupies the same volume as the rock before it is drilled. If a drilled formation has a 10% pore volume, or porosity, the rock added is 9 bbl and the volume of fluid in the pore space is 1 bbl. If the fluid in the pore space is liquid, no volume change will occur when the formation is ground into small pieces and enters the drilling fluid. The pit levels remain constant except for the volume of the drill string added to drill the hole. If the fluid space is filled with gas, the pit levels decrease as the gas is liberated at the surface. The pit levels decrease only by the volume removed from the system by the solids-control equipment.

The 9 bbl of rock remaining in the system increase the drilled-solids concentration. Most drilling-fluid systems require maintaining drilled-solids concentration at some predetermined value.

How much fluid would be required to dilute these 9 bbl of rock to a concentration of 4%vol?

Answer:

- The requirement will be that the 9 bbl of rock will be 4% of the volume of new drilling fluid built, or 9 bbl = 0.04 (new drilling fluid built).
- The volume of new drilling fluid built would be 225 bbl.
- The 225 bbl of new drilling fluid would contain 9 bbl of drilled solids and some volume of clean drilling fluid, that is, clean drilling fluid + 9 bbl = new drilling fluid built = 225 bbl.
- So, the clean drilling fluid volume would be 216 bbl. The volume of liquid added would depend on the other ingredients in the drilling fluid, such as barite, deflocculants, filtration control additives, other chemicals, and low-shear-rate viscosifiers.

Solids-control processes are designed to remove drilled solids so that such large quantities of additional fluid will not be required to keep drilled solids at their prescribed values with dilution only. If the 216 bbl of clean drilling fluid cost \$50/bbl, the drilling fluid would cost \$10,800 plus eventual disposal costs. For comparison, 100 feet of $9\frac{7}{8}$-inch hole would be about 10 bbl, and few drilling budgets can afford to tolerate over \$10,000 for drilling fluid for 100 feet of hole.

The clean drilling fluid added would also increase the system volume by 216 bbl. This volume would need to be sent to a reserve pit or discarded.

Normally, drilling fluid is processed through equipment to remove drilled solids when the drilling fluid reaches the surface.

15.2 REMOVAL EFFICIENCY

The term *solids removal equipment efficiency* is frequently used to describe solids-control equipment performance. This term may be somewhat confusing. American Petroleum Institute (API) Recommended Practice (RC) 13C, "Solids Control," refers to the solids-removal process in terms of system performance.

Solids-control equipment is designed to remove drilled solids, which are not dry when they are removed from the system. For example, the underflow discharge from a properly operating 4-inch desilter or hydrocyclone contains around 35%vol solids and 65%vol drilling fluid. The liquid concentration of the discard from a shale shaker depends on the screens. An API 200 screen discards a much wetter discard stream than an API 20 screen. The liquid drilling fluid that accompanies these separated solids comes from the drilling-fluid system. After screening, however, the solids in the discarded drilling fluid cannot be expected to have the same solids distribution as the drilling fluid in the tanks. Quantities of barite in the liquid phase of the discard from a fine screen may not be in the same concentration as barite in the pits. The discard obviously contains more drilled solids than the drilling fluid in the pits. The barite, or weighting agent, will probably also have a different concentration in the liquid phase of the discard than in the drilling fluid in the pit. Measuring the quantity of barite in the discard will not reveal the amount of drilling fluid discarded. The drilling fluid accompanying the drilled-solids discard will, however, contain drilled solids that have remained in the system after the fluid was originally processed by the solids-control equipment.

Efficiency is defined by the ratio of output to input. For example, if a 100-hp motor drives a rotary table and produces 85 hp to rotate a drill string, the efficiency of the system is 0.85, or 85%. Drilled solids removal efficiency would imply a ratio of output (or discard) to input (circulating volume).

Consider the situation in which 6 of the 9 bbl of rock described in the previous example reach the surface and are removed by the solids-control equipment. Not all of these 6 bbl of solids will be removed from the well bore in the same interval of time that they were drilled. Drilling at 20 ft/hr, solids from the 100 feet of hole would enter the system during a 5-hour period. Assume that these solids arrive at the surface during a 10-hour period (or longer). Circulating at 10 bbl/min, 6000 bbl of drilling fluid would reach the surface and be processed through the solids-control equipment. If this drilling fluid contained 4%vol drilled solids in addition to the 9 bbl of rock, a total of 249 bbl (240 bbl + 9 bbl) of solids would pass through the equipment.

What would be the removal efficiency of all of the drilled solids removed? Assume that the 6 bbl of drilled rock were discarded in a slurry containing 35%vol solids. This means that the total discard volume would be 17.1 bbl (6 bbl of drilled rock and 11.1 bbl of drilling fluid). The 11.1 bbl of drilling fluid with 4%vol drilled solids would have 0.44 bbl of drilled solids. So, the input would be 249 bbl of solids, and the output would be 0.44 bbl. Clearly, the resulting 0.2% removal efficiency reveals nothing.

If only the removal of the drilled rock entering the system is considered in the process calculation, no credit would be taken for the resident drilled solids present in the drilling fluid. The input volume of drilled solids would be the 9 bbl of drilled rock, and the output volume would be the 6 bbl. The system performance would be at 67% efficiency. This is the solids removal equipment efficiency of the system, also referred to as solids removal equipment performance or drilled-solids removal system performance.

Only solids removed that decrease the solids concentration in the drilling fluid are considered in calculating solids removal equipment efficiency (SREE). If a valve is opened and 200 bbl of drilling fluid are removed from a system, some drilled solids are obviously removed with the liquid. However, the concentration of drilled solids in the remaining drilling-fluid system does not change. Removal of the 200 bbl of drilling fluid provides space for clean drilling fluid to be added. The addition of the clean drilling fluid will decrease the concentration of drilled solids in the system, but this is a very expensive method—that is, dilution—of maintaining a low concentration of drilled solids. The calculation of SREE considers only the drilled-solids removal that decreases the drilled-solids concentration in the system compared with the new drilled solids introduced during that interval.

15.3 REASONS FOR DRILLED-SOLIDS REMOVAL

Many years ago, a controversy raged concerning the effect of drilled solids on the cost of a well. Many thought that drilled solids were beneficial as an inexpensive substitute for weighting agents. As oil well drilling encountered more and more difficult conditions, hole problems finally became undeniably associated with excessive drilled solids. Frequently, production horizons near the surface were normally pressured and could be drilled with unweighted drilling fluids. Usually, these drilling conditions were relatively trouble free, and a poor-quality drilling fluid was used for drilling. Of course, drilling performances and well productivity could be enhanced with better-quality drilling fluids, but those effects were difficult to quantify. As these areas graduated from unweighted drilling fluids to weighted drilling fluids, better drilling-fluid properties were required to prevent trouble. The primary problem was that large quantities of drilled solids were intolerable. The drilling trouble costs could easily be traced to failure to limit drilled-solids concentration. This provided the impetus for most drilling rigs to upgrade their surface systems handling drilling fluids. The benefits of a clean drilling fluid have been well stated in previous chapters and have been well validated.

Some rigs now process all drilling fluid sequentially in accordance with good practices, as discussed in Chapter 5 on tank arrangements. Drilling fluid type does not affect proper rig plumbing. Dispersed or non-dispersed, fresh- or saltwater, clay-based or polymer-based, any drilling fluid must be treated sequentially to remove smaller and smaller drilled solids.

The cost of solids-control equipment was justified initially economically as an insurance policy to prevent catastrophes. Subsequently, more expensive drilling fluids required lower drilled-solids concentrations. Polymer additives that adhere to active solids require significantly lower concentrations of drilled solids to prevent loss of too much polymer. Environmental concerns also dictate minimization of waste fluid; this requires careful attention to mechanical removal of drilled solids.

15.4 DILUTING AS A MEANS FOR CONTROLLING DRILLED SOLIDS

One way that drilled solids can be kept at a manageable level is to simply dump some of the drilling fluid containing the drilled solids and replace it with clean drilling fluid. One half of the drilled solids is eliminated if one half of the system is dumped and replaced with clean fluid. Generally this is too expensive, so mechanical equipment is used. Traditionally, large volumes of drilling fluid were dumped from the system by aggressively dumping sand traps. This makes room available for clean drilling fluid needed for dilution without calling it "dilution."

15.5 EFFECT OF SOLIDS REMOVAL SYSTEM PERFORMANCE

Assume that the surface system contains 1000 bbl of drilling fluid, the targeted drilled-solids level is 4%vol, and 100 bbl of drilled solids report to the surface. For reference, 100 bbl is the volume of 1029 feet of a 10-inch-diameter hole.

If this volume of 100 bbl of drilled solids remains in the drilling fluid, the pit levels remain constant. The drilling-fluid system has increased to 1100 bbl because new hole of 100 bbl was drilled. The volume of the rock represented by the new hole is virtually the same whether it is ground into cuttings or is solid rock. The 4%vol drilled-solids concentration before drilling means that the drilling fluid contained 0.04×1000 bbl, or 40 bbl of drilled solids when drilling started. After drilling, these 40 bbl plus the 100 bbl of new drilled solids would be in the 1100-bbl drilling-fluid system, or (140 bbl/1100 bbl) $\times$ 100, or 12.7%vol drilled solids.

To reduce the concentration of drilled solids to 4%vol by only adding clean drilling fluid would require adding enough clean drilling fluid to reduce the 100 bbl of drilled solids to a volume concentration of 4%.

$$100 \text{ bbl} = (0.04) \text{ new drilling fluid built}$$

$$\text{new drilling fluid built} = 2500 \text{ bbl.}$$

This newly built drilling fluid would consist of drilled solids plus clean drilling fluid, or

$$2500 \text{ bbl} = \text{clean drilling fluid} + 100 \text{ bbl}'$$

$$\text{clean drilling fluid} = 2400 \text{ bbl.}$$

Observe that this volume is independent of the original volume of the system.

If the tanks are full when the drilling starts, the 2400 bbl of clean drilling fluid cannot be contained in the mud tanks. The only volume available for the clean drilling fluid is the volume of discard removed from the system. Since nothing is removed from the system, the volume added must be removed to return the pit levels to the original level. The excess drilling fluid (2400 bbl) would need to be removed from the drilling-fluid system to keep the pits from overflowing. Not only would the cost of the clean drilling fluid be prohibitive, but this fluid must also eventually go to a disposal site.

The intent of this analysis is to build the basis for the concept of an appropriate removal efficiency to build the minimum quantity of new drilling fluid and also minimize the volume of discarded fluid. The assumption is that the drill pipe will be returned to the position at which it started drilling the interval. Although the rock does not change in volume, the pit levels will rise because of the volume of the steel added as the drill string enters the well.

Evaluation of SREE is very difficult to do with only short-term tests. Capturing equipment discards for only 15 minutes to 2 hours will not provide sufficient data to calculate SREE. The quantity of solids entering the drilled solids removal equipment is usually unknown and impossible or very difficult to determine. Drilled solids do not arrive at the surface in the same order in which they were drilled. Fluid flow in the annulus is usually laminar. The center part of the annular flow moves faster than the portion of the flow adjacent to the formation or the drill pipe. A 30-minute sample from the shale shaker discards may contain samples that were drilled 2 to 4 hours apart, even in a well bore drilled to gauge. With rugosity and borehole enlargements, lag times are extended. Mud loggers often observe formation cuttings that were drilled several days prior to the sample's being taken from the end of the shale shaker. This is the reason that material balances are difficult to obtain with only snapshots. Circulating a borehole clean before drilling an interval, drilling a known volume of solids, and circulating all of the cuttings from the well bore will permit estimating a known volume of solids arriving to the surface. Another variable in this analysis is the porosity of the formation. Solids analysis requires being informed of the initial condition of the quantity, or volume, of solids that arrive at the surface.

15.6 FOUR EXAMPLES OF THE EFFECT OF SOLIDS REMOVAL EQUIPMENT EFFICIENCY

If the mechanical equipment does not remove a significant portion of the drilled solids reporting to the surface, it can become very expensive to maintain a reasonable level of undesirable drilled solids. Dilution, then, becomes a major portion of the solids-management strategy. Calculations indicate the performance of the solids-removal equipment.

This set of calculations is simply a material balance of the volumes added and the volumes discarded. The calculations will be based on a drilling-fluid processing plant system in which the solids-removal section is removing either 100, 90, 80, or 70% volume of the drilled solids arriving at the surface. Obviously 100% removal efficiency is not currently possible while retaining the liquid phase. Removal percentages listed are used simply to demonstrate the method and concepts of solids-removal efficiency minimization.

The average drilled-solids concentration in the discard stream has been selected as 35%vol. The underflow from a decanting centrifuge and the mud cleaner discharge stream will contain 55–63%vol solids. The underflow from desilters and desanders will vary around 35%vol solids. The discharge stream from a shaker can vary from 45%vol, for very coarse screens, to 20%vol solids for very fine screens. The average for all of these devices is assumed to be 35%. With very coarse screens, most of the liquid is removed from the large solids as they travel down the shaker screen. So, the concentration of solids is much higher than it is with screens with smaller openings, although the total volume of solids discarded is usually smaller. Screens with smaller openings remove more solids as well as more liquid. The reason is related to the ratio between the surface area and the volume of the cuttings. For example, a golf ball would retain very little drilling fluid when removed from the fluid. Grind the ball into very small pieces and the volume of liquid would increase greatly. The volume of the solid (golf ball) would not change, but the surface area would change greatly. More liquid is required to wet the increased surface area.

To examine (1) the quantity of discards, (2) the clean drilling fluid needed to dilute the remaining drilled solids, and (3) the excess volume of drilling fluid built, four SREEs will be discussed: 100%, 90%, 80%, and 70%.

15.6.1 Example 1

100% Removal Efficiency

In the 100% removal efficiency case, all of the drilled solids reaching the surface are removed. In addition to drilled solids, the solids-control equipment also removes drilling fluid clinging to those solids. With a 35%vol concentration of solids, the discard will contain 65%vol drilling fluid. The pit levels will decrease by the volume of drilled solids and

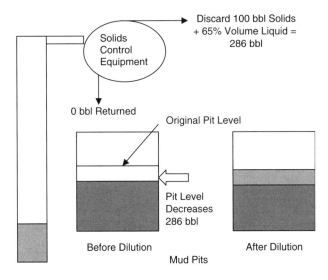

Figure 15.1. 100% drilled-solids removal efficiency.

drilling fluid removed from the system, as shown in Figure 15.1. (This makes the assumption that the rock porosity and compressibility are zero.) No drilling fluid is needed for dilution because no drilled solids are retained in the system. However, 286 bbl of drilling fluid are needed to replace the material removed from the system in order to keep the pit levels constant. The drilled-solids concentration will decrease. In other words, a level of 4%vol drilled solids will not and cannot be sustained. In the preceding example, the 40 bbl of drilled solids originally in the drilling fluid will now be distributed into 1100 bbl of drilling fluid (the 1000 bbl of initial volume plus the 100 bbl of hole drilled). The drilled-solids concentration will decrease to 3.6%vol, or (40 bbl / 1100 bbl) × 100. The ratio of clean drilling fluid needed per barrel of hole drilled is 2.86 [286 bbl / 100 bbl]. In this case, the 4%vol drilled solids in the system will be diluted and the concentration of drilled solids will be reduced.

15.6.2 Example 2

90% Removal Efficiency

In the 90% removal efficiency situation, the volume of newly drilled solids removed from the system is 90 bbl. The discards will again be

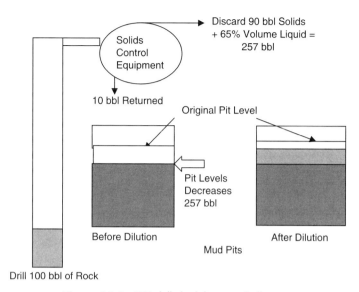

Figure 15.2. 90% drilled-solids removal efficiency.

assumed to contain 35%vol solids. The total volume of discard would be [90 bbl / 0.35], or 257 bbl (Figure 15.2). The solids returned to the system (10 bbl) must be reduced to a 4%vol concentration. The 10 bbl of newly drilled solids must become 4%vol of newly built mud. So the new mud built would be 250 bbl [10 bbl / 0.04]. The 250 bbl would be composed of 240 bbl of clean drilling fluid and 10 bbl of drilled solids. Since the volume removed by the solids-control equipment was 257 bbl, this would be the volume available in the mud pits to hold the dilution fluid. Since 240 bbl is needed to dilute the remaining 10 bbl of drilled solids, an additional 17 bbl must be added to bring the pit level. This is almost a *balanced* system, that is, no excess drilling fluid is needed to dilute the drilled solids returning to the system. If the discarded volume exactly matches the required volume needed for dilution, the minimum quantity of drilling fluid will be built. The optimum removal efficiency for any targeted drilled-solids level may be calculated by mathematically equating the removal volume to the dilution volume required. The ratio of clean drilling fluid needed per barrel of hole drilled is 2.57 [257 bbl / 100 bbl]. Again, the 4%vol drilled solids in the system will be decreased because more dilution was needed to keep the pit levels constant than was needed to dilute the 10 bbl of retained solids.

The concentration of drilled solids would be less than the targeted 4%vol. The volume of drilled solids would be the original 40 bbl plus the 10 bbl retained in the system by the solids-control equipment. The system volume would be the original 1000 bbl plus the 257 bbl of new hole drilled, or 1100 bbl. The drilled-solids concentration would be 3.98%vol.

15.6.3 Example 3

80% Removal Efficiency

For the case of the 80% removal efficiency, 229 bbl of drilled solids and drilling fluid will be discharged (Figure 15.3). Although this is only 21 bbl less than the 90% removal efficiency, the dilution volumes are significantly higher. The dilution of the 20 bbl of returned drilled solids to a 4%vol level requires the addition of 480 bbl [20 bbl/0.04] to the system. The reconstituted 500 bbl of drilling fluid will contain 20 bbl of drilled solids and 480 bbl of clean drilling fluid. Since only 229 bbl of space is available, 271 bbl of drilling fluid must be discarded. The total discard will therefore be the 229 bbl from the solids-removal equipment and the 271 bbl of drilling fluid. The volume of clean drilling fluid needed per barrel of hole drilled is 4.8%. The excess drilling fluid generated is 271 bbl.

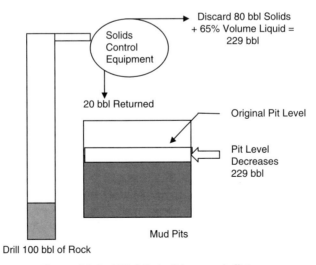

Figure 15.3. 80% drilled-solids removal efficiency.

15.6.4 Example 4

70% Removal Efficiency

For the case of 70% removal efficiency, 70 bbl of drilled solids will be discarded in a 200-bbl slurry that contains 130 bbl of drilling fluid (Figure 15.4). So the pit levels will decrease by the removal of 200 bbl from the system. The 30 bbl retained in the system must be diluted to 4%vol. The new drilling fluid contains 4%vol drilled solids, or mathematically: drilled solids = (4%)(new drilling fluid). The volume of retained drilled solids that must be diluted is 30 bbl, so the volume of new drilling fluid would be (30 bbl / 0.04), or 750 bbl. This new drilling fluid will contain 30 bbl of drilled solids; so, 720 bbl of clean drilling fluid must be added to the system. Only 200 bbl of volume is available in the drilling fluid system, so 520 bbl of drilling fluid must be discarded from the system or stored. The volume of clean drilling fluid needed per barrel of hole drilled is 720 / 100, or 7.0. Note that the excess drilling fluid generated is 520 bbl.

15.6.5 Clean Fluid Required to Maintain 4%vol Drilled Solids

The volume of clean drilling fluid required is a function of the targeted drilled-solids concentration, 4%vol for these cases, and the SREE.

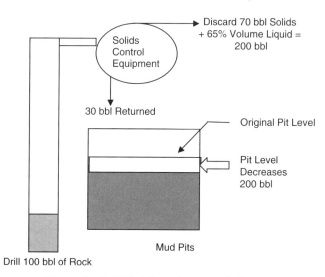

Figure 15.4. 70% drilled-solids removal efficiency.

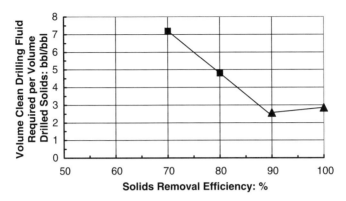

Figure 15.5. Dilution required to maintain 4%vol drilled solids.

The line designated by the triangles in Figure 15.5 indicates that the volume of clean drilling fluid required to dilute the drilled solids remaining in the drilling fluid is less than the volume of fluid required to return the pit levels back to their original values. In this case the targeted drilled-solids concentration will be decreasing. The line designated by the squares indicates that more fluid was required to dilute the solids remaining in the pits after processing through the solids-control equipment. The pit levels would increase so much that excess drilling fluid would need to be removed from the system. The intersection of these two lines would indicate the smallest quantity of drilling fluid required to maintain a concentration of 4%vol drilled solids in the system. This intersection is predictable and the calculation is presented in the next section.

15.7 SOLIDS REMOVAL EQUIPMENT EFFICIENCY FOR MINIMUM VOLUME OF DRILLING FLUID TO DILUTE DRILLED SOLIDS

The minimum volume required to dilute solids remaining after processing by the solids-control equipment depends on the drilled-solids concentration in the drilling fluid. If all of the drilled solids are removed from the system, the clean drilling fluid added to return the pit levels back to the original level will dilute the solids already in the drilling fluid. As noted earlier, more clean drilling fluid will be needed to return the pits to the original level with 100% removal than 90% removal of drilled solids. The smallest volume required will occur when the system is balanced, as previously defined. The same solids-removal efficiency that provides the minimum quantity of new drilling fluid to be built will also be the removal efficiency that generates the minimum discard volume. This would be a condition in which the volume of clean drilling fluid required to dilute the solids remaining after processing through the removal equipment is exactly the volume discarded by the equipment.

optimum solids removal efficiency

$$= \frac{\left(1 - \dfrac{\text{target drilled solids}}{\text{conc. in drilling fluid}}\right)}{1 - \left(\begin{array}{c}\text{target drilled} \\ \text{solids conc.}\end{array}\right) + \left(\begin{array}{c}\text{target drilled solids} \\ \text{concentration}\end{array}\right) \Big/ \left(\begin{array}{c}\text{drilled solids} \\ \text{conc. in discard}\end{array}\right)}$$

The derivation for this equation is presented in the following section. For the preceding case, in which the targeted drilled-solids concentration is 4%vol, and the drilled-solids concentration in the discard is 35%vol, the optimum solids-removal efficiency would be:

$$\text{optimum solids-removal efficiency} = \frac{(1 - 0.04)}{1 - 0.04 + (0.35/0.04)} = 89.4\%$$

This agrees with the preceding calculation, which indicates that 90% removal efficiency is almost an optimum value.

15.7.1 Equation Derivation

$$\text{volume of discard} = \frac{\text{(solids removal efficiency)(solids to surface)}}{\text{(drilled solids concentration in discard)}}$$

- Volume of clean drilling fluid:

  volume of new drilling fluid built = volume of clean drilling fluid
  + volume of retained drilled solids.

- The volume of new drilling fluid built and the volume of retained drilled solids may be expressed in terms of the discard concentration and the targeted drilled-solids concentration in the drilling fluid.
- The volume of new drilling fluid built requires that the drilled-solids concentration in the new fluid be the targeted concentration, or

  drilled solids volume = (targeted drilled-solids concentration)
  × (new drilling fluid built).

- The second term on the right side of the equation relates to the volume of retained drilled solids, which is determined by the solids-removal efficiency:

    volume of retained drilled solids = (1 − solids removal efficiency)
    × (drilled solids to the surface).

- These expressions may now be substituted into the expression for the volume of clean drilling fluid that needs to be added to the drilling fluid system, which is:

$$\text{volume of new drilling fluid built} = \begin{pmatrix} \text{volume of} \\ \text{clean drilling} \\ \text{fluid needed} \end{pmatrix} + \begin{pmatrix} \text{volume of} \\ \text{retained} \\ \text{drilled solids} \end{pmatrix}.$$

$$\frac{\text{volume of drilled solids to surface}}{\text{target drilled solids concentration}}$$

$$= \begin{pmatrix} \text{volume of clean} \\ \text{drilling fluid needed} \end{pmatrix} + (1 - \text{solids-removal efficiency})$$

× (volume of drilled solids to surface)

- Solving this equation for the volume of clean drilling fluid that must be added to dilute drilled solids remaining in the system:

$$\frac{\text{volume of clean drilling fluid}}{\text{volume of drilled solids to surface}}$$

$$= \frac{1 - \begin{pmatrix} \text{solids removal} \\ \text{efficiency} \end{pmatrix} - (1 - \text{SRE}) \begin{pmatrix} \text{target drilled solids} \\ \text{concentration} \end{pmatrix}}{\text{target drilled solids concentration}}$$

15.7.2 Discarded Solids

Solids discarded by the solids-removal equipment contain some of the original resident drilled solids and the new drilled solids that have just entered the system. For example, in the case of 70% removal efficiency, 70% of the newly drilled solids are discarded in a slurry of drilling fluid. The target drilled-solids concentration in the preceding examples was 4%vol. For the case of 70% removal efficiency (Figure 15.4), 70 bbl of drilled solids will be discarded in a 200-bbl slurry that contains 130 bbl of drilling fluid. The 130 bbl of drilling fluid will contain 5.2 bbl of resident solids. The total quantity of drilled solids will be 75.2 bbl. The SREE relates only to removed solids that decrease the solids concentration in the system. If a volume of 130 bbl of drilling fluid is dumped from the system, the remaining drilling fluid still has 4%vol drilled solids, whereas the removal of 70 bbl by the equipment reduces the total solids concentration in the system.

15.8 OPTIMUM SOLIDS REMOVAL EQUIPMENT EFFICIENCY (SREE)

Equating the volume of clean drilling fluid needed to the volume of discard results in the minimum volume of clean drilling fluid needed and, as a consequence, the minimum volume of drilling fluid disposal. For that reason the resulting SREE required is called the optimum solids-removal efficiency. It is independent of the volume of drilled solids reaching the surface, or the volume of the drilling-fluid system.

If the SREE is less than the value required to achieve the minimum volume, the amount of fluid increases rapidly as the removal efficiency decreases. Figures 15.6 and 15.7 present the results of calculations similar to the ones for the targeted 4%vol drilled solids concentration previously described. As the targeted drilled-solids concentration is increased, the volume requirements decrease for any particular solids-removal efficiency. For example, at 70% SREE, maintaining 4%vol drilled solids requires adding 7 bbl of clean drilling fluid for every bbl of drilled solids, whereas maintaining 10%vol drilled solids requires adding 2.7 bbl of clean drilling fluid for every bbl of drilled solids.

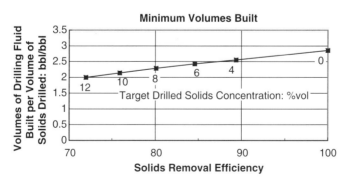

Figure 15.6. Minimum volumes required for various target drilled-solids concentrations for 35%vol discarded solids concentration.

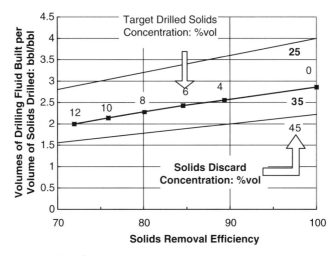

Figure 15.7. Clean fluid required decreases as the discard concentration increases.

The consequences of permitting 10%vol drilled solids, however, will usually completely eliminate any economic advantage of the lower volume requirement. The most effective, economical procedure is to improve solids-removal processing so that the efficiency approaches the minimum value for any targeted drilled-solids concentration. The targeted solids concentrations must be carefully selected. Artificially large values result in trouble costs and rig downtime; artificially low values will significantly increase drilling fluid and disposal costs (Figure 15.8).

Frequently, some rules of thumb are discussed in the literature about how much drilling fluid is required per barrel of drilled solids. In Figure 15.9 a common value of 3 is shown. This value depends on the targeted drilled-solids concentration and the SREE.

Some evaluation techniques involve not only accounting for the new solids drilled but also the resident drilled solids in the system before and after drilling. In the case of the 80% removal efficiency described, the discarded slurry contained 80 bbl of new drilled solids with 149 bbl of the drilling fluid (a total of 229 bbl). The 149 bbl of drilling fluid contained 4%vol drilled solids, or 6 bbl. Since this 6 bbl did not reduce the drilled-solids concentration in the system (it did reduce the total volume, but not the concentration), it was not part of the removal efficiency of the equipment.

To account for the 4%vol resident drilled solids, the volume of the system must be known. In this case, assume that a volume of 1000 bbl was available in the system before the drilling started. The drilling

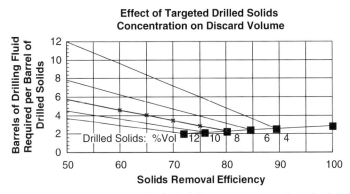

Figure 15.8. Effect of targeted drilled-solids concentration on discard volume.

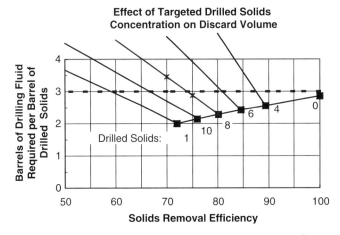

Figure 15.9. Material blance of 80% equipment solids removal efficiency.

fluid had a 4%vol drilled-solids concentration, or 40 bbl of drilled solids. If new rock of 100 bbl is drilled, the system would now have a volume of 1100 bbl, with 100 bbl of new drilled solids and 40 bbl of the original drilled solids. The drilled-solids concentration in the system would now be 140 bbl/1100 bbl, or 12.7%vol. Since the new rock occupies the same volume before and after drilling, the pit levels remain constant (except for the volume of drill pipe that enters the borehole). The total volume of drilled solids in the system is now 140 bbl.

These new drilled solids arrive at the surface and 80 bbl of them are discarded in a 35%vol slurry, or a total of 229 bbl of material leave the system. The pit levels drop by 229 bbl.

How many drilled solids are left in the system now? The discard was 80 bbl of new drilled solids and 6 bbl of the original drilled solids, or 86 bbl. This leaves 140 bbl − 86 bbl, or 54 bbl of drilled solids in the system. The system volume is now 1100 bbl − 229 bbl, or 871 bbl. The system contains 20 bbl of the new drilled solids and 40 bbl − 6 bbl, or 34 bbl, of the original drilled solids, for a total of 54 bbl of drilled solids. The drilled-solids concentration at this time is 54 bbl / 871 bbl, or 6.2%vol.

These 54 bbl of drilled solids must be diluted to the requisite 4%vol drilled solids concentration. Mathematically, this could be stated: 54 bbl = 0.04 (total new drilling fluid volume). So, the new drilling fluid volume must be 1350 bbl. Currently the system volume is 871 bbl, so 479 bbl of clean drilling fluid must be added to dilute the remaining 54 bbl of drilled solids. This is almost the same (within roundoff errors) as the 480 bbl originally calculated in the preceding section.

Another way to evaluate this would be to observe that the original 4%vol drilled solids that was in the drilling fluid originally did not have to be diluted. Only the new 20 bbl of drilled solids that remained in the system had to be diluted to 4%vol.

In the field, the SREE is not known. The discard volumes and solids concentrations in the system can be measured. From these numbers, the SREE can be calculated.

15.9 SOLIDS REMOVAL EQUIPMENT EFFICIENCY IN AN UNWEIGHTED DRILLING FLUID FROM FIELD DATA

Situation: NoProfit Drilling Company is drilling 100 bbl of hole daily in a formation with 15% porosity. For four consecutive days, 400 bbl of discards and fluid were captured each day in discard tanks. The pit levels remained constant, but some drilling fluid was jetted to the reserve pits daily to keep the pits from overflowing. The unweighted drilling fluid weighed 9.4 ppg daily and contained 2%vol bentonite.

Since no barite is contained in the drilling fluid, the volume of low-gravity solids (V_{LG}) is the same as the volume of total suspended solids (V_s). Assume SG of the low-gravity solids to be 2.6.

The equation for determining V_{LG} is:

$$V_{LG} = 62.5 + 2.0\ V_s - 7.5(MW)$$

where MW = mud weight. Or, $V_s = V_{LG} = 7.5(MW) - 62.5 = 8\%vol$.

Part of this 8%vol low-gravity solids concentration was 2%vol bentonite, so the content of low-gravity drilled solids was 6%vol.

The pit levels decrease by the quantity of material removed from them. If no fluid or solids are removed from the system, the pit levels remain constant (except for the increase in volume of the drill pipe entering the hole). In this case, the volume decrease is 400 bbl daily, or 1600 bbl. This must also be the total volume of clean mud added to the system if the pit levels are returned to their original position.

The %vol drilled solids in the mud remained at 6% daily. So, the drilled solids retained must be 6% of the volume of the newly built mud, or

volume of retained solids = (0.06)(volume of new mud built).

The new drilling fluid built daily comprises the clean mud added and the drilled solids that remain in the drilling fluid after it has been circulated through the solids-removal equipment.

volume of new drilling fluid built in the drilling fluid system
= drilled solids retained + clean drilling fluid added.

The quantity of drilled solids retained can be substituted into that equation, resulting in:

volume drilling fluid mud built
= (0.06)(volume new drilling fluid built) + clean drilling fluid.

The volume of clean drilling fluid must be exactly the volume that was discarded, or 1600 bbl. This gives an equation with one unknown:

volume of new mud built = (0.06)(volume of new mud built)
+ 1600 bbl
= (1600)/(1 − 0.06) = 1702 bbl.

Since the drilled solids retained are 6% of the volume of the new mud built, the drilled solids retained = (0.08)(1702 bbl) = 136 bbl.

With 15% porosity, the 400 bbl drilled resulted in the addition of 340 bbl of solids to the system. If 136 bbl of drilled solids were retained, 204 bbl were discarded. This gives a ratio (or SREE) of 204 bbl / 340 bbl, or 0.60. The SREE is 60%.

For comparison, the procedure in API RP 13C used to calculate the same SREE is presented in the Appendix.

15.9.1 Excess Drilling Fluid Built

Normal discards from fine screens on linear motion shale shakers and from hydrocyclones contain about 35%vol solids, as shown in Figure 15.6. If all of these are drilled solids, the volume of drilling fluid discarded with the drilled solids can be calculated. The statement that the volume of discarded solids is equal to 35% of the discarded volume can be written:

discarded drilled solids
$= 0.35$(volume of fluid discarded with drilled solids)

204 bbl $= (0.35)$(volume of total fluid discarded with drilled solids)

volume discarded with drilled solids $= 583$ bbl.

The 583 bbl of waste discard would contain 204 bbl of drilled solids and 379 bbl of drilling fluid.

Since a total of 1600 bbl were discarded, 1017 bbl of good drilling fluid were removed from the system along with the 583 bbl of waste drilled solids and drilling fluid. The 1017 bbl of good drilling fluid was the excess clean fluid added to dilute the retained drilled solids and could be pumped to a storage pit. However, eventually this excess drilling fluid must go to disposal. Ideally, the amount of clean drilling fluid added to the system will be exactly the volume discarded with the drilled solids (583 bbl).

Consider a case in which the discard from the hydrocyclones and the shale shaker was very wet (meaning that a large volume of liquid was discarded with the drilled solids), so that the discard contained 20%vol drilled solids instead of 35%vol.

discarded drilled solids
$= 0.20$ (volume of fluid discarded with drilled solids)

50.5 bbl $= (0.20)$(volume of fluid discarded with drilled solids)

or volume of fluid discarded with drilled solids $= 253$ bbl. Again, since a total of 400 bbl were discarded daily, 400 bbl – 253 bbl, or 147 bbl, of good drilling fluid were also discarded.

A conclusion should be obvious at this point. Efforts to eliminate all dripping of drilling fluid from the end of a shale shaker are futile and not needed when the SREE is around 60%. This drilling fluid will need to be discarded eventually, and shale shakers do a better separation when the effluent is still wet. This will usually point the way toward using shaker screens with smaller openings. Coarse screens allow more drilled solids to pass through.

15.10 ESTIMATING SOLIDS REMOVAL EQUIPMENT EFFICIENCY FOR A WEIGHTED DRILLING FLUID

After drilling 1000 ft of hole with a 12.5-lb/gal drilling fluid circulated at 25 bbl/min, the hole was circulated clean. This required four hole volumes to eliminate all solids in the discard. Assuming that the formation averaged about 13%vol porosity, a multiarmed caliper indicated that a volume of 97.3 bbl of new hole was drilled. The drilling fluid was freshwater-based mud weighted with barite and contained 2%vol bentonite, no oil, and 5%vol drilled solids. While drilling this interval, 1350 sacks (sx) of barite (100 lb/sx) were added to the system, and the drilled solids remaining in the system were diluted as required to control their concentration at the targeted 5%vol. Some drilling fluid was pumped to the reserve pits, and all discards of the solids-control equipment were captured in a container to be shipped back to shore. One drilling-fluid technician reported that 200 bbl were hauled to shore, and another reported that 180 bbl were captured.

These data represent information available in most field operations. Certainly, knowledge of the volume of solids reaching the surface is necessary for any calculation. These data will be assumed to be available from a hole caliper and circulation of the hole clean after drilling a particular interval. But it is one of the most difficult parameters to determine. Solids do not report to the surface in the same order that they were drilled, nor do they report in a predictable period of time. The preceding problem was deliberately set up to remove all of the drilled interval. Next, it is needed to know (1) the volume of clean drilling fluid added to the active system to dilute the cuttings from the hole, and (2) the volume of clean drilling fluid added to the active system to dilute the drilled solids remaining in the system. With a weighted mud, the number of sacks of barite and an analysis of solids concentrations in the drilling fluid allow a calculation of the clean drilling fluid added. Similarly, if the liquid volume added (water, oil, or synthetics) is known, the volume of clean drilling fluid can be calculated. Finally, if all of the discard volumes are captured in a disposal tank or container, the volume of discard can be measured.

Calculation of SREE does not require knowledge of the volume of the circulating system if the other information is available. The system has reached a stable drilled solids concentration, and the changes to the system are the primary concern. In actual practice, the system is dynamic, with continuous additions of small amounts of drilling fluid ingredients and continuous discards from the solids-removal equipment. At the drilling rig, sand traps are dumped with a variety of quantities of good drilling fluid. For this reason, these calculations should involve a reasonably long interval of hole to include all of the solids reaching the surface.

15.10.1 Solution

Volume of New Drilling Fluid Built While Drilling the Interval

Assuming a density (SG) of low-gravity solids of 2.6 g/cc and a barite density of 4.2 g/cc, the content of low-gravity solids can be calculated from the equation:

$$V_{LG} = 62.5 + 2\,V_s - 7.5\,(MW)$$

where

V_{LG} = content of low-gravity solids, %vol
V_s = total solids content, %vol
MW = mud weight, lb/gal.

Rearranging to solve for total solids:

$$V_s = [V_{LG} + 7.5(MW) - 62.5]/2.0.$$

For a 12.5-ppg drilling fluid containing 5%vol drilled solids and 2%vol bentonite (V_{LG} = 7%vol), the total solids are 19.1%vol. Since the bentonite and drilled solids account for 7%vol, the remaining 12.1%vol is barite.

A barrel of barite, SG 4.2, weighs 1470 lb:

$$(4.2)(8.34\ \text{lb/gal})(42\ \text{gal/bbl})/100\%\text{vol}.$$

The 1350 sx barite, or 13,500 lb, added during the drilling of this interval are equivalent to 91.8 bbl of barite. Assume that the MW and drilled solids were maintained at the stated levels during the drilling of the interval.

Stated in an equation:

$$\text{barite volume in the drilling fluid} = (12.1\%)(\text{volume of new drilling fluid built})$$

$$91.8 \text{ bbl barite} = (0.121)(\text{volume of new drilling fluid built}).$$

The volume of new drilling fluid built is 759 bbl. The volume of drilled solids would be 5% (759 bbl) or 38 bbl.

Solids Removal Equipment Efficiency (SREE)

SREE is calculated from the ratio of the volume of drilled solids discarded (97.3 bbl − 38.0 bbl) to the volume of drilled solids arriving at the surface (97.3 bbl).

$$\text{SREE} = [59.3 \text{ bbl}/97.3 \text{ bbl}] = 61\%.$$

The volume of clean drilling fluid added to dilute the 38 bbl of retained drilled solids must be exactly the volume discarded if the pit levels remained constant.

The new drilling fluid built consists of the retained drilled solids and the clean drilling fluid added to dilute those solids to 5%vol. This is stated mathematically:

$$\text{new drilling fluid volume built} = \text{clean drilling fluid added} + \text{drilled solids remaining}.$$

The volume of clean drilling fluid added would therefore be (759 bbl − 38 bbl), or 721 bbl.

Volume of Drilling Fluid Created from Adding the Clean Drilling Fluid

The volume of clean drilling fluid added was 721 bbl (as calculated from the amount of barite added). The pit levels would decrease by the volume discarded. The discarded volume is 57.3 bbl of drilled solids and the associated drilling fluid. After this decrease, the pit levels increase by 721 bbl. This could have been added as water and a blend of ingredients or, as most common, as individual components during the drilling process.

Excess Drilling Fluid Generated

The pit levels decrease only by the quantity of drilling fluid removed from the system. If nothing were removed, the pit levels would not change (except by the volume of drill pipe added to the system). The 97.3 bbl of drilled solids have been added to the system, but 97.3 bbl of new hole means that the pit levels stay constant. The pit levels will drop by the amount of fluid and solids removed from the pits and will rise by the volume of new material added to the system.

The volume of clean drilling fluid added was 721 bbl. The volume of material removed was either 200 bbl or 180 bbl, depending on which drilling-fluid technician is correct. This would indicate that a volume of 521 or 541 bbl of excess drilling fluid was built while drilling this interval.

As we previously noted in the case of unweighted drilling fluid,

- it is obviously futile to seek to prevent all drilling fluid from dripping from the end of the shale shaker, nor is the attempt needed when SREE is around 60%;
- the drilling fluid will need to be discarded eventually, and shale shakers do a better separation when effluent is wet; and
- wet effluent will usually point the way to using shaker screens with smaller openings. Coarse screens allow more drilled solids to pass through.

15.10.2 Inaccuracy in Calculating Discard Volumes

The volume of discard was either 200 bbl or 180 bbl. In either case the discard tanks had to contain 59.3 bbl of drilled solids that came to the surface from the drilling operation. In one case, the drilled-solids concentration of newly drilled solids in the discard would be 59.3 bbl/200 bbl, or 29.7%vol. In the other, it would be 59.3 bbl/180 bbl, or 33.9%vol.

In the 200-bbl case, 141 bbl of good drilling fluid were discarded; and in the 180-bbl case, 121 bbl of good drilling fluid were discarded. In other words, the total difference was only 20 bbl of good drilling fluid. So either 521 or 541 bbl of good drilling fluid would be sent to storage. The difference was relatively insignificant. Note that the 20-bbl error did not in any way affect the calculation of the SREE.

15.11 ANOTHER METHOD OF CALCULATING THE DILUTION QUANTITY

The dilution required to compensate for the incorporation of 37.9 bbl of drilled solids was the 759 bbl of new mud, less the volume of the drilled solids in the new drilling fluid built, or 721 bbl. This is 19 bbl of dilution per bbl of incorporated solids (721.0 bbl/37.9 bbl). The calculation for the dilution, that is, the volume of new mud that must be prepared, is

$$V = (100 - \%)/\%$$

where $V=$ the volume of new mud (dilution) required, in bbl/bbl of incorporated solids, and $\% =$ the concentration of drilled solids.

The volume of new mud plus the incorporated cuttings, which is the total volume increase, is simply the volume increase factor (VIF) multiplied by the volume of incorporated solids:

$$\text{VIF} = 100/\%.$$

At 5%vol drilled solids, V, the dilution volume, is 19 times the volume of incorporated solids, and the volume increase is 20 times that.

15.12 APPENDIX: AMERICAN PETROLEUM INSTITUTE METHOD

Using the current API technique for the problem: The dilution corresponding to total incorporation of the drilled solids (no separation) would be the volume of solids drilled divided by the drilled-solids fraction; in this case:

$$97.3 \text{ bbl}/0.05, \text{ or } 1946 \text{ bbl}.$$

The dilution factor (DF) is the ratio of the volume of new mud actually prepared to that which would have been required with no drilled-solids removal. In this instance, a volume of 759 bbl of new mud has been built and the DF is 759 bbl/1946 bbl, or 0.39. The required dilution was 39% of what it would have been if none of the drilled solids had been separated.

15.12.1 Drilled Solids Removal Factor

Observe the relationship between DF and the drilled-solids removal of 61% calculated above. The drilled solids removal factor (DSRF) is defined as

$$\text{DSRF} = 100(1 - \text{DF}).$$

It is numerically equal to the SREE previously calculated. In this case:

$$\text{DSRF} = 100(1 - 0.39) = 61\%.$$

15.12.2 Questions

1. What is the volume of new drilling fluid built while diluting drilled solids remaining in the drilling fluid?
2. What volume of drilled solids was discarded?
3. What is the solids removal equipment efficiency?
4. What volume of drilling fluid resulted from the clean drilling fluid being added?
5. How much excess drilling fluid, containing 5%vol drilled solids, was generated that had to be stored?
6. What is the cost of the error in the mud engineer's estimate of material captured for discard? (i.e., does this number have to be very accurate?)

15.13 A REAL-LIFE EXAMPLE

15.13.1 Exercise 1

Never-Wrong Drilling Company is drilling a $12\frac{1}{4}$-inch hole from a casing seat at 4000 feet to the next casing seat at 10,000 feet. The 12-ppg water-based drilling fluid has an 18%vol solids content with a methylene blue test of 18 lb/bbl. After drilling the well, a caliper indicates that the borehole washed out to an average diameter of 14 inches. While drilling this 5000-foot interval, a volume of 7163 bbl of clean drilling fluid was metered into the system to maintain a constant solids concentration of 18%vol.

This is a fairly common example. From these data, the SREE can be calculated. First, the hole volume, or volume of solids reaching the surface, is calculated as 1143 bbl.

$$\text{hole volume} = \frac{(14.0 \text{ inches})^2}{1027} = 1143 \text{ bbl}$$

This assumes that the rock drilled has no porosity. The value of solids could be reduced by 20% to a value of 914 bbl, but the accuracy of the diameter measurement is insufficient to justify such a minutia.

Second, the concentration of drilled solids in the drilling fluid must be determined. For this calculation, the density (SG) of the barite will be assumed to be 4.2 g/cc and that of the low-gravity solids 2.6 g/cc.

$$\begin{bmatrix} \text{volume fraction of} \\ \text{low-gravity solids} \end{bmatrix} = 62.5 + 2.0 \begin{bmatrix} \text{volume fraction} \\ \text{of solids} \end{bmatrix}$$
$$+ 7.5 \text{ [mud weight, ppg]}$$

This calculation indicates that the concentration of low-gravity solids is 8.5%vol. Since 2%vol is bentonite, the drilled-solids concentration in this 12-ppg drilling fluid is 6.5%vol. The total solids concentration is 18%vol, so the barite concentration must be 9.5%vol. The solids-removal equipment will discard some drilled solids and also retain some in the system. The solids remaining in the tanks must be diluted with clean drilling fluid to produce a 6.5%vol drilled-solids concentration. The liquid added to the mud system to dilute the returned solids will be the *dilution volume*.

$$\text{drilled solids returned} = (\text{drilled solids concentration})$$
$$\times (\text{dilution volume})$$

This equation says that the drilled solids in the new drilling fluid added to the tanks will be diluted to the drilled-solids concentration (6.5%vol). The total fluid added to the mud system (which is the dilution volume) will consist of two components, the clean drilling fluid and the drilled solids returning to the system.

Third, calculate the volume of drilled solids returned to the system by the solids-control equipment. The volume of clean drilling fluid added during the drilling of this 6000 feet of hole was 7163 bbl. The drilled-solids concentration in the drilling fluid was maintained at 6.5%vol. From the preceding equation: 7163 bbl = drilled solids returned [(1/0.065) −1], or drilled solids returned to the system = 497 bbl.

Fourth, calculate the removal efficiency of the surface equipment on this rig. Since 1143 bbl arrived at the surface during the drilling of this interval, 646 bbl were discarded. The removal efficiency would be the ratio of the solids discarded and the solids arriving × 100, or 56.5%.

Costs!!

Without commenting on the hole problems associated with drilled solids, the cost of this inefficient removal system will perhaps be surprising. For this example, the drilling fluid will be assumed to cost $90/bbl and the disposal costs will be assumed to be $40/bbl.

How much drilling fluid was discarded during this 6000-foot interval? This would be the sum of the volume discarded with the solids-removal equipment and the excess volume needed for dilution of the drilled solids.

This system discarded 646 bbl of drilled solids. These drilled solids were not dry; they carried with them some of the drilling fluid. For 200-mesh shaker screens and other equipment, a concentration of 35%vol drilled solids in the discard is a reasonable number. This means that the volume of discard would be (646 bbl/0.35), or 1846 bbl.

The pit levels would have decreased by 1846 bbl during the drilling of this interval. During this part of the well, a volume of 7163 bbl of clean fluid was added to the 497 bbl of drilled solids returned to the pits by the solids-control equipment. The total drilling-fluid volume built would be 7660 bbl. Since only 1846 bbl of fluid were removed from the pits, an excess volume of (7660 − 1846), or 5814 bbl, was built.

The new drilling fluid would cost ($90/bbl × 7660 bbl), or $689,000. The discarded volume would be the sum of the discarded volume from the solids-control equipment and the excess volume generated to dilute the returned drilled solids, or 1846 bbl + 5814 bbl, or 7660 bbl. This cost would be $40/bbl × 7660, or $306,000. Total drilling-fluid cost for operating this system for 6000 feet of drilling would be $995,000, or $166/ft of hole.

15.13.2 Exercise 2

In this inefficient drilled-solids removal system, the decision is made to drill with a polymer drilling-fluid system that required a 4%vol drilled-solids concentration. The cost of reducing the drilled-solids concentration by only 2.5% is startling.

With 56.5% removal efficiency and 1143 bbl of drilled solids reporting to the surface, 646 bbl will be discarded and 497 bbl will be returned to the pits (as before). The returned drilled solids must be reduced to 4%vol in the pits, so the dilution volume will be (497 bbl/0.04), or 12,430 bbl. Assuming the same volume concentration of drilled solids in the equipment discard of 35%, as before, the equipment discard (and the decrease in pit volume) will be 646 bbl/0.35, or 1850 bbl. The excess drilling fluid generated is (12,430 bbl − 1850 bbl), or 10,580 bbl.

Now the costs:

- New drilling fluid added to the system = 12,430 bbl × $90/bbl, or $1,119,000
- Disposal for this interval = (1850 bbl + 10,580 bbl) × $40/bbl, or $497,000
- Total cost for this 6000-foot interval = $1,616,000, or $269/ft of hole. Decreasing the drilled-solids concentration from 6.5 to 4%vol with this inefficient removal system increases the cost by $621,000.

15.13.3 Exercise 3

This exercise involves the cost benefit of increasing SREE to 80% for the 4%vol drilled-solids concentration:

With 80% removal efficiency and 1143 bbl of drilled solids reporting to the surface, 914 bbl would be discarded and 229 bbl returned to the pits. The drilling fluid needed to dilute the 229 bbl to 4%vol would require adding (229 bbl/0.04), or 5725 bbl of new drilling fluid. This 5725 bbl would consist of 229 bbl of drilled solids and 5496 bbl of clean drilling fluid. Assuming the 35%vol concentration of drilled solids in the discard, the discard volume would be 914 bbl/0.35, or 2611 bbl. The excess volume of drilling fluid is (5725 bbl − 2611 bbl), or 3114 bbl.

The drilling-fluid cost for 80% removal efficiency and 4%vol drilled solids would be (5725 bbl × $90/bbl) + [(3114 bbl + 2611 bbl) × $40], or $744,000. Increasing the SREE from 56.5 to 80% decreases the cost of the new fluid and disposal from $1,616,000 to $744,000 for this 6000 feet of hole. The $872,000 difference could justify significant changes in the drilling-fluid system.

15.13.4 Exercise 4

This exercise demonstrates the effect of a slight increase in the drilled-solids concentration:

If the 80% removal efficiency were achieved and a 6%vol drilled-solids level would not create hole problems, another significant cost reduction is possible. With 80% removal efficiency and 1143 bbl of drilled solids reporting to the surface, 914 bbl would be discarded and 229 bbl returned to the pits.

The drilling fluid needed to dilute the 229 bbl to 6%vol would require adding (229 bbl / 0.06), or 3817 bbl, of new drilling fluid. This volume would consist of 229 bbl of drilled solids and 3588 bbl of clean drilling fluid. Assuming 35%vol drilled solids in the discard, the discard volume would be (914 bbl / 0.35), or 2611 bbl. The excess volume of drilling fluid is (3817 bbl − 2611 bbl), or 1206 bbl.

The drilling-fluid cost for 80% removal efficiency and 6%vol drilled solids would be (3817 bbl × \$90 / bbl) + [(1206 bbl + 2611 bbl) × \$40], or \$496,000. With an 80% removal efficiency, increasing the drilled-solids target concentration from 4 to 6%vol decreases the cost of the new fluid and disposal from \$744,000 to \$496,000 (or by \$248,000) for this 6000 feet of hole.

15.13.5 General Comments

Exercise 1 assumed that the clean drilling fluid volume was known. Frequently, drilling fluid is not metered into the pits, but the new fluid is built by adding liquid and solids to the tanks to keep the pit levels constant and maintain the drilled-solids concentration at some target value.

Since the barite concentration in the drilling fluid was also kept at the same value, 9.5%vol, it has been used as a tracer to determine how much clean drilling fluid was added. In Exercise 1, the dilution fluid contains 9.5%vol barite, or 680 bbl. Barite weighs 1471 lb / bbl, so this would be equivalent to 1,000,000 lb of barite used (or 100,000 sx). The new drilling fluid built contained 7163 bbl of clean fluid and some volume of drilled solids that was returned to the pits by the solids-removal equipment.

Barite may be used as a tracer to determine the volume of clean drilling fluid added to the system. But using measured volumes of barite in the discard is fraught with problems. The liquid removed by fine-mesh shale shaker screens does not contain the same concentration of barite as does the drilling fluid in the pits. In one field test, between 100 and 300 lb / hr of excess barite was discarded from a continuous-cloth screen. During that field test, a deficiency of as much as 100 lb / hr of barite was observed in the discard from panel screens. In this case, barite went through the screen, which decreased the barite in the discard below the concentration values of the drilling fluid in the pits.

CHAPTER 16

WASTE MANAGEMENT

William Piper
Piper Consulting

Tim Harvey
Oiltools, Inc.

Hemu Mehta
Kem-tron Technologies, Inc.

16.1 QUANTIFYING DRILLING WASTE

Drilling waste consists of waste drilling fluid, drilled cuttings with associated drilling fluid, and, to a lesser extent, miscellaneous fluids such as excess cement, spacers, and a variety of other fluids. The amount of drilling waste depends on a number of factors. These include hole size, solids-control efficiency, the ability of the drilling fluid to tolerate solids, the ability of the drilling fluid to inhibit degradation or dispersion of drilled cuttings, and the amount of drilling fluid retained on the drilled cuttings.

One simple expression states the amount of wet drilled solids to be discarded as:

$$S = \varepsilon \times HV/F_s$$

where

S = volume of wet drilled solids, in bbl
ε = efficiency of solids control, expressed as a fraction
HV = hole volume, in bbl
F_s = fraction of solids in the discard stream.

The fraction of the solids in the discard stream varies from a maximum of about 50% to a lower value of about 25–30%. There is always some amount of drilling fluid associated with drilled cuttings being discarded. Solids-control systems, no matter how good, cannot totally separate the drilling fluid from the drilled cuttings. By the same token, rarely can all of the drilled cuttings be separated from the circulating system. This means that, with time, drilled solids will build up in the circulating system.

16.1.1 Example 1

If 16 m^3 of hole volume is drilled, if the solids-control efficiency is 70%, and if the fraction of solids in the discard stream is 0.5 (50%), how much drilled-solids discard is generated?

Sixteen cubic meters of earth (as drilled solids) is removed from the hole, and 70% of that volume is removed from the drilling fluid by the solids-control system. That means that 11.2 m^3 of "dry" drilled solids would be removed (16 m^3 × 0.7). But, since removing dry material without drilling fluid is impossible, some drilling fluid will be attached. In this case, a fraction of 0.5 indicates that an equal volume of drilling fluid is removed. Thus, two times the dry amount is removed (one part fluid and one part drilled solids). Therefore, the amount of drilled cuttings removed from the system is 22.4 m^3 (11.2/0.5).

16.1.2 Example 2

From a drilling rate of 50 feet per hour in a 12¼-inch hole, dry cuttings are being generated at 7.3 bbl/hr (1.16 m^3/hr). Tests indicate that the solids-control system is removing 15.3 bbl/hr of drilled solids with associated drilling fluid. Tests on the discard indicate that the solids fraction in the discard stream is 0.33, or 33%. What is the solids-control efficiency?

The solids-control efficiency can be estimated by using the formula

$$\varepsilon = S/(HV/F_s).$$

In this case, hole volume (HV) is 7.3 bbl/hr, solids removal (S) is 15.3 bbl/hr, and the fraction of solids in the discard (F_s) is 0.33. Thus, the efficiency is 70%:

$$15.3/(7.3/0.33).$$

Solids buildup in the drilling fluid is the portion of solids not removed and leads to dilution and discard of whole dirty fluid. The solids buildup is calculated by multiplying the hole volume by 1 minus the efficiency. Whole, dirty drilling fluid is discarded when it reaches the limit of drilled-solids tolerance for the fluid type. Typically, 5, 7, and 10% are used in water-based systems. Five percent indicates that the drilling fluid type is very intolerant to solids. If 7% is used, then the system is of about average tolerance. When 10% is used, then the system is reasonably tolerant for water-based fluid. A quick way of estimating the amount of drilling-fluid discard is to divide the solids buildup in the fluid by the drilled-solids tolerance (or the maximum percentage allowable of low-gravity solids).

$$L = HV \times (1 - \varepsilon)/T,$$

or

$$L = DS_L/T$$

where

L = liquid discard, in bbl
HV = hole volume, in bbl
ε = efficiency of solids control, expressed as a fraction
T = tolerance of the fluid system to solids contamination, expressed as a fraction
DS_L = solids buildup in the drilling fluid.

The total waste or discard for water-based drilling-fluid systems is the sum of the waste solids with associated fluid (S) and the liquid discard (L). As the solids-control efficiency increases, the waste solids increase slightly, but the liquid discard decreases dramatically. For this reason, small improvements in solids-control efficiencies can significantly reduce the total amount of waste for disposal.

16.1.3 Example 3

If 16 m^3 of hole volume is drilled and the solids-control efficiency is 70%, how much solids buildup in the circulating system will occur?

Since the solids-control efficiency is 70%, the amount of drilled solids not removed must be 30% $(1 - \varepsilon)$. Therefore, the amount of drilled solids buildup in the circulating system is 4.8 m^3 (16 × 0.3).

16.1.4 Example 4

If 189 m³ of hole volume is drilled and the solids-control efficiency is 70%, how much drilling fluid will need to be discarded if the tolerance of the system is 7%? In this example, 189 m³ of hole volume represents the approximate volume of 4000 feet of 17½-inch hole, which is a large but reasonable hole section. Since 30% of the hole volume is not removed by the solids-control equipment, 56.7 m³ of cuttings (189 × 0.3) will build up in the system. Since the tolerance of the system is 7%, the amount of fluid discard is 810 m³ (56.7/0.07).

Table 16.1 indicates the amount of waste generated under similar conditions with varying solids-control efficiencies. Hole volume is constantly 6000 barrels (954 m³). The solids content in the liquid discard is 50%. Tolerance of the fluid is 5%, meaning that the fluid system is sensitive to buildup of low-gravity solids.

In Table 16.1, notice that the hole volume is *not* the amount of waste fluid and cuttings. Yet calculated hole volume is still frequently reported as the total drilling waste. It can also be seen from the table that as the solids-control efficiency increases, the total waste or discard volume decreases. The converse is also true.

Data for total waste in the table are simply compared with the calculated hole volume. The larger the ratio, the lower the solids-control efficiency. In soft rock drilling conditions when water-based fluid is used, this ratio is frequently in the range of 8 or higher.

Table 16.2 illustrates the amount of waste generated under similar conditions but with solids tolerance at 10% rather than 5%. The total waste in Table 16.2 is much lower than that in Table 16.1 because the drilling-fluid system is much more tolerant to the accumulation of drilled solids. The system is considered more tolerant because the drilling fluid

Table 16.1

Efficiency (%)	Solids and associated LD	LD	TW	Ratio TW:HV
30	3,600	84,000	87,600	14.6
50	6,000	60,000	66,000	11.0
70	8,400	36,000	44,400	7.4

LD = liquid discard; TW = total waste; HV = hole volume.

Table 16.2

Efficiency (%)	WS	FD	TW	Ratio TW:HV
30	3,600	36,000	39,600	6.6
50	6,000	30,000	36,000	6.0
70	8,400	18,000	26,400	4.4

WS = waste solids; FD = fluid discharge; TW = total waste; HV = hole volume.

system is more inhibitive, the solids are less dispersive, or the effects upon the fluid properties can be tolerated.

When an oil-based fluid (or a synthetic fluid with similar inhibitive characteristics) is used, the system is both more tolerant and more inhibitive. Typically, whole fluid is not discarded from an oil-based system, except when fluid is lost or during cleaning, etc. Measured discard rates are on the order of three or four times the hole volume, or less. This indicates that the solids-control efficiencies obtained while drilling with oil-based fluid are high and that the sensitivity of oil-based fluid to solids contamination is high (probably greater than 10%).

This can have significant implications. Modern water-based fluids contain inhibitive additives, fluid loss agents, lubricants, etc., that are used to adjust fluid properties. Many of these additives contain organic materials. When large amounts of the drilling fluid are discarded into the sea, the chemical loading from the organic materials can be high. Synthetic fluid has potentially higher organic materials per barrel discharged, but less fluid is discharged. Therefore, the resulting chemical loading from organic materials can be less for synthetic-based fluid. All of these factors are used to determine the ultimate fate of the environment in the area of disposal.

16.1.5 Example 5

If 189 m^3 of hole volume is drilled with water-based fluid and the solids-control efficiency is 30%, how much total waste will be generated? Assume that the drilling fluid will need to be discarded if the tolerance of the system is 7% and the solid fraction in the discard is 0.5 (50%). What is the ratio of total discard to hole volume?

Use the following formula for solid and associated fluid discard:

$$S = \varepsilon \times HV/F_s$$

where S = volume of wet solids, in bbl; ε = efficiency; HV = hole volume, in bbl; F_s = solids fraction in the discard. The solids discard is 113.4 m³ (189 × 0.3/0.5).

Use the following formula for dirty, whole fluid discard:

$$M = HV \times (1 - \varepsilon)/T$$

where M = drilling fluid, in m³, and T = tolerance. Dirty, whole fluid discard is 1890 cubic meters (189 × 0.7/0.07).

The total discard is the sum of S and M. Thus, the total discard for this case is 2003 m³. The ratio of total discard to hole volume is 10.6 (2003/189).

16.1.6 Example 6

If 189 m³ of hole volume (as in example 5) is drilled with oil-based fluid and the solids-control efficiency is 80%, how much drilled-solid waste (S) will be generated? Assume that the fraction of solids in the discard is 0.5. How much solids buildup in the system can be expected? If 400 m³ of drilling fluid exists, then what would the concentration of drilled solids in the fluid be?

The drilled-solids discard would be 302.4 m³ (189 × 0.8/0.5).

The drilled-solids buildup in the system would be 37.8 m³ (189 × 0.2). If this amount is incorporated into 400 m³ of fluid, then the concentration is 9.5%.

Examples 5 and 6 illustrate the difference in waste generation between using water-based fluid and nonaqueous fluid (NAF). Since the water-based fluid is sensitive to drilled-solids contamination and is not very inhibitive, the total waste amount is high (2003 m³). By using NAF, inhibition of the formation solids is increased and NAF sensitivity to contamination is decreased. The result is less total discard being generated (302 m³). However, the fluid on cuttings being discarded is NAF. And half of the total discard (by volume) is fluid.

16.2 NATURE OF DRILLING WASTE

It is obvious, from the preceding discussion, that drilling waste contains a large amount of base fluid, whether that fluid is diesel oil, mineral oil, olefin, ester, or water. A more detailed discussion about the nature or characteristics of the waste should consider the place of disposal. In a broad sense, this can be accomplished by considering that all waste must be disposed in the water, on land, or in the air. For example, the characteristics of drilling waste when discharged offshore (disposal in water) will be viewed from the potential effects between the waste and water. These are effects to the seabed, to the water column itself, and to the air/water interface at the surface. In this scenario, diesel oil is an obvious contaminant. Diesel oil creates a sheen on the water surface, disperses in the water column, and creates a toxic effect in cuttings piles on the seabed. For this reason, diesel oil-based drilling fluids and the cuttings generated while using them are not discharged into the sea.

While it is beyond the scope of this text to fully discuss the nature of drilled cuttings, it is important to at least identify some of the common characteristics. Water-based fluids are generally considered relatively benign. The main concern is with the smothering effect of potential cuttings piles, although the creation of piles can be somewhat moderated by the manner of discharge, water depth, and strength of prevailing currents. There is also a concern for entrained oil, either from the formation or from surface additions. With modern emulsifiers, it is possible to entrain fairly large amounts of oil (say, 3–4%) without detection by standard rig site testing. There is also a concern for toxicity, as defined by the standard toxicity test run in the Gulf of Mexico. This is not truly a test of toxicity, but simply an indicator with a discharge/no discharge implication. Modern drilling fluids formulated for high inhibition can run close to the boundary of this test. Another concern is with heavy metals. With the use of barium sulfate (barite) to increase the drilling-fluid density, there is little direct concern with barium solubility or the biological availability of barium. However, there is concern for trace heavy metals within barite, such as mercury and cadmium.

All of the water-based considerations are also considerations with NAFs. In addition, there are specific concerns with the NAF itself. Generalized concerns associated with offshore discharges and NAFs include

- benthic smothering
- toxicity (aquatic or in sediments)
- sheen or entrained oil
- biodegradability (aerobic and anaerobic)
- bioaccumulation
- dispersibility
- persistence
- taint (alteration of flavor or smell of fish)
- heavy metals.

Most of these concerns are addressed by some sort of stock (base fluid) limitation and by limiting the amount of fluid to be discharged. Some areas restrict the type of base fluid that can be discharged based on biodegradation rate. There may also be limits on the amount of fluid retained on the cuttings when discharged. In this manner, any fluid on cuttings discharged (whole fluid is not discharged) will biodegrade rapidly and any effects will be short term.

The preceding discussion applies to discharges at sea when no special environmental condition exists. Special environmental conditions might be reefs, oyster beds, kelp beds, subsistence fishing grounds, or sites near shore. In freshwater environments such as lakes and rivers (or enclosed brackish waters), discharges may also pose a hazard due to simple sedimentation.

When considering land disposal options, the concerns are of a different nature. The concern with oil is still present, but to a much less extent. The *type* of oil is also important. Oil can be incorporated into dirt or soil and will biodegrade. The major concerns are about the concentration of oil remaining after biodegradation and potential plant toxicity of some portions of diesel oil. Some types of NAF will biodegrade to very low concentrations and do not exhibit toxicity to plants. Salts are a major concern. Salt is toxic to plants even at fairly low concentrations. Associated with the salt is the concern over sodium from sodium chloride. Sodium replaces calcium and magnesium in clays, causing a condition known as sodicity. Sodic soils collapse, causing a low permeability to water and a hard surface. Since water cannot infiltrate the soil matrix, there is no water available to support plant life. Further, salt inhibits the transport of water via osmosis to the plant.

Heavy metal content is the third major concern with drilled cuttings disposed onshore. While barium from barite has low solubility and bioavailability, there is still a concern with the concentration of barium in dirt or soil. Other heavy metals of potential concern that are found in drilled cuttings are lead and zinc, although these are found to be a problem far less often.

16.3 MINIMIZING DRILLING WASTE

Waste minimization or reuse of resources that can become waste are key strategies in waste avoidance and a sound waste management plan. Two general approaches to waste minimization have developed. They can be called total fluid management (TFM) and environmental impact reduction (EIR).

16.3.1 Total Fluid Management

One of the largest sources of drilling waste for onshore operations is location water. This happens to be the source that can be reduced most. Most wastewater originates from drilling-fluid usage, storm water, rig wash water, or cooling water. The volume of location water requiring handling and disposal could be as much as 30 times the hole volume!

Table 16.3 shows data collected from a project in Louisiana for wells drilled during 1995 and 1996. The first well established the baseline. Water reduction schemes began on the following wells. Some of the wells were concurrent because two rigs were used for the project.

While the hole volume remained fairly constant throughout the project, waste volume was reduced to about one-half the original amount. The later wells (#5 and #7) had waste volumes consistent with multiples of 15 times hole volume. One well (#6) achieved a waste volume multiple of 8 times hole volume. This well began using oil-based fluid earlier than the rest, which reduced total waste volume. However, this was an economic failure due to low penetration rate and higher cost of drilling.

The approaches taken to reduce wastewater generation were based on reuse of as much water as possible. They included the following techniques:

1. Single-pass systems, such as cooling water, brake water, and seal water, were eliminated. These should be contained by enclosed

Table 16.3
Waste Handled on Similar Wells

Well	Days	HV (bbl)	Waste Handled (bbl)	Ratio Waste:HV
1	105	2,792	102,000	36.7
2	85	2,748	80,000	28.9
3	94	2,671	72,000	26.8
4	69	2,634	59,000	22.5
5	76	2,820	41,000	14.5
6	102	3,142	25,000	7.9
7	85	2,807	53,000	18.9

HV = hole volume.

systems. Recycling these fluids is inexpensive and can save a large amount of fluid.

2. Storm water was reused. Storm water can be reused for fluid makeup water, although the drilling personnel may not like it much. It can also be used for rig wash water. Rig wash water (which falls into the same ditches as the storm water) should be reused until it is too dirty to be used as wash water. It is surprising how many times wash water can be used effectively.

3. The dirtiest water (such as drilling-fluid waste) was used for slide wash water. Desanders and desilters generate copious amounts of drilling-fluid waste (usually calculated at two or three parts liquid to one part solid), yet still require washing to the disposal pit. Shaker slides and centrifuge slides almost always require wash water. Slide wash water does not need to be clean, and the introduction of any clean water into the waste solids and fluid chemicals is an unnecessary addition of water that becomes difficult to separate during disposal.

4. Liquid waste was not generated needlessly. The use of rig vacuums rather than washing is increasing precisely because of the expense involved with disposal of waste liquids. Pistol-grip shutoff valves on hoses are a great idea. When the floor hand is called for a connection, the hose that is thrown down will shut off automatically rather than run the whole time during connections. High-pressure/low-volume washers are a favorite with rig crews, because they clean better with less effort. They also save liquid waste volume. Vacuums and washers are usually a breakeven cost unless the disposal cost is high, but pistol grips always pay off.

5. Wastes that were to be handled in different ways were separated. For instance, do not combine oil-based wastes with water-based wastes, unless they will be handled together. In this project, all liquid from the reserve pit was injected, so all liquid went to the reserve pit.

Preplanning was an integral part of the Louisiana project. Thinking about the operation and developing a plan to handle the waste streams always pays off. If you are going to dump the sand trap, then where will that waste stream go (especially if your pits are aboveground)? How will you handle the low-contaminant but large volume associated with surface hole and keep it separated from the oil-based fluid or highly treated fluid later? Once the location is built, it is usually too late to consider these things.

As can be seen, the TFM technique can be used to reduce waste volume or amount. However, the reduction in waste is mostly by reducing the amount of water in the waste. While this may not seem important to some, it is nonetheless saving a valuable resource and will prove cost-effective as disposal costs continue to escalate.

16.3.2 Environmental Impact Reduction

Another form of minimization strategy is to evaluate the environmental impact of the project and attempt to reduce it. In the EIR method, all fluids are evaluated for their chemical components. Certain environmental data are collected on each of the chemicals. The data might include parameters of

- toxicity
- biodegradation potential
- persistence
- bioaccumulation
- heavy metal concentrations

A review of the chemicals to be used would be made, and those chemicals with the least environmental impact would be selected.

A simple example of this is prequalifying a drilling-fluid system. In the prequalification, every chemical to be used is examined for the desired environmental characteristics and approved for use. In addition to each chemical individually, the entire system would be approved. Only approved chemicals, and only at the maximum approved concentration, would be allowed. This is, of course, a very complex system. Many fluid programs contain contingency chemicals that are used under only certain circumstances for a small portion of the hole.

16.4 OFFSHORE DISPOSAL OPTIONS

There are a limited number of options for dealing with drilling waste generated in an offshore environment. The drilling waste can be discharged to the ocean (direct discharge), injected into the ground beneath the sea, or taken to shore for commercial disposal or a land-based disposal option.

16.4.1 Direct Discharge

Direct discharge is the most common mode of disposal for cuttings and waste drilling fluids generated during offshore drilling operations. Directly discharging drilling waste is inexpensive and simplifies operations. But, in recent years, increasing attention has been paid to environmental risks posed by this activity. It is now recognized that there may be some long-term liability associated with discharging, even if water-based fluid is used.

Direct discharge is needed when drilling with water-based fluids due to the large quantity of waste associated with this activity. Besides generating cuttings and associated fluid, drilling with water-based fluid generates relatively large volumes of waste fluid. This is due, in part, to the relative intolerance of water-based fluid to solids buildup in the active system. It is very hard to control the desired properties of water-based fluid with a content of low-gravity solids (drilled solids) greater than 10%, and penetration rate will be adversely affected above 5%. Due to the sensitivity of water-based fluid to low-gravity drilled solids, a "dump and dilute" strategy may be adopted. Dumping whole, dirty fluid may create eight times the amount of waste created by the drilled cuttings.

Switching to oil- or synthetic-based drilling fluid will reduce the total amount of drilling waste generated. These NAFs are much more solids tolerant, and cuttings are not degraded as much as with WBM. The net result is that dumping of whole fluid is usually not needed.

In the last 20 years, cuttings generated while drilling with diesel-based fluid systems have been considered to be unsuitable for discharge. Other NAFs with drilling-performance benefits approaching diesel-based systems have been developed. While using some of these systems, cuttings may be discharged with certain limitations in many parts of the world. With NAFs that discharge cuttings, the associated fluid will be lost. Most of these fluids are supplied on a rental basis. The cost of any fluid not returned will be charged to the operator, which has become a major cost.

16.4.2 Injection

Injection involves making a suitable slurry out of the waste generated during drilling operations. This solids-laden slurry is pumped into the formation at pressures exceeding the fracture. Since this is the only disposal method not involving the surface or the sea bottom, waste streams that would be undesirable to surface-dispose of could be safely disposed of by injection. However, in the United States, only exempt waste can be injected. Nonexempt waste must be taken to an appropriate and approved UIC (underground injection control) well.

The use of injection as a method of disposal in offshore drilling operations has a history of mixed results. The first obstacle is having a conduit into a receiving formation. The slurry to be injected can be pumped down the inside of tubing or casing (through tubing injection), or it can be pumped down an annulus between casing strings into an open formation. Whichever method is used, the receiving formation must be isolated from the surface and other zones, especially up the cemented annulus. In many drilling operations, a conduit does not exist. The previous annuli have been cemented to surface, another well bore is not available, or a potential conduit simply does not exist. At other times, a satisfactory receiving formation does not exist.

Weak or unconsolidated sands are preferred as injection target. The sand can fluidize and repack, accommodating very large volumes of waste. Dedicated injection wells are preferred to annuli because the wells can be worked over if plugged by cuttings. Careful geotechnical surveys should be done to choose an appropriate zone. The zone must have a good seal above it, and the pipe must have a premium cement job.

If injection is to be used, then slurry must be made from the fluid and cuttings. The fluid and waste fluids are collected into slurry tanks to make the slurry. These slurry tanks are circulated with special slurry pumps designed to break up particles into natural grain sizes. If more fluid is needed because the slurry is too thick, then seawater can be added. The fluid may be too thin if excessive amounts of water are collected in the waste stream or if the mixing is insufficient. Frequently a shaker is used prior to the slurry tanks to prevent large particles or junk from entering the slurry tanks. In addition, there is a suction line screen to protect the pumps. There are usually two or three slurry tanks of 100- to 150-bbl capacity. This large volumetric capacity may be needed to handle large hole drilling.

As slurry is made in one tank, fluid and cuttings waste are diverted to another slurry tank. When the slurry is sufficiently mixed in the first tank and reaches an acceptable consistency, it is transferred to the slurry holding tank. This tank, or possibly series of tanks, is designed to hold the injection batch volume. The volume is based on the desired radius of injection. Care must be taken not to intersect nearby well bores, which may not be cemented to surface.

Weight (or space) limitations on the rig or platform frequently pose problems. Generally, batches of about 300 bbl of slurry are made (this is limited mainly by space and weight criteria). If 100 bbl of seawater are pumped before and after the slurry, then the total volume of each injection would be 500 bbl. Total slurry and equipment weight may be between 125 and 150 tons, which includes the slurry holding tanks in addition to the slurry skid. The footprint for the equipment may be as much as 40 feet wide by 90 to 120 feet long, depending on the equipment required. However, with a little innovation, the weight and space problems can be overcome.

The final problem is rate of waste generation. Offshore there is not a great deal of space in which to collect waste. For example, if 8½-inch hole is being drilled, then drilling 200 feet per tour would generate about 30 bbl of waste cuttings. An additional 50 bbl of oily wash water might also be generated. Total slurry volume would be about 100 bbl per tour. However, total slurry volume for 400 feet of 12¼-inch hole in a tour would be about 300 bbl. Similarly, 500 feet of 17½ inches may require handling as much as 1000 bbl of total slurry. For the large jobs, it is advantageous to have some storage (buffer) for generated waste. On the other hand, the cuttings from an 8½-inch hole can be collected in boxes to be slurried and injected later while drilling the 6-inch hole section. This may save rental cost on the slurry equipment during the slower drilling intervals and more fully utilize the rental equipment.

16.4.3 Collection and Transport to Shore

If discharge is not allowed or desired and injection is not possible, then the drilling waste must be collected and transported to shore for treatment or disposal. Generally, commercial disposal operations have been established in areas with supporting infrastructure. If no commercial operations have been established, then company-operated land disposal options must be implemented.

16.4.4 Commercial Disposal

Commercial disposal processes have been established in two main areas: the U.S. Gulf Coast and various countries bordering the North Sea. These processes will be described in order to establish a baseline of current commercial practices. Just because the commercial process exists does not mean that individual operators approve of the process or that long-term liabilities do not exist. However, it is possible that in a remote area with no onshore infrastructure for handling waste fluid and cuttings, some sort of commercial activity might be mimicked or encouraged.

At the present time in the Gulf Coast, all waste coming to shore from offshore drilling activities (approximately 6 million bbl per year) is handled by commercial facilities. It is possible for companies to permit, own, and operate their own disposal facilities, but it has not been done to date.

In general, fluid (mostly oil base) and cuttings returned to shore for disposal are separated into solid and liquid fractions. The liquid fraction is generally injected in a disposal well. The solid fraction is generally disposed of similar to land farming or converted to a dirtlike product that can be used.

One commercial disposal service provider in the Gulf Coast area operates several strategically placed waterfront waste receiving/transfer stations in Louisiana and Texas. The company sends material from these stations to four processing/disposal facilities in Texas. The waste fluid and cuttings received at any of the receiving/transfer facilities is offloaded into Coast Guard–approved (double hull construction) hopper barges for temporary storage. They can dump cuttings from boxes, receive waste from trucks into sumps, and pump off boat tanks. The filled hopper barges (possibly filled with multiple operators' waste) are transported from the transfer stations to the main/central facility to be offloaded for processing, recycling, and/or disposal.

At the processing facility, reclaimed components from spent drilling fluids may be conditioned into recycled drilling fluids and reused for other drilling operations, although this is a minor amount. Nonreusable liquids and solids are transported via truck to one of the company's injection wells. Slurries, containing up to 15% solids, are injected into porous geological formations below fracture pressure (these are typically depleted, subpressured wells). Typically, 20% of the fluid waste goes to reuse and 80% is injected.

A portion of the solids reclaimed at the processing facility are reused as daily cover in municipal sanitary landfills, under authority of the Texas Railroad Commission and the Texas Commission on Environmental Quality (TCEQ). The solids (dirt) used in this manner must meet criteria for landfill cover stipulated by the TCEQ.

Another company uses a process of collection and transfer similar to the one described, but differs in that it uses a salt cavern as a disposal site. Mud and cuttings are pumped into the salt cavern via a large-diameter tubing string. The brine in the salt cavern is displaced to a saltwater storage system up the annulus between the tubing string and casing. The brine is then injected into a permitted saltwater disposal well. The oil contained in the cuttings floats to the top of the heavier brine in the cavern, forming an oil blanket. This oil blanket prevents erosion of the cavern ceiling.

Commercial disposal processes in the North Sea area tend to focus on oil removal from the cuttings. The oil is intended to be reused, either as drilling fluid makeup or some other purpose, such as a fuel source. Currently, only about 30% of the recovered oil (fluid) is being reused. The other 70% is being used for fuel material. The cuttings (after de-oiling) are sent to landfills. Thermal desorption is used to de-oil the cuttings (refer to the Treatment Techniques section later in this chapter for a discussion of thermal desorption).

Commercial operations in the developed areas of the Gulf Coast and the North Sea rely on certain infrastructure or systems that have been established. For instance, landfills are lined and daily cover is used. In many overseas areas, landfills may not be lined and the contents may be burned on a periodic basis. This poses considerable liability if an attempt is made to use them anyway. Also, injection wells to handle liquid being brought to shore are rarely available overseas.

16.5 ONSHORE DISPOSAL OPTIONS

Onshore disposal options aim at incorporating drilling waste into either the surface (or rooting zone) or beneath the rooting zone. The former is called land application. The latter is called burial.

16.5.1 Land Application

The term *land application* refers to a disposal technique of incorporating drilling wastes (cuttings and associated drilling fluids) into the top few inches or feet of soil so that the resulting soil is still good for agricultural use. The same basic thing is meant by other terms, such as *land farming*, *land application*, and *land treatment*. Occasionally, the user might intend some fine points of difference. For instance, if the land is to be tilled, then *land farming* might be used, whereas in *land application*, no tilling is intended. No such distinction will be made here. It is assumed that some sort of mixing of waste and soil to form a waste/soil mixture will be required.

The technique of land spreading is ideally suited to the large amounts of solids (cuttings) and fluid generated from the drilling process.

The relatively small amounts of contaminants are easily incorporated into waste/soil mixtures, sometimes resulting in an improved soil. This is especially true if the soil has been neglected. By land spreading, the waste is actually *becoming* soil.

Land application occurs when drilling waste is mixed into the top few inches of soil. The drilling waste is spread in a thin, calculated layer and the waste is tilled into the soil, usually using farming plows. Several natural mechanisms help reduce the contaminant concentration in the resulting waste/soil mixture. Rapid biodegradation of the oil occurs when the oily cuttings are exposed to air, water, and naturally occurring microorganisms in the soil. By managing the sodicity of the clays in the soil and drilling waste, and the salinity in the drilling waste, soil structure can be maintained. The structure controls how air and water pass through the soil; for example, dispersed soils reduce hydraulic conductivity, while flocculated soils increase it. There is also a reduction in concentration of potential contaminants due to mixing with soil.

The contaminants most commonly found in drilling waste that must be addressed are salts, oil, and heavy metals. *Salts* in this case refers to the large family of chemicals generated by mixing an acid with a base. Salt content is determined by measuring the electrical conductivity in the extracted water of a saturated paste. All salt measurements must be related back to the saturated condition from the condition of the waste/soil mixture at the *in situ* condition.

Oil and heavy metals are measured by determining the weight of oil or heavy metal in the sample after all the water is removed. This is called the *dry weight basis*. The oil referred to in this test is total petroleum hydrocarbon (TPH).

Occasionally, lead, zinc, or chromium may be of concern, but lead and chromium have generally been replaced in oil field use. Barium levels in waste cuttings from high-weight fluid systems (say, 17 or 18 ppg) may be as high as 300,000 mg/l. While barium from barite has very, very low solubility and is not bioavailable, many areas of the world have regulations that limit the concentration of barium that can be incorporated into the soil.

Deuel and Holliday have described acceptable limits of contaminants commonly found in E&P (exploration and production) wastes in agricultural soils. These limits were examined by the state of Louisiana and adopted, with slight modification, as the regulatory limits known as 29b. These limits are shown in Table 16.4.

Table 16.4
Waste/Soil Mixture Limits

Parameter	Limit, La. 29b
Oil and grease (%weight)	1
pH	6–9
Electrical conductivity, mmhos/cm	4
Metals (mg/kg):	
Barium	20,000 or 40,000
Arsenic	10
Cadmium	10
Chromium	500
Lead	500
Mercury	10
Selenium	10
Silver	200
Zinc	500

By knowing the levels of potential contaminants in the waste, one can estimate the amount of dilution with dirt needed to reduce the contaminant level to the acceptable range. Many times, dilution with dirt is all that is needed. A simple method is to divide the waste concentration by the established limit for the waste/soil mixture and subtract 1. While not mathematically rigorous, this gives a close approximation of the amount of dirt required to dilute the concentrations down to the acceptable level.

Sometimes large amounts of oil are encountered, and it is desirable to allow some oil to biodegrade before mixing with dirt to achieve the final waste/soil mixture. Oily cuttings can easily be 30% by volume (about 15% by weight) when they are returned to shore. Most oils biodegrade readily to a natural endpoint. Diesel biodegrades to about two thirds of its original content. In warm climates, where moisture conditions are kept reasonable (neither flooded nor dry), biodegradation rates are very rapid. The cuttings should be spread in a thin layer at about 3–5% and kept moist. If the conditions are correct, then the oil content should be reduced to 1% within 3 months.

Managing salinity is complicated. Salt is toxic to plants at fairly low levels. However, salts tend to flocculate soils containing clay, making them more permeable to water infiltration. Sodium has the opposite effect, dispersing them. Thus, sodium chloride salt may flocculate the soil initially, but as the sodium adsorbs onto the clay matrix it may cause the clay to collapse and reduce water infiltration, precipitating a condition of sodicity, which is a major problem in agricultural chemistry.

Since oil-based fluids are dominated by calcium content, sodicity is not a problem with land application of oil-based cuttings. However, water-based fluid is dominated by sodium content. Thus, any salt content in water-based cuttings will need some sort of amendment, which may take the form of calcium additives. Another approach is to incorporate soil humus into the soil by mixing horse manure or sawdust with the waste/soil mixture. Soil humus adsorbs large amounts of sodium, allowing the clay in the soil to remain flocculated.

If the soil is flocculated, then rainfall or irrigation will remove salt by infiltration. For this reason, land application works better in wetter climates, but irrigation can replace rainfall. Also, groundwater should be reasonably deep to prevent salt from reaching it.

Heavy metal content cannot be reduced. It must be diluted with dirt to the acceptable level. The method to achieve this is to plow the waste mixture into the land surface. Most plows till only about 6 inches deep, but with sufficient power, tilling 2 feet deep can be accomplished.

Several best practices should be observed:

- Land application is meant to be a onetime event. A limited amount of waste material can be applied to a given area. A commonly cited value is 1000 metric tons of material per hectare, although this depends on the nature of the waste material. If the drilling-fluid system being used contains a high amount of barite, then the amount that can be supplied will be severely limited.
- Waste/soil mixtures that are to be biodegraded should be kept moist, but not saturated. Both water and air are required for biological degradation of oil. Land application in desert environments should not be used unless water can be obtained. Percolation and drainage should be considered in very wet locations.
- Make sure that the waste brought to the site is acceptable for land application. Highly saline solutions, such as zinc bromide, are not acceptable.
- Obtain equipment for spreading and tilling that can adequately do the required job.
- Get help from someone who has performed land farming or land application and understands the objectives.

Land farming ensures the proper chemical balance in the final waste/soil mixture, but further enrichment by fertilizer addition is usually needed.

The main advantage of land application is that, if done right, the waste is incorporated into the land and the land can be returned to its original status (e.g., for growing crops). While it cannot be said that liability is completely eliminated, proper land spreading can certainly minimize any long-term liability.

Not understanding that a limited amount of waste material can be managed in a given area is the biggest pitfall in land farming. Time and again, so-called land farms are set up and waste is dumped into the area in an uncontrolled manner. Land farming is meant to be a mixing of waste and soil in the top foot or two of soil. If 3 or 4 feet of waste (or the equivalent volume in barrels) is brought to a site and dumped, then this is not land farming. These types of misapplication usually end up being long-term liabilities.

16.5.2 Burial

Burial is a method of disposal in which cuttings are mixed with dirt to achieve physical and chemical properties in the resultant waste/soil mixture that are suitable for burial. The waste/soil mixture will be placed in a burial cell. The top of the burial cell should be below the common rooting zone of 3 feet. Since the material is out of the rooting zone, the chemical properties are less strenuous than with land application. The bottom of the burial cell should be at least 5 feet above the seasonal high water table. Burial is the most common method of disposing of cuttings collected in a reserve pit while drilling onshore. However, burial practices have developed over time with little scientific input. Thus, the *method* of burial varies widely among operating areas and contractors. The most common practice is to bury the solids in the existing reserve pit after the water is allowed to evaporate. However, this practice should be examined closely in view of waste/soil mixtures and placement of the burial cell.

In order to apply scientific knowledge to the practice and *responsibly* bury cuttings, several issues need to be addressed. These are:

- Chemical content of the buried cuttings
- Depth or placement of the burial cell
- Moisture content or condition of buried cuttings
- Leakage or leaching from the burial cell

Chemical Content of Buried Cuttings

Chemical content of the buried material is important to prevent potential contaminants from affecting surrounding soil or groundwater. The level of potential contaminants can be higher than would normally be tolerated in good soil (i.e., soil not adversely affecting crop growth), because the burial cell is out of the rooting zone. But it must not be so high that leaching to groundwater or the rooting zone can occur. Table 16.5 lists the chemical criteria and maximum levels suggested for each material.

Most of the parameters in Table 16.5 cannot be met without mixing the cuttings with dirt or soil prior to burial. Thus, burial *in situ* will not, generally, meet the above criteria. Mixing the drilled cuttings with dirt will increase the volume of material to be buried but decrease the concentration of potential contaminants.

Depth or Placement of the Burial Cell

Burial can be considered entombment of waste/soil mixtures. The waste/soil mixture contains oil, salt, or heavy metals in excess of the

Table 16.5
Burial Limits

Parameter	Limit, La. 29b
Oil and grease (%weight)	3
pH	6–9
Electrical conductivity (mmhos/cm)	12
Moisture content (%)	50
Metals (mg/kg):	
Barium	40,000
Arsenic	10
Cadmium	10
Chromium	500
Lead	500
Mercury	10
Selenium	10
Silver	200
Zinc	500
Depth of cell (ft)	5
Height above water (ft)	5

surrounding dirt and soil. This means that, over time, equilibrium is established with the surrounding soil, or there exists a transition from soil properties to waste/soil mixture. The concentration of potential contaminants is within the waste/soil mixture but gradually decreases to background levels in the surrounding soil.

If the waste/soil mixture to be buried were spread as soil, then it would adversely affect some crop growth. This is because the chemical concentrations exceed established limits for crop growth. But the material is not spread; it is buried in a cell. Thus it is important that the burial cell be placed below the rooting zone of future plants. The rooting zone for most plants is 3 feet. If the top of the burial cell is 5 feet below the surface, then 2 feet of buffer zone protects the rooting zone from contamination from the buried material.

The cell should also be placed 5 feet over any groundwater to prevent migration to the groundwater. The zone over the groundwater should be discontinuous. This means that there should be some sort of clay barrier between the waste material and the groundwater, especially in sandy soil.

If the burial cell is placed in this manner, future problems with the cell are unlikely. The waste/soil mixture should not adversely affect crops and should not contaminate groundwater.

Placement of the cell should be considered, too. The cell should not be placed near sensitive resources, like water wells or streambeds. In hilly areas, the cell should not be placed where leaching out of the hillside could occur. Other natural environments may make burial difficult or impossible. Sites in swamps, tundra, or desert that blows sand may not be possible.

Moisture Content or Condition of the Buried Cuttings

Moisture content is expressed as the weight of water contained in the cuttings divided by the mass of the dry sample. This means that the moisture content is 100% when the weight of water contained in the sample equals the weight of dry cuttings. Pit contents have more than 100% moisture content without dewatering. Moisture content is important because dissolved solids could potentially migrate to groundwater. By limiting moisture content, the carrier fluid is limited and the potential for contamination is minimized.

Moisture content is also important because burial of cuttings with high moisture content can result in so-called slumping of the burial cell: When wet cuttings are buried, moisture content will be lost to the surrounding dirt. As this water is lost, the water in the cuttings matrix is replaced with air. As more air is incorporated into the matrix, the strength of the matrix is reduced. Eventually the weight of the covering dirt/soil causes the cuttings matrix to collapse.

Excessive amounts of water associated with the cuttings can be pumped off and processed for reuse or disposed of. Smaller amounts of moisture can be incorporated into the waste/soil mixture by mixing dirt with the cuttings. Clays will contain about 60% moisture at saturation and will drain to about 38% at field capacity. Sands will contain about 40% moisture at saturation and drain to about 7% at field capacity. Cuttings containing high amounts of clay will probably not require much drying or extra work. Cuttings containing high amounts of sand, though, will probably require water to be extracted and/or require mixing with dirt/soil. In tropical or subtropical climates, collection areas may need to be covered to prevent additional moisture being absorbed into the cuttings.

Leakage and Leaching from the Burial Cell

Leaking from a burial cell refers to whole fluid escaping. Improper cell construction or poor practices usually cause this. If the cell is properly constructed, leaking should not occur.

Leaching refers to potential contaminants dissolving in water (usually rainwater) and escaping the cell with the solvent (water). This can occur when sufficient rain saturates the soil and water moves gravitationally down to the burial cell. Soluble chemicals may dissolve in the water and may move laterally or upward while trying to establish an equalized concentration. If the cell is constructed with discontinuous zones (such as clay barriers), movement of lightly contaminated water will not occur or will be very limited.

The disadvantage of limiting leaching is that the main mechanism for removing salt from cuttings is through leaching. This means that high concentrations of salt in cuttings should not be buried. High concentrations of salt can be minimized by leaching prior to burial (as in land application) or by mixing with dirt/soil.

Pit design has a dramatic impact on burial operations. There are several types of pit design. They are:

- In-ground
- Partially aboveground

- Aboveground
- Perched (a variation of aboveground)

The in-ground pit is dug into the earth. The top of the pit is level with the surface of the adjacent drilling pad. This type of pit makes it easy to operate. Wash water can be collected in ditches surrounding the rig and gravimetrically drained into the pit. If the ditches do not drain on their own, then they can be flushed with water easily. These types of pits are also easy to close. If the pit is deep enough, then it can be dried and dirt can be backfilled over the top. The pits need to be deep enough so that the contents (after drying) can be covered with 3–5 feet of dirt. Groundwater in the area needs to be deep enough that the bottom of the pit is at least 5 feet above it. A hazard with this type of pit is lack of control over what goes into the pit from the ditches. Accumulated hydraulic fluid or used motor oil in the ditches might be flushed into the pit and jeopardize its exempt status.

The partially aboveground pit is a shallow hole dug into the earth using the excavated material for berm walls. This increases the holding capacity of the pit and decreases the amount of digging. It might also be used in areas with shallow groundwater. Since the ditches surrounding the rig do not drain gravimetrically into the pit, sumps are placed strategically between the pit and the drilling pad. Wash water collects in the sumps and is pumped or jetted into the pit in batches. If the sand trap is dumped, then it is usually dumped into the sumps. Whole fluid discard is usually jetted rather than dumped. One positive is that there is better control over what gets into the pit. A disadvantage is that the pit is usually not deep enough for direct burial. This means extra handling and cost during the closure operations.

The aboveground pit is constructed where it is impractical to dig a pit. This can occur in very soft, sandy conditions or in very hard surface conditions. The aboveground pit has similar operational advantages and disadvantages as the partially aboveground pit. Burial is obviously not possible using an aboveground pit. In soft, sandy conditions, a large burial cell or trench is usually dug especially for burial during closure operations. In very hard surface conditions, burial becomes a difficult problem. If the pit could not be dug due to very hard surface conditions, then a burial cell is probably not economical either. Construction of an aboveground pit is fairly cheap and simple, but closure costs are slightly higher.

The perched pit is a special case of the aboveground pit. This type of pit is used in hilly terrain. It is constructed by making a berm out of fill material on the outside of a hill. In this case, the pad is located on the cut portion of a cut-and-fill location. The inside wall of the pit is the hillside. The outer wall is the constructed berm. These types of pits are cheap and easy to construct. If conditions are ideal, then operations will run smoothly, as well. But all too often, problems occur and the berm wall breaks. Then, the contents of the pit can run down the hill and collect in whatever is at the bottom. This is usually a stream, creek, or other water body. For this reason the perched pit is usually not recommended.

16.6 TREATMENT TECHNIQUES

Treatment techniques differ from disposal techniques in that they modify or separate the properties of the waste, but the waste must still be placed somewhere. In burial, discharge, land application, and injection, the waste is placed on the land, in the sea, or deep in the earth. Treatment techniques are aimed at removing oil, reducing the mobility of contaminants, or otherwise modifying the properties of the waste material.

Many treatment techniques have been tried. Most have aimed at reducing the amount of oil retained on the cuttings, so as to allow the cuttings to be discharged at sea. Others have aimed at encapsulating the cuttings and contaminants to prevent their leaching into the environment.

Not all of the treatment techniques will be discussed. However, it is important to discuss three types of treatment. The first is dewatering. Dewatering is the mechanical and chemical separation of liquid and solids from drilling fluid and cuttings. The second is thermal desorption, which is the heat separation of oil from cuttings. The third is stabilization. This technique is the attempt to encapsulate contaminants in the cuttings to prevent leaching.

16.6.1 Dewatering

Dewatering is the art and science of chemically enhanced centrifuge separation. Dewatering is the final step of a closed loop system and follows the separation process after shakers, hydrocyclones, and centrifuges. While high-speed centrifuges remove particles 2 to 3 microns and larger, dewatering can remove all colloidal particles down to clear water.

Dewatering has become common in many instances, especially as technology has advanced and the units have become more compact and less expensive. At first, dewatering units were introduced where only stringent environmental conditions existed. However, dewatering has now become economical where freshwater is scarce or disposal sites for off-spec fluid are too far from the drill site, making transportation costs expensive. Dewatering units have also found application in drilling through fragile clay formations.

The presence of oils or lubricants does not affect the dewatering of water-based fluids. Once the colloidal solids are removed, the oils, lubricants, or organics separate out with the liquid and tend to float on top of the water. All water-based fluids can be dewatered, although some are easier to dewater than others. Dewatering oil-base fluids is not easy and requires a prior treatment with a demulsifier to break the oil–water emulsion. However, today, even cement-contaminated fluids can be dewatered on location in order to reduce disposal costs by requiring only the solids to be removed from the site.

Dewatering units can be compact systems located on the rig, or they can be mobile, trailer-based systems parked next to the rig. Dewatering units are also sometimes set up in a central fluid plant serving a cluster of rigs. A dewatering unit on location allows a portion of the off-spec fluid to be dumped into a storage tank and then processed into solids and fluid. Depending on the dewatering process and the chemical treatment applied, the fluid can be recycled as is or further treated before recycling. The water generated from the first step generally tends to be clear but sometimes requires pH adjustment prior to being recycled. However, further treatment may be required to bring the water up to local standards before disposal.

Use of an on-location dewatering unit during operations has become mandatory in some parts of the world, such as the Arctic, jungles, and rain forests and in close proximity to urban environments. Additionally, on-location dewatering units may be mandated when drilling near freshwater sources, near sensitive fishing areas, or where concern for protecting ocean species is very strong or regulated.

With the pH of fluid typically being between 7.0 and 10.0, the colloidal particles in the fluid tend to be negatively charged. The negative charges repel the particles, preventing them from clumping together to form larger particles. To remove these submicron colloids is difficult, even with a high-speed centrifuge with 2000 g force. Therefore, to remove these tiny particles in the fluid, it is first necessary to treat the fluid with chemicals to agglomerate the solids to make them large enough to be removed by a high-speed centrifuge.

The process of agglomeration to create large, dense clusters requires three steps:

1. Destabilize the submicron particles so they no longer repel each other. This is easily achieved by lowering the pH from 7.0–10.0 to approximately 5.5.
2. Coagulate or bring together the fine solids—create an attraction between the particles.
3. Flocculate, bundle, or wrap together to create large dense clusters.

These three steps can be accomplished by sequentially adding three chemicals; or sometimes only one or two chemical additives can accomplish all three steps. To maximize dewatering efficiency and effectiveness, enough time between the addition of each of the three chemicals should be allotted. This enables the chemicals to fully react with the particles and therefore ensures that all chemical treatment affects primarily the solids, with negligible amounts of chemical remaining in the liquid phase. This helps generate a reusable fluid (water) or one that can comply with local norms if being disposed of.

Typical dewatering systems include:

1. A holding tank with mixing to create a homogeneous waste fluid that is ready to be processed.
2. Small storage tanks for the chemical additives with controllable feed pumps to calibrate minute levels of treatment—one tank each for acid and coagulant and two tanks for flocculant. Since flocculant makedown takes time, having two allows one for makedown while the other one is in use.
3. Pumps to feed the treated fluid under steady pressure and at constant flow to a centrifuge.
4. A manifold with inline mixers that will allow the fluid time to react with the chemical additives before reaching the centrifuge.
5. A high-speed centrifuge in which the desired g force can be attained to remove/discard the coagulated solids and discharge clear fluid.
6. A storage tank for clear fluid exiting the centrifuge before recycling to the active system or disposal.

7. Skimmers to remove any oil or lubricants present. The oil remains with the cleaned water and floats above the water phase, where it can be removed.

A simple explanation of the dewatering process is that it diverts fluid through a manifold system prior to pumping to the centrifuge. It is in the manifold that the chemicals are mixed sequentially, inline with the fluid. As the fluid flows through the manifold, the acid is first added, the fluid continues through the manifold to where coagulants are mixed in, and finally it arrives at where the flocculant is added. The chemicals should be properly metered in order to allow for minimum dosage. With a properly designed manifold, one that is sufficiently long and includes inline mixers, the chemicals will have time to react individually with the fluid, producing the desired result of larger clustered solids that are more easily removed by centrifugal force.

When the fluid exits the manifold system, it is ready move into the high-speed centrifuge. The particles in the fluid have been clumped together into relatively large ones, with a high enough density for a centrifuge to remove. After being processed by the centrifuge, the clear effluent (water) is stored in a tank and the solids are conveyed to a separate pit or tank for disposal. A pump takes the fluid back to the fluid system or to a separate storage tank.

Using clean chemistry, buffered phosphoric acid, and coagulants and flocculants approved by the National Sanitation Foundation can usually generate clean water that is good for recycling or disposal with minimal treatment. Depending on the nature of solids from the formation and fluid additives, with the proper treatment chemistry it is possible to generate solids with minimal chemical content that can be disposed of like other cuttings.

Dewatering oil-base fluids generally follows the same procedures as that for water-based fluids. Oil-based fluid, however, must first be treated to break the emulsion. This can be accomplished by adding acid and additional water or by use of a demulsifier. The oil on separation rises to the surface, where it is removed by a skimmer. The remaining solids in the water are then treated in the same manner as water-based fluids. Careful treatment with adequate dilution can ensure that the remaining oil content in the sludge is well within acceptable environmental norms. Figure 16.1 shows a schematic of a dewatering set-up. The process can be quite complex.

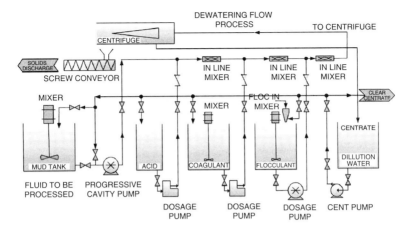

Figure 16.1. Dewatering flow process.

16.6.2 Thermal Desorption

Thermal desorption describes a treatment technique in which the oily cuttings are heated and the volatile liquids are driven off, resulting in two phases. The liquid phase containing water and oil is separated. The solids component is generally de-oiled to less than 0.5% oil by weight. This is considered a treatment technique, because something must still be done with both phases. It is important to understand that this technique does not address anything except oil. The salinity remains with the solids, as does the heavy metal content (barium, zinc, lead, etc.).

In thermal desorption, oily cuttings are fed to a heating unit. Many types of heating units exist, but the purpose is to efficiently transfer heat to the drilled cuttings to where oil and water are driven off. The water and oil are separated. Depending on quality, the recovered oil is used to further fuel the thermal desorption process, as makeup fluid for the fluid system; or it can be sold to industries needing boiler fuel (cement kilns and power plants).

In order to apply thermal desorption to the treatment of oily cuttings from offshore, the cuttings are transported to shore. Large-scale thermal desorption has not yet been applied at an offshore location. As of this writing, the maximum throughput for any thermal plant on an offshore location has been about 5 to 6 tons per hour. At the dock, the cuttings are transported to the treatment site by truck, box, or other transport method. Cuttings arriving at the desorpton facility are weighed in and kept segregated by operator (generator). The cuttings are screened for foreign materials before being fed into the feed hopper of the desorption unit. All of this is simplified if the thermal desorption unit is located near the receiving dock.

At this point, the individual processes vary considerably. There are single and twin screws, hollow-flight augers, hollow-paddle augers, rotary kilns, and hammer mills. They may be direct fired, as in a rotary kiln type of incinerator; indirect fired as in calciners; indirectly heated by circulating a high-temperature silicone oil through the hollow augers; or, in the case of the hammer mills, use the heat generated by friction to vaporize the liquid components. There is a perceived advantage to being able to control the temperature and the amount of oxygen in the process. This is partially to reduce fire hazard potential and partially to prevent cracking of the hydrocarbons.

The use of nitrogen to purge the units (called a nitrogen blanket) effectively excludes oxygen from the process. For cost reasons, the "steam quench" method of preventing fire is frequently used in lieu of a nitrogen blanket. The hazard from having volatile vapors above the flashpoint in the presence of oxygen is obvious. The propensity of hydrocarbons to crack and form polycyclic aromatic hydrocarbons and other known carcinogens in the presence of oxygen is a serious concern. Further, cracked hydrocarbons do not retain their original fluid properties and become unsuitable for reuse as base oil in a drilling fluid In addition to process type, it is important to understand how the presence of water in the cuttings reduces the process rate while increasing fuel requirements.

Two problems must be addressed as the hot, de-oiled cuttings leave the desorption unit. The cuttings will be around 500°C and must be cooled rapidly before exposure to air. This is usually addressed by having the cuttings pass through an enclosed cooling auger. A second problem is that fine solids are stripped from the unit with the vapor phase. There will usually be a centrifuge to remove solids from the liquid phase before oil/water separation. Bag houses to control dust have proven to be susceptible to catching fire and their use has generally been discontinued.

The processing rate for a typical unit is about 100 metric tons per day. This is equivalent to about 350 bbl of cuttings per day. Feed consistency is critical to the overall processing rate. High water content will require the feed to be diluted, so to speak, with processed cuttings containing no water or set aside under cover to allow the material to drain before processing. A water content below 10% is desired, while most cuttings will contain more than that. Oil content is a concern for those units that allow oxygen in the process. Many of the rotary kiln desorption units restrict the feed rate to control the process or attempt to maintain the concentration of oil in the feed at about 10%.

The main advantage of thermal desorption is in removing almost all of the oil from the waste solids. If disposal criteria demand extremely low hydrocarbon content, the only way to obtain it is by thermal desorption. A second advantage is in reclaiming oil that can theoretically be reused. However, this is dependent on the process, the temperature control, and the ability to control oxygen.

The primary disadvantages are cost and safety. Most reputable companies have addressed the safety concerns, but some fly-by-night companies still exist, especially in developing countries. Other disadvantages include:

- Processed solids must still be handled and disposed of (see the preceding sections on commercial options, land farming, and burial).
- The quality of the recovered oil varies, depending on the process.
- There are high capital costs.
- Maintenance is relatively high.
- The technique is highly dependent on operator experience and expertise.
- Air emissions must be controlled and monitored.

16.6.3 Solidification/Stabilization

Solidification continues to be sold around the world as an inexpensive alternative treatment technique to other methods of waste management, particularly in countries with relatively lax environmental regulations. This treatment technique is not recommended unless a thorough investigation of the process to be used is conducted.

Solidification (also referred to as encapsulation, fixation, and stabilization) is a technique in which material is added to reduce free water and possibly reduce or slow potential contaminant release. This is considered a treatment technique because something must still be done with the solidified material. In the past, this technique was used prior to burial or road spreading (as surface material) of the solidified material. It has also been used to make reusable material, such as bricks or blocks.

Probably the most promising application has been to make roadbed material (the dirt used under the surface of a road).

Typically, fly ash or cement has been used to solidify cuttings. However, many products have been marketed claiming successful results. The main problem with solidification has been that there have been no well-defined processes or quality control techniques used. One could argue that adding dirt to cuttings is a solidification technique, but obviously this form of dilution would not be of much advantage except to reduce free water.

In any solidification process, the solidified material must first pass some sort of leachate test. As a process control, the amount of solidification material added to the cuttings should always be equivalent to the amount required to pass the leachate test. Because there is such variability in the consistency of the feed material, this has rarely been achieved. The second result that must be obtained is that the solidified material must resemble whatever material is being made. If the product desired is a brick, then a good brick should be made. If the product desired is road-grade material, then the material should pass the tests for good road-grade material.

Solidification leachate standards are given in Louisiana's regulations (29b). If no regulation exists in the area where this technique is to be applied, then La. 29b is a good standard to follow. It assumes that the material is to be solidified and buried. The standard calls for a minimum compressive strength of 20 psi to be developed in a test block. The test block is then crushed, and sized particles of the block are immersed in water to test the leaching of oil and salt. The leachate limits for oil and chloride are less than 10 mg/L and 500 mg/L, respectively. There are other tests as well. It should be noted that almost no solidification process can be economically applied if this standard is met.

Assuming that the leachate standards can be met, then the second criterion is to meet standards set for the particular product. Again, no standard has ever been used in previous work, so there is no history to recall. But there is a road-grade material standard published by the American Association of State Highway and Transportation Officials. In the standard, there are specifications for materials used for embankments and subgrades (M 57-80) and for aggregate and soil-aggregate subbase, base, and surface courses (M 147-65), as well as a classification of soils and soil-aggregate mixtures for highway construction purposes (M 145-91). These specifications discuss desirable particle or aggregate size, liquid or moisture content, and plasticity (relating to clay content and wet/dry cycling and expansion/contraction). A quick review of the standards would lead most readers to conclude that cuttings would probably not make good roadbed material.

But this technology continues to be sold, especially in developing countries and remote areas. The attractiveness seems to lie in the view that roads can be made from cuttings and that waste can be converted to a usable product. The promise has not come to fruition in the more developed countries.

Another seemingly promising technology has been conversion of the waste material into bricks or other construction slabs or blocks. Again, the technology has not proven reliable or economical if leachate standards are met. Salt readily leaches out of bricks and concrete, as can be seen in any tropical climate (salt is used to speed up setting time in building construction). In Venezuela, a wall was built of concrete blocks as part of an encapsulation/solidification process. Years later oil could be seen and smelled at considerable distance from the wall. Unfortunately, this result is the rule rather than the exception.

While the advantage seems obvious (if elusive), there are some disadvantages:

- Poor or uncontrolled application of the technology
- Leaching of contaminants
- Increases in volume of waste (if it cannot be used as a product)
- Labor intensiveness (may not be a problem if labor is cheap and unemployment is high)
- Long-term liability with the product (would you want to live in a house made from solidified cuttings bricks?)
- And the main disadvantage: the unfulfilled promise that this technology really works.

This treatment technique is not recommended unless a thorough investigation of the process to be used is conducted.

16.7 EQUIPMENT ISSUES

Anytime drilling waste is handled, equipment must be used. This section deals with some of the equipment issues in moving, storing, and handling drilling waste.

16.7.1 Augers

Augers (screw conveyors) are commonly used to move drilled cuttings and associated fluid. They can be arranged to collect the cuttings (usually relatively dry oil-based cuttings) from the individual pieces of solids-control equipment and convey them to another area of the drilling rig where they are used to load cuttings boxes (skips). The standard screw conveyor is composed of an auger or screw housed in a flanged, U-shaped trough with bolt-on covers. It is powered by an electric motor and equipped with an appropriate gearbox. The motor must be sized to provide enough horsepower and torque to permit the transport of cuttings at a rate at least equal to the maximum rate at which they are delivered to the screw. The feed and discharge ends are fitted with flanges and ports to allow the cuttings to flow into and out of the conveyor without plugging. For multiple screw sections, hanger bearings are used to support the ends of the screw sections where they are joined. Operating parameters such as loading, housing enclosures, and length of the section are all considered in determining the required bearing type.

Augers are used to convey relatively dry cuttings, such as those generated while drilling with oil-based fluid. They are generally not used for conveying liquid discharges other than the associated liquid with the cuttings. The conveyors are usually sloped at a shallow angle so that liquid can drain into a catch tank located under the lower end. The cuttings should flow directly from the feed inlets into the screw in a manner that prevents their building up and plugging the inlet. The most common arrangement has the conveyor running perpendicular to the flow down the shaker screens, so that each shaker has its own inlet to the screw. The inlet opening should be as wide as required to maintain a steady flow without plugging.

Because of the variety of rig designs and operating conditions, particular care must be taken to ensure that the sizes of the conveyor, motor, and gearbox are adequate for the application.

Augers are available in different diameters and lengths. Common diameters are 9 inches to 18 inches. Augers are typically supplied in 10- or 12-foot lengths that can be linked together to form up to 50- or 60-foot sections. It is imperative that individual lengths align precisely within a section. A motor is supplied for each section. Auger runs greater than 50 or 60 feet require multiple sections. Angles can be formed where one section meets another.

Utilization of correctly sized augers for the drilling conditions is critical. The auger must provide the capacity to transport cuttings at the maximum rate at which they will reach the surface. Attempted use of augers that are too small for the drilling conditions can lead to bottlenecks, breakdowns, and the need to interrupt, or shut down, the drilling process.

The estimated peak volume of cuttings and associated fluid to be handled by an auger, where all of the cuttings are being collected by one auger and the solids-control efficiency is high, can be estimated as double the gage hole volume at the maximum instantaneous penetration rate. A typical maximum instantaneous penetration rate could be 300 ft/hr. Table 16.6 indicates several examples of volume of cuttings that need to be handled by an auger.

Auger sizing is dependent on the required handling volume, auger speed, and trough loading. For cuttings handling, it is recommended to use relatively low auger speeds and minimize trough loading. Higher speeds may increase capacity, but higher wear may also occur. Higher loading may increase likelihood of plugging. Table 16.7 gives an example of auger sizes based on speed and trough loading.

Typically, 18-inch-diameter augers are installed and used when all of the cuttings are discharged into a single run of augers and when 17½-inch

Table 16.6
Handling Volume for 300 ft/hr Drilling Rate

Hole Size (in.)	Handling Volume (ft^3/hr)	Handling Volume (m^3/hr)
8½	250	7
12¼	500	14
17½	1000	28

Table 16.7
Auger Size Based on Volumetric Rate and Loading Conditions

Volume (m^3/hr)	Auger Size, 30% loading, high speed (in.)	Auger Size, 30% loading (in.)	Auger Size, 15% loading, (in.)	Maximum rpm
7		9	12	50
14	9	12	14	50/100
28	12	14	18	45/90

hole cuttings are collected. By running a 14-inch-diameter auger with higher loading rates, the same capacity could be achieved, but with higher likelihood of plugging. A 12-inch-diameter auger would require both higher loading and higher speed and would pose greater danger of plugging incidents.

Occasionally, a single auger run cannot collect all of the cuttings from the solids-control equipment. In this case, a smaller branch auger can be used to collect isolated cuttings. For instance, a 9-inch auger might be used to bring centrifuge cuttings to the main auger run, where the centrifuge is isolated from the shale shakers. Also, if smaller hole size and slower rates of penetration are anticipated before the auger is required, then smaller augers can be used. This condition is typical where oil-based drilling fluid is used in lower hole sections and the oily cuttings are collected for disposal.

Augers can represent significant safety hazards. The bolt-on covers must be kept in place when the auger is turning. However, when they are being used to convey drilled cuttings, this is not always possible. Grates can be used to cover open sections. If neither grate nor cover is possible, then a barrier fence with a warning sign posted should enclose the exposed section. A remote cutoff switch should be located within reach.

16.7.2 Vacuums

Vacuum transfer systems provide an alternative to augers. The initial applications were on jackups, primarily in the Gulf of Mexico, where the cramped areas required multiple conveyors to get around the legs and other obstacles. It was also easy to close the discharge chute, making a sump to collect the cuttings that is suitable for the vacuum nozzle.

A major advantage of vacuum transfer systems is that they permit more flexibility in siting components on the rig. The individual components (collection troughs, conveyors, collection boxes, etc.) are connected via hoses. This facilitates the routing of flow through and around congested areas on the rig and makes it possible to place equipment at different elevations when this is advantageous. The use of these systems has grown in recent years as the demand for total containment of wastes (zero discharge) has increased.

Vacuum transfer systems come in various types. Typical components include one or more vacuum blowers, a filtration device to protect the blower, vacuum hoses, and cuttings boxes with vacuum hose ports or vacuum lids that seal on the box lip. The vacuum pump may be skid mounted and can be either electric or diesel powered. Noise reduction devices and sound insulation should be used to reduce noise levels to below 80 dB, if possible. Many of the units commonly in use exceed noise levels set by the Occupational Safety and Health Administration and require hearing protection to work around them.

A vacuum pump is connected via hoses and/or pipes to a filtration unit, which in turn is connected to a collection box (or hopper, although these are rarely used). A hose or pipe connects to either a single point or to a manifold with multiple suction points at the waste source. The vacuum must be continuous to permit transport of the cuttings. The cuttings box allows the cuttings to fill in from one of the openings while the vacuum is pulled through the other side. While cuttings and waste are traveling through the hoses, sufficient velocity must be maintained to prevent them from settling in the line. Once the waste reaches the cuttings box, the material velocity slows, allowing the transported material to drop out into the collection chamber. Once the box is full, the hoses are switched, or the valve positions on the manifold are changed, to fill other boxes. As boxes are filled, they are lifted away and replaced with empty boxes.

16.7.3 Cuttings Boxes

Boxes are the primary method of transporting waste drilling fluid and cuttings to shore around the world. Boxes were developed as an easy method of collecting and transporting cuttings given the weight restrictions of offshore cranes on earlier drilling rigs. Boxes are typically placed near the solids-control equipment, where cuttings can be moved relatively short distances and collected in the box. When a box gets full, it is removed and an empty box is shuffled into position. When a sufficient number of full boxes are ready, they are backloaded onto a workboat and returned to shore. Empty boxes from the dock facility replace the returned boxes.

In the United States, cuttings boxes (small tanks) must be Coast Guard approved. In other developed areas there are similar requirements. These standards have been developed for safety reasons. Some of the box requirements are:

1. Boxes should have a cover that can be locked down and sealed shut. Lid gaskets should be checked frequently.

2. The tank must be of sufficient strength to survive a drop of 1 meter without leaking or deforming. Skids, frames, and braces should be in good shape. If the tank is rusty or shows signs of abuse, it is probably not maintained properly.
3. Since the tank is sealed, it must have pressure relief to prevent explosion and vacuum relief to allow the lid to be opened when the contents cool and form a vacuum. A rupture disc is usually installed on the lid to allow emergency venting. Check the rupture disc inspection date. A small valve may be used for vacuum venting.

In developing countries, homemade boxes are sometimes encountered. Transporting and emptying boxes is always dangerous. The use of homemade boxes or poorly maintained boxes can be hazardous.

Removing settled solids from the box has always been a problem. The boxes have undergone a long period of settling in conjunction with severe vibration (both on the rig and during transport). Typically, the top-filled boxes are lifted and turned over to remove the solids. Sometimes the boxes have a side-emptying hatch that can be opened, to eliminate turning the box over. Some of the boxes have a side-mounted flushing valve to help flush settled cuttings out of the box. In any event, emptying the box is very dangerous and a nasty, sloppy job.

Boxes typically come in two sizes: 25 bbl and 15 bbl. The 25-bbl boxes can approach 15 tons when full, especially if highly weighted fluid is being used. The 15-bbl boxes were designed for situations in which the larger boxes could not be handled by the rig crane. The 15-bbl boxes weigh between 7 and 10 tons when full. Boxes can usually be stacked two high when empty, but must not be stacked when full. An area of 4 by 6 feet per box can be used for planning purposes. For larger jobs, as many as 25 to 30 boxes per group (three groups overall) will be used. Rig space needs to be available to accommodate all the boxes.

Box Alternatives

An alternative to box use is the use of bulk systems. An example of a bulk system is the collection and transport of inland water cuttings by shale barges. A shale barge typically holds 4000 bbl of fluid and cuttings in four compartments. The compartmentalized tank is basically mounted on a barge. However, shale barges are meant for relatively calm waters and are not intended for offshore use. The forces developed by the heavy cuttings in the barge tanks due to wave action can break the tank walls and/or sink the barge.

Bulk systems for offshore use are basically seagoing vessels. The tanks used for storage are mounted in the hull. The vessels and the tanks are designed to overcome the "sloshing" of the cuttings due to wave action. The problem with putting cuttings in tanks within vessels has historically been that the cuttings settle into a solid mass that is very difficult to remove.

Two relatively new systems have been developed for moving cuttings in larger volumes. The first is a hydraulic-driven submersible pump. The second is a pneumatic system.

The hydraulically driven, modified submersible pump is usually mounted on a movable arm, although it may be fixed in an offshore application. The pump can also be mounted on a tank. A hydraulic motor is used rather than an electric motor. This makes the combination of motor and pump much lighter (and safer for submersible operations). Whereas a 50-hp electric pump is very heavy, a 150-hp hydraulic pump is much lighter. This means that much more power to slurry and pump the cuttings can be applied.

The system is capable of pumping fluid or cuttings, with relatively low liquid content, over considerable distances. The original use of the system was to move cuttings from barges into sealed dump trucks at a dock facility. It was also used to pump fluid and cuttings from earthen pits for downhole injection. One early system was mounted on a 150-bbl tank so that it could be used to collect cuttings and transfer them as well. The pump ran on a track and could be moved to any part of the tank to facilitate slurrying the cuttings, even if they had settled and appeared hard packed. The system is fairly reliable from a technical standpoint. Of course, as with most systems, the state of reliability depends on the operator's expertise and the maintenance of the equipment.

Typically, for an offshore application, it would be desirable to pump the cuttings from a collection tank to boat tanks. The boat would take the cuttings to shore. Another pump at the shore base would pump the cuttings out of the boat tanks into transport vehicles to be transferred to the disposal site. The boat tanks must have specially designed entryways to allow the pump to be inserted into the tanks and moved around to reslurry the settled cuttings. The shore transportation might be avoided if the disposal site is near enough to the shore base.

In the pneumatic system, a *cuttings blower* blows the cuttings from the cuttings trough to temporary storage containers. The blower uses pulses of air provided from an air compressor. The units are automated and the pulses controlled by relays and timers. The operator adjusts only the timer, which would vary only if there were a great change in the rate of penetration.

The temporary storage containers are a series of pressure tanks with 95-bbl capacity. The number of tanks depends on the amount of cuttings generated and needing to be stored. In this character the tanks are similar to boxes, except they are larger. A number of tanks are manifolded together and can be located in a variety of locations as space and deck load permit and as required for cuttings storage. The size of each unit is $8 \times 8\frac{1}{2} \times 20$ feet tall. Each is located on a hydraulic load cell that is used to determine the weight of cuttings produced and when it is necessary to switch to an empty tank. The system can be manually operated or automatic. In automatic mode, a valve closes and directs the flow to a new tank when the original tank is full (by weight). In manual mode, an operator simply changes the hoses.

The system is designed so that tanks are also mounted on a boat. When the boat arrives at the rig site, the cuttings in the rig's tanks are transferred to the boat's tanks. The contents of the tanks are removed in less than an hour. The boat returns to shore, where its contents are transferred to shore-based transportation. The tank can also be lifted and moved if crane capacity allows the lift.

Alternatives to boxes offer the advantage that larger quantities of waste fluid and cuttings can be transferred, thus reducing the logistical requirements. However, they are fairly new and have limited usage.

16.7.4 Cuttings Dryers

Cuttings drying is sometimes referred to as secondary drying of cuttings. Drilled cuttings with associated fluid from the rig solids-control equipment have been passed over a second drying shaker for a number of years. The recovery of oil-based drilling fluid coupled with a 10–25% reduction in disposal volume is usually easily justified and has become standard procedure in areas where so-called pitless drilling or closed loop systems are the norm. Generally, the secondary drying shaker is a four-panel screening device running at 7.0–7.3 g's at the screen surface. Drilled cuttings from a drying shaker typically test between 8 and 12% base oil (NAF) by wet weight. This retention-on-cuttings (ROC) figure is significantly higher than the current minimum allowed for offshore discharge in the United States and, increasingly,

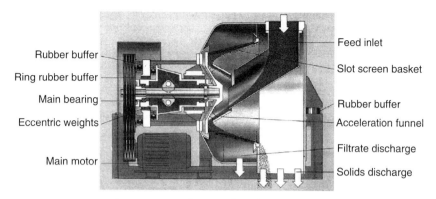

Figure 16.2. Schematic of one horizontal dryer.

elsewhere around the world. Accordingly, there has been increased interest in lowering the ROC figure by the use of different types of drying equipment.

Figure 16.2 shows an example schematic for one typical horizontal dryer. Cuttings removed in the primary solids control section enter in the feed inlet. Processed cuttings are removed after the dryer removes oil (or NAF). The oil or NAF is recovered and returned to the drilling fluid system.

The quest for improved technology to reduce the ROC figures is ongoing with a number of competing machines and technologies. At this time the market is composed of the following commercial products:

- Horizontal and vertical basket centrifuges
- Basket centrifuges that move cuttings by vibration
- Basket centrifuges that move cuttings with a scroll
- Basket centrifuges that move cuttings with a so-called pusher rod
- Basket centrifuges that move cuttings with vibration and a pusher rod
- Perforated rotating vacuum cylinders

Of these competing technologies, basket centrifuges (vertical and horizontal) using centrifugal force and a flighted scroll to effect separation make up approximately 80% of the units in use today. Typical rotation speeds from 300 to 870 rpm provide a g force between 48 and 375 g's. Retention time of the drilled solids within the screen basket, screen geometry, and basket rpm vary widely among units. Generally speaking, more rpm means more g force, drier drilled cuttings, and more capacity.

Legislation has been the driving force behind the use of cuttings dryers, coupled with the use of synthetic 16-18 IO—base fluids. The basket centrifuge is thus far the only piece of equipment on the market that can reliably reduce ROC figures below 4% while handling the volume and variety of solids presented to it during the drilling process. Fast drilling generally produces large cuttings with small surface area and thus a lower ROC value. Slow drilling with very fine drilled cuttings will have a large surface area that is proportionately more difficult to bring to the desired level of dryness.

Figure 16.3 shows cuttings prior to being processed by a cuttings dryer. Figure 16.4 shows cuttings after processing. Visually, the cuttings appear drier and test results confirm a significant reduction in retention on cuttings.

By bringing the average ROC to below 4%, a number of issues are much more easily dealt with. For example, the pneumatic transfer/ conveyance of cuttings is much more reliable. As a prelude to thermal desorption, the process rate through a thermal plant can be increased. As a prelude to land farming or composting, the amount of amendments and the time required to reduce total petroleum hydrocarbons below

Before Dryer ≈ 10%

Figure 16.3. Cuttings before drying.

After dryer 2%

Figure 16.4. Cuttings after drying.

1% is significantly reduced. On land, the volume of liquids attached to drilled cuttings to be disposed of offsite can be reduced by between 25 and 45%.

Figure 16.5 shows the reduction in retention on cuttings when centrifuge cuttings are processed by a cuttings dryer. The volume reduction indicated can be as high as almost 50%.

The vast majority of offshore rigs today were designed before secondary drying was mandated. Frequently, waste management, including solids control, is treated as an afterthought. Accordingly, few offshore rigs are amenable to the installation of equipment required for secondary drying.

One of the issues related to secondary drying is the amount of fine solids that accompany the recovered drilling fluid. The effluent from dryer operations is always laden with low-gravity solids that are passed either over a fine screen or through one or more centrifuges. Passing the recovered drilling fluid through one or more high-speed centrifuges should allow the cleaned drilling fluid being returned to the active system to have a content of low-gravity solids below the target in the operator's drilling program.

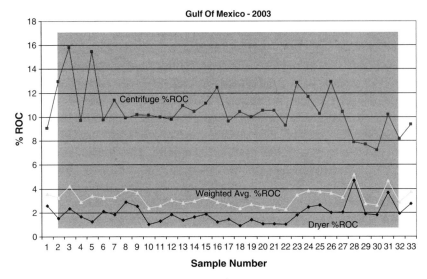

Figure 16.5. Reduction in ROC.

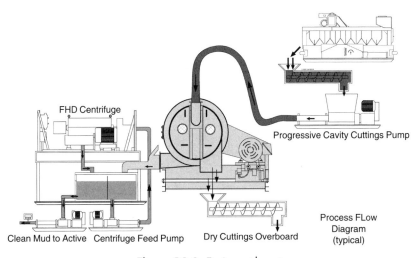

Figure 16.6. Equipment layout.

Figure 16.6 shows a schematic of fluid removed by a cuttings dryer being further processed by a centrifuge. The intent is to remove fines from the fluid prior to returning the fluid to the drilling system. By removing LGS cleaner fluid is returned, but some fluid is lost again.

Figure 16.7. A typical horizontal cutting dryer installation on a jackup rig with two high-speed centrifuges processing in series removing low-gravity solids. All pumps and process tanks are tucked beneath the process equipment.

Figure 16.7 shows a picture of an installation of a cuttings dryer with associated processing equipment and tanks. An idea of the size of the processing area can be envisioned in this picture.

Another benefit of any secondary drying operation is the ability to catch all drilling fluid that may flow off the end of the shaker due to screen blinding, surges in circulation, etc. The potential for excess fluid, particularly on ultra-deepwater operations, should be evaluated and planned for. For instance, "boosting the riser" frequently will overwhelm the screening capacity of a deepwater drilling rig.

The single most difficult aspect of operating a reliable cuttings dryer operation is getting the volume and variety of drilled cuttings with associated drilling fluid to the dryer in a continuous stream.

The second most problematic aspect of dryer operations is dealing with the discharge "on stream," that is, balancing all equipment, fluid levels, and feed streams. Dryers and their associated centrifuges or fines units are fed with a variety of equipment types. Without delving into the various feed methodologies, suffice it to say that simplicity reigns.

The simpler systems are generally more reliable and more easily brought back online when something untoward occurs.

Any secondary drying installation for an offshore rig must provide the capability to easily switch between any of the following modes:

- Bypass all drilled solids, cement, and displacement fluids overboard.
- Zero discharge—everything from the well bore is diverted to either cuttings boxes, cuttings tanks, or cuttings barges for further processing. In some instances, cuttings are ground and injected down the annulus on-stream, or as created.
- Normal dryer operations in which all well-bore discharges are passed through a cuttings dryer prior to final disposal.

REFERENCES

Duel, L. L., Jr. and Holliday, G. H. Soil Remediation for the Petroleum Extraction Industry. Pennwell Publishing, 1994.

CHAPTER 17

THE AC INDUCTION MOTOR

Michael Kargl
Martin Engineering Co.

Wiley Steen
Consultant

This chapter will discuss electrical and magnetic properties along with theory of operation of the induction motor. Adjustable speed drives (ASDs) and their control of induction motors will be discussed, along with various motor applications and considerations regarding alternating current (AC) induction motors on oil rigs.

17.1 INTRODUCTION TO ELECTRICAL THEORY

The properties of electricity are voltage, current, and resistance. Voltage (also known as electromotive force, or emf) is the strength of a circuit that causes current to flow through the resistance in the circuit. It is a force that causes electrons to move from one atom to the next. Voltage is analogous to pressure in the mechanical system. Voltage is measured in volts.

Current, measured in amperes, is the measurement of electrons flowing through a conductor. One ampere is 1 coulomb passing an imaginary plane in 1 second. A coulomb is a quantity of electrons (6.24×10^{18}). The amp, in the mechanical system, is analogous to flow rate, such as gpm.

Resistance is measured in ohms and is a measure of a circuit's resistance to current flow. Resistance is that property that opposes the movement of electrons from one atom to the next. As resistance goes up, current flow decreases if the voltage is constant. As resistance goes down, current flow increases. Resistance varies with temperature. As temperature increases, resistance increases.

In direct current (DC) circuits, Ohm's law defines the relationship between voltage, current, and resistance. Ohm's law states that a potential difference of 1 volt applied across a 1-ohm resistance will cause a current of 1 amp to flow through the resistance. The formula is as follows:

$$V = (I)(R)$$

where

V = voltage, in volts
I = current, in amps
R = resistance, in ohms.

Note that varying the voltage or the resistance changes the current directly.

Work is a force applied through a distance or moment arm. *Power* is work per unit time and is usually measured in watts (joules per second). Watts are calculated in a DC circuit by multiplying the current, in amps, by the voltage, in volts, that is, watts = (current)(voltage).

The preceding formulas are also useful for AC circuits, but they are modified because of the constant change in magnitude of the voltage and current waveforms in three-phase AC circuits. In a DC circuit, the magnitude of the voltage and current is constant with respect to time. In an AC circuit, the voltage and current constantly change sinusoidally with respect to time (see Figure 17.1). The voltage and current start at zero, increase to maximum positive value, fall back to zero, increase to maximum negative value, and finally return to zero. This cycle constitutes one hertz (Hz), or cycle per second. In North America, most power is generated at 60 Hz.

Common circuit voltages are 120, 240, and 480 VAC (voltage alternating current). These values are actually root-mean-square values, or effective values. The maximum circuit voltage for a 120-VAC circuit is actually $120/0.707 = 169.7$ volts.

In a three-phase circuit, there are three voltage and current sinusoidal waveforms. The three waveforms are called phases, thus the term *three-phase*. The waveforms are spaced 120 electrical degrees apart. Electrical degrees and mechanical degrees differ, as shown in Figure 17.1. One

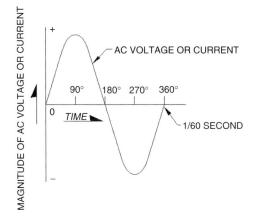

Figure 17.1. AC Waveform at 60 Hz.

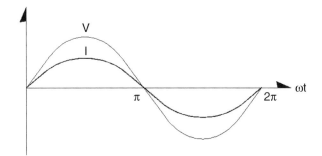

Figure 17.2. Voltage and current are in phase.

sinusoidal waveform (cycle) constitutes 360 electrical degrees, whereas one full rotation of a circle equals 360 mechanical degrees.

The relationship between the voltage waveform and the current waveform defines the power factor of the electrical circuit (more on power factor later). The voltage and current are in phase in a pure resistive AC circuit (see Figure 17.2).

An inductor is a device that stores energy in the magnetic field set up by the current through a coil. *Inductance* is that property that opposes the change in current. Inductance is measured in henrys. The inductor causes the voltage waveform to lead the current waveform by 90 electrical degrees (see Figure 17.3). Inductance in a circuit not only causes the current to lag the voltage, but it also reduces the current to a smaller value than it would have been if there were no inductance present.

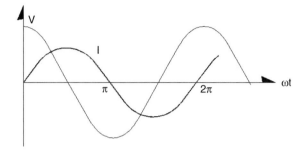

Figure 17.3. The voltage leads the current by 90 degrees in a pure inductive AC circuit.

The inductive reactance, in ohms, is calculated as follows:

$$X_L = 2\pi f L$$

where

f = frequency, in Hz
L = inductance, in henrys

A *capacitor* is a device that stores electrical energy. It is typically constructed of two conducting parallel plates that are separated by a nonconductor (dielectric). The plates are continuously charged and discharged. Capacitance delays the voltage waveform and causes it to lag the current waveform. Or, put another way, capacitance causes current to lead the voltage by 90 electrical degrees (see Figure 17.4). Capacitance is measured in farads. One farad can store one coulomb per volt. Capacitance also presents impedance to alternating current. This is known as capacitive reactance, in ohms, and is calculated as follows:

$$X_C = 1/(2\pi f C)$$

where

f = frequency, in Hz
C = capacitance, in farads

Now we can define power in AC circuits. In a single-phase circuit, watts = (volts)(amps)(power factor). See subsequent discussions for a description of the power factor. In a three-phase circuit, watts = (volts)(amps)(power factor)(1.732). The 1.732 factor for three-phase circuits is used to adjust the measurements taken on a single line or phase (the volts

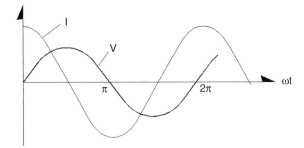

Figure 17.4. The voltage lags the current by 90 degrees in a pure capacitive circuit.

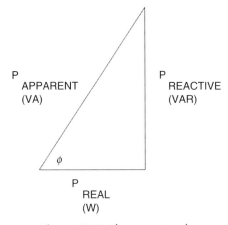

Figure 17.5. The power triangle.

and amps used in the formula are measured on one line or phase) to account for the net effect of the three phases. This three-phase power is known as the *real power* (see Figure 17.5). For a more detailed analysis, see Sidebar 1.

Sidebar 1: Three-Phase Circuits

To illustrate three-phase power, imagine three coils spaced 120 degrees apart. Three voltages spaced 120 degrees from each other will be produced when a magnetic field cuts through the coils. This is effectively how three-phase power is generated. There are two basic connections for three-phase power: the wye connection and the delta connection.

Sidebar 1: Three-Phase Circuits (continued)

The wye connection consists of three coils with one common connection (see Figure A). The voltage measured across a single coil, or winding, is known as the *phase voltage*. The voltage measured from line to line is known as the *line voltage*. In a wye-connected motor, the line voltage is higher than the phase voltage by a factor of the square root of 3, or 1.732. Phase current and line current are the same in a wye-connected motor.

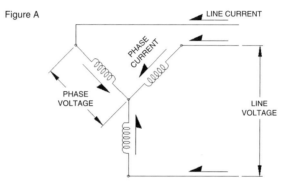

1) LINE VOLTAGE (V_L) = 1.732 X PHASE VOLTAGE (V_P)
2) LINE CURRENT = PHASE CURRENT

Three coils in Figure B have been arranged to form a delta-connected three-phase circuit. The phase voltage and the line voltage are the same in a delta-connected motor (or circuit). However, the line current and the phase current are different. The line current in a delta-connected motor is higher than the phase current by a factor of the square root of 3, or 1.732.

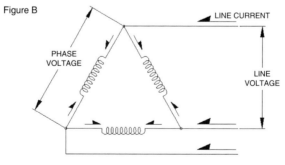

1) LINE VOLTAGE = PHASE VOLTAGE
2) LINE CURRENT (I_L) = 1.732 X PHASE CURRENT (I_P)

> An error can easily be made with three-phase power calculations. The confusion is rooted in the fact that there are actually two formulas to calculate three-phase power. If line voltage and line current are used, the power (watts) in a motor circuit is expressed as
>
> $$(1.732)(V_L)(I_L)(\text{power factor}).$$
>
> If the phase voltage and the phase current are used in the calculation, the *apparent power* is determined, not the real power. The apparent power in a motor circuit (volt-amps [VA]) is
>
> $$(3)(V_P)(I_P)(\text{power factor}).$$
>
> Power factor and apparent power are discussed in Sidebar 2 on the power triangle and power factor.

Figure 17.5 shows the reactive power as lagging. The reactive power could just as easily be leading, in which case the reactive power would be shown as negative and pointing down. The power triangle illustrates the fact that voltage and current waveforms can be out of phase.

The reactive power is shown perpendicular to the real power because it is out of phase by 90 electrical degrees. The inductive reactance and the capacitive reactance must have different signs. Inductive reactance is conventionally denoted as positive and capacitive reactance is conventionally designated as negative.

The apparent power, in VA, is the product of the measured voltage and the current. Alternatively, the apparent power may be calculated by multiplying the impedance (Z) by the current squared. The impedance can be calculated as follows:

$$Z = (R^2 + (X_L - X_C)^2)^{1/2}$$

where

R = resistance, in ohms
X_L = inductive reactance, in ohms
X_C = capacitive reactance, in ohms

The cosine of the angle between real power and apparent power in Figure 17.5 is known as the *power factor*. For a more detailed analysis, see Sidebar 2.

> **Sidebar 2: The Power Triangle and Power Factor**
>
> In AC circuits, the power is not necessarily the product of current and voltage, although it would be in an entirely resistive circuit. There are three types of power, as shown in the power triangle (Figure 17.5): apparent power, real power, and reactive power.
>
> 1. Apparent power is the product of line voltage and line current. It is obtained by measuring line voltage and line current and multiplying the two measured values.
> 2. In an inductive circuit such as a motor circuit, there are magnetic characteristics that result in drawing a larger apparent power, in VA, than real power, in watts. Remember that this induction causes the voltage to lead the current by 90 electrical degrees, as shown in Figure 17.5. This reactive load effectively delivers less power to the load. Reactive power is measured in volt-amps-reactive (VARs).
> 3. Figure 17.5 is essentially a vector diagram that shows the components of power in an AC circuit. The VAR component represents the reactive power, while the watt component represents real power. The combination of the two represents VA, the apparent power. The apparent power is the square root of the sum of the squares of the reactive power and the real power.
> 4. The power factor is equal to the real power divided by the apparent power. The power factor is also the cosine of the angle between the real power and the apparent power (the angle is typically designated φ). The cosine of φ is the power factor.

Electrical utility companies charge for apparent power, but only real power is dissipated. It is possible to reduce power bills by changing the power factor (φ) without changing the real power. This is done by changing the reactive power and is commonly known as *power factor correction*. This concept is best explained in an example:

A 10-hp, 460-volt (V), 60-Hz induction motor draws 22 amps (A) and has an efficiency of 90%. Find the capacitance that could be connected in parallel that would improve the power factor to 95%.

$$\text{Real power} = \frac{10 \text{ hp} \times 746 \text{ W/hp}}{0.90 \text{ eff.}} = 8289 \text{ W}$$

where W = watts. The apparent power is determined from the measured values:

$$460 \text{ V} \times 22 \text{ A} = 10{,}120 \text{ W}.$$

Therefore, the reactive power $= (10120^2 - 8289^2)^{1/2} = 5805$ W.

The power angle, φ, = arcos $(8289/10120) = 35$ degrees. At the desired power factor = 0.95, the new power angle = arcos $0.95 = 18.2$ degrees. The new reactive power would be (tan 18.2)(8289) = 2725 VARs. The difference in reactive power is $(5805 - 2725) = 3080$ VARs.

The required capacitance, C (in microfarads [µf]), equals the change in reactive power divided by $2\pi f V^2$:

$$C = 3080/(2\pi)(60)(460)^2 = 38.6 \, \mu f$$

A 40-µf capacitor would be installed in each of the three phases.

It should be noted that it is not advisable to raise the no-load reactive power to a point at which unity power factor can be achieved (95% maximum is recommended). This will cause overexcitation and could damage the motor or injure personnel. Overexcitation results in high transient voltages, currents, and torque.

17.2 INTRODUCTION TO ELECTROMAGNETIC THEORY

Current flow in a conductor causes a magnetic field to form around the conductor. The *right-hand rule* states that pointing the right thumb toward the direction of current flow causes the fingers to show the direction of the magnetic field. In a DC circuit, current flows out the positive terminal and into the negative terminal.

A bar magnet has two poles: a north pole and a south pole. Invisible magnetic lines of flux exit one pole and enter the other pole. The magnetic field, defined by the location and strength of the flux lines, can also be established electromagnetically. Electromagnetic induction is the generation of a voltage by movement of a conductor through lines of magnetic force, or flux lines.

A voltage is generated in a conductor as it moves through a magnetic field. The generated voltage is

(flux density)(length of the conductor linked by the flux)
  × (velocity of the conductor through the magnetic field).

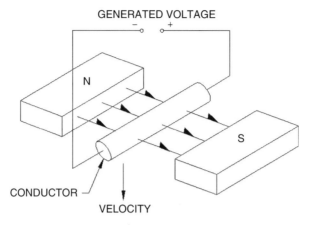

Figure 17.6. Voltage generated in a conductor.

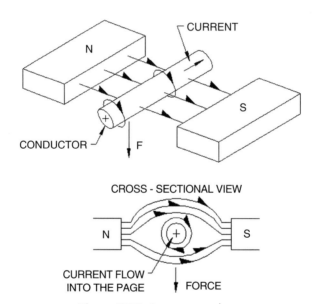

Figure 17.7. Force on a conductor.

Figure 17.6 shows voltage generated in a conductor. A force is exerted on a current-carrying conductor placed in a magnetic field. Figure 17.7 shows this concept graphically. The force in the magnetic circuit is equal to

(flux density)(current)(length of the conductor in the magnetic circuit).

If current-carrying conductors are arranged so that they may rotate on an axis (shaft) that is centered in a magnetic field, torque will be produced on the axis. The torque is proportional to the number of conductor turns and the current. *Ampere-turns* is a common motor-winding design term and is derived from this concept. The electromagnetic torque is increased proportionally by an increase in the number of winding turns and/or an increase in the current.

17.3 ELECTRIC MOTORS

An electric motor is a device used to convert electrical energy into mechanical energy. The *stator* is the stationary part of the motor. The *rotor* is the rotating part of the motor. Induction motors are usually rated three-phase. The motor's construction is unique because the rotor receives power through induction. There is no physical connection to the rotor.

An induction motor is thus a motor that has no physical connection to the rotor. Current in the rotor is induced by the magnetic field in the stator. Induction motors have the following performance characteristics:

1. At a given frequency, speed is constant.
2. Maximum current is realized during starting when the percentage slip is 100%.
3. Total torque is three times the phase torque.
4. Starting torque is directly proportional to rotor resistance and line voltage.
5. Torque is inversely proportional to the rotor winding resistance.
6. Torque is directly proportional to slip.

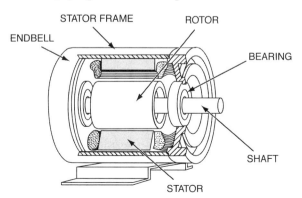

The AC induction motor is prevalent for the following reasons:

1. Electric utility companies generate and deliver three-phase power.
2. The three-phase induction motor is simple in construction and easy to maintain.
3. Three-phase induction motors cost less to operate per horsepower and are less expensive than other motors of the same horsepower.
4. The three-phase induction motor does not require complicated motor starters and usually can be started across the line (application of full-line voltage).
5. Various speed/torque ratings are readily available and speed control is now common with the use of ASDs.

Two formulas that are commonly used for three-phase motor applications are as follows:

$$hp = (1.732)(V)(I)(\text{efficiency})(\text{power factor})/746$$
$$hp = \text{torque (ft-lb)} \times \text{rpm}/5250.$$

where

V = voltage, in volts
I = current, in amps

17.3.1 Rotor Circuits

The rotor circuit may be either one of two types regardless of stator winding: the squirrel-cage rotor or the wound rotor.

The simplest rotor circuit is the squirrel cage. There are no windings in this type of rotor. It has a large number of parallel connected conductor bars made of aluminum or copper or alloys of these metals. These bars are embedded into partially enclosed slots on the rotor periphery. Two end rings permanently join all bars. It is important to note that the rotor bars (conductors) are located some distance from the mechanical axis or shaft centerline.

There are many motor applications in which little or no load is connected to the motor at startup. The load is applied after the motor has reached operating speed. For these applications, the squirrel-cage rotor is suitable. Squirrel-cage rotors are constructed so that there is no ability to change the rotor resistance. The rotor may be built with relatively high or relatively low starting torque. Use of rotor bar materials with high or low resistivities provides high starting torque at the expense of additional heating. The impedance of the squirrel-cage rotor circuit at standstill is sufficient to limit starting current to five to seven times normal operating current. Starting torque is 150–250% of full-load torque.

Some applications require a motor with a wound rotor rather than a cage rotor. A wound rotor has a polyphase winding similar to that in the stator circuit. The winding leads are connected to shaft-mounted slip rings. These slip rings connect the rotor windings to an external circuit. The rotor can be connected to external resistance during starting to improve the power factor of the rotor circuit and to increase the starting torque. There is a limit to the increase in torque that can be obtained with additional resistance.

17.3.2 Stator Circuits

When voltage is applied to the stator windings, a rotating magnetic field is set up automatically. The windings (coils) in the stator are connected to form three separate windings, or phases. Each phase consists of one third of the total number of windings in the stator. These three composite windings are known as phases A, B, and C.

Each phase is placed in the stator so that it is 120 electrical degrees from the other phases (see Figure 17.8). Since each phase reaches its peak value 120 degrees apart, a rotating magnetic field is produced in the stator. Electric coils consisting of several turns of wire are placed in the stator at strategic locations to establish a desired number of magnetic poles. The number of magnetic poles will define the synchronous speed of the motor. The synchronous speed (frequency in Hz) = 120 × frequency/number of poles. A two-pole motor, which has a synchronous speed of 3600 rpm at 60 Hz, has its poles located 180 mechanical degrees apart. See Figure 17.9 for a schematic of magnetic poles located in a stator. For instance, a four-pole motor, which has a synchronous speed of 1800 rpm at 60 Hz, has its poles located 90 mechanical degrees apart; a six-pole motor, which has a synchronous speed of 1200 rpm at 60 Hz, has its poles located 60 mechanical degrees apart.

When voltage is applied to the stator windings, a rotating magnetic field is automatically effected. The stator magnetic flux lines cut across the rotor conductors and induce a voltage in them. Because the rotor conductors are shorted at each end with the shorting rings, the rotor bars

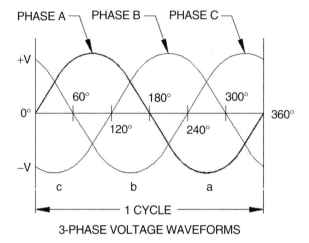

Figure 17.8. Three-phase voltage waveforms.

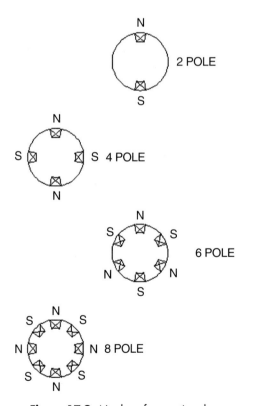

Figure 17.9. Number of magnetic poles.

constitute an electrical circuit. With the induced voltage, a current flow is present. This current flow effects a second magnetic field surrounding the rotor bars. The interaction of the two magnetic fields produces a torque around the mechanical center (shaft).

The stator field rotates at synchronous speed (120 × frequency/number of poles). In order for there to be a change in the flux linkage, the rotor must turn slower than synchronous speed. The difference in speed is small but necessary, known as *slip*. Expressed as a percentage, it is calculated as follows:

$$\text{slip} = \frac{\text{synchronous rpm} - \text{full-load rpm}}{\text{synchronous rpm}}$$

The slip is typically 2–5%. Multiply the figure determined from the formula above by 100.

17.4 TRANSFORMERS

The transformer is an inexpensive piece of stationary equipment that operates at a high efficiency. Transformers make it possible to generate energy at any level, say 20,000 volts, and transform it to energy at high voltage, such as 100,000 volts. Power lines transmit energy at high voltage to minimize losses. Power losses are proportional to the square of the current. At high transmission voltages, the current is low.

Transformers are also used to transform energy at high voltage to energy at a lower voltage. Utility substations receive high-voltage energy and distribute it at a lower voltage. The voltage is usually transformed down once again at a location closer to the load, such as within residential neighborhoods or at industrial plant feeder circuits. By means of a transformer, electrical energy in one circuit is transformed to electrical energy of the same frequency in another circuit.

In its simplest form, a transformer consists of two coils of wire, a primary and a secondary, insulated from each other and wound on a common laminated steel core. Energy enters the transformer through the primary coil and exits via the secondary coil. When the primary coil is connected to a voltage source and the secondary coil is connected to a load (such as lamps or motors), the energy is transferred from the primary to the secondary coil through the magnetic field in the core. Thus, the direction of energy flow is from primary to secondary.

The transformer usually transforms the primary voltage to a different voltage in the secondary. The secondary circuit supplies voltage to the load. Transformers sometimes are used for isolation only (no voltage change), for purposes of protecting the load or protecting the circuit from the load should the latter have the ability to generate undesirable harmonics or voltage surges.

When the high-voltage coil is the primary and the low-voltage coil is the secondary, the transformer is a *stepdown transformer*. Conversely, in a *stepup transformer*, the primary is the low-voltage coil and the secondary is the high-voltage coil.

The vast majority of transformers used in commercial power applications receive power from and deliver power to circuits having approximately constant potentials. Such transformers are called *constant-potential transformers*. They are used to deliver relatively large amounts of energy.

An ideal transformer is one in which resistance of the windings is negligible and the core has no losses. When the primary coil is energized with an AC voltage, the impressed potential causes an alternating current to flow in the primary winding. Assuming an ideal transformer, as defined, the effective resistance is zero, the circuit is primarily reactive, and the current waveform lags the voltage waveform by 90 electrical (time) degrees.

The current flowing in the turns of the primary winding (N1, where N = number of turns) magnetizes the core, produces a flux that is proportional to the current, and is in time phase with the current. The flux intercepts the secondary winding. A voltage is induced in the secondary. The ratio of the voltages, which is commonly referred to as the turn ratio, is

$$V2/V1 = N2/N1.$$

Transformers have very high efficiencies, from 98 to 99.5%. This would be necessary for the power factors of the primary (input) and the secondary (output) to be approximately equal. The power input and output can be determined by:

$$V1/N1 = V2/N2$$

or

$$I1/I2 = V2/V1.$$

As the voltages are approximately the same turn ratio,

$$I1/I2 = N2/N1.$$

In other words, the currents in the primary and secondary of a transformer are inversely proportional to the number of turns in the primary and secondary windings.

Transformers are designed to transfer power between circuits having a variety of voltages, and so a variety of transformer ratios are required. Common ratios for distribution circuits are from 2300 to 230/115 and from 2300 to 460/230 volts.

Certain voltages have become standard for different sections of transmission and distribution systems. Standard circuit voltages for various parts of distribution systems are 2300, 4600, 6900, 11,500, 13,800, 22,000 and 33,000 volts. These may be considered as standard generator voltages. The standard voltages at which power is transmitted over great distances are 44,000, 66,000, 88,000, 110,000, 132,000, 154,000, 220,000 and 440,000.

Heavy power circuits require high voltages, especially if they are many miles long. Standard service voltages for lighting and small motors are 230 and 115 volts. Other common low voltages are 460 and 575. Large motors may be operated at voltages of 2300, 4600, or 6600.

17.5 ADJUSTABLE SPEED DRIVES

As discussed earlier, a motor's synchronous shaft speed is determined by the product of 120 and the frequency divided by the number of magnetic poles. The number of magnetic poles is a motor attribute—it is a fixed physical characteristic. If it is desired to vary a motor's speed, the line frequency must vary. The speed of a motor is directly proportional to the power supply frequency. Varying the frequency to a motor is done by an ASD, which is a motor controller.

For many years, DC motors were the only choice for speed and torque control. In some applications, this still applies today. DC machines have application where precision control of speed and torque is needed. The problem with DC motors and DC drives is that there are high initial and maintenance costs.

The ASD actually goes by many names. "Drive" to some people may mean a mechanical speed controller such as a gearbox or pulley system. People working in the electrical fields may refer to an ASD as a variable frequency drive or a silicon-controlled rectifier power module. By whatever name, ASDs provide the required level of current and voltage in a form that the motor can use. The ASD provides user controls for adjusting the minimum and maximum frequency or speed settings and typically provides overload protection in that the user can set the motor current rating to protect it from prolonged high current.

There are different types of ASDs, but all have the same function: to change the fundamental frequency (usually 50 or 60 Hz) in order to change the motor speed proportionally. The basic components of an ASD are a rectifier, inverter, and controls. An ASD rectifies the incoming AC power to DC and then the inverter converts the DC back to AC at an adjustable fundamental frequency. Inductors or capacitors are also used to condition the rectifier output and minimize ripple harmonics.

As previously stated, the motor shaft speed is proportional to the power supply frequency. On the other hand, motor torque is proportional to the magnetic field strength. The magnetic field is proportional to the voltage and inversely proportional to the frequency. Therefore, the motor torque can be changed by adjusting the voltage at any given frequency.

Operating an induction motor to deliver constant torque at various speeds requires a variable voltage and a variable frequency power supply. The ASD must supply a constant "boost," or voltage/frequency ratio. Torque versus rpm load characteristics would typically fall within one of the following load profiles:

1. Torque is constant while speed varies. Typical applications include conveyors, positive displacement pumps, punch presses, and extruders.
2. Torque is proportional to rpm. Typical applications include low-speed mixing. This application is known as a variable torque load profile.
3. Liquids and gases when moved require a pressure or voltage proportional to the velocity squared. As centrifugal pumps deliver volumes proportional to rpm, the pressure or voltage will be proportional to the speed squared. As the motor develops torque to maintain a pressure, the torque will also be proportional to the speed squared. Therefore, the power will be proportional to the speed cubed. Reducing the flow to 50% of maximum reduces the hp to only 12.5% of that needed at full flow.
4. Maximum power and maximum torque may be required at low operating speeds. Some machine tools have this load characteristic. A material may initially be highly viscous, but as it is moved, the viscosity of the material drops.

5. It is common for applications to require constant torque or variable torque up to the fundamental frequency, and constant hp for operation above the fundamental frequency. Constant torque applications include hoists and cranes. Variable torque applications include fans and centrifugal pumps. Constant hp applications include high-speed machine tools.

The most common type of ASDs used with induction motors are

- Current source inverter (CSI)
- Voltage source inverter (VSI)
- Pulse-width modulation (PWM)

When a large inductor is connected in series with a rectifier output, it becomes a current stabilizer or current source, thus the name *current source inverter*. The CSI inductance maintains constant DC current for the inverter. CSIs are used for very large motors and high-inertia loads.

When a large capacitor is connected in series with the rectifier output, it becomes a VSI. The PWM inverter has replaced the VSI for the most part and is the most common ASD.

The PWM inverter rectifies the AC into DC segments of constant amplitude every half cycle. A full cycle consists of one half positive DC segments and one half negative DC segments. This modulating process of the power produces very high frequency DC voltage pulses. These pulses are rectangular when looked at in an exploded view or on an oscilloscope. The high frequency of the voltage pulses is referred to as the *carrier frequency*. The rate of rise of the voltage pulses is on the order of tenths of microseconds. The rate of change of the voltage is significantly higher than the rate of change for a normal sine wave. Voltage spikes can be produced due to this high rate of change. The PWM ASD current output is also in the form of a sine wave, with numerous small irregularities. This distortion amounts to a total harmonic distortion on the order of 5–10%.

ASDs provide several benefits:

- Energy savings
- Improved process control
- Wider operating speed ranges
- Increased productivity

- High acceleration and deceleration
- Higher production rates
- Longer equipment life
- Reduced maintenance
- Soft starting

A word about disadvantages. ASDs are more complicated than standard motor starters. They also are more expensive than standard motor starters. Voltage spikes and harmonic distortion may require special filters and/or isolation transformers. One must be careful not to adjust line frequencies and equipment operating speeds to a point where vibration is excessive. Electrical noise can increase with use of an ASD.

17.6 ELECTRIC MOTOR APPLICATIONS ON OIL RIGS

Continuous-duty electric motors are an integral part of a drillings rig's solids-control and processing systems. Centrifugal pumps that feed hydrocyclones, circulate mud for mixing, and transfer mud to and from reserve and also into the trip tank are powered by electric motors. Shale shakers, mud cleaners, centrifuges, and pit agitators are also driven by electric motors.

Continuous-duty electric motors meet well-defined performance standards. Motors are designed with conductor, frame, and insulating materials to continuously deliver rated horsepower and not exceed the insulation's temperature limits. A *service factor rating* defines the ability of the motor to continuously withstand prolonged overload conditions while remaining within the temperature limitations of the insulating material.

17.6.1 Ratings

A statement of operating limits of each commercial machine is provided by the manufacturer in the form of data stamped on a nameplate fastened to the machine frame. These data usually include the hp output, speed, voltage, and current assigned to the machine by the manufacturer. These data constitute the rating of the machine, and the various parameters are referred to as the rated voltage, rated current, etc.

The rating of a machine is an arbitrarily specified safe operating limit for the machine, determined in accordance with accepted standards and procedures. The rating specifies the operating limit that the machine cannot ordinarily exceed for a considerable length of time without some damage occurring to it, or at least causing an accelerated rate of wear in one or more of its parts. Unless stated otherwise, generators and motors are rated for continuous service. Their specified loads may be carried for an unlimited period of time. Machines that operate on intermittent, varying, or periodic duty are given a short time rating such as 5, 10, 15, 30, 60, or 120 minutes.

The rated output of a generator is expressed in kilowatts (kW) available at the machine terminals. A generator whose rating is 25 kW at 125 V and 1000 rpm will, when operated at 1000 rpm, supply an output current of $(25,000/125) = 200$ A. The value of the current that appears on the nameplate is 200 amperes.

The rated output of a motor can be given in kilowatts but is more often given in horsepower available at the shaft when the specified voltage is impressed and the motor runs at rated speed. The motor input is greater than the motor output because there are certain friction and other heat losses that must be supplied from the electrical input in addition to the useful work, which the motor does.

Heating is the main factor affecting the ratings of motors and generators. There is no definite load a machine can carry in the sense that a 5-gallon bucket holds a certain maximum amount of liquid and no more. A machine may exceed its rating by 25% or even 50%, but if the excess load is carried for a considerable period of time, the temperature will rise to a value that will result in permanent damage to the insulation.

Voltage regulation, in the case of a generator, and speed regulation, in the case of a motor, also influence rating. The full-load and no-load voltages of a generator are often specified, as are the full-load and no-load speeds of a motor.

17.6.2 Energy Losses

I^2R losses (both stator and rotor), friction losses, hysteresis, and eddy currents are electrical-energy phenomena converted into heat that must be radiated away by currents of air.

Heat energy can flow only if a difference in temperature of the heated part or surface and the surrounding air exists. There is a definite temperature limit that no winding can exceed without incurring permanent damage to the machine (generator or motor). Heat losses should be kept to a minimum. Good design allows for the transfer of heat away from the motor frame—typically through conduction or convection.

Generators and motors are sometimes enclosed and provided with a forced ventilating system. Many others are open and have free access to the air for cooling. Dirt and dust accumulating in the air ducts may impair ventilation to such a degree that a machine may overheat even though its rated load is not exceeded.

Altitude affects heating. Air is a better cooling medium at sea level than at elevations. Standard ratings apply to altitudes of 3300 feet (1000 m) or less.

17.6.3 Temperature Rise

Excessive temperatures deteriorate and finally destroy the insulating properties of the materials used to insulate the windings of motors and generators. The highest temperature to which an insulating material may be subjected continuously is the maximum rated temperature for that material. Exceeding the maximum rated temperature of the insulating material will result in premature breakdown of that material.

The criterion for sizing and selection of any motor is its ability to deliver startup power under the process load and to then provide power that drives the equipment throughout operation. Adequate torque must be developed to overcome inertia during startup. The load must then be accelerated to the desired operating speed and full-load power requirements supplied without overheating. These parameters depend on motor design and the full-load rating (output hp).

Electric-motor operating efficiency is the ratio of output power to input power. The power loss is the difference between the power into the motor and the power out of the motor. This power loss is caused by:

- Heat from the electrical resistance of motor windings and rotor
- Windage losses from cooling fans or rotor fins
- Magnetic and core losses from currents induced in the laminations of frame and stator
- Friction losses from shaft bearings

The motor's internal heat is a function of load conditions, motor design, and ventilation conditions. Heat produced internally by the motor raises operating temperature and adversely affects insulation used to isolate electrical conductors from each other and from the motor frame. Insulation materials are rated based on thermal capacity, or the ability to withstand heat effects. High-quality insulation systems with high thermal capacity can withstand relatively high temperature increases and deliver a long motor service life at rated performance. Because motors may be operating properly and still be too hot to touch, it is important to check the manufacturer's guidelines.

17.6.4 Voltage

Motors are rated for operation at specific voltages. Motor performance is affected when the supply voltage varies from the motor's rated voltage. Motors generally operate satisfactorily with voltage variations within ±10%. However, equipment connected to the motor may not always function properly with such variations.

Surge voltage is any higher-than-normal voltage that temporarily exists on one or more of the power lines of a three-phase motor. A surge causes a large voltage rise during an extremely short period of time. Surges are of concern because the higher voltage is impressed on the first few turns of the motor windings. The winding wire insulation may be destroyed, causing the motor to fail. Frequent voltage surging can result from line switching of large generators.

Undervoltage at the motor terminals can result when large current demands are placed on the generator, such as starting the top drive motor. Operation below 10% of the marked motor voltage will generally result in excessive overheating and torque reduction. Overheating prematurely deteriorates the insulation system. Torque reduction may result in the motor stalling or, in the case of shale shaker vibrators, in poor performance.

Figure 17.10 provides general guidelines for the effects on induction motors of voltage variation and the effects of voltage unbalance on motor performance.

17.7 AMBIENT TEMPERATURE

The National Electrical Manufacturers Association (NEMA) full-load motor ratings are based on an ambient temperature of 40°C (104°F), at a

EFFECT OF VOLTAGE UNBALANCE ON MOTOR PERFORMANCE

When the line voltages applied to a polyphase induction motor are not equal, unbalanced currents in the stator windings will result. A small percentage voltage unbalance will result in a much larger percentage current unbalance. Consequently, the temperature rise of the motor operating at a particular load and percentage voltage unbalance will be greater than for the motor operating under the same conditions with balanced voltages.

Should voltages be unbalanced, the rated horsepower of the motor should be multiplied by the factor shown in the graph below to reduce the possibility of damage to the motor. Operation of the motor at above a 5 percent voltage unbalance condition is not recommended.

Alternating current, polyphase motors normally are designed to operate successfully under running conditions at rated load when the voltage unbalance at the motor terminals does not exceed 1 percent. Performance will not necessarily be the same as when the motor is operating with a balanced voltage at the motor terminals.

MEDIUM MOTOR OPERATING FACTOR DUE TO UNBALANCED VOLTAGE

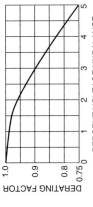

Percent Voltage Unbalance = $100 \times \dfrac{\text{Max. Volt. Deviation from Avg. Volt.}}{\text{Average Volt.}}$

Example: With voltages of 460, 467, and 450, the average is 459, the maximum deviation from the average is 9, and the

Percent Unbalance = $100 \times \dfrac{9}{459}$ = 1.96 percent

Reference: NEMA Standards MG 1-14.35.

EFFECT OF VOLTAGE VARIATION ON INDUCTION MOTOR CHARACTERISTICS

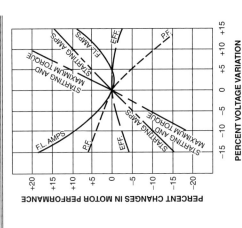

POWER SUPPLY AND MOTOR VOLTAGES

NOMINAL POWER SYSTEM VOLTAGE, VOLTS	MOTOR UTILIZATION (NAMEPLATE) VOLTAGE, VOLTS
120	115
208	200
240	230
480	460
600	575
2400	2300
4160	4000
6900	6600

Reference: NEMA Standards MG-10.

Figure 17.10. General guidelines for the effects on induction motors of voltage variation and the effects of voltage unbalance on motor performance. © 2000, Electrical Apparatus Service Association, Inc., St. Louis, MO. Reprinted with permission.

Table 17.1
Effect of Ambient Temperature on Electric Motors

Ambient Temperature (°C/°F)	Ambient Temperature Factor
−20/−4	1.27
0/32	1.19
20/68	1.10
40/104	1.00
60/140	0.88

maximum altitude of 1000 meters (3300 ft) above sea level. Variance in ambient temperature requires rerating of motor hp requirements. See Table 17.1 for a determination of hp requirements at varying ambient temperatures.

Machines intended for use at altitudes above 1000 meters (3300 ft), at an ambient temperature of 40°C (104°F), should have temperature rises at sea level not exceeding the values calculated from the following: when altitude is expressed in meters:

$$T_{RSL} = T_{RA}[1 - (\text{Alt} - 1{,}000)/10{,}000]$$

when altitude is expressed in feet:

$$T_{RSL} = T_{RA}[1 - (\text{Alt} - 3{,}300)/33{,}000]$$

where

T_{RSL} = test temperature, in degrees Celsius at sea level

T_{RA} = temperature rise, in degrees Celsius from tables
(see NEMA standard MG-1, 1993, section II, par 14.04.3)

Alt = Altitude above sea level at which machine will be operated.

NEMA performance ratings are also based on operation at voltage within 10% of the nameplate voltage and a frequency within 5% of the nameplate frequency. If both frequency and voltage vary, the combined total variance is not to exceed 10%.

A 50-Hz motor should not be operated at 60 Hz unless it is specifically designed and marked for 60-Hz operation. A 60-Hz, three-phase induction motor may be operated at 50 Hz if the voltage and hp are reduced to 83% of the 60-Hz values. It should be noted that the speed and the slip would also be reduced by 83%. The shaft torque would remain the same.

17.8 MOTOR INSTALLATION AND TROUBLESHOOTING

When replacing a motor, its exact dimensions, as well as speed, hp, and torque characteristics should be determined and duplicated if the same performance is desired. When replacing a motor, the entire system should be inspected for internal and external degradation. Neither the motor mounting nor the mechanical coupling should exhibit signs of wear.

The power supply and connections should not be damaged. These should be frequently checked for proper frequency, voltage, and voltage balance between phases. Poor and broken connections of one of the supply lines are a major cause of voltage unbalance. Overload relays for each phase should protect against extreme (greater than 5%) voltage unbalance.

Misalignment between the motor and the driven machine (e.g., centrifugal pump) can cause bearing failures and shaft breakage. Excessive vibrations frequently indicate misalignment. All motor feet must be fastened to a flat, preferably machined, surface. Otherwise, the frame can bend when the motor is tightened down, which twists the motor frame and causes misalignment. Care should be taken to evenly tension mounting bolts. If torque values are specified, follow the manufacturer's recommendations. The mounting should be inspected frequently. If necessary, retighten bolts with the proper torque. If vibration is detected, one or more of the motor feet may have to be shimmed. A few thick shims are preferable to many thin shims if it is necessary to align motor and machine (pump) shafts. Misaligned couplings create bearing loading in both the motor and the machine, causing high-speed distortions and also increasing power consumption.

Motor bearings should be greased using manufacturer's specified greases in concert with manufacturer's specified lubrication frequency and quantities (sealed ball bearings cannot be lubricated after manufacture). Relubrication is necessary to replenish grease that has broken down by oxidation or been lost by evaporation and centrifugal force.

Inspect and keep cooling and ventilation vents clear of obstructions.

If a motor burns out, the windings should be inspected for signs of single phasing, short circuiting, overloading, and voltage unbalance. Any cause of winding damage should be identified and corrected. If a motor burns out, the circuit supplying the voltage should also be inspected for broken or shorted wires, burnt contacts, or voltage unbalance.

17.9 ELECTRIC MOTOR STANDARDS

Generally, standards for electric motors are based on those of NEMA, the International Electrotechnical Commission (IEC), and the Institute of Electrical and Electronic Engineers (IEEE). Most countries typically have their own standard or recognized standard.

NEMA designs (A, B, C, D, and E) classify motors according to specific torque characteristics for effective startup and operation of equipment under particular loading and operating situations. Design B motors are commonly used on drilling rigs. These are general-purpose motors suitable for normal startup required by pumps, fans, and low-pressure compressors (see Figure 17.11).

In Europe and Asia, national standards for electric motors are, in general, based on those of the IEC, which facilitates coordination and unification of motor standards. IEC standards for dimensions, tolerances, and output ratings are contained in IEC Publications 72 and 72A. IEC standards for rating, performance characteristics, and testing of rotating machinery for nonhazardous locations are contained in a series of IEC Publications No. 34, while IEC standards dealing with apparatus for explosive gas atmospheres are contained in a series of Publications No. 79.

The IEC recommendations hold international applicability. The European standards are identical in all European Community (EC) countries in regard to their contents and are published as national standards. Before existence of these standards, each country had its own national certifying authority. Today, with certification from a recognized national *notified body*, the motors are acceptable in all EC countries and most other European and Asian countries as well. Electrical equipment certified to conform to these standards may be installed and used in any EC member state. Participating non-EC states may require additional testing standards.

While there are many similarities and even direct interchangeabilities between U.S.- and IEC-recognized standards, specific applications must be considered. Motors may be acceptable under all standards but not necessarily certified under all standards. The IEC flameproof motor is essentially the same as the U.S. explosion-proof motor. Each design withstands an internal explosion of a (specified) gas or vapor and prevents ignition of the specified gas or vapor that may surround the motor. However, construction standards are not identical. The U.S. standard is generally more stringent, and acceptability can be based on approval by local authorities.

GENERAL SPEED-TORQUE CHARACTERISTICS
THREE-PHASE INDUCTION MOTORS

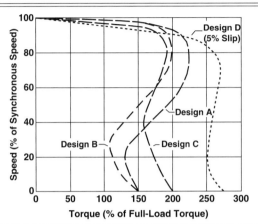

NEMA DESIGN	LOCKED ROTOR TORQUE	BREAKDOWN TORQUE	LOCKED ROTOR CURRENT	% SLIP	RELATIVE EFFICIENCY
B	70–275%*	175–300%*	600–700%	0.5–5%	Medium or High
	Applications: Fans, blowers, centrifugal pumps and compressors, motor-generator sets, etc., where starting torque requirements are relatively low.				
C	200–250% *	190–225%	600–700%	1–5%	Medium
	Applications: Conveyors, crushers, stirring machines, agitators, reciprocating pumps and compressors, etc., were starting under load is required.				
D	275%	275%	600–700%	5–8%, 8–13%, 15–25%	Medium
	Applications: High peak loads with or without flywheels, such as punch presses, shears, elevators, extractors, winches, hoists, oil-well pumping, and wire-drawing machines.				
E	75–190%*	160–200%*	800–1000%	0.5–3%	High
	Applications: Fans, blowers centrifugal pumps and compressors, motor-generator sets, etc., where starting torque requirements are relatively low.				

Based on NEMA Standards MG 10, Table 2-1. NEMA Design A is a variation of Design B having higher locked-rotor current.
*Higher values are for motors having lower horsepower ratings.

Figure 17.11. © 2000, Electrical Apparatus Service Association, Inc., St. Louis, MO. Reprinted with permission.

The U.S. totally enclosed purged and pressurized, or inert gas filled, motors are manufactured to similar standards as those of IEC pressurized motors. Each operates by first purging the motor enclosure of any flammable vapor and then preventing entry of the surrounding (potentially explosive or corrosive) atmosphere into the motor enclosure by maintaining a positive gas pressure within the enclosure.

IEC type "e" (Increased Safety) motors are nonsparking motors with additional features that provide further protection against the possibilities of excess temperature and/or occurrence of arcs or sparks.

NEMA and IEEE standards and testing are more comprehensive than the IEC standards. In general, motors designed to NEMA/IEEE standards should be suitable for application under IEC standards from a rating, performance, and testing viewpoint. Mounting dimensions and tolerances should always be verified.

17.10 ENCLOSURE AND FRAME DESIGNATIONS

Motors operate best in areas free of airborne particles and corrosives and should have sufficient cool airflow to dissipate heat developed during operation. Poor ventilation causes many industrial motor failures. Motors should also be protected or shielded from damage by liquids. Rarely, if ever, do all these conditions exist. NEMA has defined various enclosures suitable for different operating environments. A list of these is given in Sidebar 3.

Sidebar 3: Widely Used Electric-Motor Enclosures

- **Encapsulated** motors have coated windings to protect them from moisture, dirt, and abrasion.
- **Open** motors (IEC class IP 00) have ventilating openings for passage of external air over and around the windings for cooling purposes.
  1. **Drip-proof** motors (IEC class IP 12) are open motors protected from entry of liquids or solids falling on the motor at angles up to 15° from vertical.
  2. **Guarded** motors (IEC class IP 22) are open motors with ventilating openings of such size and shape to prevent fingers or rods from coming in contact with rotating or electrical parts.
  3. **Splash-proof** motors (IEC class IP 46) are open motors protected from entry of liquids or solids falling on the motor or coming in contact with the motor in a straight-line path at angles up to 100° from vertical.
- **Totally enclosed** motors (IEC class IP 44) are constructed to prevent free exchange of air between the inside and the outside of the motor case but are not airtight.
  1. **Dust-ignition–proof** motors are totally enclosed motors designed to exclude the entry of combustible dusts into the enclosure or bearing chamber. They are also designed to operate under any normal or abnormal operating condition (including being heavily blanketed with dust) such that external

surface temperatures of the motor casing do not exceed the motor's maximum operating temperature.
2. **Explosion-proof** motors (IEC "flameproof") are totally enclosed motors built to contain the flames and pressures resulting from repeated internal (within the motor casing) explosions. They are also designed to operate under any normal or abnormal condition such that external surface temperatures of the motor casing do not exceed the motor's maximum operating temperature. Explosion-proof (or flameproof) motors are used almost exclusively on drilling rigs.
3. **Totally enclosed blower-cooled** motors have an independently powered external frame cooling fan.
4. **Totally enclosed fan-cooled** motors have a shaft-mounted fan that directs air across the external frame (applications include dusty, dirty, corrosive atmospheres).
5. **Totally enclosed nonventilated** motors are, however, not self-equipped for cooling.

IEC Publication 72 and its extension, 72A, provide standards for dimensions, tolerances, and output ratings. Publication 72A addresses larger machines. There are small, but significant, differences between the IEC and NEMA frame dimensions (Table 17.2). In most cases, machines built to either series can be adapted by special machining or shimming. For example, the IEC bolt hole is larger than the NEMA bolt hole. Couplings can normally be obtained or machined that accommodate the shafts of either series.

One potential point of confusion between NEMA and IEC dimensional nomenclature is the different letter symbols used to indicate basic mounting dimensions (Table 17.3). For example:

- IEC 112M28 = 112-mm foot height, M frame length, 28-mm shaft diameter
- IEC 18M1-1/8 = 18 in. mounting foot height, M frame length, 1 ⅛-inch shaft diameter

In IEC nomenclature, when a flange exists on the drive end, the flange number is added directly following the shaft diameter. For example: 112M28F215 = 112-mm mounting foot height, M frame length, 28-mm shaft diameter, 215-mm pitch circle diameter flange.

Table 17.2
IEC Versus NEMA Mounting Dimensions (mm), IEC/NEMA[a]

Frame Size	H/D	A/2E	B/2F	C/BA	K/H
90S/143	90/86.9	140/139.7	100/101.6	56/57.2	10/8.6
90L/145	90/88.9	140/139.7	125/127	56/57.2	10/8.6
112S/180	112/114.3	190/190.5	114/114.3	70/69.9	12/10.4
112M/184	112/114.3	190/190.5	140/139.7	89/88.9	12/10.4
132S/213	132/133.4	216/215.9	140/139.7	89/88.9	12/10.4
132M/215	132/133.4	216/215.9	178/177.8	89/88.9	12/10.4
160M/254	160/158.8	254/254	210/209.5	108/108	15/13.5
160L/256	160/158.8	254/254	254/254	108/108	15/13.5
180M/284	180/177.8	279/279.4	241/241.3	121/120.6	15/13.5
180L/286	180/177.8	279/279.4	279/279.4	121/120.6	15/13.5
200M/324	200/203.2	318/317.5	267/266.7	133/133.4	19/16.8
200L/326	200/203.2	318/317.5	305/304.8	133/133.4	19/16.8
225S/364	225/228.6	356/355.6	286/285.8	149/149.4	19/16.8
225M/365	225/228.6	356/355.6	311/311.1	149/149.4	19/16.8
250S/404	250/254	406/406.4	311/311.2	168/168.1	24/20.6
250M/405	250/254	406/406.4	349/349.2	168/168.1	24/20.6
280S/444	280/279.4	457/457.2	368/368.3	190/190.5	24/20.6
280M/445	280/279.4	457/457.2	419/419.1	190/190.5	24/20.6
315S/504	315/317.5	508/508	406/406.4	216/215.9	28/-
315M/505	315/317.5	508/508	457/457.2	216/215.9	28/-
355S/585	355/368.3	610/584.2	500/508	254/254	28/-
355M/586	355/368.3	610/584.2	560/558.8	254/254	28/-
400S/684	400/431.8	686/685.8	560/558.8	280/292.1	35/-
400M/685	400/431.8	686/685.8	630/635	280/292.1	35/-

IEC = International Electrotechnical Commission; NEMA = National Electrical Manufacturers Association.

[a] For foot-mounted motors, the IEC frame designation consists of the shaft height followed by the shaft extension expressed in millimeters, while the NEMA system uses a specific number for each frame.

17.10.1 Protection Classes Relating to Enclosures

The ingress protection code (IP), published as European Standard EN 60 529, provides for enclosures that protect against

- Contact with live or moving parts,
- Entry of solid foreign matter, and
- Entry of water.

This classification system is now the main system used by NEMA. See Figure 17.12 for a comparison of the old and new enclosure designation

Table 17.3
IEC Versus NEMA Mounting Dimensions Nomenclature

IEC Letter	NEMA Letter	Dimension
H	D	Distance from shaft centerline to foot bottom
A	2E	Distance between centerlines of foot mounting holes (end view)
B	2F	Distance between centerlines of foot mounting holes (side view)
C	BA	Distance from shoulder on shaft to centerline of mounting holes in the nearest feet
K	H	Diameter of holes or width of slots in the feet
D	U	Diameter of shaft extension
M	AJ	Pitch circle diameter of fixing holes in face, flange, or base
N	AK	Diameter of spigot on face, flange, or base
S	BF	Diameter of threaded or clearance hole in face, flange, or base

systems (the table references starters but is applicable to any electrical enclosure). The IP system uses the letters "IP" followed by two digits (Table 17.4). The first digit of the code indicates the degree that persons are protected against contact with moving parts and the degree that equipment is protected against solid foreign bodies (tool, wires, etc.) intruding into an enclosure. The second digit of the code indicates the degree of protection to the equipment from moisture entry by various means such as dripping, spraying, or immersion.

17.11 HAZARDOUS LOCATIONS

Construction and installation of all electrical equipment placed in a flammable or potentially explosive location must receive careful consideration. In some drilling and production sites where the occurrence of explosive mixtures of flammable materials and air cannot be prevented, special construction measures for prevention and/or containment of ignition sources are warranted. Such areas are classified by hazardous ratings, which will be discussed later in this chapter.

Hazardous locations are those where potentially explosive atmospheres can occur due to local and/or operational conditions. Leaks inevitably occur during manufacture or movement of volatile, or slightly volatile, liquids. Such leakage, in the form of gas, vapor, or mist, may

STARTER ENCLOSURES

TYPE	NEMA ENCLOSURE
1	General Purpose—Indoor
2	Driproof—Indoor
3	Dust-tight, Raintight, Sleet-tight—Outdoor
3R	Raintight, Sleet Resistant—Outdoor
3S	Dust-tight, Raintight, Sleet-tight—Outdoor
4	Watertight, Dust-tight, Sleet Resistant—Indoor & Outdoor
4X	Watertight, Dust-tight, Corrosion-Resistant—Indoor & Outdoor
5	Dust-tight, Drip-Proof—Indoor
6	Occasionally Submersible, Watertight, Sleet Resistant—Indoor & Outdoor
6P	Watertight, Sleet Resistant—Prolonged Submersion—Indoor & Outdoor
12	Dust-tight and Driptight—Indoor
12K	Dust-tight and Driptight, with Knockouts—Indoor
13	Oiltight and Dust-tight—Indoor
	HAZARDOUS LOCATION STARTERS
7	Class I, Group A, B, C, or D Hazardous Locations—Indoor
8	Class I, Group A, B, C, or D Hazardous Locations—Indoor & Outdoor
9	Class II, Group E, F, or G Hazardous Locations—Indoor
10	Requirements of Mine Safety and Health Administration

CONVERSION OF NEMA TYPE NUMBERS TO IEC CLASSIFICATION DESIGNATIONS

(Cannot be used to convert IEC Classification Designations to NEMA Type Numbers)

NEMA ENCLOSURE TYPE NUMBER	IEC ENCLOSURE CLASSIFICATION DESIGNATION
1	IP10
2	IP11
3	IP54
3R	IP14
3S	IP54
4 and 4X	IP56
5	IP52
6 and 6P	IP67
12 and 12K	IP52
13	IP54

Note: This comparison is based on tests specified in IEC Publication 529.
Reference: Information in the above tables is based on NEMA Standard 250-1991.

Figure 17.12. © 2000, Electrical Apparatus Service Association, Inc., St. Louis, MO. Reprinted with permission.

combine with oxygen from the atmosphere to form mixtures of explosive concentrations. Ignition of such mixtures by an electrical spark or arc or by contact with an excessively hot surface may result in an explosion. Different techniques are used to minimize the risk of explosion, including

Table 17.4
Scope and Protection According to IP Protection Classes*

Digit Protection	First Digit (Physical)	Foreign Body	Second Digit (Water)
0	No protection	No protection	No protection
1	Protection against back and body contact	Protection against solid foreign bodies, 50 mm (2.08 in.) diameter	Protection against water drops falling vertically
2	Finger contact	Solid foreign bodies, 12.5 mm (0.52 in.) diameter	Water drops falling 15° from vertical
3	Tool contact	Solid foreign bodies 2.5 mm (0.1 in.) diameter	Water spray at angles up to 60°
4	Wire contact	Solid foreign bodies 1.0 mm (0.04 in.) diameter	Water spray from all directions
5	Wire contact	Dust	Water jets
6	Wire contact	Dust-tight	Strong water jets
7	—	—	Intermittent immersion in water
8	—	—	Continuous immersion in water

*If a code character is not necessary, it should be replaced by the letter "X."

explosion-proof construction, purging, pressurization, encapsulation, oil immersion, and intrinsic safety.

When oxygen reacts with other elements or compounds, heat is usually liberated. Because of this, the temperature rises and causes the reaction to proceed at a more rapid rate. Generally, the term *combustion* refers to the vigorous and rapid reaction with oxygen attended by liberation of energy in the form of heat and light.

Reactions other than those involved with oxygen can also liberate heat and light. For example, a jet of acetylene burns brilliantly in chlorine:

$$C_2H_2 + Cl_2 \rightarrow\rightarrow 2HCl + 2C.$$

Hydrogen also burns brilliantly in chlorine:

$$H_2 + Cl_2 \rightarrow\rightarrow 2HCl.$$

Various substances must be heated to different temperatures before they will ignite and continue to burn in air without being supplied with additional heat from an outside source. Some substances (sand and clay, for example) will not ignite at any temperature because the elements they contain have already combined with as much oxygen as they are capable of. Hence, further reaction with oxygen is not possible. Some substances ignite at very low temperatures—white phosphorus, for instance, ignites at 35°C (95°F). Gasoline ignites at a lower temperature than kerosene; kerosene ignites at a lower temperature than motor oil; and ether ignites at a lower temperature than alcohols.

The combustion of some substances is accompanied by the production of flames, which are burning gases. When wood or coal (especially soft coal) is heated to its ignition point or below, combustible gases are released. These combustible gases usually ignite at a lower temperature than the residue of wood or coal, and their combustion produces the effect known as flame.

Hydrogen burns with an almost colorless flame, as opposed to flames produced by wood, which are generally yellow. Flames are usually colorless if solid particles are not present in the burning gases and are produced by the decomposition of substances into the gases as they burn. The hydrogen flame can be yellow if a small quantity of sodium chloride is vaporized and mixed with the burning hydrogen.

Kindling temperature is the temperature at which a substance bursts into flames and combustion proceeds without further application of heat. Kindling temperature varies considerably with the state of division of the substance (for instance, the stem of a match), its surface area, porosity, and so forth. Finely divided particles offer much more surface area than the same weight of a substance in one large mass. Iron and lead can both be produced in small enough particles (large surface area per unit mass) that they ignite without preliminary heating when poured from a container into air.

Flashpoint is the lowest temperature at which the vapors above a volatile, combustible substance (such as any petroleum product) ignite momentarily in air due to a spark or small flame applied near the liquid surface. It has also been described as the lowest temperature at which a liquid will give off sufficient vapor to ignite momentarily on application of a flame. The degree of flammability of a substance is expressed mainly by its flashpoint.

An *ignitable mixture* is one within the flammable range (between upper and lower limits) capable of flame propagation away from the source of ignition when ignited. Some evaporation occurs below the flashpoint, but not in quantities sufficient to form an ignitable mixture. This term applies mostly to flammable and combustible liquids, although there are certain solids (such as camphor or naphthalene) that slowly evaporate or volatilize at ordinary room temperatures. Also, liquids such as benzene freeze at relatively high temperatures and therefore have flashpoints while in the solid state.

The term *propagation of flame* is used to describe the spread of flame from the ignition source through a flammable mixture. A gas or vapor mixed with air in proportions below the lower limit of flammability may burn at the source of ignition. In other words, it may burn in the zone immediately surrounding the source of ignition without propagating (spreading away) from the source of ignition. However, if the mixture is within the flammable range, the flame will spread throughout when a source of ignition is supplied. The term *flame propagation,* therefore, can be used to distinguish between combustion that takes place only at the source of ignition and that which travels (propagates) through the mixture.

The *ignition temperature* (or autoignition temperature) of a substance, whether solid, liquid, or gas, is the minimum temperature required to initiate or cause self-sustained combustion in the absence of any ignition source, such as a spark or flame. To avoid the risk of explosion, the temperature of any part or surface must always be below the ignition temperature. Ignition temperatures observed under one set of conditions may alter significantly with changing conditions, such as:

- Percentage composition of the vapor or gas/air mixture
- Shape or size of the space in which ignition occurs
- Rate and duration of heating
- Reactivity of any other materials present

Thus, ignition temperatures should be viewed as approximations. There are many differences in ignition temperature test methods, including size, shape, and composition of ignition chambers, method and rate of heating, residence time, and method of flame detection. Reported ignition temperatures are affected by the test methods employed.

Since ignition temperature is the temperature at which ignition may occur due to contact with a hot surface, it follows that motor selection must be based on the maximum surface temperature that will never exceed the autoignition temperature of any potentially explosive mixture likely to exist. The National Electrical Code mandates that motors be marked to indicate the maximum temperature when they are placed in service with combustible materials (see Table 17.5).

Table 17.5
National Electrical Code Maximum External Surface Temperatures

Temperature Class	Maximum Surface Temperature	Ignition Temperature of Combustible Material
T1	450°C/842°F	> 450°C/842°F
T2	300°C/572°F	> 300°C/572°F
T2A	280°C/536°F	> 280°C/536°F
T2B	260°C/500°F	> 260°C/500°F
T2C	230°C/446°F	> 230°C/446°F
T2D	215°C/419°F	> 215°C/419°F
T3	200°C/392°F	> 200°C/392°F
T3A	180°C/356°F	> 180°C/356°F
T3B	165°C/329°F	> 165°C/329°F
T3C	160°C/320°F	> 160°C/320°F
T4	135°C/275°F	> 135°C/275°F
T4A	120°C/248°F	> 120°C/248°F
T5	100°C/212°F	> 100°C/212°F
T6	85°C/185°F	> 85°C/185°F

Table 17.6
Flashpoint and Autoignition Temperatures for Some Common Materials

Gas/Vapor/Liquid	Flashpoint	Autoignition Temperature	Class	Explosion Group
Acetone	−20°C/−4°F	465°C/869°F	T1	IIA
Benzyl alcohol	93°C/200°F	436°C/817°F	T2	IIA
Benzene	−11°C/12°F	498°C/928°F	T1	IIA
Gasoline (petro)	−43°C/45°F	280°C/536°F	T2A	IIA
Hydrogen sulfide	Gas	260°C/500°F	T2B	IIB

Ignition temperature depends on the type and concentrations of gases and vapors present. Table 17.6 compares the flashpoint and ignition (or autoignition) temperatures for some common materials.

17.12 MOTORS FOR HAZARDOUS DUTY

A summary of hazardous-location designations as outlined in the U.S. National Electrical Code is given in Table 17.7.

IEC standards that address equipment for use in explosive atmospheres are contained in a series of Publications 79-0 through 79-10.

Table 17.7
Hazardous (Classified) Location Reference Guide
(specification must include class, division, and group)

Class I	Areas containing flammable gas or vapor
Class II	Areas containing combustible dust
Division 1	Explosion hazard may exist under normal operating condition or due to maintenance, leakage, or breakdown of equipment
Division 2	Explosion hazard may exist under abnormal operating conditions such as rupture of containers or failure of ventilation equipment
Class I group	
A	Atmospheres containing acetylene
B	Atmospheres containing hydrogen and the like
C	Atmospheres containing ethylene and the like
D	Atmospheres containing acetone, methanol, propane, and the like
Class II group	
E	Atmospheres containing combustible metal dust such as magnesium or aluminum
F	Atmospheres containing combustible carbonaceous dust such as coal
G	Atmospheres containing combustible dust such as flour, grain, wood, and plastic

For additional information on the properties and group classification of class I and class II materials, see *Manual for Classification of Gases, Vapors and Dusts for Electrical Equipment in Hazardous (Classified) Locations*, National Fire Protection Association, 497M.

Motor classification and applicability differ considerably from U.S. standards and practices. A summary of some of these differences follows.

IEC classifies equipment into two broad categories:

- Group I: Underground mines
- Group II: Use in other industries

This discussion is restricted to motors in the group II classification, and specifically to groups IIA, IIB, and IIC, which relate to the gas or vapor involved. A general comparison to the U.S. National Electrical Code is as follows:

IEC	U.S.
Group II A	Group D
Group II B	Group C
Group II C	Groups A and B

The IEC classifies hazardous locations into zones according to the probability of a potentially explosive atmosphere occurrence. The degree of danger varies from *extreme* to *rare*:

- *Zone 0:* An explosive gas/air mixture is continuously present or present for long periods of time. No electric motors may be used in these areas.
- *Zone 1:* An explosive gas/air mixture is likely to occur in normal operations.
- *Zone 2:* An explosive gas/air mixture is not likely to occur in normal operations and, if it does occur, will exist for only a short time.

The following is a comparison of IEC and U.S. designations:

IEC Designation	U.S. Designation
Group II, Zone 0	Rotating equipment generally not recommended
Group II, Zone 1	Class I, Division 1
Group II, Zone 2	Class I, Division 2

Construction features and test requirements for motors used in hazardous locations are defined by the IEC as:

- Flameproof enclosures
- Pressurized enclosures
- Increased safety protection, "e"

Common IEC symbols are:

- Flameproof: "d" or (Ex) d
- Pressurized: "p" or (Ex) p
- Increased Safety: "e" or (Ex) e

Table 17.8 provides a detailed description of these terms and symbols.

17.13 EUROPEAN COMMUNITY DIRECTIVE 94/9/EC

The European Community Directive 94/9/EC became mandatory on July 1, 2003, and will be enforced on all equipment traded within the

Table 17.8
IEC Nomenclature Applied to Motors for Hazardous Areas

Protection, IEC or European Standard	Basic Principle	Applications
Flameproof Enclosure, "d"	Parts that can ignite an explosive atmosphere are enclosed to withstand pressure developed during an explosion and prevent transmission of the explosion to explosive atmospheres around the enclosure	Switch gear, control and indicating equipment, control boards, motors, transformers, light fixtures, and other spark-producing parts
Increased Safety, "e"	Additional measures against internal or external arcs or sparks, or excessive temperatures not produced in normal service	Terminal and connection boxes, control boxes and housing, squirrel-cage rotors, and light fittings.
Pressurized Apparatus, "p"	Entry of surrounding atmosphere is prevented by maintaining a protective gas at pressure higher than the surrounding atmosphere	As above, usually for large equipment and contained rooms
Intrinsic Safety, "i"	Internal electric circuits are incapable of causing explosions in the surrounding atmosphere	Measurement and control equipment
Oil Immersion, "o"	Electric apparatus or its parts are immersed in oil to prevent ignition of a surrounding or covering atmosphere	Transformers (rarely used)
Powder Filling, "q"	Enclosure is filled with a finely granulated material so that an internal arc will not ignite the surrounding atmosphere; also, ignition will not be caused by flame or excessive temperature of enclosure surfaces	Transformers, capacitors, heater strip connection boxes, and electronic assemblies
Molding, "m"	Parts that can ignite an explosion are enclosed (encapsulated) in a resin to prevent ignition of an explosive atmosphere by internal sparking or heating	Only small capacity switch gear, control gear, indicating equipment, and sensors

European Union (EU). Several markets outside of the EU will simultaneously apply the same requirements.

Previous European regulations dealing with electrical equipment for use in potentially explosive atmospheres were based on application or supplemental directives, which were often modified to maintain the state of the art. European Directive 76/117/EEC provides a general outline for the present system. This system promotes free movement of goods and machinery within the EU.

Previously, electrical equipment for use in potentially explosive atmospheres was evaluated by a European organization such as KEMA, located in Arnhem, Netherlands. This organization evaluated the equipment with respect to a protection method (shale shaker vibrators are flameproof type "d") to the applicable European Norm standard (in the case of type "d" equipment, EN 50 014 and EN 50 018). Upon successful completion of the evaluation, the European organization would issue an Ex Certificate, which specifies that the product may be marked with the Ex symbol.

This system, sometimes referred to as the Old Approach, has been used for about 15 years and has proven successful, with some drawbacks. The main drawback has been the slow and tedious process to change requirements to keep up with technology. The new European Directive 94/9/EC supersedes the Old Approach. The new directive allows the continued use of the Ex Certificate only up to June 30, 2003. Thereafter, the Ex Certificate will not be accepted and the goods and machinery will not be allowed to move freely within the EU.

The New Approach will effect harmonization of requirements for equipment for use in potentially explosive environments. The new directive makes no direct references to standards but sets out the essential health and safety requirements to be met (Annex II) and also introduces the CE (Conformité Européene) marking.

The CE marking is a declaration that the equipment complies with all applicable directives. In the case of shale shaker vibrators for use in potentially explosive atmospheres, the applicable directives include the low-voltage directive, 72/23/EC; the electromagnetic compatibility, or EMC, directive 89/336/EC; the CE marking directive 93/68/EC; and the ATEX (ATmosphere EXplosive) directive 94/9/EC.

The new classification scheme becomes a bit complicated. There now are two groups: group I for underground mines and group II for surface applications. Within each group, there are categories that designate the degree of the hazard similarly to the zone designation system.

Further, the classification is divided into gas (G) applications and dust (D) applications.

Table 17.9 outlines the New Approach classification system for group II surface applications. The equipment type and intended use is specified in the markings provided on the product. An example of a New Approach marking would be "CE #### EX II 2 G, DEMKO 01 ATEX 0135585 EEx d IIB T4." The CE mark is followed by a four-digit number indicating the notified body that qualifies and audits the manufacturing system, followed by the Ex mark, the Group (II), the category (2) and the substance present (G = gas). DEMKO is the notified body (located in Herlev, Denmark) that issued the Certificate 01 ATEX 0135585. The product employs protection method "d" and is intended for gas group IIB with an operating temperature of T4 (135C).

17.14 ELECTRIC MOTORS FOR SHALE SHAKERS

With the exception of specialized motors for centrifuge feed, practically all AC electric motors encountered in drilling-fluid operations are integral-horsepower, across-the-line start, horizontal squirrel-cage motors. Across-the-line motors are the simplest and lowest cost. The motor is connected directly to the input power through a starter switch. Full current and torque are realized at startup. This is acceptable with solids-control and processing equipment; however, it is suggested that centrifugal pumps be started with the discharge valve partially closed to restrict initial pump output and load demand on the motor.

Shale shaker motors are generally three-phase induction motors that are explosion proof, having NEMA design B or similar characteristics (Table 17.10). The number of magnetic poles in a shale shaker motor can be four (1800 rpm synchronous shaft speed at 60 Hz), six (1200 rpm), or two (3600 rpm). The motor should have independent, third-party markings indicating its suitability in explosive or potentially explosive environments. It is recommended that these motors be suitable for Class I, Division 1, Groups C and D, and Group IIB atmospheres. The motor also should have the proper operating temperature or code designation for the anticipated ambient temperature.

A 50-Hz motor driving a shale shaker vibrator should not be operated at 60 Hz, since the centrifugal force output will increase by 44%. This will likely damage the bearings, the vibrating screens, and the shale shaker. A 60-Hz motor driving a shale shaker vibrator can be operated at 50 Hz with the understanding that the centrifugal force output will decrease

Table 17.9
Classification by Group and Category According to Intended Use (Surface Industry)

Area	Category of Equipment	Presence or Duration of Explosive Atmosphere	Inflammable Substances	Level of Protection Faults to	Comparison with Present Practice
Equipment Group II (surface)	1	Continuous presence Long Period Frequent	Gas, vapors, mist, dust	Allow for very high level of protection: 2 types of protection or 2 independent faults	Group II Zone 0 (gas) Zone 20 (dust)
	2	Likely to occur	Gas, vapors, mist, dust	High level of protection: 1 type of protection Habitual frequent malfunction	Group II Zone 1 (gas) Zone 21 (dust)
	3	Unlikely to occur Present for a short period	Gas, vapors, mist, dust	Normal protection: Required level of protection	Group II Zone 2 (gas) Zone 22 (dust)

Table 17.10
Electric Motor Specifications for Shale Shakers

	U.S. Designation	IEC Designation
Terminology	Explosion-proof	Flameproof
Hazardous location rating	Class I, Division 1 group D	Eexd Gas Group IIA
Hazardous location rating if hydrogen sulfide is encountered	Class I, Division 1 groups C and D	Eexd Gas Group IIB

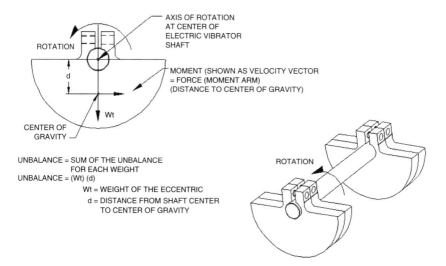

Figure 17.13. Eccentric weight unbalance.

by 31%. If, at 60 Hz, the centrifugal force is 1000 lb, the centrifugal force will only be 690 lb at 50 Hz.

For a given frame size, higher-speed motors will have high hp ratings, low slip, high starting torque, and low bearing life. Conversely, lower-speed motors will have lower hp ratings, high slip, low starting torque, and long bearing life.

Electric industrial vibrators are rated in centrifugal force output, frequency, unbalance (static moment), and hp. Centrifugal force is caused by torque resultant from the offset eccentric weight acting through the moment arm (the distance from the shaft center to the center of gravity of the weight) (see Figure 17.13). This torque is referred to as unbalance or static moment. The unbalance provides the amplitude at which the vibrating screen will move.

Two counterrotating shale shaker motors will produce a linear force that should be located through the center of gravity of the shaker basket (see Chapter 7 on Shale Shakers). The resultant motion is perpendicular to a plane drawn between the rotating shafts directed through the center of gravity of the machine (see Figure 17.14). The shale shaker motor should be selected to meet or exceed the desired stroke of the machine, centrifugal force, and acceleration (g's). Adequate hp is required to perform the work and to ensure synchronization. Synchronization results in opposing forces from two counterrotating vibrators that cancel each other and double directional forces.

Stroke, which is independent of motor speed, is the peak-to-peak displacement imparted to the machine. Dampening may occur in the system affecting the total stroke. Stroke is a function of the unbalance (or static moment) of the motor and of the total weight of the shaker basket, including the weight of the motors and the live load. The stroke

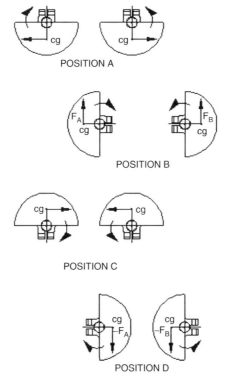

NOTE: FORCE VECTOR SHOWN PARALLEL TO VELOCITY VECTOR

Figure 17.14. Movement of two counterrotating shale shaker motors.

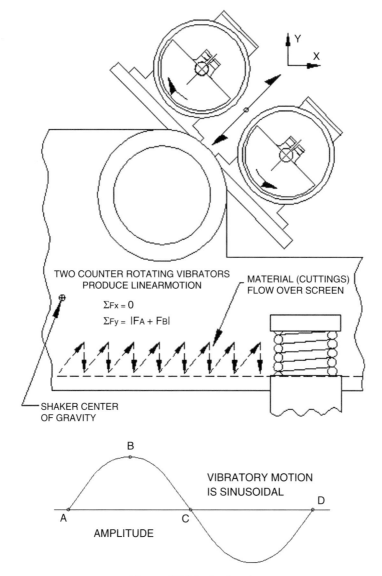

Figure 17.14. Continued.

equals two times the motor unbalance, multiplied by the number of motors, divided by the total weight. Note that the motor unbalance is a function of the eccentric weight setting. For example, the motor unbalance is 50% of the maximum unbalance if the eccentric weights are set at 50%.

The centrifugal-force output of the vibrating motor (lb) is equal to the shaft speed squared times the unbalance (in.-lb), divided by 35,211. Once again, the vibrating motor's centrifugal-force output is a function of the eccentric weight setting. For example, the centrifugal force is 50% of the maximum centrifugal force if the eccentric weights are set at 50%. Typical acceleration rates for vibrating screens are 4 to 8 g's.

17.15 ELECTRIC MOTORS FOR CENTRIFUGES

Most centrifuges use the same NEMA design B explosion-proof motors used for centrifugal pumps: either 1500 rpm at 50 Hz or 1800 rpm at 60 Hz. Centrifuges may draw up to seven times the full-load current for approximately 15 seconds at startup. It is considered good practice to limit centrifuge startups to two starts per hour to protect the motor because the current draw closely approaches the limit.

Oilfield centrifuges do not always use a direct drive between the motor and centrifuge. Direct drive requires expensive, variable-speed motors that have restricted availability of replacement parts and repair facilities. Most oilfield centrifuges are connected to the motor by a fluid clutch or a hydraulic drive, which uses a system of adding or subtracting motor or hydraulic oil to increase/decrease slippage between the driver and driven coupling halves. Some centrifuges use a variable-speed electric motor startup system that brings the centrifuge slowly up to operational rotating speed. It is important that all personnel understand the manufacturer's recommended startup and shutdown procedures.

Early centrifuges were capable of generating 500 to perhaps 1500 g's of acceleration. Today, machines commonly generate 2000 to 3000 g's. The advent of higher g centrifuges is attributed to improvements in bearing design and manufacturing procedures, including hard surfacing with tungsten carbide and precision robot welding.

17.16 ELECTRIC MOTORS FOR CENTRIFUGAL PUMPS

The fluid volume necessary to be moved by many centrifugal pumps is related to the rig circulation rate and the specific rig plumbing. Centrifugal pumps should be sized by the particular application and the maximum anticipated flow rate. Piping friction losses—if lines are reasonably short, with few turns or restrictions, and flow velocities between 5 and 10 feet per second—are readily estimated. The pressure, or head, which should be delivered to each piece of equipment (for hydrocyclones, typically 75 feet of head) is specified by the equipment manufacturer.

Horsepower requirements for centrifugal pumps, when pumping water or fluids of water-like viscosity, are well established and published with the performance curve for each design. See Chapter 18 on centrifugal pumps.

17.17 STUDY QUESTIONS

1. Before testing, a four-wire ohmmeter is used to measure a three-phase motor winding resistance between two phases. The resistance is measured as 1.239 ohms. The ambient temperature is recorded as 25°C. After several hours of testing in a 40°C ambient, the motor is shut off and the same phase resistance is quickly measured and determined to be 1.651 ohms. What is the winding temperature when the motor was shut off? Assume that the winding is copper. What is the significance of the four-wire ohmmeter? Was the test procedure correct?

Answer: Resistance is proportional to temperature. As the temperature of the circuit increases, the circuit resistance increases. The equation that relates resistance to temperature is

$$R_i = \frac{R_f(234.5 + T_i)}{(234.5 + T_f)}$$

where

R_i = initial resistance, in ohms
R_f = final resistance, in ohms
T_i = initial temperature, °C
T_f = final temperature, °C

Rearranging the equation gives

$$\frac{R_f(234.5 + T_i)}{R_i} - 234.5$$

$$T_f = \frac{1.651(234.5 + 25)}{1.239} - 234.5 = 111.3°C \text{ (winding temperature)}$$

A four-wire ohmmeter is used to negate the resistance of the equipment test leads. If a two-wire ohmmeter is used, the resistance of these leads must be accounted for.

Technically, if one wanted to determine the winding temperature, this procedure is not correct. Although live winding test equipment is available (variation of a Wheatstone bridge), most motor temperature testing is still done with an ohmmeter. The winding temperature calculated above is accurate at the exact time of the resistance measurement. A small amount of time expired before the resistance measurement actually occurred. After the motor is shut off, the time it takes to obtain the resistance is recorded. The resistance is also recorded at 30 sec and at 2 min. Additional readings may be taken. Using linear regression, the resistance values are plotted with respect to time and the curve extended back to time zero. The winding temperature is more accurately indicated as the temperature at time zero.

2. For a single-phase circuit, calculate the electrical angle between the apparent power waveform and the real power waveform if the circuit voltage is 230 V, the circuit current is 7 A, and the circuit power is 1.3 kW.

Answer: The cosine of the electrical angle between the apparent power waveform and the real power waveform is the power factor.

$$\text{watts} = (\text{volts})(\text{amps})(\text{powerfactor})$$

therefore

$$\text{power factor} = \text{watts}/(\text{volts})(\text{amps})$$
$$= 1300/(230)(7) = 0.81$$

The angle, φ, $= \arccos 0.81 = 35.9°$.

3. How many mechanical degrees apart are the poles in an eight-pole winding?

Answer: In an eight-pole winding, there are four north and four south magnetic poles around the winding periphery (360°). The mechanical angle between poles would be $(360/8) = 45°$.

4. In a single-phase transformer, the primary current is 10 A with 1000 primary coil turns. How many secondary coil turns would be required for a 2:1 stepdown transformer if the primary voltage were 460 V?

Answer: In a 2:1 stepdown transformer, the secondary coil voltage is one-half the primary coil voltage. The secondary voltage is ½(460) = 230 V. The secondary coil turns = (1000) 230/460 = 500 turns.

5. If a four-pole shale shaker vibrator has a synchronous speed of 1800 rpm and a centrifugal-force output of 10,000 lb, what line frequency is needed to operate the vibrator at 7500 lb? Neglect slip.

Answer: The electric vibrator unbalance or static moment is calculated as

$$\text{unbalance} = (10,000)(35,211)/1800^2 = 108.7 \text{ in-lb.}$$

The shaft speed at the lower centrifugal force is then calculated:

$$\text{rpm} = [(35211)(7500)/108.7]^{1/2} = 1559 \text{ rpm.}$$

The frequency is calculated as Hz = (1559)(4)/120 = 52 Hz.

6. A 50-hp, 1800-rpm centrifugal pump is connected to an ASD with a 3:1 turndown ratio. What is the hp at maximum turndown?

Answer: A turndown ratio of 3:1 would result in operation at $\frac{1}{3}$(1800) = 600 rpm at maximum turndown. Horsepower at 600 rpm = 50(600/1800)^3 = 5.6 hp.

7. Why would a 60-Hz induction motor generally be derated to 83% when operating at 50 Hz?

Answer: The motor hp is proportional to the shaft speed. The torque does not change when a 60-Hz motor is operated at 50 Hz. Assume that the motor is a four-pole motor that has a synchronous speed at 60 Hz of (120)(60)/4 = 1800 rpm. At 50 Hz, a four-pole motor has a synchronous speed of (120)(50)/4 = 1500 rpm. The hp at 50 Hz must then be 1500/1800 = 0.83 = 83%. The same is true for any motor with a different number of magnetic poles. Note: Since the frequency is also directly proportional to the rpm, the ratio of the line frequencies can also be used to determine the new rating, that is, 50/60 = 0.83 = 83%.

8. What does the designation CE 0539 EX II 2 G, DEMKO 03 ATEX 12345 EEx d IIB T4 designate?

Answer:
- CE = European Community mark designating compliance with all applicable directives
- 0539 = notified-body number for the organization responsible for initial quality system review and ongoing follow-up
- EX = European Commission mark for Ex (explosive) equipment
- II = equipment group: surface installations
- 2 = equipment category: zone 1
- G = gas or vapor applications
- DEMKO = notified body that conducted type testing
- 03 = year certificate issued: 2003
- ATEX = acronym for ATmosphere EXplosive
- 12345 = certificate number.
- EEx = indicates compliance with CENELEC (the European Committee for Electrotechnical Standardization) standards
- d = protection method: "d" is flameproof construction
- IIB = gas group IIB
- T4 = operating temperature: T4 = 135°C

9. Describe how the centrifugal force is decreased by 31% when a 60-Hz shale shaker vibrator is operated at 50 Hz.

Answer: Given that the centrifugal force $= \dfrac{(\text{unbalance , in.-lb})(\text{rpm})^2}{35211}$

it is apparent that for a given unbalance, the centrifugal force is proportional to the square of the speed. When a 60-Hz shale shaker vibrator is operated at 50 Hz, the centrifugal force is $(50^2/60^2) = 0.69$. The centrifugal force reduction is then $(1 - 0.69) = 0.31 = 31\%$.

10. Prove that a shale shaker stroke is independent of the shale shaker vibrator speed.

Answer: The force required to move the shale shaker is defined by Newton's second law of motion, which states that the force (F_1) equals mass times acceleration (ma). The electric vibrator generates a force (F_2) by accelerating a mass at some distance from the rotating axis. These forces must be equal, $F_1 = F_2$:

$$F_1 = \text{ma} = (\text{wt}/g)(\omega^2)(s/2)$$

$$F_2 = (WR)(\omega^2/g)$$

$$F_1 = F_2$$

$$(wt/g)(\omega^2)(s/2) = (WR)(\omega^2/g)$$

$$(wt)(s/2) = WR$$

$$(wt)(s) = 2(WR)$$

$$s = \frac{2(WR)}{wt}$$

where

- F_1 = force needed to drive the shale shaker
- F_2 = force generated by a rotating unbalance
- WR = eccentric weight unbalance, in.-lb
- wt = weight of the shale shaker including the electric vibrators above the springs
- g = acceleration due to gravity
- ω = angular acceleration
- s = stroke, in.

Therefore, the stroke is directly proportional to the eccentric weight unbalance and inversely proportional to the total system weight on the springs. The stroke is not related to the frequency of operation. Note that this conclusion assumes that the vibrating shale shaker is not operating at or near its natural or resonant frequency.

CHAPTER 18

CENTRIFUGAL PUMPS

Todd H. Lee
National Oilwell

This chapter discusses fundamentals involved in hydraulic design, proper sizing, and operation of centrifugal pumps. It is not intended to fully educate the reader in the art of hydraulics design, but rather to give a basic understanding of hydraulic performance of a centrifugal pump. Two basic components of a centrifugal pump that are related to hydraulic performance are the impeller and casing.

This discussion is limited to one particular type of centrifugal pump—the radial flow pump. Radial flow pumps are designed to produce a flow pattern through the impeller radially outward and perpendicular to the pump shaft. This discussion is limited to a single-stage pump with an end suction and tangential discharge case design (Figures 18.1 and 18.2).

18.1 IMPELLER

Centrifugal pumps are often referred to as kinetic energy machines. Rotation of the impeller causes fluid within the impeller to rotate at a high velocity, imparting kinetic energy to the fluid. This concept is described mathematically by the equation:

$$H_i = (u_2 \times c_{u2})/g$$

where

H_i = theoretical head developed by the centrifugal pump, in ft
u_2 = rotational velocity of the impeller at the outer diameter, in ft/sec
c_{u2} = rotational velocity of the fluid as it leaves the impeller, in ft/sec
g = gravitational constant, in ft/sec^2.

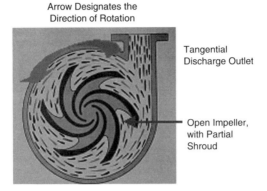

Figure 18.1. Open impeller incased by tangential casing.

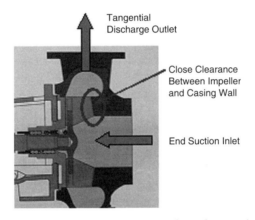

Figure 18.2. Tight tolerance casing serves as front of open style impeller.

There are three basic impeller designs:

- a closed impeller that has a shroud (rotating wall) on both the front and the back of the impeller,
- a semi-open impeller that has a shroud on one side and is closely fitted to the stationary wall of the casing on the other side, and
- an open impeller (see Figure 18.1) that may or may not have part of a shroud on one side and is closely fitted to the casing wall on the other side (Figure 18.2).

As fluid approaches the pump suction, it is assumed to have very little to no rotational velocity. Note: *Prerotation of fluid in suction piping can and often does exist, but will be disregarded in this discussion.* When fluid enters rotating passages of the impeller, it begins to spin at the rotating velocity of the impeller. Fluid is forced outward from the center of the impeller, and its rotating velocity increases in direct proportion to the increasing impeller diameter. The rotating velocity of the impeller can be calculated at any diameter by the equation:

$$u = (D)(N)/229$$

where

u = rotational velocity, in ft/sec
D = diameter at which the velocity is being calculated, in.
N = impeller rotating speed, in rpm
$1/229$ = constant to convert rpm × in. to ft/sec.

The exit velocity of the fluid (c_{u2}) approaches the rotating velocity of the impeller (u_2) at D_2 but does not equal u_2 in normal operation. The main reason $c_{u2} < u_2$ is the "backward" sweep of the impeller vane (see Figure 18.1). The exit velocity of the fluid, c_{u2}, can be calculated from the design parameters of the impeller, but that derivation is beyond the intended scope of this discussion.

The discussion thus far has been of *theoretical* head (H_i), which does not account for losses that occur as fluid moves through the impeller during normal operation. Losses in the impeller that normally occur are friction, eddy currents, fluid recirculation, entrance losses, and exit losses. Additional losses will occur in the casing.

It should be noted that head produced by a centrifugal pump is a function of fluid velocity and is not dependent (normally) on the fluid being pumped. For example, a pump that will produce 100 feet of head on water (8.34 lb/gal) will also produce 100 feet of head on gasoline (6.33 lb/gal). Note: Fluids with viscosity greater than 20 centipoises will decrease output head produced by a centrifugal pump. For viscous fluids, corrections can be made to predict actual pump performance.

18.2 CASING

The function of the pump casing is to

1. direct fluid into the eye of the impeller through the suction inlet,
2. minimize fluid recirculation from impeller discharge to impeller suction, and

3. capture fluid discharge from the impeller in the case volute to most efficiently utilize work performed by the impeller and direct fluid away from the impeller. (Figures 18.3 and 18.4.)

The impeller performs useful work and increases the head of the fluid. The casing consumes part of the work imparted to the fluid and creates head losses due to friction, eddies, and other flow characteristics. A good casing design will minimize the losses, as opposed to a bad casing design. However, no casing design will increase pump head above what exists at the discharge of the impeller. Typically a centrifugal pump casing is designed so that the suction flange is one or two pipe sizes larger than the

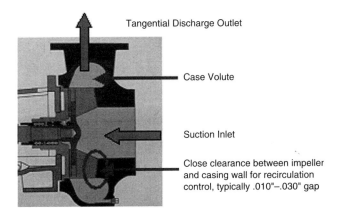

Figure 18.3. Tight tolerance casing serves as front of open style impeller.

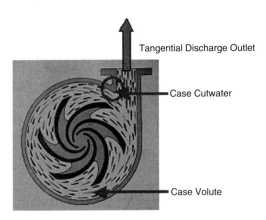

Figure 18.4. Casing volute and cutwater areas.

discharge flange. This is done to manage velocity of the fluid as it approaches the impeller inlet and also to minimize friction losses ahead of the pump. Excessive losses on the suction side of a pump can cause severe and rapid damage to the pump impeller and casing.

As fluid discharges from the impeller into the case volute, it has increased in head value by the amount of work imparted by the impeller. Since fluid will naturally flow in the direction of least resistance, it will tend to flow back (recirculate) toward the suction inlet, where fluid entering the impeller is at a relatively low head. In order to prevent this recirculation, a restriction must be created between the impeller and casing to minimize flow back to the suction and cause the fluid to flow out the discharge. In this example, with an open impeller, that restriction is the very close clearance (gap) between the impeller and the casing wall. This gap is typically 0.010 to 0.030-inch wide. Two major factors in determining gap size are (1) desired performance (the smaller the better) and (2) minimum allowable clearance to prevent impeller rubbing during pump operation. This works well when the pump is in a new condition, but eventually the impeller and casing will begin to wear and the gap size will increase, allowing more fluid to recirculate to the suction side of the impeller. Eventually pump performance will deteriorate to the point that an adjustment to the impeller location will have to be made (if possible in the pump design) or the impeller and casing will have to be replaced in order to restore the pump's original performance.

As shown in Figures 18.3 and 18.4, fluid is discharged from the impeller into the case volute. One aspect of a good casing design is that the volume of the volute is sized to match the volume flow through the impeller so that fluid velocity in the volute is somewhat less than fluid velocity exiting the impeller. This reduction in velocity is where a portion of the overall pump head is generated. If velocity reduction is too great, excessive shock losses and eddy currents in the volute will degrade pump head output. If velocity reduction is too small, excessive friction losses will occur in the volute that will also degrade pump head output.

In Figure 18.4, the part of the casing called the *cutwater* can be seen. This is where fluid is guided into the discharge outlet and led away from the pump. The cutwater must be accurately located relative to the impeller and angled to minimize flow disruption as fluid exits the casing. A cutwater that is too close to the OD of the impeller will cause a pressure pulse to be created as the impeller vane passes by the cutwater. This momentary high-pressure pulse will disrupt flow and cause pump output to degrade. Conversely, a cutwater that is too far from the impeller will allow too much fluid to pass by the outlet and simply recirculate within the case volute and thereby reduce pump output. The angle of the cutwater must match the flow path of the fluid as it exits the casing. If it does not, eddy currents will be created that degrade pump performance.

In summary, the impeller of a centrifugal pump is designed to impart kinetic energy to a fluid, raising the head value of the fluid to accomplish a specific goal. The pump casing design must be matched to the impeller to most efficiently guide fluid into and away from the impeller. When these components are properly designed, fluid will flow smoothly through the machine with minimal recirculation and losses caused by eddy currents and friction that can degrade pump output head.

18.3 SIZING CENTRIFUGAL PUMPS

Many factors affect performance of a centrifugal pump and must be considered during pump selection. This chapter describes conditions that affect the centrifugal pump and is followed by details that will assist in eliminating negative conditions that cause pump failure.

During rig design, the centrifugal pump is often considered a low-cost product that does not warrant a great deal of engineering consideration. Many times the centrifugal is sized and ordered based on existing packages utilized on other rigs. This can cause serious problems because each rig has unique operating conditions and piping designs. Centrifugal pumps are used for a variety of applications and feed other, much more expensive equipment. If centrifugal pumps are not properly sized, they and other equipment can be adversely affected. Proper sizing, design, and installation of centrifugal pumps can directly affect the efficiency and operating cost of the rig

Centrifugal pumps are available in a variety of materials, configurations, sizes, and designs. Normally, a single size, configuration, and speed can be selected to best meet the intended application. Accurate centrifugal pump selection can occur only with knowledge of system details. It is imperative to obtain accurate information such as fluid temperature, specific gravity, pipe diameter, length of pipe, fittings, elevations, flow required, head required at end of transfer line, type of driver required, and type of power available. Without all of this information, assumptions have to be made that could cause pump failure, high maintenance costs, downtime, and/or improper performance.

18.3.1 Standard Definitions

Terms associated with centrifugal pumps have been defined in a variety of ways, but for our purpose the terms will be referenced as defined below:

- *Flow rate:* volume of liquid going through a pipe in a given time. If a hose stream will fill a 10-gal bucket in 2 minutes, then the flow rate is 5 gpm. The bottom axis of most pump curves is flow rate and is usually measured in gallons per minute or cubic meters per hour (m^3/hr).
- *Friction loss:* resistance to movement of fluid within the pipe, or head loss caused by turbulence and dragging that results when fluid comes into contact with the ID of pipe, valves, fittings, etc. This value is normally measured in feet per 100 feet of pipe and is noted in Tables 18.1 through 18.14 under the column heading "Friction Loss in Feet Head per 100 Ft of Pipe." These tables are based on Schedule (SCH)-40 new steel pipe. Other piping material or older scaled or pitted pipe will have higher friction losses. Consult engineering handbooks for piping other than SCH 40 new steel pipe.
- *Head:* distance, in feet, that water will rise in an open-ended tube connected to the place where the measurement is to be taken. Throughout this chapter, head is measured in terms of distance (ft) and not pressure (psi, for example). Units of psi vary with the weight of the fluid, but head in feet or meters is constant regardless of fluid weight.
- *Net positive suction head available ($NPSH_A$):* amount of head that will exist at the suction flange of the pump above absolute zero. Friction losses, atmospheric pressure, fluid temperature/vapor pressure, elevation, and specific gravity affect this value. This value must be calculated.
- *Net positive suction head required ($NPSH_R$):* amount of inlet head above absolute zero required by the centrifugal pump to operate properly. This value varies with pump size and flow rate and is normally represented on the pump curve.
- *Total differential head (TDH):* amount of head produced by a centrifugal pump in excess of pump inlet head. This value is found on the left axis of most pump curves.
- *Total discharge head:* sum of the inlet head + total differential head of a centrifugal pump, measured in feet or meters.
- *Velocity (V ft/sec):* refers to the average speed that liquid travels through a pipe. Velocity is measured in feet per second (ft/sec). Velocities measured in ft/sec can be found in Tables 18.1 through 18.14,

(text continued on p. 479).

Tables 18.1–18.14
Friction Loss for Water in Feet Head per 100 Feet of Pipe Length

	Table 18.1				Table 18.2		
¾-inch Nominal	Steel Schedule-40 Pipe ID: 0.824 Inches €/D: 0.00218			1-inch Nominal	Steel Schedule-40 Pipe ID: 1.049 Inches €/D: 0.00172		
Flow Rate (gpm)	V (ft/sec)	$V^2/2g$ (ft)	Friction Loss in Feet Head per 100 Ft of Pipe	Flow Rate (gpm)	V (ft/sec)	$V^2/2g$ (ft)	Friction Loss in Feet Head per 100 Ft of Pipe
2.0	1.20	0.02	1.21	3	1.11	0.02	0.77
2.5	1.50	0.04	1.80	4	1.48	0.03	1.30
3.0	1.81	0.05	2.50	5	1.86	0.05	1.93
3.5	2.11	0.07	3.30	6	2.23	0.08	2.68
4.0	2.41	0.09	4.21	7	2.60	0.10	3.56
4.5	2.71	0.11	5.21	8	2.97	0.14	4.54
5.0	3.01	0.14	6.32	9	3.34	0.17	5.65
6.0	3.61	0.20	8.87	10	3.71	0.21	6.86
7.0	4.21	0.28	11.8	12	4.45	0.31	9.62
8.0	4.81	0.36	15.0	14	**5.20**	**0.42**	**12.8**
9.0	**5.42**	**0.46**	**18.8**	16	5.94	0.55	16.5
10	6.02	0.56	23.0	18	6.68	0.70	20.6
11	6.62	0.68	27.6	20	7.42	0.86	25.1
12	7.22	0.81	32.6	22	8.17	1.04	30.2
13	7.82	0.95	37.8	24	8.91	1.23	35.6
14	8.42	1.10	43.5	26	9.65	1.45	41.6
15	9.03	1.27	49.7	28	10.39	1.68	47.9
16	9.63	1.44	56.3	30	11.1	1.93	54.6
17	10.23	1.63	63.1	32	11.9	2.19	61.8
18	10.8	1.82	70.3	34	12.6	2.48	69.4
19	11.4	2.03	78.0	36	13.4	2.78	77.4
20	**12.0**	**2.25**	**86.1**	38	14.1	3.09	86.0
22	13.2	2.72	104	40	14.8	3.43	95.0
24	14.4	3.24	122	42	15.6	3.78	104.5
26	15.6	3.80	143	44	16.3	4.15	114
28	16.8	4.41	164	46	17.1	4.53	124
30	18.1	5.06	187	48	17.8	4.93	135
				50	18.6	5.35	146
				55	20.4	6.48	176

Table 18.3				Table 18.4			
1 ¼-inch Nominal	Steel Schedule-40 Pipe ID: 1.380 Inches ε/D: 0.00130			1 ½-inch Nominal	Steel Schedule-40 Pipe ID: 1.610 Inches ε/D: 0.00112		
Flow Rate (gpm)	V (ft/sec)	$V^2/2g$ (ft)	Friction Loss in Feet Head per 100 Ft of Pipe	Flow Rate (gpm)	V (ft/sec)	$V^2/2g$ (ft)	Friction Loss in Feet Head per 100 Ft of Pipe
6	1.29	0.03	0.70	6	0.95	0.01	0.33
7	1.50	0.04	0.93	7	1.10	0.02	0.44
8	1.72	0.05	1.18	8	1.26	0.02	0.56
9	1.93	0.06	1.46	9	1.42	0.03	0.69
10	2.15	0.07	1.77	10	1.58	0.04	0.83
12	2.57	0.10	2.48	12	1.89	0.06	1.16
14	3.00	0.14	3.28	14	2.21	0.08	1.53
16	3.43	0.18	4.20	16	2.52	0.10	1.96
18	3.86	0.23	5.22	18	2.84	0.13	2.42
20	4.29	0.29	6.34	20	3.15	0.15	2.94
22	4.72	0.35	7.58	22	3.47	0.19	3.52
24	**5.15**	**0.41**	**8.92**	24	3.78	0.22	4.14
26	**5.58**	**0.48**	**10.37**	26	4.10	0.26	4.81
28	**6.01**	**0.56**	**11.9**	28	4.41	0.30	5.51
30	**6.44**	**0.64**	**13.6**	30	4.73	0.35	6.26
32	**6.86**	**0.73**	**15.3**	**32**	**5.04**	**0.40**	**7.07**
34	**7.29**	**0.83**	**17.2**	**34**	**5.36**	**0.45**	**7.92**
36	**7.72**	**0.93**	**19.2**	**36**	**5.67**	**0.50**	**8.82**
38	**8.15**	**1.03**	**21.3**	**38**	**5.99**	**0.58**	**9.78**
40	**8.58**	**1.14**	**23.5**	**40**	**6.30**	**0.62**	**10.79**
42	**9.01**	**1.26**	**25.8**	**42**	**6.62**	**0.68**	**11.8**
44	**9.44**	**1.38**	**28.2**	**44**	**6.93**	**0.75**	**12.9**
46	**9.87**	**1.51**	**30.7**	**46**	**7.25**	**0.82**	**14.0**
48	**10.30**	**1.65**	**33.3**	**48**	**7.56**	**0.89**	**15.2**
50	**10.7**	**1.79**	**36.0**	**50**	**7.88**	**0.97**	**16.4**
55	**11.8**	**2.16**	**43.2**	**55**	**8.67**	**1.17**	**19.7**
60	12.9	2.57	51.0	**60**	**9.46**	**1.39**	**23.2**
65	13.9	3.02	59.6	**65**	**10.24**	**1.63**	**27.1**
70	15.0	3.50	68.8	**70**	**11.03**	**1.89**	**31.3**
75	16.1	4.02	78.7	**75**	**11.8**	**2.17**	**35.8**
80	17.2	4.58	89.2	**80**	**12.6**	**2.47**	**40.5**
85	18.2	5.17	100.2	85	13.4	2.79	45.6
90	19.3	5.79	112	90	14.2	3.13	51.0
95	20.4	6.45	124	95	15.0	3.48	56.5
100	21.5	7.15	138	100	15.8	3.86	62.2
				110	17.3	4.67	74.5
				120	18.9	5.56	88.3
				130	20.5	6.52	103

	Table 18.5				Table 18.6		
2-inch Nominal	Steel Schedule-40 Pipe ID: 2.067 Inches €/D: 0.00087			2½-inch Nominal	Steel Schedule-40 Pipe ID: 2.469 Inches €/D: 0.000729		
Flow Rate (gpm)	V (ft/sec)	$V^2/2g$ (ft)	Friction Loss in Feet Head per 100 Ft of Pipe	Flow Rate (gpm)	V (ft/sec)	$V^2/2g$ (ft)	Friction Loss in Feet Head per 100 Ft of Pipe
10	0.96	0.01	0.25	16	1.07	0.02	0.24
12	1.15	0.02	0.34	18	1.21	0.02	0.30
14	1.34	0.03	0.45	20	1.34	0.03	0.36
16	1.53	0.04	0.58	22	.47	0.03	0.43
18	1.72	0.05	0.72	24	1.61	0.04	0.50
20	1.91	0.06	0.87	26	1.74	0.05	0.58
22	2.10	0.07	1.03	28	1.88	0.05	0.66
24	2.29	0.08	1.20	30	2.01	0.06	0.75
26	2.49	0.10	1.39	35	2.35	0.09	1.00
28	2.68	0.11	1.60	40	2.68	0.11	1.28
30	2.87	0.13	1.82	45	3.02	0.14	1.60
35	3.35	0.17	2.42	50	3.35	0.17	1.94
40	3.82	0.23	3.10	55	3.69	0.21	2.32
45	4.30	0.29	3.85	60	4.02	0.25	2.72
50	4.78	0.36	4.67	65	4.36	0.30	3.16
55	5.26	0.43	5.59	70	4.69	0.34	3.63
60	5.74	0.51	6.59	75	5.03	0.39	4.13
65	6.21	0.60	7.69	80	5.36	0.45	4.66
70	6.69	0.70	8.86	85	5.70	0.50	5.22
75	7.17	0.80	10.1	90	6.03	0.57	5.82
80	7.65	0.91	11.4	95	6.37	0.63	6.45
85	8.13	1.03	12.8	100	6.70	0.70	7.11
90	8.60	1.15	14.2	110	7.37	0.84	8.51
95	9.08	1.28	15.8	120	8.04	1.00	10.0
100	9.56	1.42	17.4	130	8.71	1.18	11.7
110	10.52	1.72	20.9	140	9.38	1.37	13.5
120	11.5	2.05	24.7	150	10.05	1.57	15.4
130	12.4	2.40	28.8	160	10.7	1.79	17.4
140	13.4	2.78	33.2	170	11.4	2.02	19.6
150	14.3	3.20	38.0	180	12.1	2.26	21.9
160	15.3	3.64	43.0	190	12.7	2.52	24.2
170	16.3	4.11	48.4	200	13.4	2.79	26.7
180	17.2	4.60	54.1	220	14.7	3.38	32.2
190	18.2	5.13	60.1	240	16.1	4.02	38.1
200	19.1	5.68	66.3	260	17.4	4.72	44.5
220	21.0	6.88	80.0	280	18.8	5.47	51.3
240	22.9	8.18	95.0	300	20.1	6.28	58.5

Centrifugal Pumps

Table 18.7				Table 18.8			
3-inch Nominal	Steel Schedule-40 Pipe ID: 3.068 Inches ϵ/D: 0.000587			3½-inch Nominal	Steel Schedule-40 Pipe ID: 3.548 Inches $\epsilon\epsilon/D$: 0.000507		
Flow Rate (gpm)	V (ft/sec)	$V^2/2g$ (ft)	Friction Loss in Feet Head per 100 Ft of Pipe	Flow Rate (gpm)	V (ft/sec)	$V^2/2g$ (ft)	Friction Loss in Feet Head per 100 Ft of Pipe
25	1.09	0.02	0.19	35	1.14	0.02	0.17
30	1.30	0.03	0.26	40	1.30	0.03	0.22
35	1.52	0.04	0.35	45	1.46	0.03	0.27
40	1.74	0.05	0.44	50	1.62	0.04	0.33
45	1.95	0.06	0.55	60	1.95	0.06	0.46
50	2.17	0.07	0.66	70	2.27	0.08	0.60
55	2.39	0.09	0.79	80	2.60	0.11	0.77
60	2.60	0.11	0.92	90	2.92	0.13	0.96
65	2.82	0.12	1.07	100	3.25	0.16	1.17
70	3.04	0.14	1.22	110	3.57	0.20	1.39
75	3.25	0.17	1.39	120	3.89	0.24	1.64
80	3.47	0.19	1.57	130	4.22	0.28	1.90
85	3.69	0.21	1.76	140	4.54	0.32	2.18
90	3.91	0.24	1.96	150	4.87	0.37	2.48
95	4.12	0.26	2.17	**160**	**5.19**	**0.42**	**2.80**
100	4.34	0.29	2.39	**170**	**5.52**	**0.47**	**3.15**
110	4.77	0.35	2.86	**180**	**5.84**	**0.53**	**3.50**
120	**5.21**	**0.42**	**3.37**	**190**	**6.17**	**0.59**	**3.87**
130	**5.64**	**0.50**	**3.92**	**200**	**6.49**	**0.66**	**4.27**
140	**6.08**	**0.57**	**4.51**	**220**	**7.14**	**0.79**	**5.12**
150	**6.51**	**0.66**	**5.14**	**240**	**7.79**	**0.94**	**6.04**
160	**6.94**	**0.75**	**5.81**	**260**	**8.44**	**1.11**	**7.04**
170	**7.38**	**0.85**	**6.53**	**280**	**9.09**	**1.28**	**8.11**
180	**7.81**	**0.95**	**7.28**	**300**	**9.74**	**1.47**	**9.26**
190	**8.25**	**1.06**	**8.07**	**320**	**10.4**	**1.68**	**10.48**
200	**8.68**	**1.17**	**8.90**	**340**	**11.0**	**1.89**	**11.8**
220	**9.55**	**1.42**	**10.7**	**360**	**11.7**	**2.12**	**13.2**
240	**10.4**	**1.69**	**12.6**	**380**	**12.3**	**2.36**	**14.6**
260	**11.3**	**1.98**	**14.7**	400	13.0	2.62	16.2
280	**12.2**	**2.29**	**16.9**	420	13.6	2.89	17.8
300	13.0	2.63	19.2	440	14.3	3.17	19.4
320	13.9	3.00	22.0	460	14.9	3.46	21.2
340	14.8	3.38	24.8	480	15.6	3.77	23.0
360	15.6	3.79	27.7	500	16.2	4.09	25.0
380	16.5	4.23	30.7	550	17.8	4.95	30.1
400	17.4	4.68	33.9	600	19.5	5.89	35.6
420	18.2	5.16	37.3	650	21.1	6.91	41.6
440	19.1	5.67	40.9				
460	20.0	6.19	44.6				

	Table 18.9				Table 18.10		
4-inch Nominal	Steel Schedule-40 Pipe ID: 4.026 Inches €/D: 0.000447			5-inch Nominal	Steel Schedule-40 Pipe ID: 5.047 Inches €/D: 0.000357		
Flow Rate (gpm)	V (ft/sec)	$v^2/2g$ (ft)	Friction Loss in Feet Head per 100 Ft of Pipe	Flow Rate (gpm)	V (ft/sec)	$v^2/2g$ (ft)	Friction Loss in Feet Head per 100 Ft of Pipe
40	1.01	0.02	0.12	70	1.12	0.02	0.11
50	1.26	0.02	0.18	80	1.28	0.03	0.14
60	1.51	0.04	0.25	90	1.44	0.03	0.17
70	1.76	0.05	0.33	100	1.60	0.04	0.20
80	2.02	0.06	0.42	120	1.92	0.06	0.29
90	2.27	0.08	0.52	140	2.25	0.08	0.38
100	2.52	0.10	0.62	160	2.57	0.10	0.49
110	2.77	0.12	0.74	180	2.89	0.13	0.61
120	3.02	0.14	0.88	200	3.21	0.16	0.74
130	3.28	0.17	1.02	220	3.53	0.19	0.88
140	3.53	0.19	1.17	240	3.85	0.23	1.04
150	3.78	0.22	1.32	260	4.17	0.27	1.20
160	4.03	0.25	1.49	280	4.49	0.31	1.38
170	4.28	0.29	1.67	300	4.81	0.36	1.58
180	4.54	0.32	1.86	320	5.13	0.41	1.78
190	4.79	0.36	2.06	340	5.45	0.46	2.00
200	5.04	0.40	2.27	360	5.77	0.52	2.22
220	5.54	0.48	2.72	380	6.09	0.58	2.46
240	6.05	0.57	3.21	400	6.41	0.64	2.72
260	6.55	0.67	3.74	420	6.74	0.71	2.98
280	7.06	0.77	4.30	440	7.06	0.77	3.26
300	7.56	0.89	4.89	460	7.38	0.85	3.55
320	8.06	1.01	5.51	480	7.70	0.92	3.85
340	8.57	1.14	6.19	500	8.02	1.00	4.16
360	9.07	1.28	6.92	550	8.82	1.21	4.98
380	9.58	1.43	7.68	600	9.62	1.44	5.88
400	10.10	1.58	8.47	650	10.4	1.69	6.87
420	10.6	1.74	9.30	700	11.2	1.96	7.93
440	11.1	1.91	10.2	750	12.0	2.25	9.05
460	11.6	2.09	11.1	800	12.8	2.56	10.22
480	12.1	2.27	12.0	850	13.6	2.89	11.5
500	12.6	2.47	13.0	900	14.4	3.24	12.9
550	13.9	2.99	15.7	950	15.2	3.61	14.3
600	15.1	3.55	18.6	1000	16.0	4.00	15.8
650	16.4	4.17	21.7	1100	17.6	4.84	19.0
700	17.6	4.84	25.0	1200	19.2	5.76	22.5
750	18.9	5.55	28.6	1300	20.8	6.75	26.3
800	20.2	6.32	32.4				

Centrifugal Pumps

	Table 18.11				Table 18.12		
6-Inch Nominal	Steel Schedule-40 Pipe ID: 6.065 Inches €/D: 0.000293			8-Inch Nominal	Steel Schedule-40 Pipe ID: 7.981 Inches €/D: 0.000226		
Flow Rate (gpm)	V (ft/sec)	$V^2/2g$ (ft)	Friction Loss in Feet Head per 100 Ft of Pipe	Flow Rate (gpm)	V (ft/sec)	$V^2/2g$ (ft)	Friction Loss in Feet Head per 100 Ft of Pipe
100	1.11	0.02	0.08	60	1.03	0.02	0.05
120	1.33	0.03	0.12	180	1.15	0.02	0.06
140	1.55	0.04	0.16	200	1.28	0.03	0.08
160	1.78	0.05	0.20	220	1.41	0.03	0.09
180	2.00	0.06	0.25	240	1.54	0.04	0.11
200	2.22	0.08	0.30	260	1.67	0.04	0.13
220	2.44	0.09	0.36	280	1.80	0.05	0.14
240	2.66	0.11	0.42	300	1.92	0.06	0.16
260	2.89	0.13	0.49	320	2.05	0.07	0.18
280	3.11	0.15	0.56	340	2.18	0.07	0.21
300	3.33	0.17	0.64	360	2.31	0.08	0.23
320	3.55	0.20	0.72	380	2.44	0.09	0.25
340	3.78	0.22	0.81	400	2.57	0.10	0.28
360	4.00	0.24	0.90	450	2.89	0.13	0.35
380	4.22	0.28	1.00	500	3.21	0.16	0.42
400	4.44	0.31	1.10	550	3.53	0.19	0.51
420	4.66	0.34	1.20	600	3.85	0.23	0.60
440	4.89	0.37	1.31	650	4.17	0.27	0.70
460	**5.11**	**0.41**	**1.42**	700	4.49	0.31	0.80
480	**5.33**	**0.44**	**1.54**	750	4.81	0.36	0.91
500	**5.55**	**0.48**	**1.66**	**800**	**5.13**	**0.41**	**1.02**
550	**6.11**	**0.58**	**1.99**	**850**	**5.45**	**0.46**	**1.15**
600	**6.66**	**0.69**	**2.34**	**900**	**5.77**	**0.52**	**1.27**
650	**7.22**	**0.81**	**2.73**	**950**	**6.09**	**0.58**	**1.41**
700	**7.77**	**0.94**	**3.13**	**1000**	**6.41**	**0.64**	**1.56**
750	**8.33**	**1.08**	**3.57**	**1100**	**7.05**	**0.77**	**1.87**
800	**8.88**	**1.23**	**4.03**	**1200**	**7.70**	**0.92**	**2.20**
850	**9.44**	**1.38**	**4.53**	**1300**	**8.34**	**1.08**	**2.56**
900	**9.99**	**1.55**	**5.05**	**1400**	**8.98**	**1.25**	**2.95**
950	**10.5**	**1.73**	**5.60**	**1500**	**9.62**	**1.44**	**3.37**
1000	**11.1**	**1.92**	**6.17**	**1600**	**10.3**	**1.64**	**3.82**
1100	**12.2**	**2.32**	**7.41**	**1700**	**10.9**	**1.85**	**4.29**
1200	13.3	2.76	8.76	**1800**	**11.5**	**2.07**	**4.79**
1300	14.4	3.24	10.2	**1900**	**12.2**	**2.31**	**5.31**
1400	15.5	3.76	11.8	2000	12.8	2.56	5.86
1500	16.7	4.31	12.5	2200	14.1	3.09	7.02
1600	17.8	4.91	15.4	2400	15.4	3.68	8.31
1700	18.9	5.54	17.3	2600	16.7	4.32	9.70
1800	20.0	6.21	19.4	2800	18.0	5.01	11.20
1900	21.1	6.92	21.6	3000	19.2	5.75	12.8
2000	22.2	7.67	23.8	3200	20.5	6.55	14.5

Table 18.13

10-Inch Nominal — Steel Schedule-40 Pipe, ID: 10.020 Inches, ∈/D: 0.000180

Flow Rate (gpm)	V (ft/sec)	$V^2/2g$ (ft)	Friction Loss in Feet Head per 100 Ft of Pipe
240	0.98	0.01	0.04
260	1.06	0.02	0.04
280	1.14	0.02	0.05
300	1.22	0.02	0.05
350	1.42	0.03	0.07
400	1.63	0.04	0.09
450	1.83	0.05	0.11
500	2.03	0.06	0.14
550	2.24	0.08	0.16
600	2.44	0.09	0.19
650	2.64	0.11	0.22
700	2.85	0.13	0.26
750	3.05	0.15	0.30
800	3.25	0.17	0.33
850	3.46	0.19	0.37
900	3.66	0.21	0.41
950	3.87	0.23	0.46
1000	4.07	0.26	0.50
1100	4.48	0.31	0.60
1200	4.88	0.37	0.70
1300	**5.29**	**0.44**	**0.82**
1400	**5.70**	**0.50**	**0.94**
1500	**6.10**	**0.58**	**1.07**
1600	**6.51**	**0.66**	**1.21**
1700	**6.92**	**0.74**	**1.36**
1800	**7.32**	**0.83**	**1.52**
1900	**7.73**	**0.93**	**1.68**
2000	**8.14**	**1.03**	**1.86**
2200	**8.95**	**1.25**	**2.23**
2400	**9.76**	**1.48**	**2.64**
2600	**10.6**	**1.74**	**3.08**
2800	**11.4**	**2.02**	**3.56**
3000	**12.2**	**2.32**	**4.06**
3200	13.0	2.63	4.59
3400	13.8	2.97	5.16
3600	14.6	3.33	5.76
3800	15.5	3.71	6.40

Table 18.14

12-Inch Nominal — Steel Schedule-40 Pipe, ID: 11.938 Inches, ∈/D: 0.000151

Flow Rate (gpm)	V (ft/sec)	$V^2/2g$ (ft)	Friction Loss in Feet Head per 100 Ft of Pipe
350	1.00	0.02	0.03
400	1.15	0.02	0.04
450	1.29	0.03	0.05
500	1.43	0.03	0.06
550	1.58	0.04	0.07
600	1.72	0.05	0.08
650	1.86	0.05	0.10
700	2.01	0.06	0.11
750	2.15	0.07	0.12
800	2.29	0.08	0.14
850	2.44	0.09	0.16
900	2.58	0.10	0.17
950	2.72	0.12	0.19
1000	2.87	0.13	0.21
1100	3.15	0.15	0.25
1200	3.44	0.18	0.30
1300	3.73	0.22	0.34
1400	4.01	0.25	0.40
1500	4.30	0.29	0.45
1600	4.59	0.33	0.51
1700	4.87	0.37	0.57
1800	**5.16**	**0.41**	**0.64**
1900	**5.45**	**0.46**	**0.70**
2000	**5.73**	**0.51**	**0.78**
2200	**6.31**	**0.62**	**0.93**
2400	**6.88**	**0.74**	**1.10**
2600	**7.45**	**0.86**	**1.28**
2800	**8.03**	**1.00**	**1.47**
3000	**8.60**	**1.15**	**1.68**
3200	**9.17**	**1.31**	**1.90**
3400	**9.75**	**1.48**	**2.13**
3600	**10.3**	**1.65**	**2.37**
3800	**10.9**	**1.84**	**2.63**
4000	**11.5**	**2.04**	**2.92**
4500	**12.9**	**2.59**	**3.65**
5000	14.3	3.19	4.47
5500	15.8	3.86	5.38

(continued)

10-Inch Nominal	Table 18.13 (continued) Steel Schedule-40 Pipe ID: 10.020 Inches ∈/D: 0.000180			12-Inch Nominal	Table 18.14 (continued) Steel Schedule-40 Pipe ID: 11.938 Inches ∈/D: 0.000151		
Flow Rate (gpm)	V (ft/sec)	$V^2/2g$ (ft)	Friction Loss in Feet Head per 100 Ft of Pipe	Flow Rate (gpm)	V (ft/sec)	$V^2/2g$ (ft)	Friction Loss in Feet Head per 100 Ft of Pipe
4000	16.3	4.12	7.07	6000	17.2	4.60	6.39
4500	18.3	5.21	8.88	6500	18.6	5.39	7.47
5000	20.3	6.43	10.9	7000	20.1	6.26	8.63

No allowance has been made for age, differences in diameter, or any abnormal condition of interior surface. Any factor of safety must be estimated from the local conditions and the requirements of each particular installation.

Recommended flow rates **(boldface)** for suction and discharge pipes are to avoid sanding at lower flow rates and to avoid too much friction at higher flow rates.

where one can look up the velocity for recommended flow rates in a pipe size or the flow rate in a pipe for any velocity desired. For example, 10 ft/sec in an SCH 40 4-inch-diameter pipe will flow 400 gpm.

18.3.2 Head Produces Flow

Most water used at home and in industry comes from tank towers or standpipes. Figure 18.5 shows water flowing through a straight pipe of constant diameter lying on level ground. If a clear sight tube is installed on the pipe near the open end, it can be used to measure head at that place. By closing the end of the pipe, flow will stop, and the water in the sight tube will rise to a level equal to that of the standpipe. When the end of the pipe is opened fully, the flow will be the most the standpipe head can deliver, and the water in the sight tube will drop to the bottom. Almost all of the head is consumed while pushing the water through the pipe and overcoming friction.

The velocity head is used first, to speed the water from a standstill in the standpipe up to the velocity in the pipe as it enters. The velocity head (amount in ft) depends on only the velocity of the flow (in ft/sec), not on diameter or gpm. It is the same amount at any point in the pipe (constant diameter) even to the open end, where it shows as the strength of the flow stream. It is usually small, 3–6% of the total head for pipes of 100 feet or more in length. It is shown in the friction loss tables (Tables 18.1–18.14) under the column headed "$V^2/2g$."

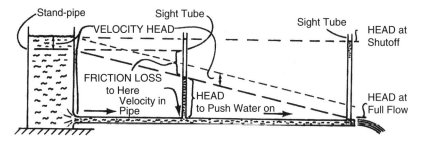

Figure 18.5. Head pressure measured by a standpipe.

A sight tube installed somewhere near the halfway point will show the (pressure) head remaining at that point that pushes the water on to the end. The difference of the height in the sight tube from the height at shutoff is the velocity head plus the friction loss from the standpipe to that point.

The use of a standpipe to supply fluid for pipe friction problems is the clearest way to demonstrate how pipe friction tables are made. While a standpipe illustration explains the system head and flow, it is not a practical method of producing head in most applications, and pumps are substituted for standpipes. Pumps can be sized to produce the proper amount of head to achieve the desired flow rate and overcome the friction losses and elevation in a system.

18.4 READING PUMP CURVES

A centrifugal pump curve comprises a grid depicting head and flow rate and a series of lines that illustrate pump performance characteristics. Figure 18.6 is a typical pump curve, and each set of lines will be reviewed individually.

Figure 18.6 is a curve for a Magnum pump, size 8 (suction) × 6 (discharge) × 14 (maximum impeller size) inches, operating at 1750 rpm. (Note: The suction will always be equal to or greater than the pump discharge size.) Impeller sizes, from 10 to 14 inches in diameter, can pass a spherical solid up to 1⅜ inches in diameter. All this information is located above the grid. Performance characteristics of this pump will change if the speed is altered, and this curve simply shows performance when the pump is driven at 1750 rpm.

Figure 18.6 utilizes a left and bottom axis. Other curves may include a top and right axis. The left axis denotes the scale for TDH. This is the amount of head, in feet; the pump will produce in excess of suction head. The bottom axis denotes the scale for capacity, in U.S. gpm. In order to

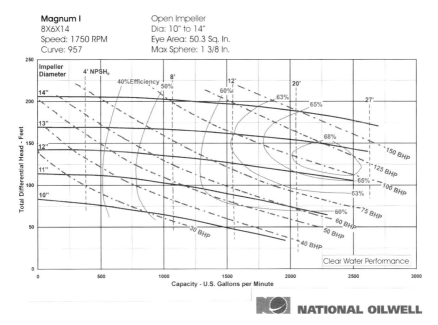

Figure 18.6. Centrifugal pump curve.

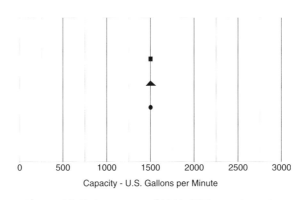

Figure 18.7. Designation of 1500 GPM operating points.

read a curve, review each set of lines individually. In Figure 18.7, several points have been marked on a line, which all depict a flow rate of 1500 gpm. If a flow rate of 1200 gpm were desired, it would be necessary to estimate the position on the grid

Figure 18.8 shows several points marked on the line that all depict a TDH of 100 feet. If the pump had an inlet suction head of 20 feet and were sized to produce 100 feet of head, then the pump discharge head

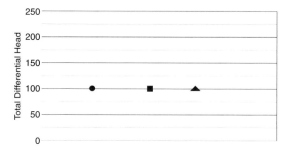

Figure 18.8. Designation of 100 feet total differential head.

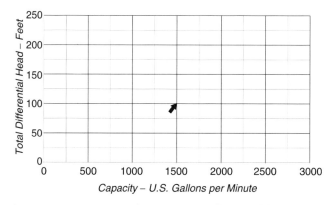

Figure 18.9. Designation of 1500 GPM 101 feet total differential head.

(total discharge head) would be equal to 120 feet. If TDH were 100 feet and fluid had to be lifted 5 feet above the suction liquid level, the total discharge head would be 95 feet.

Suppose that 8-inch-diameter SCH 40 new pipe will be used and that it lies on level ground. Table 18.12 shows that at 1500 gpm, friction loss per 100 feet of pipe will be 3.37 feet. Pipe 3000 feet long will have 30 times (3000/100) as much friction loss, or 101 feet:

$$(30)(3.37) = 101.$$

Therefore, 8-inch pipe 3000 feet long will require 101 feet of head to flow 1500 gpm.

A curve can be marked at 1500 gpm at 101 feet of head as shown in Figure 18.9.

Other lines on the grid represent pump performance characteristics (Figure 18.10).

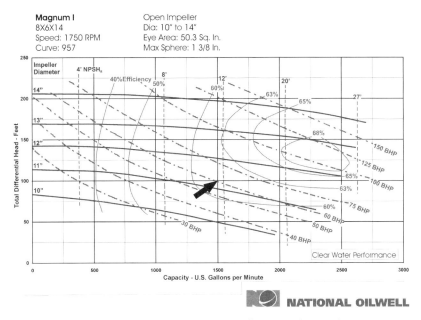

Figure 18.10. Impeller diameter, HP, efficiency and NPSHr lines.

The solid curved lines in Figure 18.10 that extend from the left axis to the right and are labeled 10", 11", 12", etc., designate the diameter of the impeller, the TDH produced by the impeller at this speed, and the flow rate produced by the impeller. The pump discharge head can be altered by changing rpm or by changing impeller diameter. If a fixed-speed driver is utilized, such as 1750 rpm, the only way to vary pump head is to alter the impeller diameter. If the requirement is for 1500 gpm at 101 feet of head and the operating speed is 1750 rpm, the impeller would have to be trimmed to 11¼ inches. Impellers can be sized to ⅛-inch increments, but for common installations a ¼-inch increment is sufficient.

The dash-dot curves in Figure 18.10 running diagonally from upper left to lower right labeled 30, 40, 50 BHP (brake horsepower), etc., designate hp required to transfer clear water. In this example, the pump requires 60 hp for clear water. This value must be corrected for fluids with an SG other than 1.0 (this will be discussed later). Solid curves running from top to bottom in a circular pattern and labeled 40, 50, 60%, etc., designate the pump efficiency. The higher the efficiency level, the lower the power operating cost. Concentric casing pumps have lower efficiency levels than other styles of pump; however, concentric-style pumps last longer and have less downtime and maintenance operating

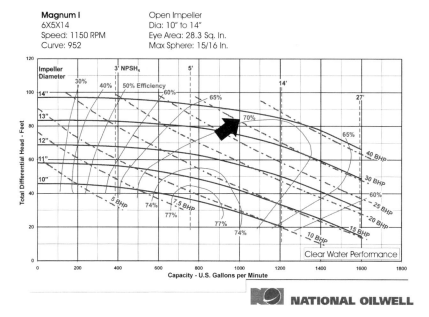

Figure 18.11. Exercise 1.

costs when transferring abrasive fluids. In the preceding example, the pump is approximately 61% efficient. Dashed lines running from top to bottom designate minimum $NPSH_R$ for the pump to operate properly. This is explained in the next section. In this example, the pump has an $NPSH_R$ of approximately 11–12 feet. Look at Figure 18.11, then do Exercise 1 in the Exercises section at the end of this chapter.

18.5 CENTRIFUGAL PUMPS ACCELERATE FLUID

Standard centrifugal pumps are not self-priming and require the fluid end to be primed prior to activation. This can be accomplished by installing the pump in a location that provides a flooded suction or by using a device to prime the pump. Once the pump casing is full of fluid, it can then be energized. Running a pump dry or restricting suction flow can severely damage the fluid end, mechanical seal, or packing. The designs of self-priming pumps result in turbulent flow patterns, which cause excessive wear during pumping of abrasive fluids and increase operating costs. The drilling industry avoids using self-priming pumps due to increased downtime and costs.

Once a pump is primed and then activated, suction head at the eye of the impeller drops. Actual positive suction head required at the eye of the impeller to prevent cavitation varies by pump size and flow and is noted on pump curves as $NPSH_R$. When this suction-head drop occurs, atmospheric pressure pushes on the liquid surface and forces it into the pump suction. As fluid enters the pump, the impeller accelerates it. The diameter of the impeller and the rpm at which the impeller is rotated directly affect the velocity of the fluid. The casing of the pump contains this velocity and converts it into head. Casing size and impeller width control the volume that the pump is able to produce.

18.5.1 Cavitation

Consider a pump suction located 5 feet above liquid level in the suction tank. Head at the suction flange will be less than atmospheric pressure. Absolute zero pressure, a perfect vacuum, is -14.7 psi, or -34 feet, of water at sea level. If the suction of the centrifugal pump is 35 feet above liquid level in the suction tank, a vacuum will exist and no fluid will enter the pump. Each centrifugal pump has a minimum suction head required above absolute zero pressure that must exist at the suction to keep the pump full of liquid. A system head that produces a suction pressure less than this value will cause cavitation. When a pump is severely cavitating, it sounds like it is pumping gravel. If a pump suction line is too small or too long or has too many valves or elbows, the friction loss in the suction line may reduce head to a value below $NPSH_R$.

If the supply tank is at sea level and is vented, inlet pressure to the pump will be

$$(34 \text{ feet} - \text{vapor pressure in feet})/(SG)$$
$$\pm \text{ liquid level above/below pump centerline}$$
$$- \text{ suction head friction losses.}$$

The sum of this calculation is the $NPSH_A$. If $NPSH_A$ is greater than $NPSH_R$, the pump will function as designed. If $NPSH_A$ is equal to or less than $NPSH_R$, the pump will cavitate. It is advisable to maintain $NPSH_A$ at least 3 feet above $NPSH_R$ to allow for calculation errors or system changes. This formula is discussed in greater detail later in the chapter.

Cavitation severely reduces life of the pump. As fluid enters the pump, the pressure at the eye of the impeller drops. If insufficient inlet pressure ($NPSH_A$) is present, fluid transforms from a liquid state to a gas (boils).

Figure 18.12. Results of clear fluid cavitation.

Gas forms low-pressure bubbles and as these bubbles travel from the ID to the OD of the impeller, pressure increases. Eventually the pressure increases enough to collapse the low-pressure bubbles. When this occurs, the bubbles implode, and space once occupied by the bubbles fill with fluid. Fluid fills this space with such force that it actually fractures adjacent metal. As this process repeats, it will knock out sections of the fluid end and can even knock a hole through the stuffing box, impeller, or casing. Cavitation can be caused by improper suction or discharge conditions.

Figures 18.12 and 18.13 show the result of severe cavitation in a pump that was transferring clear fluid. Notice that the fractured metal has sharp corners.

Figure 18.14 shows the result of cavitation in a pump that was handling abrasive fluids. After the fracture occurs, sharp corners are worn smooth by abrasive fluid. The damaged part will look as if a spoon were used to scoop sections of metal from the part.

18.5.2 Entrained Air

Entrained air in transferred fluid causes excessive turbulent flow patterns and can vapor-lock the pump. Air bubbles do not collapse like

Figure 18.13. Results of clear fluid cavitation.

Figure 18.14. Results of abrasive fluid cavitation.

low-pressure bubbles. As an air bubble enters the pump, it moves from the ID toward the OD of the impeller. Increased pressure at the OD of the impeller pushes the bubble back into the ID, where it combines with other air bubbles to become a larger bubble. This process continues to occur until the bubble at the ID of the impeller becomes large enough to impede suction flow, which can cause cavitation, and/or it becomes as large as the suction inlet and prevents fluid from entering the pump, resulting in vapor lock. Once the pump is shut down, the bubble will normally escape through the discharge, but when the pump is restarted, the process repeats itself. Recirculation of air bubbles from the ID to the OD and back again also causes turbulent fluid flow patterns that will result in excessive pump wear.

Air, entrained in fluid, can enter a pump through a loose flanged or threaded connection, through the pump packing when the pump has a high lift requirement, or through an air vortex formed in the suction tank.

18.6 CONCENTRIC vs VOLUTE CASINGS

Turbulent flow is detrimental to a centrifugal pump during handling of abrasive fluids. The drilling industry has standardized centrifugal pumps with concentric casings and wide impellers, a design that has proven to offer less turbulence and greatest pump life. The walls of a concentric style of casing (Figure 18.15) are an equal distance from the impeller throughout the impeller circumference, resulting in a smooth flow pattern. A volute style of casing (Figure 18.16) has a cutwater point that disturbs the fluid flow pattern, creating an eddy.

Wide impellers and larger casing cavities utilized by concentric pumps (Figure 18.17) reduce the effect of the fluid velocity when it exits the impeller OD. This wider area allows fluid to smoothly blend with recirculating fluid within the casing, thus reducing turbulent flow patterns that exist in volute pumps. These characteristics also reduce the sandblasting effect on the ID of the casing that is present with narrow impellers and close-proximity casing walls (Figure 18.18). Smooth flow patterns and wider recirculation area extend fluid end life and reduces operating costs of the centrifugal pump.

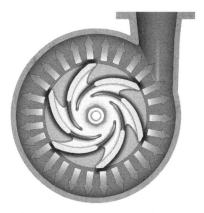

Figure 18.15. Concentric style casing.

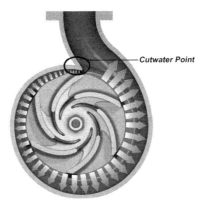

Figure 18.16. Volute style casing.

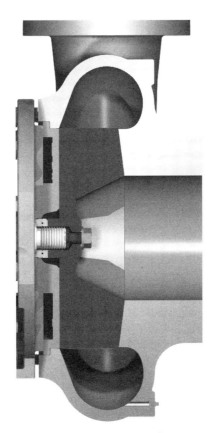

Figure 18.17. Concentric style casing.

Figure 18.18. Volute style casing.

18.6.1 Friction Loss Tables

When selecting pipe size, both friction losses and line velocity must be considered. Friction losses should be kept to a minimum in order to reduce hp requirements. Discharge piping should be sized to achieve a flow rate of 5 to 12½ ft/sec. Suction piping should be sized to achieve a flow rate of 5 to 8½ ft/sec. A minimum line velocity of 5 ft/sec is recommended because at lower velocities, solids within the liquid can settle in the piping. When settling occurs, it becomes difficult to open and close valves. Exceeding 12½ ft/sec on the discharge line will cause excessive wear of valves, elbows, and tees. Exceeding 8½ ft/sec on the suction line will cause excessive wear of the pump fluid end due to turbulent flow patterns that will occur as the fluid impacts the impeller.

Tables 18.1–18.14 provide line velocities and friction losses that occur when fluid travels through new SCH 40 steel pipe. Scaled pipe and piping with different IDs will have different values, and corrections to shown values must be made. A design factor of 15–20% minimum should be added to values in these tables. The values in boldface designate recommended optimum flow rates based on discharge line velocities.

To determine line velocities and friction losses from Tables 18.1 through 18.14, first locate the proper pipe diameter. In the "Flow Rate (gpm)" column, locate the maximum anticipated flow rate. Line velocities will be located in the "V (ft/sec)" column. Friction loss values will be located in the column "Friction Loss in Feet of Head per 100 Ft of Pipe." For example, for a 6-inch nominal pipe size at 1000 gpm, line velocities are 11.1 ft/sec and friction losses per 100 feet of pipe will be 6.17 feet of head.

18.7 CENTRIFUGAL PUMPS AND STANDARD DRILLING EQUIPMENT

Hoppers, mud guns, desanders, desilters, degassers, and triplex pumps requiring supercharging all have one thing in common: they require 76–80 feet of inlet head to operate as designed. Exceptions do exist, and the equipment manufacturer should be consulted. This simplifies the job of sizing centrifugal pumps. Since most applications in drilling systems require 80 feet of head at the inlet of the equipment, knowledge of volume needed by each piece of equipment is required. Following are standard flow rates when equipment has an 80-foot inlet head:

- 6-inch hopper with standard 2-inch nozzle: 550 gpm
- 4-inch hopper with $1\frac{1}{2}$-inch nozzle: 300 gpm
- Mud gun with ¾-inch nozzle: 85 gpm per gun
- Mud gun with 1-inch nozzle: 150 gpm per gun
- Desander/desilter 4-inch cone: 60 gpm per cone
- Desander/desilter 10-inch cone: 500 gpm per cone
- Degasser: 600 gpm

These are general standards. All values should be verified prior to sizing the centrifugal pump. Considering both suction and discharge conditions when sizing a centrifugal pump is very important.

18.7.1 Friction Loss and Elevation Considerations

Once the head requirement at the end of the transfer line and volume required is known, it is time to determine elevation and friction losses that need to be overcome. Elevation is the distance above or below the centerline of the pump. Therefore, if a centrifugal is mounted on deck 1 and the transfer line ends on deck 2, which is 20 feet above the pump centerline, then the discharge elevation is 20 feet. This 20 feet of elevation must be added to the discharge head required. Additionally, the suction supply tank elevation must be considered. If minimum liquid surface is 8 feet above the pump centerline, then 8 feet of positive head will feed the pump and can be subtracted from the discharge head requirement. However, if the minimum liquid surface level is 8 feet below the pump centerline, a negative head is created and this must be added to the discharge head requirement.

Friction loss is the amount of resistance to flow, measured in feet of head, that occurs when fluid flows through pipe, valves, elbows, etc. This resistance varies with flow rate and pipe diameter

Take, for example, a contractor who wishes to operate a two-cone desander equipped with 10-inch cones and anticipates the maximum mud weight to be 16 lb/gal. The pump is mounted on the same deck as the desander and is 150 feet away. The inlet to the desander is 10 feet above the deck. The supply tank minimum liquid surface level is 8 feet above the pump centerline. Remember that two 10-inch cones will flow 1000 gpm and that the desander requires 80 feet of inlet head:

> 80 feet required by desander
> + 10 feet of discharge elevation
> − 8 feet of positive suction elevation
> + ? discharge friction loss
> + ? suction friction losses
> = TDH required at discharge of centrifugal pump.

Since friction losses are unknown, refer to Table 18.15 (values taken from Tables 18.1 through 18.14).

At 1000 gpm, 4- and 5-inch lines have line velocities that exceed the maximums recommended and should therefore not be used. A 10-inch line has a velocity that does not meet minimum velocity requirements, and

Table 18.15
Excerpt from Tables 18.1–18.4

Line Size Schedule-40 Steel Pipe	Velocity (ft/sec)	Head Loss (ft/100 ft)
4	25.5	64.8
5	16	15.8
6	11.1	6.17
8	6.41	1.56
10	4.07	0.50

Chart based on 1000-gpm flow rate.

settling may occur. Line velocities for 6-inch pipe are excessive for the suction side of the pump. This means that optimum-size suction piping for this application is 8 inches. Discharge piping could use either 6-inch or 8-inch pipe, but it would be most economical to utilize 6-inch piping.

Since suction and discharge line sizes have been selected, friction losses can be calculated. In this example, assume that the discharge line is new 6-inch SCH 40 and has one butterfly valve, six ells, one running tee, and one branched tee and is 150 feet long. The suction line is new 8-inch SCH 40 and has one elbow, one branched tee, and one butterfly valve and is 30 feet long. Each fitting causes friction losses that can be measured and compared to equivalent feet of pipe (see Table 18.16).

Discharge line:

- (1) 6-inch butterfly valve = 22.7 feet
- (6) 6-inch ells = 91.2 feet
- (1) 6-inch running tee = 10.1 feet
- (1) 6-inch branched tee = 30.3 feet

Actual feet of pipe = 150 feet. Total = 304.3 equivalent feet of 6-inch pipe.
  Suction line:

- (1) 8-inch ells = 20 feet
- (1) 8-inch branched tee = 39.9 feet
- (1) 8-inch butterfly valve = 29.9 feet

Actual feet of pipe = 30 feet. Total = 119.8 equivalent feet of 8-inch pipe. Now calculate friction losses using Table 18.17. Friction loss values are based per 100 feet of pipe. Divide the equivalent feet of pipe by 100 to determine the multiplier. Tables 18.1 through 18.14 are based on new steel pipe, and even if new steel pipe is utilized, a 20% design factor should be added to the friction loss values. For pipe other than new SCH 40, refer to engineering handbooks for friction loss values.

Three hundred four equivalent feet of 6-inch discharge line flowing 1000 gpm would have a friction loss of 18.76 feet (3.04 feet × 6.17 = 18.76). With a 20% design factor, the value is 22.5.

One hundred twenty equivalent feet of 8-inch suction line flowing 1000 gpm would have a friction loss of 1.87 feet (1.20 × 1.56 = 1.87). With a 20% design factor, the value is 2.24.

Discharge elevation above pump centerline is 10 feet; supply tank liquid surface level is 8 feet above the pump centerline; and the required

Table 18.16
Friction Loss in Pipe Fittings in Terms of Equivalent Feet of Straight Pipe

Nominal Pipe Size	Actual Inside Diameter	Gate Valve (f.o.)	90° Elbow	Long Radius 90° or 45° std. elbow	Std. Tee (thru flow)	Std. Tee (branch flow)	Close Return Bend	Swing Check Valve (f.o.)	Angle Valve (f.o.)	Globe Valve (f.o.)	Butterfly Valve
1½	1.61	1.07	4.03	2.15	2.68	8.05	6.71	13.4	20.1		
2	2.067	1.38	5.17	2.76	3.45	10.3	8.61	17.2	25.8	7.75	7.75
2½	2.469	1.65	6.17	3.29	4.12	12.3	10.3	20.6	30.9	9.26	9.26
3	3.068	2.04	7.67	4.09	5.11	15.3	12.8	25.5	38.4	11.5	11.5
4	4.026	2.68	10.1	5.37	6.71	20.1	16.8	33.6	50.3	15.1	15.1
5	5.047	3.36	12.6	6.73	8.41	25.2	21	42.1	63.1	18.9	18.9
6	6.065	4.04	15.2	8.09	10.1	30.3	25.3	50.5	75.8	22.7	22.7
8	7.981	5.32	20	10.6	13.3	39.9	33.3	58	99.8	29.9	29.9
10	10.02	6.68	25.1	13.4	16.7	50.1	41.8	65	125	29.2	29.2
12	11.938	7.96	29.8	15.9	19.9	59.7	49.7	72	149	34.8	34.8
14	13.124	8.75	32.8	17.5	21.8	65.6	54.7	90	164	38.3	38.3
16	15	10	37.5	20	25	75	62.5	101	188	31.3	31.3
18	16.876	16.9	42.2	22.5	28.1	84.4	70.3	120	210	35.2	35.2
20	18.814	12.5	47	25.1	31.4	94.1	78.4	132	235	39.2	39.2

f.o. = full open.
Calculated from data in Crane Co., Technical Paper 410.

Table 18.17
Excerpt from Tables 18.1–18.14

Line Size Schedule-40 Steel Pipe	Velocity (ft/sec)	Head Loss (ft/100 ft)
4	25.5	64.8
5	16	15.8
6	11.1	6.17
8	6.41	1.56
10	4.07	0.50

Chart based on 1000-gpm flow rate.

desander inlet head is 80 feet. Therefore, $80 + 10 - 8 + 22.5 + 2.24 = 106.74$ TDH required. Knowing the flow rate of 1000 gpm and TDH required at the pump to be 107 feet, an individual can begin the pump selection process.

Centrifugals must be sized using maximum values anticipated, to ensure that they can perform without cavitation and that the driver is adequately sized. If there is a possibility that the contractor would add a third cone and flow at 1500 gpm, the pump must be sized to handle up to 1500 gpm (this would also affect line velocities and friction losses). Motors must be sized for maximum mud weight. For this example, assume 1000 gpm to be the maximum flow rate and 16 lb/gal mud to be maximum mud weight. A pump to produce 1000 gpm at 107 feet TDH is required. This operating point is marked on attached curves (see pages 496–501).

Following are possible pump selections for this application:

5 × 4 × 14	11.50″ impeller	39 hp (water)	29′ $NPSH_R$	70% efficiency
6 × 5 × 11	10.75″ imp	39 hp (water)	9′ $NPSH_R$	71% eff.
6 × 5 × 14	10.75″ imp	39 hp (water)	10′ $NPSH_R$	70% eff.
8 × 6 × 11	10.75″ imp	40 hp (water)	10′ $NPSH_R$	67% eff.
8 × 6 × 14	11.25″ imp	50 hp (water)	7′ $NPSH_R$	54% eff.
10 × 8 × 14	12.25″ imp	71 hp (water)	16′ $NPSH_R$	40% eff.

With six different pumps that meet the criteria, it is important to select the pump that best meets the application. First consider where on the curve the operating point is located:

- 5 × 4 × 14: Located at the end of the curve. $NPSH_R$ is very high. If this pump is used, cavitation is likely due to insufficient $NPSH_A$. Even given

(*text continued on p. 502*).

496 *Drilling Fluids Processing Handbook*

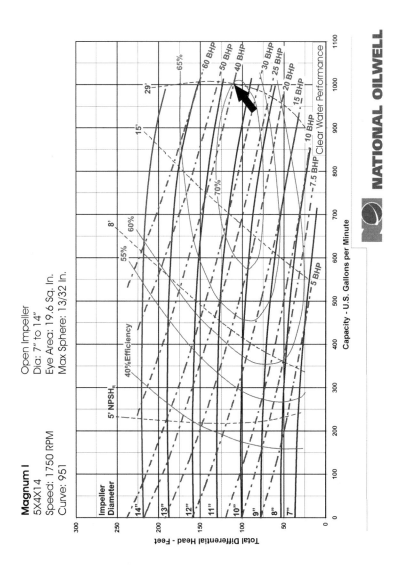

Centrifugal Pumps

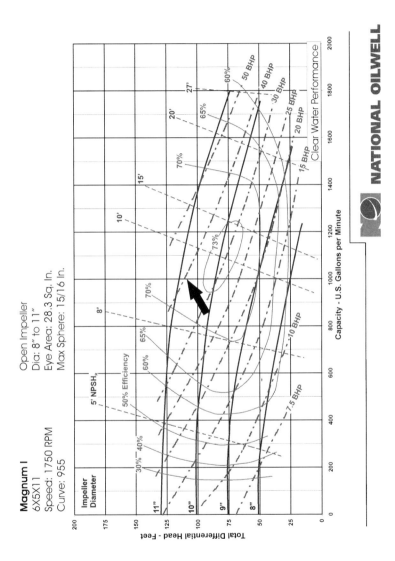

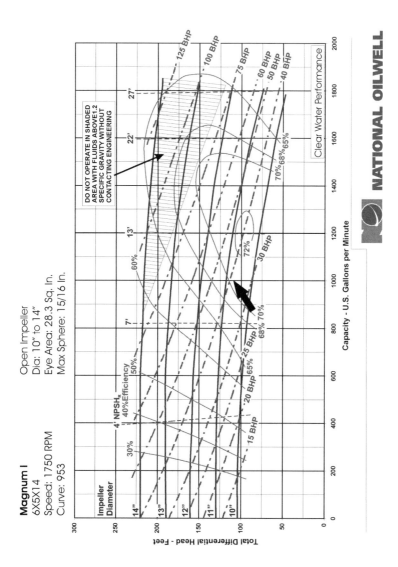

Centrifugal Pumps

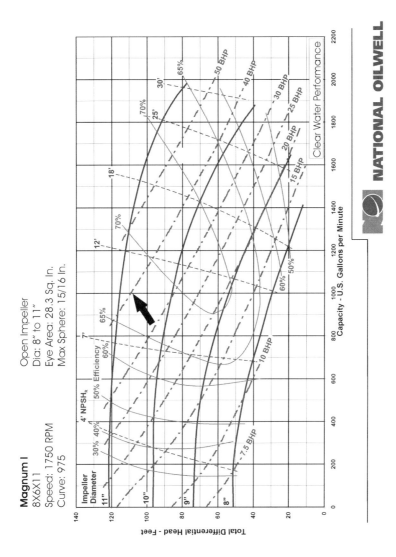

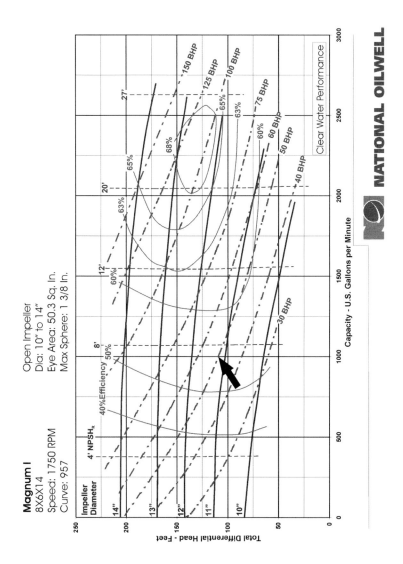

Centrifugal Pumps

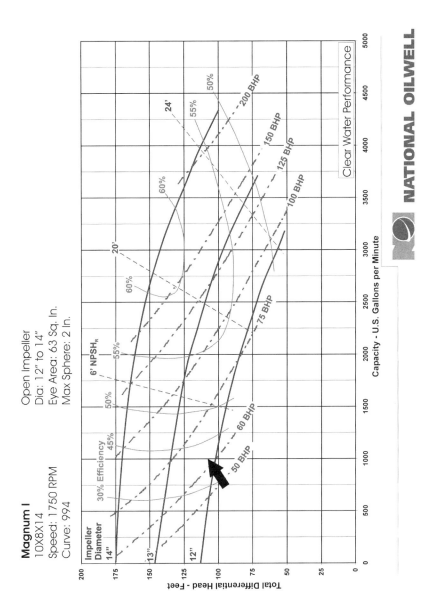

sufficient $NPSH_A$, if additional cones are added, mud temperatures rise, or if the desander cone wears, the pump could not handle the increased volume required. This would be an *unacceptable* sizing choice.
- $6 \times 5 \times 11$: Most efficient pump for the application. An impeller diameter of 10.75 inches is near the maximum impeller diameter for this pump. If there were any miscalculation of friction losses or if the pump were to be relocated farther from the equipment, there would not be a way to significantly increase discharge head. This would be a *fair* choice.
- $6 \times 5 \times 14$: Located in the center of the curve near the best efficiency point. Discharge head can be increased or decreased by changing impeller size. $NPSH_R$ is low. The $6 \times 5 \times 14$ costs less than larger pumps. This would be an *excellent* choice.
- $8 \times 6 \times 11$: Comments are the same as for $6 \times 5 \times 11$.
- $8 \times 6 \times 14$: Located left of the best efficiency point. This pump would perform well in the application but would require more energy than the $6 \times 5 \times 14$, resulting in higher operating costs. Some contractors may still wish to use this pump if they are utilizing a majority of $8 \times 6 \times 14$'s because they could reduce their spare-parts requirements. However, a different-size pump would only require an additional impeller and maybe casing held in stock. If space is available for spares, this is not a good reason to make this selection. This would be a *good* choice for this application.
- $10 \times 8 \times 14$: Located at the very far left of the curve, and a 10-inch suction line would have a line velocity below the recommended minimum. Horsepower requirements are much higher than for other pumps. Although this pump would function in this application, it would be a *bad* choice.

The $6 \times 5 \times 14$ is the best pump for this application.

Knowing the SG of the fluid is necessary to determine the hp required. Maximum mud weight in this example is 16 lb/gal. To determine SG, use the following formula:

$$SG = \frac{lb/gal}{8.34}; \text{therefore, } 16/8.34 = 1.92\,SG$$

where 8.34 lb = weight of 1 gallon of water.

Horsepower required:

$$SG \times hp \text{ for water} = 1.92 \times 39 = 74.88\,hp.$$

A more accurate hp formula is as follows:

$$hp = \frac{\text{gpm} \times \text{feet of head} \times \text{SG}}{3960 \times \text{eff}}.$$

therefore

$$74.11 = \frac{1000 \times 107 \times 1.92}{3960 \times .70}$$

Motors are available in 75 hp, but it is advisable to select a 100-hp motor (if using an electric driver). A 100-hp motor would offer flexibility for the package, compensate for any errors, increase flow rates against pipe or equipment wear, and allow the contractor to exceed 16 lb/gal if future requirements dictate.

Normally a motor with a 1.15 service factor (SF) would be utilized. This means that a 75-hp motor is capable of producing

$$(75)(1.15) = 86.25\,\text{hp}.$$

However, the SF of a motor is intended for intermittent service, and running in the SF continuously will shorten the life of the motor.

If the pump delivers insufficient head, the equipment being fed may not be operating properly and could be damaged. If the pump delivers excess head, flow rates will increase and may cause motor overload. Excessive inlet head to some equipment can also cause premature wear, equipment failure, and/or improper operation.

18.8 NET POSITIVE SUCTION HEAD

NPSH is extremely important to the operation of a centrifugal pump. Factors that affect NPSH are atmospheric pressure, suction line friction loss, elevation, fluid temperature and SG. To calculate NPSH_A:

$$\text{NPSH}_A = \frac{P_{AF} - P_{VF}}{\text{SG}} \pm Z - S_{HF}$$

where

- P_{AF} = atmospheric pressure, feet—at sea level is 34 feet of water. Therefore, if the supply tank is a vented or open-air tank at sea level, the atmosphere will apply 34 feet of pressure to the fluid surface. When the pump casing is filled with liquid and then activated, pressure at the eye

of the impeller drops. Atmospheric pressure, being greater than this value, pushes liquid into the pump suction. The pump does not suck fluid; fluid is pushed into the pump by the atmosphere.
- P_{VF} = vapor pressure, in feet—the amount of head required to maintain fluid in a liquid state. This value (for water) can be found in Table 18.18.
- SG = specific gravity of fluid
- S_{HF} = suction head friction losses, calculated as shown previously
- Z = elevation, or the liquid level above or below the pump centerline, in feet. If the supply tank is mounted on the same level as the pump, the elevation is the number of feet the fluid is above the centerline. Therefore, if the tank is 9 feet tall and when full has 8 feet 9 inches of liquid, then 8 feet can be added to the $NPSH_A$ equation as long as the tank is not going to be drained (centerline of the pump is 9 inches). If there is a desire to have the ability to drain the tank, this value should not be added to the equation. If the fluid level is below the pump centerline, this distance must be subtracted from the equation

Table 18.18
Properties of Water

Temperature		Vapor Pressure	
F	C	psi	ft
40	4.4	0.12	0.28
50	10	0.18	0.41
60	15.6	0.26	0.59
70	21.1	0.36	0.82
80	26.7	0.51	1.17
90	32.2	0.70	1.61
100	37.8	0.95	2.19
110	43.3	1.28	2.94
120	48.9	1.69	3.91
130	54.4	2.22	5.15
140	60	2.89	6.68
150	65.6	3.72	8.56
160	71.1	4.74	10.95
170	76.7	5.99	13.84
180	82.2	7.51	17.35
190	87.8	9.34	21.55
200	93.3	11.50	26.65
212	100	14.70	33.96

(the lowest possible liquid surface level should be used when calculating).

To continue the previous example, the suction line will be 8 inches; flow rate, 1000 gpm; fluid temperature, 180°F; location is sea level; tank is open vented; liquid level above pump centerline is 8 feet; and there is not a desire to drain the tank. The 8-inch-diameter. suction line has one elbow, one branched tee, and one butterfly valve and is 30 feet long:

Atmospheric pressure = 34 feet of water
Vapor pressure = ? feet
SG = 1.92
Suction line friction losses previously calculated = 2.23 feet
Elevation = 8 feet

To determine vapor pressure of the fluid, refer to the properties of water in Table 18.18.

Water-based drilling mud in 180°F will require 17.9 feet of head to maintain fluid in a liquid state:

$$\text{NPSH}_A = \frac{P_{AF} - P_{VF}}{SG} \pm Z - S_{HF}.$$

Therefore

$$\text{NPSH}_A = \frac{34 - 17.35}{1.92} + 8 - 2.23 = 14.44$$

Review the curve in Figure 18.19 to determine NPSH_R for the pump at the operating point (shown by the arrow).

In Figure 18.19, NPSH_R at the operating point is slightly less than halfway between 7 and 13 feet and would therefore equal 10 feet NPSH_R. Because NPSH_A is 14.44 and NPSH_R is 10 feet, the pump will still have 4.44 feet of positive head that will prevent fluid from cavitating. If NPSH_A were less than or equal to NPSH_R, the pump would cavitate and damage would occur.

Fluid temperature/vapor pressure is the most common factor overlooked during pump sizing. However, in this example it was the most significant factor in determining NPSH_A.

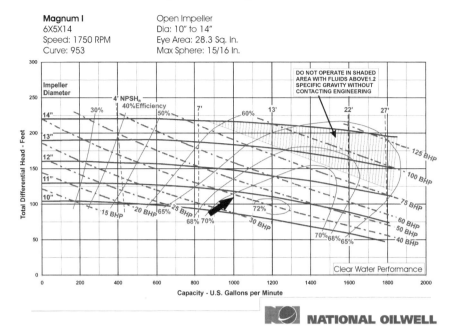

Figure 18.19. Centrifugal pump curve.

Note that NPSH values are used only to determine whether adequate head will be maintained on the suction side of the pump to prevent cavitation. It does not, however, have any bearing on TDH required by the system.

18.8.1 System Head Requirement (SHR) Worksheet

A worksheet is a useful tool for sizing a centrifugal pump. Following a worksheet will simplify the sizing process and allows the user to calculate TDH required. An SHR Worksheet problem is given in Exercise 2 at the end of the chapter.

18.8.2 Affinity Laws

If there is a known operating point and a different operating point is required, the following algebraic formulas can be used to accurately predict what changes should be made to alter flow or head and what the resulting horsepower requirements will be. A pump's performance can be altered by changing speed or by changing impeller diameter.

Note: *Speed formulas are very reliable. Impeller diameter formulas are accurate only for small variations in diameter.*

Speed Formulas	or	Impeller Diameter Formulas (valid for small variations in dia. only, max 1")
Flow:		
$\dfrac{gpm_1}{gpm_2} = \dfrac{rpm_1}{rpm_2}$	or	$\dfrac{gpm_1}{gpm_2} = \dfrac{Dia._1}{Dia._2}$
Total Differential Head:		
$\dfrac{TDH_1}{TDH_2} = \dfrac{(rpm_1)^2}{(rpm_2)^2}$	or	$\dfrac{TDH_1}{TDH_2} = \dfrac{(Dia_1)^2}{(Dia_2)^2}$
Horsepower:		
$\dfrac{hp_1}{hp_2} = \dfrac{(rpm_1)^3}{(rpm_2)^3}$	or	$\dfrac{hp_1}{hp_2} = \dfrac{(Dia_1)^3}{(Dia_2)^3}$

18.8.3 Friction Loss Formulas

$$\frac{\text{Friction loss}_1}{\text{Friction loss}_2} = \frac{(gpm_1)^2}{(gpm_2)^2}$$

If a particular operating point and elevation of a system are known, it is possible to calculate a new operating point by using the following friction loss formulas. Assume that a system exists that has 20 feet of elevation and the pump is transferring water at 500 gpm and the pressure gauge reads 50 psi at the pump discharge. What head is required to produce 1000 gpm?

1. First convert psi to feet:

$$\text{Head} = 50\,\text{psi} \times 2.31/1.0\,\text{SG}$$
$$\text{Head} = 115\,\text{feet}$$

2. Subtract lift of 20 feet, since this is a constant:

    115 feet head − 20 feet elevation = 95 feet of system friction loss at 500 gpm.

3. Utilize friction loss formulas to determine the new head required to produce 1000 gpm in this system:

$$\frac{\text{Friction loss}_1}{\text{Friction loss}_2} = \frac{(\text{gpm}_1)^2}{(\text{gpm}_2)^2} \quad \text{or} \quad \frac{X}{95} = \frac{1000^2}{500^2}$$

$$\text{or} \quad X = 95(1000/500)^2 = 380 \, \text{feet}$$

4. Add back lift: $380 + 20 = 400$.

It would therefore be necessary to size a pump for 1000 gpm at 400 feet to obtain the desired flow rate of 1000 gpm in the existing system. Note: It may be more economical to alter system discharge piping to reduce system friction losses than to pay power costs to produce 400 feet of head.

18.9 RECOMMENDED SUCTION PIPE CONFIGURATIONS

In addition to selecting the proper suction pipe diameter and having adequate NPSH_A, the submergence level and suction pipe configuration must be considered. Submergence level is the depth of the suction pipe inlet below the liquid surface. If an inadequate submergence level exists, an air vortex will form that extends from the liquid surface to the inlet of the suction pipe. This will introduce air into the system, resulting in either turbulent flow patterns or vapor locking of the pump. Amount of submergence required varies with velocity of the fluid. Fluid velocity is controlled by flow rate and pipe diameter. Refer to Figure 18.20 to determine submergence required based on fluid velocity (fluid velocity

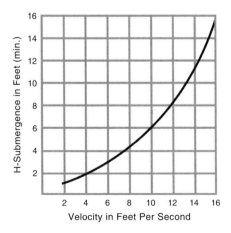

Figure 18.20. Submergence chart.

can be found in Friction Loss Tables 18.1 through 18.14, in the column "V (ft/sec)".

If a system utilizes a 6-inch suction line with a flow rate of 600 gpm, suction-line velocities will be 6.6 fps and the line will therefore require approximately 3.5 feet of liquid surface above the suction-line entrance. Once the submergence level drops below 3.5 feet, an air vortex will form, causing air to enter the pump suction, resulting in a turbulent flow pattern and/or vapor lock.

A suction-line velocity of 6.6 ft/sec is ideal. Increasing the pipe diameter to 8 inches would result in an insufficient line velocity of 3.85 ft/sec. However, most systems will require the tank to have the ability to drain lower than 3.5 feet. One solution is to install a baffle plate over the suction pipe. If a 14-inch baffle plate is installed, fluid velocities around the edge of the plate are only 1.25 ft/sec, which would allow the tank to be drained to approximately 1 foot above the suction pipe entrance. Refer to Figure 18.21 for an illustrated view of a baffle plate.

In addition to proper line size and submergence level, a suction pipe should slope gradually upward from the source to the pump suction. This prevents air traps within the suction line. There should be a straight run prior to the pump entrance of at least two pipe diameters in length to reduce turbulence. A smooth-flowing valve should be installed in the suction line that will allow the pump to be isolated for maintenance and inspection. If a suction hose is used in lieu of hard piping, the hose must be noncollapsing. Refer to Figures 18.22 and 18.23 for examples of accepted piping practices.

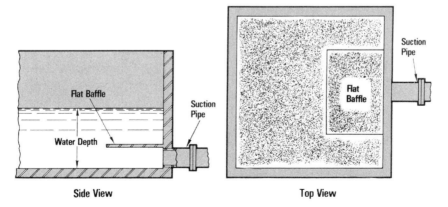

Figure 18.21. Baffle plate illustration.

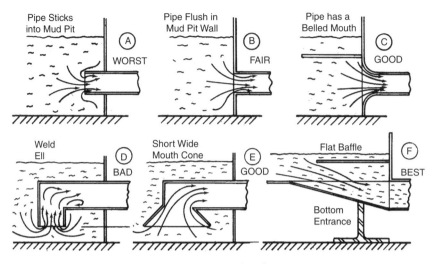

Figure 18.22. Suction line illustration.

18.9.1 Supercharging Mud Pumps

Triplex mud pumps are often operated at speeds at which head in the suction tank is insufficient to maintain fluid against the piston face during the filling stroke. If fluid does not remain against the face, air is sucked in from behind the piston, causing a fluid void. If a void is formed, the piston strikes the fluid when the piston reverses direction during the pressure stroke. This causes a shock load that damages the triplex power end and fluid end and lowers expendable parts life. Supercharging pumps are used to accelerate fluid in the suction line of a triplex mud pump during the filling stroke, allowing fluid to maintain pace with the piston. A properly sized supercharging pump will accelerate fluid so that fluid voids and shock loads do not occur.

Triplex mud pumps normally have shock loads at speeds greater than 60 strokes per minute (spm) (when not supercharged). Without proper equipment, this would go unnoticed until the pump exceeded 80 strokes per minute, but meanwhile the shock load is damaging the pump. Supercharging requires an oversized pump with wide impellers to adequately react to rapid changes in flow required by the triplex mud pump. When sizing a centrifugal pump for a mud pump supercharging application, the pump should be sized for $1\frac{1}{2}$ times the required flow rate. Therefore, if the triplex mud pump maximum flow rate is 600 gpm, the centrifugal pump should be sized for 900 gpm. High-speed piston

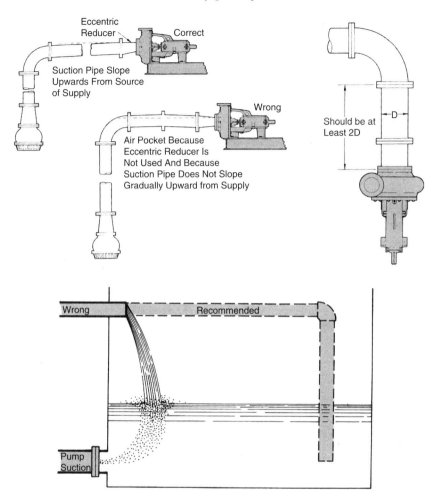

Figure 18.23. Piping recommendations.

and plunger pumps that stroke above 200 spm should be designed with a supercharging pump that produces 1 ¾ to 2 times the required flow rate.

Supercharging is one of the few applications in which the centrifugal pump does not have steady flow. The flow pulsates. Small impellers operating at 1750 rpm have a tendency to slip through the fluid when acceleration is needed. This is similar to car tires slipping on wet pavement. Even though it sometimes appears that the small impeller running at 1750 rpm is providing enough head, shock loading may be occurring. Supercharging pumps should have larger impellers running at either 1150 (60 cycles) or 1450 rpm (50 cycles) and should normally be sized to produce 85 feet of head at the triplex suction inlet. Supercharging pumps should be located as close to the supply tank as possible. Mounting supercharging pumps near the triplex and away from the supply tank transfers suction problems from the triplex to the centrifugal pump. If the centrifugal pump does not have a favorable supply with short suction run, it will have an insufficient supply to accelerate fluid.

Piping for supercharging pumps and triplex pump suctions should be oversized for the flow rate. Pipe should be sized so the change in line velocity during pulsations will not be over 1.5 ft/sec during the change from low flow rate to high flow rate during the triplex pulsation cycles.

Example:

Triplex will be sized for 600 gpm. What pipe size should be used?

The triplex does not have a constant flow rate but varies as it goes through its different crank positions:

- High flow rate is 107% of average $= 600 \times 1.07$. High flow rate $=$ 642 gpm.
- Low flow rate is 86% of average $= 600 \times 0.86$.
- Low flow rate $= 516$ gpm.
- Line velocity in 6-inch pipe at 642 gpm $= 7.29$ ft/sec.
- Line velocity in 6-inch pipe at 516 gpm $= 5.86$ ft/sec.
- Change in line velocity $= 7.29 - 5.86$.
- Change in line velocity $= 1.43$ ft/sec.

Since the change in line velocity in 6-inch pipe is less than 1.5 ft/sec, this pipe size can be used for supercharging a triplex with an average flow of 600 gpm.

18.9.2 Series Operation

There are times when a single centrifugal pump will not meet the head requirements of an application. Two pumps can be operated in series to achieve the desired discharge head, in which the discharge of one pump feeds the suction of the second pump. The second pump boosts the head produced by the first. Therefore, if an application required 2900 gpm at 200 feet of head, one option would be to run two $10 \times 8 \times 14$ pumps in series. Each pump could be configured with a

13-inch impeller to produce 2900 gpm at 100 feet of head. When operated in series, the pumps would produce 2900 gpm at 200 feet of head.

This type of configuration is most commonly used for extremely long discharge runs. When running pumps in series, it is important not to exceed flange safety ratings. Additionally, it is not required to place pumps within close proximity of each other. If an application had a 6-mile discharge line the first pump could be located at the supply source and the second pump could be located 3 miles away.

18.9.3 Parallel Operation

Parallel operation is discouraged for centrifugal pumps. If an application exists that requires high volume and low head and volume required is greater than can be produced by a single pump, two pumps are sometimes used in a *parallel* configuration to meet the demand. Two pumps that produce the same TDH can be configured so that each pump has an individual suction but both pumps feed into the same discharge line. If the pumps are identical, head in the discharge line is equal to that of the pumps, but the volume is double what a single pump can produce. However, two centrifugal pumps will never have the exact same discharge head, and as wear occurs one pump will produce less head than the other and the stronger pump will overpower the weaker pump and force fluid to backflow into the weaker pump. For this reason, parallel operation is not normally recommended.

18.9.4 Duplicity

Two pumps can be configured in parallel but only one pump is operated at a time, thus providing a primary and a backup pump. The two pumps are separated by a valve in each discharge line that prevents one pump from pumping through the other. This type of configuration is perfectly acceptable and, in crucial applications, encouraged.

18.10 STANDARD RULES FOR CENTRIFUGAL PUMPS

1. Never install a suction line smaller than the pump suction, as this can cause NPSH deprivation.
2. Never close or throttle the pump suction valve while the pump is operating.

3. Prime the pump and purge all air from the casing before starting, to prevent seal damage and vapor locking.
4. Never run the pump dry.
5. Pump discharge should always be above the pump suction, to prevent air from being trapped in the casing.
6. Ensure adequate $NPSH_A$.
7. Ensure that the suction line is adequately submerged, to prevent an air vortex from forming.
8. Ensure that fluid transferred is free of air.
9. Ensure that suction and discharge piping always lead upward. Avoid lines that go up over and back down, as this will cause air to become trapped in the line. If this is unavoidable, provide a means for bleeding air out of the line.
10. When sizing centrifugal pumps, work in feet, not psi.
11. Ensure that the pump driver is turning in the proper direction.
12. Oil-lubricate bearings for speeds above 2400 rpm.
13. Oil-lubricated pumps must be operated in a level horizontal position.
14. Do not allow the pump to support suction and discharge piping.
15. Ensure that the pump and driver are properly aligned after all piping and positioning has been completed.
16. Size the pump to produce head to meet the system requirements or provide means of throttling the discharge line. Proper sizing is desirable.
17. Size the pump for the maximum flow rate, temperature, and head that may be required.
18. The suction line should be as short as possible.
19. The suction line should have a straight run at least two pipe diameters long directly in front of the pump suction.

18.11 EXERCISES

18.11.1 Exercise 1

Answers are in the Appendix at the end of this chapter.

Examine the curve and operating point marked by the arrow in Figure 18.11. Determine pump size, speed, gpm, feet of head, impeller diameter, hp for water, efficiency point, and $NPSH_R$. Fill in the answers below and then check the answers with those in the chapter Appendix 1.

Pump size   _____
Speed rpm   _____
gpm   _____
Feet of head   _____
Impeller dia   _____
hp Required   _____
Efficiency   _____
$NPSH_R$ _____

18.11.2 Exercise 2: System Head Requirement Worksheet

Answers are in the Appendix at the end of this chapter.

Conditions:

Liquid Pumped _____ Flow Rate (gpm) _____ Specific Gravity _____ Fluid Temperature _____ °F Calculated Feet of Head (line 6) _____

1. Suction: Pipe Size _____ inches.
   1a. Vertical distance (liquid surface to pump centerline ±). Positive number if above pump centerline or negative number if below pump centerline _____ feet.
   1b. Total length of suction line _____ feet.
   1c. Straight pipe equivalent of suction fittings:

Type	Qty.	Equiv. Ft per Fitting [cf. Table 18]		Total Equiv Ft. of Straight Pipe
Elbow	_____	× _____	=	_____
Tee Running	_____	× _____	=	_____
Tee Branched	_____	× _____	=	_____
Swing Check	_____	× _____	=	_____
Globe Valve	_____	× _____	=	_____
Butterfly Valve	_____	× _____	=	_____
_____	_____	× _____	=	_____

(1c) Sum total = _____

1d. Add (1b) and (1c) = _____ equivalent feet of straight suction pipe.
1e. Convert to friction loss head: $\dfrac{(1d)}{100} \times$ head loss [cf. Tables 18.1 to 18.14] $\times$ 1.2 [design factor] = _____ feet of head (S_{HF})

2. Discharge: Pipe Size _____ inches.
   2a. Vertical distance (centerline of pump to highest point in discharge system ±) _____ feet.
   2b. Total length of discharge line _____ feet.
   2c. Straight pipe equivalent of discharge fittings:

Type	Qty.		Equiv. Ft per Fitting [cf. Table 18]		Total Equiv Ft. of Straight Pipe
Elbow	____	×	____	=	____
Tee Running	____	×	____	=	____
Tee Branched	____	×	____	=	____
Swing Check	____	×	____	=	____
Globe Valve	____	×	____	=	____
Butterfly Valve	____	×	____	=	____
_____	____	×	____	=	____
(2c) Sum total = ____					

2d. Add (2b) and (2c) = _____ equivalent feet of straight discharge pipe.
2e. Convert to friction loss head: $\dfrac{(2d)}{100} \times$ head loss [Tables 18.1 to 18.14] $\times$ 1.2 [design factor] = _____ feet of head (discharge friction loss [D_{HF}]).

3. Head required at discharge point _____ psig $\times \dfrac{2.31}{SG}$ = _____ feet of head.

4. Total friction head (H_f) = (1e) + (2e) _____.
5. Total elevation head (H_e) = (2a) − (1a) _____.
6. TDH required at pump discharge = $H_f + H_e +$ line 3 = _____ feet head required. (Note: NPSH$_A$ must also be considered. See pages 507 to 509 for NPSH$_A$ calculation method.

18.11.3 Exercise 3

Answers are in the Appendix at the end of this chapter.

Application: Hopper Feed Pump	System Data
Equipment	6″ hopper with 2″ nozzle
Hopper required inlet head	80 feet
Hopper flow rate	550 gpm
Fluid temperature	130° F
Hopper elevation above pump centerline	22 feet
Discharge line	215′ of 6″ SCH 40 pipe
Discharge fittings	(1) BF [blastfire] valve, (1) running tee, (2) 90's, (4) 45's
Suction line	20′ of 6″ SCH 40 pipe
Suction fittings	(1) BF valve, (1) 90
Liquid surface level	$-10'$ (below pump centerline)
Fluid maximum weight	18 lb/gal
Equipment location	Sea level

Utilizing the System Head Requirement Worksheet and curves on page 496–501, determine the following values:

Suction friction losses = _____
Discharge friction losses = _____
Pump flow rate = _____
TDH required = _____
Optimum pump size = _____
Pump speed = _____
Impeller diameter = _____
Horsepower required = _____
$NPSH_R$ = _____
$NPSH_A$ = _____

18.11.4 Exercise 4

Answers are in the Appendix at the end of this chapter.

An $8 \times 6 \times 14$ pump with an 11-inch impeller is operating at 1000 gpm at 103 feet and requires 48 hp when pumping water. A contractor wants to increase discharge head to 115 feet. What will be the required impeller diameter, hp, and resulting flow rate?

Answers: _____ Imp dia. _____ gpm _____ hp required

18.12 APPENDIX

18.12.1 Answers to Exercise 1

Pump size $6 \times 5 \times 14$
Speed rpm   1150
gpm   1000
Feet of head   84
Impeller dia.   13.75″
hp Req'd   31 HP for water or fluids with a specific gravity of 1.0
Efficiency   69%
$NPSH_R$ 10 Feet

18.12.2 Answers to Exercise 2: System Head Requirement Worksheet

Liquid Pumped __18 lb/gal mud__ Flow Rate (gpm) __550__ Specific Gravity __2.16__ Fluid Temperature __130__ °F

1. Suction: Pipe Size __6__ inches.
   1a. Vertical Distance (liquid surface to pump centerline ±). Positive number if above pump centerline or negative number if below pump centerline __−10__ feet.
   1b. Total length of suction line __20__ feet.
   1c. Straight pipe equivalent of suction fittings:

Type	Qty.		Equiv. Ft per Fitting [cf. Table 18]		Total Equiv Ft. of Straight Pipe
Elbow	1	×	15.2	=	15.2
Tee Running	0	×	0	=	0
Tee Branched	0	×	0	=	0
Swing Check	0	×	0	=	0
Globe Valve	0	×	0	=	0

Centrifugal Pumps 519

Butterfly Valve	1	×	22.7	=	22.7
	0	×	0	=	0
(1c) Sum total =	37.9				

1d. Add (1b) and (1c) = __57.9__ equivalent feet of straight suction pipe.
1e. Convert to friction loss head: $\frac{(1d)}{100}$ × head loss [cf. Tables 18.1 to 18.14] × 1.2 [design factor] = __1.38__ feet of head (S_{HF})

2. Discharge: Pipe Size __6__ inches.
   2a. Vertical Distance (centerline of pump to highest point in discharge system ±) __22__ feet.
   2b. Total length of discharge line __215__ feet.
   2c. Straight pipe equivalent of discharge fittings:

Type	Qty.		Equiv. Ft per Fitting [cf. Table 18]		Total Equiv Ft. of Straight Pipe
Elbow	2	×	15.2	=	30.04
Tee Running	1	×	10.1	=	10.1
Tee Branched	0	×	0	=	0
Swing Check	0	×	0	=	0
Globe Valve	0	×	0	=	0
Butterfly Valve	1	×	22.7	=	22.7
45's	4	×	8.09	=	32.36
(2c) Sum total =	95.56				

2d. Add (2b) and (2c) = __310.56__ equivalent feet of straight discharge pipe.
2e. Convert to friction loss head: $\frac{(2d)}{100}$ × head loss [Tables 18.1 to 18.14] × 1.2 [design factor] = __7.42__ feet of head (discharge friction loss [D_{HF}]).

3. Head required at discharge point _____ psig × $\frac{2.31}{SG}$ = __80__ feet of head.
4. Total friction head (H_f) = (1e) + (2e) __8.8__ .
5. Total elevation head (H_e) = (2a) − (1a) __32__ .
6. TDH required at pump discharge = $H_f + H_e$ + line 3 = __120.8__ feet head.

18.12.3 Answers to Exercise 3

Suction friction losses	= 1.38
Discharge friction losses	= 7.42
Pump flow rate	= 550.00
TDH required	= 120.80
Optimum pump size	= 5 × 4 × 14
Pump speed	= 1750
Impeller diameter	= 10.75"
Horsepower required	= 60 hp (25 hp for water × 2.16 SG)
$NPSH_R$	= 11'
$NPSH_A$	= 17.42'

$$NPSH_A = \frac{P_{AF} - P_{VF}}{Sp.Gr.} \pm Z - S_{HF}$$

$$NPSH_A = \frac{34 - 5.15}{2.16} - 10 - 1.38 = 1.98$$

Note: With only 1.98 $NPSH_A$, there is not enough $NPSH_A$ to prevent cavitation. This pump would cavitate and experience severe damage and a drastically reduced life. However, if the fluid were water with a 1.0 SG, the $NPSH_A$ would be 17.42, which would be adequate for a properly selected pump.

18.12.4 Answers to Exercise 4

Answers: 11.62' Imp dia. 1056 gpm 56 hp required

New imp dia:

$$\frac{115}{103} = \frac{X^2}{11^2} \quad \text{or} \quad (1.0566)(11) = X \quad \text{or} \quad 11.62'' = X$$

New flow rate:

$$\frac{X}{1000} = \frac{11.62}{11} \quad \text{or} \quad (1.056)(1000) = X \quad \text{or} \quad 1056 \text{ gpm} = X$$

Horsepower required:

$$\frac{X}{48} = \frac{11.62^3}{11^3} \quad \text{or} \quad (48)(1.177) = X \quad \text{or} \quad 56 \text{ hp} = X$$

CHAPTER 19

SOLIDS CONTROL IN UNDERBALANCED DRILLING

Bill Rehm
Drilling Consultant

Jerry Haston
Drilling Consultant

19.1 UNDERBALANCED DRILLING FUNDAMENTALS

Underbalanced drilling (UBD) is defined as "deliberately drilling into a formation in which the formation pressure, or pore pressure, is greater than the pressure exerted by the annular fluid or gas column" (IBD HSE Forum, IADC 2002). In this respect, "balanced" pressure drilling is a subcategory of underbalanced drilling because the annular pressure may fall below the formation pressure during pipe movement. Underbalanced drilling is used to avoid or limit lost circulation and as a method to protect reservoirs, prevent differential sticking, and increase the drilling rate.

Drilling with a hydrostatic pressure that is less than formation pressure may create a condition similar to a "kick," or well-control condition. Controlling these pressures and maintaining a safe environment require special surface pressure control equipment and a properly trained crew (Figure 19.1). The type of equipment required depends on primarily the lithology, permeability, and pressure of the formations that are to be drilled.

Figure 19.1. Spindletop.

UBD systems require surface handling equipment similar to that used on wells that are drilled overbalanced, in addition to specialized equipment. Regardless of the relationship between fluid and formation pressures, the drilling fluid must remove the cuttings from the bottom of the hole and carry them to the surface. At the surface, there must be adequate equipment to remove the solids from the fluid before it is circulated back down the drill string. In cases where the fluid is not recirculated, such as air drilling and some foam drilling, provisions must be made for handling the solids and liquids.

Solids control in UBD does not get as much attention as does solids control in overbalanced drilling. There are a number of reasons for the lack of attention. Probably the major reason is the attitude that solids control is not necessary when well-bore pressure exceeds hydrostatic pressure. This is not even close to the truth. Reservoir damage from high solids content can occur because the annulus is not underbalanced at all times. Also, excess solids decrease efficiency just as they do in overbalanced drilling. All of the existing solids-control equipment can be used in underbalanced drilling, but the system should be redesigned for each particular set of conditions.

The purpose of this chapter is to provide information regarding how well-bore solids are collected and sent to the solids-control equipment in all types of UBD. Some techniques have been developed specifically for UBD, but most are the same techniques that are used in overbalanced drilling. A description of each type of UBD will be given, followed by techniques used for proper solids control.

19.1.1 Underbalanced Drilling Methods

Underbalanced drilling may be conducted using any type of fluid, provided that the hydrostatic column of that fluid is less than formation or pore pressures. The wide variety of "fluids" include:

- Dry air, or air supplied by one or more air compressors
- Natural gas, or naturally occurring hydrocarbon (primarily methane) gas or gases
- Mist, created by injecting water into a gas or air stream
- Foam, created by adding a foaming agent (surface active agent) to injected water
- Gaseated water or gaseated oil
- Nitrogen gas, or N_2 gas supplied either as bulk cryogenic liquid or from a filter (membrane) unit
- Stiff foam, created by adding a foaming agent to a specially prepared drilling fluid injected into the air stream
- Drilling fluid, of any type, that creates a hydrostatic pressure less than the pore pressure

19.2 AIR/GAS DRILLING

The extreme underbalance that results when drilling with air or gas allows large disc-shaped cuttings to break from the formation with the impact of the bit tooth. These cuttings are degraded to dust as the turbulent air lifts them to the surface. Solids control while air drilling (including natural gas and nitrogen) consists of controlling atmospheric pollution, collecting samples, and disposing of cuttings and liquids. Normally there is no recovery or reuse of the air or gas except for a few occasions in

Figure 19.2. Gas drilling with horizontal flare.

natural-gas drilling when the gas is recycled to the gas plant. An elaborate separator/cleaning/recompressing system is required to recycle the gas. It is usually more economical to flare the gas than to recover and clean it (see Figure 19.2).

19.2.1 Environmental Contamination

Solids-control problems that result when drilling with air or gas are primarily environmental. Dust drilling creates a large cloud of fine solid particles unless some type of dust-control device is used. When "dusting," the ultra-fine particles tend to remain airborne for great distances. To properly control the dust, it needs to be wetted and settled in a tank.

The wetted fine solids from the well can create environmental problems including naturally occurring radiation (NOR), saltwater or hydrocarbon contamination, danger to wildlife, or just a mass of unsightly gummy sterile cuttings. Instead of in an earthen pit, solids and liquids should be collected in a steel tank—a "frac"[1] tank is a good choice.

---

[1] "Frac tank" is a complete description of a type of tank that was originally used in the fracturing process.

The waste can then be disposed of in accordance with appropriate environmental regulations.

Air separator and silencer systems often consist of just a vertical separator with a tangential intake and a water spray injected into the blooie line (Figure 19.3). Water, cuttings, and dust are thrown outward as the air spins when it enters the tank. The wetted dust and cuttings settle to the bottom of the tank as a damp mass. The tank also acts as a muffler and directs the sound upward. The separator must be grounded to avoid static electricity in case gas is present. Water requirements average roughly 300 gal/hr for an 8¼-inch hole, but some of the water can be recycled to the blooie line spray.

Several commercial-rental dust abatement systems are available. They range from closed-tank separator systems in Canada to a slotted pipe and steel tank in Oklahoma and Arkansas.

19.2.2 Drilling with Natural Gas

The primary requirement when drilling with natural gas is to safely flare the gas. In this case, simplest is best, subject to any special environmental requirements. Historically, a horizontal blooie line, with an igniter attached to the end, extending into an earthen pit was satisfactory. Environmental regulations now set requirements for gas handling, dust control, liquid disposal, noise abatement, etc. In Canada and Europe, and in an increasing number of U.S. states, there are specific regulations about heat radiation from flares that preclude the use of an unregulated

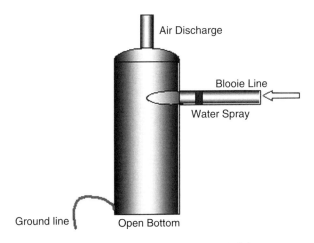

Figure 19.3. Cross section of simple air/solids separator.

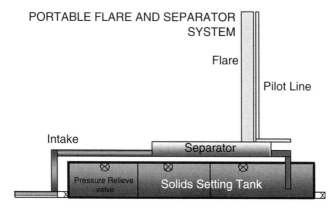

Figure 19.4. Separator with a vertical flare stack.

ground flare. Under regulated conditions, gas and cuttings with spray water have to go through a separator tank. From the separator tank, gas goes to a designed vertical flare stack. A free-water knockout tank may have to be used, depending on the amount of water required. A small backpressure, ± 5 psi, may be required to force the gas to the flare stack. A purge system with natural gas or nitrogen is also needed to keep oxygen out of the tank. The tank must be grounded to prevent static electricity. The sludge that results from wetting the dust and cuttings may be removed from the tank by hand or recirculated through the tank and over the shaker with a small circulating pump (Figure 19.4).

19.2.3 Sample Collection While Drilling with Air or Gas

Sample collecting can be done at the end of the blooie line or in a collection chamber. Catching samples at the end of the blooie line presents several problems. Cuttings are extremely fine and are hard to collect. Safely and conveniently getting to the end of the blooie line may be a problem, since the minimum safe length of a blooie line is 300 feet. When drilling with gas, the gas is usually flared. The intense heat at the end of the line prevents catching representative samples. The heat destroys some of the samples and any hydrocarbons associated with them.

A collection chamber can be installed in the blooie line. The device can be as simple as a tube or pipe welded into the bottom of the line. A tong die-welded inside the blooie line makes a satisfactory deflector. At least one valve is needed at the bottom of the tube to prevent returns

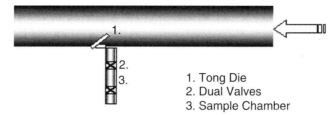

Figure 19.5. Simple sample catcher.

from continuously escaping. Two valves are better, especially when gas drilling, because cuttings can be collected in the chamber with the bottom valve closed while the top valve is open. When the sample is retrieved, the top valve is closed and the bottom valve is opened. The process is simply reversed after the sample is retrieved. The collection chamber provides samples that are much more representative of the formations being drilled than does the practice of collecting samples off the cuttings pile at the end of the blooie line. It is also readily accessible and convenient to use (Figure 19.5).

19.2.4 Air or Gas Mist Drilling

In air mist systems, the problem with cuttings is again environmental. The cuttings are finely ground and mixed with water. The simplest solution is to blow the cuttings into a steel tank and later dispose of the damp cuttings as solid waste. The water can be separated and clarified and sent to a disposal well (Figure 19.6).

Mist systems use in the range of 1000/1 to 3000/1 gas/liquid. For example, a typical 8¾-inch 8000-foot misted hole will use about 2000 standard cubic feet per minute (scfm) of air (about 3 MMscf/d) and about 4 to 5 gallons of water per minute, or a maximum of about 7000 gallons of water in a full drilling day. The cuttings will contain material from the well—oil, NOR, traces of detergent, and corrosion inhibitors. The fluid should be slightly alkaline, as part of the anticorrosion treatment, but the greater mass of damp cuttings should be pH neutral.

Mist systems will surge. A mass of damp cuttings will form in the annulus and build up a plug until enough pressure is developed below the plug to blow it out of the hole. This will cause a large surge of cuttings, water, and gas, possibly equal to a rate of 10 MMcf/d for a period of up to several minutes. The surge will have a high velocity

Figure 19.6. Mist drilling.

and impact, caused by the expanding gas and the mass of the cuttings. The separator system must be able to handle this surge. The simplest solutions utilize

- A long tank to allow the velocity to decrease,
- A spiral system in a circular tank (separator) to consume the energy of the slug,
- An open pit, or
- One of the commercial mist separator systems.

Cuttings from misting are larger than air drilling cuttings. The misting detergent will make the cuttings wet.

Misting with natural gas and diesel oil requires a closed separator in which the oil and gas are separated. This allows the oil-wet cuttings to be dumped to the bottom of the separator or recirculated to the shaker after the gas is flared. This is not a common practice. It is documented in SPE 62896 (Labat, Benoit, & Vining).

19.3 FOAM DRILLING

Foam in the hole is an emulsion of air or gas in water, but at the flowline a proper foam breaks to a mixture of droplets of water in an air stream. With proper foam breaking at the end of the flowline, there is a quickly separating mixture of gas or air with a small amount of water and a small skim of foam (Figure 19.7). During use of a shale shaker, the screen will generally appear "wet" with a skim of foam. This is the result of the chemistry of the system, and while it appears wet with foam, the water volume is very small.

19.3.1 Disposable Foam Systems

Most foam drilling is done with disposable foam. No attempt is made to recycle the water or foaming agent. Foam systems are in the range of 50/1 to 250/1 air/liquid. The liquid volumes vary from 10% of the normal amount of liquid (required to give a 120 ft/min annular velocity with rotary tools) to about 70% of that amount of liquid when using downhole motors. The amount of liquid for disposal or treatment can be relatively modest.

Figure 19.7. Foam at the flowline.

When disposable foam systems are used, the returns are taken from the flowline into a separator and then to a steel pit. Alcohol defoamer may occasionally be injected by spray near the end of the flowline. The foam breaks in the separator and finishes the last skim breaking in the pit. The water is separated and can be clarified before disposal. Most foam-system liquids are environmentally neutral at the flow line (pH of about 7 and biodegradable). The contaminants are those picked up in the hole, the foaming agent, and traces of corrosion inhibitors. Cuttings are settled on the bottom and water flows, from the top, to another pit. This can vary from a very low tech dual pit system to a complex oil separator system using a shale shaker and remote disposal of gas.

19.3.2 Recyclable Foam Systems

When large water volumes are used with downhole motors in directional holes, a recycling system may be more economical than direct disposal. Recyclable foams reuse only the water and chemicals. The primary advantage to recycling is environmental; no pit to store the used foam and water is required. Recycling systems may be less expensive on long jobs than disposable systems because a long job amortizes the added cost of the separators and solids-control equipment. The foaming agent, other chemicals for shale control, and corrosion inhibitors must be partly replaced on each cycle. Replacement of chemicals and water is a minimum of 30% per cycle and may range above 50%.

Recyclable foam systems are usually defoamed with alcohol. Alcohol evaporates when exposed to air as it passes over the shaker. Weatherford has used Transfoam^{TM}, a system in which the foam is broken at the flowline by being acidized. The pH change breaks the foam down to a liquid and a gas. In either case, the defoamed liquid, which may contain some oil, is then circulated to a separator system where the oil is skimmed from the water and pumped to a storage tank. The water is then circulated back to the pump. Foaming agents and corrosion inhibitors are injected into the water. The amount of additives required is determined by constant testing.

Foams at the flowline are about 98% gas and 2% water. If, as a result of incorrect treatment, the foam is not broken at the flowline, there will be a huge overflow volume to treat. When the formulation is correct, only a skim of foam will remain. With proper treatment, the residual foam, which has almost no mass, can be broken down to just some dampness with a small amount of alcohol.

Recyclable foam systems often use a closed and pressurized[2] separator to separate gas from the solids and liquids. The air, or gas, is sent to a flare or a free-water knockout. The oil and water are separated in the pressurized system, where the oil is skimmed and sent to a holding tank. The water and cuttings may be recycled through the separator to keep the cuttings in suspension until they are sent to the shale shaker. It is imperative that the solids be removed from recycled water because they act as a defoamer and tend to interfere with the development of proper recyclable foam.

Discarded solids are water wet and carry some small amount of residual foam. As a result, the shaker screen is always wet with a thin layer of foam, even though there is very little free-water carryover (Figure 19.8). The solids load in the water is high and the cuttings are evenly divided between fines and coarse. (Some extreme cases have been reported in which the solids were 20% by volume of the water.) In foam drilling, minimum shaker screen opening sizes are difficult to specify because of the solids and liquid volume variation.

Shaker underflow also contains a heavy load of fine solids. The most common mode of removal is the centrifuge because of the low volume of water involved. Some use has been made of 2- or 4-inch hydrocyclones.

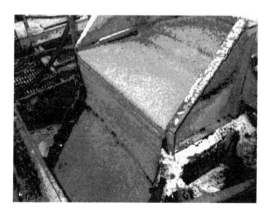

Figure 19.8. Solids on a Shaker Screen.

---

[2] There is a conflict of terminology between UBD and solids control. A *pressurized closed loop system* in UBD refers to the method of controlling gas from the well bore and sending it to a flare system. A *closed loop system* in solids control refers to a method of handling waste solids and liquid so that they are not discarded in the reserve pit.

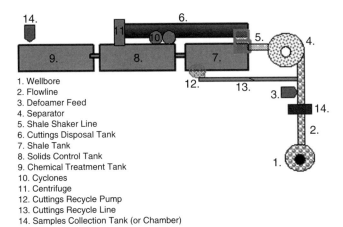

Figure 19.9. Diagram of recyclable foam solids-control system.

1. Wellbore
2. Flowline
3. Defoamer Feed
4. Separator
5. Shale Shaker Line
6. Cuttings Disposal Tank
7. Shale Tank
8. Solids Control Tank
9. Chemical Treatment Tank
10. Cyclones
11. Centrifuge
12. Cuttings Recycle Pump
13. Cuttings Recycle Line
14. Samples Collection Tank (or Chamber)

They appear very attractive from a cost basis, but there are few data available on their use with foam systems (Figure 19.9).

No major attempt has been made to save the water base of the foam after drilling is completed. The water contains very little valuable chemicals or liquids. The residual water will contain fines, have traces of foam or anticorrosion chemicals, and be slightly alkaline. The water can be clarified with commercial flocculants before disposal.

19.3.3 Sample Collection While Drilling with Foam

During use of a disposable foam system, sample collection can be done with a vessel or pail at a nipple and valve in the flowline. If a separator is used, samples can be collected from the outflow of the separator or at the shale shaker.

Cuttings samples from recyclable foam systems require a closed tank before the separator. This generally takes the form of two 10-gal tanks with discharge screens. A split of less than 5% from the bottom of the flow stream is sent to the tank. The cuttings are caught in the screen within the tank. The two tanks are used alternately (Figure 19.10).

19.4 LIQUID/GAS (GASEATED) SYSTEMS

Liquid/gas systems use 75% to 100% of the volume of drilling fluid required (for a 120-ft/min annular velocity) with typical fluid systems. These systems may be water/air, water/nitrogen, oil/gas, or oil/nitrogen.

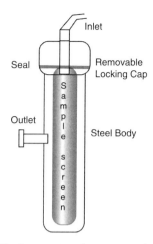

Figure 19.10. Cross section of a pressurized sample catcher.

The ratio of gas to liquid is in the range of 10/1 to 50/1. Typically, for an 8¼-inch hole, the fluid volume will be 300 gpm and the gas volume will be 1 MMcf/d (700 scfm) to 2 MMscf/d (1400 scfm).

Liquid/gas systems are essentially fluid systems with an air, nitrogen, or natural gas boost. All of the solids-control fluid problems discussed elsewhere in this manual are part of the gaseated system. There normally are no problems separating the gas from the liquid with a simple separator (Figure 19.11).

Gas/liquid separation in the well on trips and connections and during drilling is the primary problem with all gaseated systems. The separation causes volume and pressure surges that can be quite violent. In an 8¼-inch hole, using 300 gpm of fluid and 1000 scfm of gas, there can be separations in which there will be no flow at the flowline for as long as 5 minutes, and then 30 bbl of fluid will surge out in half a minute. On a connection or trip, surge can be doubled or tripled in volume, since the system has almost completely separated. As holes get larger and deeper, the surges become larger and more violent. No shaker system can economically handle this kind of volume change. In a regular rig system with open tanks, one or two extra fluid tanks need to be added to the system below the separator to handle the surge volume. The separate fluid tanks should include mixers to keep the cuttings in suspension and a centrifugal pump that will pump the average circulating volume of drilling fluid to the shaker along with cuttings. Once the drilling fluid has passed through the shaker, it can be sent to the regular fluid system and treated as is appropriate to the type of system (Figure 19.12).

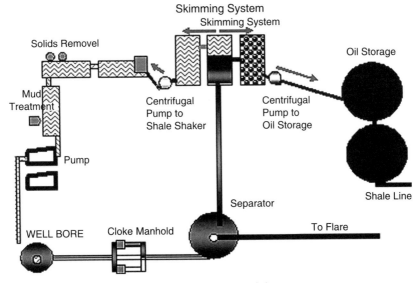

Figure 19.11. Basic separator and skimmer system.

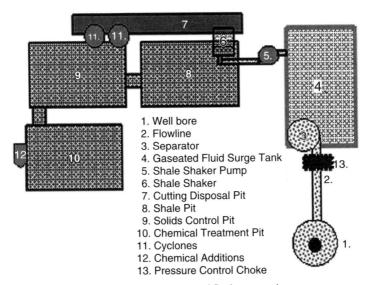

1. Well bore
2. Flowline
3. Separator
4. Gaseated Fluid Surge Tank
5. Shale Shaker Pump
6. Shale Shaker
7. Cutting Disposal Pit
8. Shale Pit
9. Solids Control Pit
10. Chemical Treatment Pit
11. Cyclones
12. Chemical Additions
13. Pressure Control Choke

Figure 19.12. Gaseated fluid surge tank.

In a commercial closed pressurized circulating system, the surge problem is controlled by the large capacity of the separator tank. Drilling fluid is supplied to the shale shaker by a pump that provides a constant controlled volume from the separator tank.

19.5 OIL SYSTEMS, NITROGEN/DIESEL OIL, NATURAL GAS/OIL

A special case of liquid/gas fluid is the diesel/nitrogen or synthetic-oil/nitrogen mixture. These systems have low viscosities and are used primarily in reentries or horizontal slim holes. They tend to require more fluid than water-based fluid systems. Annular velocities of 150–200 ft/min are common in the horizontal section of the hole. In the drilling of slim holes, hole volumes are less than they are in the drilling of conventional holes, but strong surging occurs. A closed pressurized separator system is used when drilling with an oil/gas system (Figure 19.13).

Gas is separated within the tank and sent to the flare stack. There is sometimes a free-water knockout tank between the separator and the flare stack. The oil is recycled through the system to pick up cuttings that have settled and is then pumped to the shale shaker, effectively canceling the surging effect. Since the viscosity of the oil is very low, fine screens can be used. The oil and fines are usually sent to a centrifuge for removal of the fine solids.

A buildup of ultra-fine solids in the oil can damage the reservoir because these solids form a sludge that plugs the pore spaces and can plug pipes and pumps when it settles. Eventually the oil has to be replaced to dilute the ultra-fines.

19.5.1 Sample Collection with Aerated Systems

Surging makes precise sample collection difficult. With all gaseated systems, samples need to be caught at the flowline. There is too much mixing and recirculating to get precise depth logged samples from the

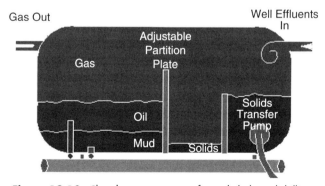

Figure 19.13. Closed separator system for underbalanced drilling.

shale shaker. The best solution is to use the closed tank and screen system shown in Figure 19.10.

19.6 UNDERBALANCED DRILLING WITH CONVENTIONAL DRILLING FLUIDS OR WEIGHTED DRILLING FLUIDS

UBD or "balanced" drilling with conventional drilling fluids is no different than conventional overbalanced drilling as far as basic solids control is concerned. Special pressure-control equipment is required to control well pressures as they occur. A rotating head, choke, and separator are basic to UBD systems (Figure 19.14).

Use of this equipment modifies the methods of feeding the shale shaker. The assumption should be made that the well will have to be drilled under pressure and fluid returned through the choke and separator. Nevertheless, the solids-control system has to be capable of handling 50–100% of the variations in flow volume.

An open, or "atmospheric," separator typically is set high enough to feed the shaker through a U-tube seal. If the underbalanced or near balanced bottom-hole pressure is strictly controlled, the output of the separator to the shaker screen will be relatively constant. Any upset to the system from gas will tend to send a surge to the separator, followed by a flow reduction. The separator will act as a buffer, but it may not be

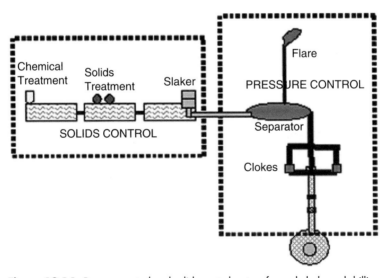

Figure 19.14. Pressure control and solids control system for underbalanced drilling.

large enough to minimize surges to the shaker screen. Some fluid may have to be bypassed to the settling tank or to a surge tank, where it will be stored temporarily as extra volume. The surge tank has been discussed under "gaseated systems."

19.7 GENERAL COMMENTS

The destination, or method of disposal, is one of the major factors in designing the solids-control system for UBD. In the best of all cases, produced oil and water can be skimmed off, cleaned, and sent into the field production system. The solids can be partly dewatered for appropriate disposal. On the other hand, if annular cuttings injection is possible, the volume of waste generated will not be as critical. In this case the waste stream should be kept as wet as possible.

If land farming or solidification is planned, the waste stream will need to be as dry as possible. Adding centrifuges or dewatering units will aid in reducing the volume of waste generated by removing a large portion of the water that would have to be dealt with in a land-farming or solidification process.

If all the cuttings have to be hauled to a disposal facility for treatment and disposal, it will be cost-effective to reduce the solids volume to a minimum. Haul-off and disposal of high chloride and oily cuttings can run as high as $15 per barrel, depending on the distance to be hauled.

In water-based fluid systems, emulsions formed between produced oil, the drilling fluid, and finely ground solids are among the most challenging problems encountered in the control of solids in UBD. Small amounts of oil will be lost in the fluid system, but larger oil flows will form an emulsion with the fluid. In some cases, the emulsion will break in a skimming tank, and the oil can be skimmed off to the sales tank. In these cases some of the cuttings can be picked up by the circulating pump and sent to the shaker. When independent pumps are used, the settling tank can be rolled and then circulated to the shaker during trips. The fluid can then be further processed. There will always be some residual settling of cuttings in the primary settling tank.

In lignosulfonate and other treated drilling-fluid systems, very little settling takes place. Removal of fine solids dispersed into the fluid system is the overriding problem. For these systems, shale shakers, fluid cleaners, and centrifuges will usually be required. Emulsion-breaking chemicals may be required to prevent the cuttings from agglomerating before being processed by the shaker or other solids-control equipment.

Breaking an emulsion in a conventional drilling fluid is difficult, expensive, and time-consuming.

In oil-fluid systems, solids do not settle easily, but any produced light oil will probably start to mix into the system. In most oil-fluid underbalanced conditions, a closed pressurized separator system is used. From the pressurized separator, the gas flows to the flare. The degassed oil fluid goes to the shaker and solids-control system.

19.7.1 Pressurized Closed Separator System

In a typical pressurized closed separator system, the water-based drilling fluid passes, under pressure, from the drilling choke into a modified production separator. Within the separator, under 2 to 5 atmospheres of pressure, the gas and free oil are separated from the drilling fluid and cuttings. Gas is sent to the flare stack, the oil is pumped to holding tanks, and the drilling fluid and cuttings are pumped to the shale shaker. If an oil-based drilling fluid is used, free water is separated from the gas and drilling fluid (see Figure 19.13, "Closed separator system for underbalanced drilling").

The type and amount of equipment that is needed for a closed loop system will be determined by several factors. Some of these are:

- Volume that the well will produce during drilling operations
- Type of fluid(s) being produced
- Hole size and footage to be drilled
- Type of drilling fluid used
- Density of drilling fluid
- Ultimate destination of waste

There are various arrangements of the closed separator system, but a description of the general process is:

1. Effluent from the well enters the separator through a cyclone arrangement that uses the velocity energy to spin the fluid and start the separation process.
2. Fluid drops into the first partition, where the solids settle. The solids may then be pumped out of the vessel by a screw pump, or removed at a hatch. *Currently, cuttings are circulated through the separator for removal in the solids-control system.*

3. Separated gas migrates to the opposite end of the vessel and then to the flare line. Tank pressurization to 2–3 atm forces the gas to the flare line.
4. Fluid goes over a second partition plate and into a section where the water and oil are gravity-separated.
5. A third partition plate protects the water suction where the drill water is sent to the drilling-fluid tanks.
6. A fourth partition plate protects the oil suction where the oil is pumped to the storage tanks for sale.

The actual separator in its final form is a complex set of pipes, tubes, electrical connections, data collection points, and heating elements. When a complete system is used, the separator service company may provide up to four men and a supervisor to service the system and operate the data collection system.

Disposal of solids from these systems faces the standard disposal problems. In general, solids-control problems increase as the fluid complexity increases. Some of the conditions that complicate the system are:

- Weighted drilling fluid
- Oil production and emulsification
- Saltwater production
- Hydrogen sulfide gas production
- Cuttings from horizontally drilled intervals (because the cuttings will be finely ground)

19.8 POSSIBLE UNDERBALANCED DRILLING SOLIDS-CONTROL PROBLEMS

19.8.1 Shale

In general, thick shale sections cause problems with UBD. They slough or cave into the hole. This is probably due to thick shale sections having some elements of laminating, geopressuring, or sensitivity to water. As a general rule, thick shale sections should not be drilled underbalanced. In the special case of air/gas drilling, shale usually remains stable as long as it is kept dry. Even the small amount of water in mist drilling will destabilize most shale. These formations need to be put behind casing within a few days. Watch out for excessive cavings and especially long, thin cuttings. Once caving starts, it cannot be controlled with underbalance operations.

19.8.2 Hydrogen Sulfide Gas

H_2S gas poses a special problem for underbalanced operations but can be controlled with a specially designed Canadian type of totally closed loop circulating system. Care and training are required in order to prevent H_2S from escaping into the atmosphere. Controlling H_2S contributes to solids-control problems. Some gas attaches to the solids and must be treated with a scavenger before the cuttings are released to the atmosphere for standard removal with the shaker and downstream equipment. During drilling of short intervals of H_2S gas-bearing zones, cuttings may be allowed to settle in the closed separator, from which they can be removed after drilling is finished. A specially designed system may be used to sweeten the H_2S gas before it is released to the flare. In 1998, Shell Canada proposed the use of a sweet-gas counterflow system to remove the H_2S from the flare gas.

19.8.3 Excess Formation Water

If excess formation water is encountered, the system must be modified to accept it. Large water flows overwhelm the chemistry of the system as well as overloading the liquid disposal system. Small amounts of formation water may require addition of a foaming agent and a corrosion control agent. Excessive amounts of water require that the system density be increased, the system changed, or the water squeezed or cased off in order to provide adequate hydrostatic pressure to control the volume of water influx.

19.8.4 Downhole Fires and Explosions

Downhole fires are one of the problems that can occur during air drilling. For years it has been suggested that one of the major causes of downhole fires in air drilling was the formation of fluid rings. It was assumed that damp cuttings and poor hole cleaning caused the rings to form. It was also assumed that as the rings became larger, the restricted annulus caused pressures to increase to the point at which spontaneous combustion of the dry gas occurred. The fire would then melt downhole tools, including drill collars. Further investigation has shown that it is impossible to develop pressure high enough to spontaneously ignite dry gas under these conditions. However, distillate will spontaneously ignite at temperatures and pressures that exist in the hole. It is not necessary for fluid rings to develop for combustion to occur. Drilling with a mist does not reduce the chances of downhole fires. The danger of downhole fires, which typically start at about 9000 feet and 150 °F bottom-hole temperature, can be avoided with the use of natural gas or nitrogen as the drilling fluid.

19.8.5 Very Small Air- or Gas-Drilled Cuttings

Evaluation of air- or gas-drilled cuttings can be difficult. The extra-small particle size of air-drilled cuttings makes the identification of index fossils very difficult. Air cuttings get smaller as the hole gets deeper. Inadequate hole cleaning and hole erosion can create additional formation identification problems. Inadequate compressor capacity is often the cause of poor hole cleaning. If additional compressor capacity is not available, air requirements may be reduced by decreasing the well annulus area. A smaller annulus imparts higher velocity for a given injection rate. Decreasing hole size or increasing drill-pipe diameter reduces the annulus area. Fine dust should be damped with water and retained in a steel pit.

An equivalent annular velocity of 3000 ft/min is adequate in most cases. When the penetration rate exceeds 60 ft/hr, or when cuttings are large or become wet, higher air volumes are needed to provide higher air velocities to effectively clean the hole during drilling (see Angel, *Air Drilling Handbook*, for the mathematics of velocity).

19.8.6 Gaseated or Aerated Fluid Surges

All gaseated or aerated fluid systems will surge because the gas and liquid are not tied together. Surging can occur during drilling and it becomes worse on connections and after a trip. The volume and pressure surge can change bottom-hole and well-bore pressure by as much as 1000 psi. Changes in circulating pressure, as read on the pump gage, gives an idea of the bottom-hole pressure change. Less air or gas, high fluid viscosity, and smaller annular size all reduce surging. There are other techniques and equipment that also help in surge control. Surges must be evened out for solids control. The easiest way to do that is with a dedicated pit and pump for the shaker.

19.8.7 Foam Control

Drilling foam is a balance between a very weak foam (wet foam) that surges and does not clean the hole and a very strong foam (dry foam) that causes a high annular pressure drop and will not break in the pit. Foam is not "soap suds." Drilling foam is chemically stabilized gas bubbles in a liquid. Proper drilling foam will carry up to 40% drilled solids and yet break quickly at the separator or flowline. Overflow of the foam in the pits is a sign of poor foam control. Good foam starts with good water. Excessive ionic solids in the water will make the foam more expensive and probably less satisfactory than is desired.

19.8.8 Corrosion Control

The potential for corrosion exists any time oxygen is induced into water. The prerequisites for corrosion control are good makeup water and a 9 pH. Any sign of red rust on the drill pipe or red color in the drilling fluid at the flowline is a sign of corrosion and should be corrected.

SUGGESTED READING

Angel, R. R. *Air Drilling Handbook*. Air Drilling Division of MI Drilling Fluids. Gulf Publishing Co., October 1958.

Angel, R. R. 1957. "Volume Requirements for Air or Gas Drilling." *Petroleum Transactions AIME*, Vol. 210.

Bennion, D. B., et al. 1995. "Advances in Laboratory Coreflow Evaluation to Minimize Formation Damage Concerns with Vertical/Horizontal Drilling Applications." Paper No. 95-105, CADE/CAODC Spring Drilling Conference, Calgary, Alberta, Canada, April 19–21.

Bennion, D. B., and Thomas, F. Brent. 1994. "Recent Investigations into Formation Damage in Horizontal Wells During Overbalanced and Underbalanced Drilling and Completion Procedures." Presented at the 6th Annual International Conference on Horizontal Well and Emerging Technologies, Houston, Texas, October 24–26.

Deis, P., et al. 1993. "Infill Drilling in the Mississippian Midale Beds of the Weyburn Field Using Underbalanced Horizontal Drilling Techniques." Paper No. 93-1105, CADE/CAODC Spring Drilling Conference, Calgary, Alberta, Canada, April 14–16.

Deis, P. V., Yurkiw, F. J., and Barrenechea, P. J. 1995. "The Development of an Underbalanced Drilling Process: An Operator's Experience in Western Canada." Presented at the 1st International Underbalanced Drilling Conference and Exhibition, The Hague, Netherlands, October 2–4.

Drillbrief Staff. 1994. "Amoco Canada Drills Horizontal Well with Natural Gas Drilling Fluid." *Drillbrief*, publication of the Amoco Production Company, June.

Eide, E., et al. 1995. "Further Advances in Coiled-Tubing Drilling." SPE 28866. Presented at the European Petroleum Conference held in London, UK, October 25–27.

Falk, K., and Scherschel, S. 1995. "An Update on Underbalanced Drilling Techniques." Paper No. 95-101, CADE/CAODC Spring Drilling Conference, Calgary, Alberta, Canada, April 19–21.

Grace, R. D., Pippin, M. 1989. "Downhole Fires During Air Drilling: Causes and Cures." *World Oil*, May 1989. 42–44.

Graham, R. A. 1995. "Horizontal Underbalanced Drilling with Coiled Tubing (a New Reservoir Management Tool)." Presented at the SPE/CIM 5th Annual One-Day Conference, "Horizontal Wells: Improvement in Horizontal Well Productivity and Profitability," Calgary, Alberta, Canada, November 21.

Guo, B., Miska, S. Z., and Lee, R. 1994. "Volume Requirements for Directional Air Drilling." IADC/SPE 27510. Paper presented at the 1994 IADC/SPE Drilling Conference, Dallas, Texas, February 15–18.

Hook, R. A., and Cooper, L. W. 1977. "Amoco's Experience Gives Right Air Drilling Techniques." *Oil & Gas Journal*, June.

IBD HSE Forum. "IADC Classification System for Under Balanced Wells."

Jones, P. W., Konopczynski, M. R., and Milligan, M. R. 1994. "Stimulation of Horizontal Wells in Shell Canada." Paper No. HWC94-70, preprint of a paper presented at the Canadian SPE/CIM/CANMET International Conference on Recent Advances in Horizontal Well Applications, Calgary, Canada, March 20–23.

Joseph, R. A. 1995. "Underbalanced Horizontal Drilling, Part 1: Planning Lessens Problems, Gets Benefits of Underbalance." *Oil & Gas Journal*, March 20.

Joseph, R. A. 1995. "Underbalanced Horizontal Drilling, Conclusion: Special Techniques and Equipment Reduce Problems." *Oil & Gas Journal*, March 27.

Labat, C. P., Benoit, D. J., and Vining, P. R. 2000. *Underbalance Drilling at Its Limits Brings Life to Old Field.* SPE 62896. Dallas, Texas, October.

Layne, A. W., and Yost, A. B. II. 1994. "Development of Advanced Drilling, Completion and Stimulation Systems for Minimum Formation Damage and Improved Efficiency: A Program Overview." SPE 27353. Presented at the SPE International Symposium on Formation Damage Control, Lafayette, Louisiana, February 7–10.

Lunan, B., and Boote, K. S. 1995. "Underbalance Drilling Techniques Using a Closed System to Control Live Wells—Western Canadian Basin Case Histories." Presented at the 3rd Annual Conference on Emerging Technology, CT-Horizontal, Aberdeen, Scotland, May 31–June 2.

MacDonald, R. 1995. "Winning with Underbalanced Drilling." Paper No. 95-104, CADE/CAODC Spring Drilling Conference, Calgary, Alberta, Canada, April 19–21.

Maclovio, Y. M. 1996. "Tecomin Oaca'n 408, First Underbalanced Drilling Application in Mexico." SPE 35320. IPC Mexico.

Rehm, W. 2002. *Practical Underbalance Drilling and Workover.* PETEX, University of Texas, Austin.

Schlumberger Dowell Staff. 1995. "Underbalanced Drilling with Coiled Tubing." Presented at the 1st International Underbalanced Drilling Conference and Exhibition, The Hague, Netherlands, October 2–4.

Shale, L. T. 1994. "Underbalanced Drilling: Formation Damage Control During High-Angle or Horizontal Drilling." SPE 27351, paper presented at the SPE International Symposium on Formation Damage Control, Lafayette, Louisiana, February 7–10.

Shale, L., and Curry, D. 1993. "Drilling a Horizontal Well Using Air/Foam Techniques." OTC 7355. Paper presented at the 25th Annual Offshore Technology Conference, Houston, Texas, May 3–6.

Tangedahl, M. J. 1995. "Well Control: Issues of Underbalanced Drilling." Presented at the 7th Annual International Conference on Horizontal Well Technologies and Applications, Philip C. Crouse & Associates, Inc., PNEC Division, Houston, Texas, November 6–8.

Taylor, J., McDonald, C., and Fried, S. 1995. "Underbalanced Drilling Total Systems Approach." Presented at the 1st International Underbalanced Drilling Conference and Exhibition, The Hague, Netherlands, October 2–4.

Teichrob, R. R. 1994. "Low-Pressure Reservoir Drilled with Air/N_2 in a Closed System." *Oil & Gas Journal*, March 21.

Wang, Z., et al. 1995. "A Dynamic Underbalanced Drilling Simulator." Presented at the 1st International Underbalanced Drilling Conference and Exhibition, The Hague, Netherlands, October 2–4.

Yee, S., Comeaux, B., and Smith, R. 1995. "Recent Advances in Underbalanced Horizontal Drilling." Presented at the 7th Annual International Conference on Horizontal Well Techologies and Applications, Philip C. Crouse & Associates, Inc., PNEC Division, Houston, Texas, November 6–8.

CHAPTER 20

SMOOTH OPERATIONS

Nace S. Peard
D F Corporation

Wiley Steen
Consultant

Mike Stefanov
BP International

Smooth operations of solids-separation equipment are accomplished with proper planning and utilization of the needed equipment to efficiently and effectively remove drilled solids from a drilling fluid. This chapter provides guidelines to ensure the smooth and efficient operations of solids-removal equipment and associated drilling-fluid equipment. This chapter is primarily directed toward rig personnel as a practical guideline for better drilling practices. Many of these suggestions and guidelines are discussed in much greater detail in other chapters of this book. This chapter consists of three sections: Derrickman's Guidelines and operations of the various solids-removal equipment; Equipment Guidelines and additional thoughts and considerations for smooth operations of the various drilling-fluid handling equipment and tankage; Solids Management Checklist and questions to consider for proper sizing, selection, and operation of a solids-management system.

The results of the smooth operation of a solids-management system are reduced well cost and reduced well problems. So many well problems have been traced back to poor solids removal over the years. Understanding the solids-removal and drilling-fluid handling equipment and the capabilities of this equipment will reduce, if not eliminate, many hole problems.

The additional information and lessons learned in the following sections will help to achieve smooth operations of solids-separation equipment and drilling fluid handling equipment needed for drilling a trouble-free well efficiently and economically.

20.1 DERRICKMAN'S GUIDELINES

This section contains comments directed specifically toward rig hands who control solids-separation operations. The information was first published in the American Association of Drilling Engineers' (AADE) *Shale Shakers and Drilling Fluid Systems* book as a short, succinct guide for rig hands. Appreciation and gratitude is due Wiley Steen and Mike Stefanov for field confirmation of the effectiveness of these suggestions. Effective use of solids-separation equipment is greatly dependent on the treatment the equipment receives on a rig every day and every tour. If some basic principles of solids-separation equipment maintenance and operations are not followed, even the most advanced techniques and high-dollar drilling equipment will be compromised.

All personnel should be thoroughly familiar with the equipment installation, including footprint, wiring pattern, and location of safety equipment. Personnel should always be aware of the potential serious hazards associated with moving equipment and electric currents. Also, to varying degrees, skin, eye, ear, and nose protection is necessary dependent on how much of an irritant the drilling fluid system is.

A carpenter's drill bit carries wood cuttings out of a drilled hole using spirals located just above on the bit itself. In oil or gas drilling, drill string spirals do not move the cuttings ("dirt," or rock drilled by the bit) out of the hole. Drilling fluid brings cuttings to the surface. Cuttings, also termed *drilled solids,* left in the mud create or contribute to many problems. Some are:

- Stuck pipe
- Bad cement jobs
- Lost circulation
- Swabbing in kicks
- Slowed drilling rates
- High mud costs
- Wear on pumps and other equipment
- Shorter bit life and more frequent trips to change bits

Good solids-removal equipment operation, including that of shale shakers, helps prevent these problems.

20.1.1 Benefits of Good Drilled-Solids Separations

- Replacement of pump fluid end parts is reduced, and pumps operate more efficiently.
- Less drill string torque and drag equates to less wear on string and less key-seating (a major potential for stuck pipe).
- Casing is more easily run. Cement jobs are better and require fewer squeeze jobs.
- Bit life is extended due to less abrasion.
- Penetration rates can increase.
- Dilutions to maintain low mud weights are reduced. This manifests not only in reduced drilling-fluid costs, but also in reduced drilling-fluid volumes and thus reduced drilling-fluid waste and waste pit volumes.
- Additions of weighting material are made with little or no difficulty.
- Downhole tools set and release with little or no interference from drilled cuttings.

20.1.2 Tank and Equipment Arrangements

1. Solids-removal equipment is arranged so that larger solids are removed before smaller solids.
2. Each piece of equipment should discharge into the compartment immediately downstream from its suction compartment.
3. Each compartment in the removal section, *except for the sand trap*, should feature backflow from the downstream compartment into the upstream compartment.
4. Except for settling tanks (sand trap), each tank should have adequate agitation.
5. Only one compartment should be used as a settling tank.
6. No solids-removal equipment of the degasser should have a settling pit for a suction pit.
7. All suction pits should be agitated.

Although all of the equipment listed below may not be needed or used, the preferred sequence, with larger-size particles being the first removed, is:
  Unweighted Drilling Fluids

- Gumbo removal
- Scalper shakers
- Main shale shakers

- Desanders
- Desilters
- Centrifuge
- Dewatering system or unit

Weighted Drilling Fluids

- Gumbo removal
- Scalper shakers
- Main shale shakers
- Mud cleaner
- Centrifuge
- Dewatering system or unit

Gumbo removal devices ("gumbo busters") are often fabricated or adapted at rig site.

20.1.3 Shale Shakers

The purpose of a shale shaker is to remove large drilled solids from the drilling fluid. The shale shaker is the first piece of solids-control equipment to treat or condition the drilling fluid. Good shaker performance is necessary if the entire system is to function at or near design efficiency or capability.

Shakers now come in a dazzling assortment of sizes, shapes, and motions. Their performance is controlled by the size(s) and shape(s) of the openings in the screen(s), the drilling-fluid properties, the amount and type of cuttings arriving at the shaker, *and the general mechanical condition of the equipment*. The shaker selected for your rig may or may not be the best for the drilling at hand. Unfortunately, if it is not, it must still be kept operational, and with intelligent, conscientious work perhaps can be made to do the job. All commercial shale shakers, however, remove cuttings—and they remove cuttings better when properly maintained and operated.

Obviously cuttings cannot be removed until the drilling fluid first brings them to the surface. Solids coming off the end of shaker screens should have sharp edges. Cuttings that "roll around" in the borehole on the way to the surface have rounded edges. Rounded edges, or round cuttings, indicate that the cuttings are not being transported directly to the surface as fast, or directly, as they should be. The driller and/or mud engineer should be advised as to the shape of the cuttings coming over the shaker in regard to round edges. Rounded-edge cuttings indicate that there are many drilled cuttings stored in the annulus. This increases the mud weight in the annulus and the pressure at the bottom of the hole. The excess pressure significantly decreases the drilling rate and cuttings removal from beneath the drill bit.

Eight general rules to assure shale shakers will work properly and remove cuttings:

1. The shale shaker should be run continuously while circulating. Cuttings cannot be separated if the shaker bed is not in motion.
2. Fluid should cover most of the screen. If only one quarter or one-third of the screen is covered, the screen is too coarse and should be replaced with a finer screen.
3. If fluid flows through a hole or tear, cuttings are not removed. Any screen with a hole or tear should be replaced immediately. With a panel screen, the hole or tear can be plugged.
4. Shaker screen replacements should be made as quickly as possible. Minimize downtime by planning your work. Locate and arrange tools and screens before starting. If possible, get help. This will decrease the amount of cuttings being kept in the mud because the shaker is not running. If possible, change screens during a connection. In critical situations, drilling may be interrupted and the pumps stopped while the screen is replaced.
5. Dilution fluid (water or oil) should not be added in the possum belly or on the shaker screen. Dilution fluid should be added downstream. Dilution-fluid (even water) additions should be metered or otherwise measured.
6. Except for cases of lost circulation (when it is necessary to retain lost circulation material), the shaker should not be bypassed, not even for a short time.
7. Large cuttings should be removed from the possum belly when mud is not being circulated. If the possum belly is dumped into the sand trap just before making a bit or wiper trip, the sand trap should also be cleaned. Otherwise, when fluid circulation starts after a trip, the large cuttings dumped into the sand trap will likely move down the pit system and plug desilters or desanders. *Note*: The possum belly and/or sand trap is not always used with synthetic-based mud or some specialized fluid systems.
8. As much as possible, flow from the well (bell nipple) should be evenly distributed among all the shakers.

20.1.4 Things to Check When Going on Tour

1. Make sure shale shaker is running properly.
2. Listen for bad bearings or motor imbalance. The vibration should be smooth and uniform.
3. Are screens installed per manufacturer's specifications? Is tension correct? Retension continuous cloth screens periodically.
4. Check for holes in screens. Wash screens during a connection for this examination. Sometimes a flashlight is needed to examine the entire screening surface. *Note*: Frequently holes appear where the overflow from the possum belly strikes the screens. It is here that the highest concentration of cuttings strikes the screens.
5. Replace screens whenever necessary. Holes in panel screens may be plugged or blanked for possible reuse.
6. Check for (sharp) cuttings, not drilling fluid, going off end of screens. If not the case, inform the driller and/or mud engineer.
7. Ensure that the fluid flow over shakers is uniform across all, or as close to uniform as possible. This may entail modifying the flowline or possum belly, including adjustment of gates on the shakers themselves.
8. Retension screens if particles cake up to form patties. This could also indicate a cracked screen frame that absorbs the acceleration force that moves solids down the screen.
9. Review the mud weights and funnel viscosity coming out the hole for the previous tour. These measurements should be made on a regular and *consistent* basis so that any gradual changes caused especially by downhole conditions can be easily detected. (For consistency, common sense applies. Samples should be taken from the same place every tour, preferably from the possum belly or flowline and from the suction tank. For weighing, the same mud balance should be used. If mud balance is adjusted, adjustment should be noted on a weight-vis chart.)
10. Check that the gas and flow sensors are properly positioned in the possum belly but do not move them. Notify the mud logger or driller if they need adjustment.

20.1.5 Sand Trap

When API 80 or coarser screens are used, the sand trap performs a valuable function as large sand-sized particles settle in it. Settled cuttings are regularly dumped overboard or to a waste pit.

The sand trap should *overflow* into the next compartment downstream. This provides an opportunity for solids to settle in the sand trap. Sand traps are very effective when water is used as the drilling-fluid makeup base in a simple system. But when a drilling-fluid system is treated to provide a large value of low-shear-rate viscosity, or high yield points, solids will not settle quickly, and sand traps are much less effective. When the shale shaker screen has large openings, many solids may settle in the sand trap; when very small openings are present on the screen, a sand trap may not be very useful.

20.1.6 Degasser

Drilling fluids will encounter gas from formations penetrated in petroleum prospective areas. Formation gases are partially dissolved under the influence of the high pressure exerted by the drilling-fluid column at depth. Also at depth, gases not in solution are compressed to occupy very small volumes. Nearing the surface, the pressure in the fluid column reduces, and gases both evolve from solution and expand to larger volumes. At the surface, these gases must be removed; otherwise, centrifugal pump operations become erratic and inefficient. In severe cases, fire hazards exist.

Neither air nor gas is effectively removed from a viscous drilling fluid by flowing through a shale shaker screen. After passing through the shale shaker and sand trap, all drilling fluid should be directed through a degasser. Either one of two types—atmospheric and vacuum—are found on most rigs. A ("poor boy") degasser, or mud-gas separator, is also installed on most rigs to remove gas from kicks. Atmospheric degassers usually sit on top of the degasser tank. A submerged pump conveys gas-cut fluid through a disc valve into a chamber where the spray formed collides at high velocity with the inner wall, driving entrapped gas out of the drilling fluid. Gas is discharged to the atmosphere at pit level, and drilling fluid is discharged into the next tank. Atmospheric degassers should discharge horizontally across the top of the subsequent tank allowing breakout of large bubbles. Their use is somewhat restricted to low-weight, low-yield point fluids.

Vacuum degassers subject thin or shallow streams of drilling fluid to a low pressure, under which the gas expands greatly to be more easily separated. The gas is drawn off by a vacuum pump and the degassed fluid pumped out through an eductor jet, which also draws additional fluid into the chamber. Vacuum degassers should discharge into a downstream tank below the fluid surface. The discharge line should turn so discharge will be directed upward to the fluid surface.

The degasser should process more drilling fluid than enters its suction compartment from the sand trap or shale shaker. This will cause some of the drilling fluid to backflow from the downstream compartment into the degasser suction compartment. The degasser suction compartment tank should equalize *overflow* from the next compartment downstream. Weigh the drilling fluid into and out of the degasser to determine whether gas is being removed. The mud weight should be compared mud weight, determined with a pressurized mud balance. "Clean mud" should be circulated through the degasser on a routine basis to avoid clogging by settled solids.

Gas exhaust piping should terminate in a nonhazardous area a safe distance from wellhead and surface pits. The exhaust line should be valved and branch in two separate directions so that gas can always be flared downwind of surface pits and wellhead.

Note: Some toxic gases, hydrogen sulfide for example, are heavier than air and will accumulate in low areas. Caution should always be exercised when approaching or working near a gas-line discharge. Also, because it is heavier than air and will settle, simply discharging hydrogen sulfide above a drilling rig (at the top of a derrick, for example) does not provide sufficient degree of safety.

20.1.7 Hydrocyclones

Hydrocyclones are simple, easily maintained mechanical devices without moving parts.

Many of the solids generated while drilling reach the surface too small for separation by shale shakers. Hydrocyclones are relied upon to separate the majority of these finer solids. Hydrocyclone units are designed to separate low-gravity solids larger than about 15 to 20 microns from an unweighted fluid system. Upstream the shale shakers remove larger particles that might cause cone plugging.

Hydrocyclones are arranged with the larger cone size unit upstream of the smaller cone size unit. The desanders function primarily to decrease the solids loading of the smaller, 4-inch desilter cones. Generally a desander size and desilter size are available as part of the rig equipment. A separate tank and centrifugal pump are needed for each size unit.

Suction into the hydrocyclone unit is taken from the tank compartment immediately upstream of the tank receiving discharge. The number of cones in use should process 100% of the flow rate of all fluids entering the suction tank of the hydrocyclones *plus* at least 100 gpm. This ensures adequate fluid processing. The suction and discharge tanks are always equalized at bottom.

A pressure gauge should always be on the inlet manifold to determine feed head (pressure) supplied by the centrifugal pump. Other than cone or manifold plugging, improperly sized or operated centrifugal pumps are by far the greatest source of problems encountered with hydrocyclones.

Most hydrocyclones, regardless of size (inner diameter at inlet), are designed to best operate at 75 feet of fluid head. Some new cones on the market may require a different head. Be certain to check with the manufacturer to determine the centrifugal pump impeller size needed to provide the proper head at the hydrocyclone manifold. A 75-foot head is equivalent to a pressure of 32.5 pounds per square inch for a 1.0 specific gravity fluid (or water). See the following chart, where head = 75 feet.

Pressure (psig)	Fluid Density or Mud Weight (ppg)
32.5	8.34
35	9.0
37	9.5
39	10.0
41	10.5
43	11.0
45	11.5
47	12.0
49	12.5
51	13.0
53	13.5
55	14.0
57	14.5
58.5	15.0
60	15.5
62.5	16.0

Remember that a centrifugal pump creates a constant head independent of mud weight. As the mud weight goes up, the centrifugal pump will maintain the same head, but the pressure will automatically increase. All feed lines should be as straight and as short as possible with a minimum of pipe fittings, turns, and elevation changes. Pipe diameters should be 6 or 8 inches for reduced friction losses and less solids settling.

A centrifugal pump should operate *only one unit* of the solids-removal system; for example, either the mud cleaner *or* the desilter *or* the desander, *not both* desilter and desander.

Cones should exhibit a spray discharge with a central air suction at apex. If cones do not operate in spray discharge,

1. too many solids are being presented, and either more efficient upstream separation or additional hydrocyclones are required. As a first step, thoroughly check shaker screens for possible tears;
2. solids have plugged the manifold or apex; or
3. feed pressure is below 75 feet of head. See the preceding chart.

If desilters are a significant distance above the liquid level in the mud tanks, a vacuum breaker should be installed on top of the discharge line. The vacuum breaker could be a simple 1-inch-diameter pipe about 12 inches or longer welded vertically onto the discharge manifold.

Hydrocyclones discard absorbed liquid with drilled solids. Solids dryness is a function of cone geometry—apex opening relative to diameter of vortex finder. Mud cleaners and/or centrifuges can process cone underflow for increased dryness. The discharge line should be above the fluid surface in the receiving tank to avoid creation of a vacuum.

Guidelines for Effective Hydrocyclone Use

1. Check cones regularly to ensure that the apex is not plugged and operation is in spray discharge.
2. Ensure that sufficient head (75 feet) is available at hydrocyclone inlet.
3. Process the total surface pit volume at least once during any bit trip.
4. When the shale shaker will not screen down below 100 microns (API 140), a desander should be used upstream of the desilter.
5. Between wells, or when drilling is interrupted, manifolds should be flushed with a fluid compatible with the drilling-fluid system, and cone internals examined for wear.

Most hydrocyclones are designed to be balanced. A properly adjusted, balanced hydrocyclone has a spray discharge to the underflow outlet and exhibits a central air suction core. A cone is balanced by pumping water through the cone at the appropriate head and adjusting the bottom opening so only a small amount of water exits the cone.

To set a cone to balance, slowly open (widen) the apex discharge while circulating water through the cone at 75 feet head (32.5 psig). When a small amount of water is discharged, the center air core is almost the same diameter as the opening, and the cone is balanced. When coarse solids are in the feed slurry, wet solids are, by design, discharged at the apex.

The discharge pattern changes from spray to "rope" when too many solids are present in the feed slurry for efficient separations. This is characterized by a slow-moving cylindrical discharge that resembles a rope. Even though rope discharge stream density will be greater than spray discharge stream density, solids separations are actually far less efficient, and spray discharge should be immediately restored.

20.1.8 Hydrocyclone Troubleshooting

Symptom	Probable Cause(s) and/or Action
Some cones continually plug at apex	Partially plugged feed inlet or discharge; remove cone and clean out lines. Check shaker for torn screens or bypassing. Possibly increase apex size. Stop dumping the possum belly into the sand trap, particularly before a trip.
Some cones losing whole mud in a stream	Plugged cone feed inlet allowing backflow from overflow manifold.
Low feed head	Check centrifugal pump operation—rpm, voltage, etc. Check for line obstructions, solids settling, partially closed valve. Check shaker for torn screens or bypassing.
Cones discharge small amounts of solids, little fluid	Increase apex size and/or install more cones.
Vacuum in manifold discharge	Install anti-siphon tube (vacuum breaker), then check for proper feed head.
Increasing solids concentration in drilling fluid	Insufficient cone capacity—more cones and/or smaller cones; solids may be too small, use finer shaker screens. Check for holes in shaker screens. Look for shaker bypasses. Check tank plumbing. Check suction and discharge line locations.
Heavy discharge stream	Overloaded cones; increase apex size and/or install additional cones.
High drilling-fluid losses	Cone apex too large, reduce discharge opening size (diameter). Reduce cone sizes.

(*continued*)

Symptom	Probable Cause(s) and/or Action
Unsteady cone discharge, varying feed head	Air or gas in feed line. Is degasser working?
Aerated mud downstream of hydrocyclone	Route overflow into trough to allow air breakout. Is overflow being discharged too deep in receiving tank?

Plugging

Obviously a plugged hydrocyclone cannot process (separate solids) drilling fluid, and maintenance of a low-level drilled-solids concentration is then not possible. On a drilling rig, someone should be assigned the task of unplugging cones. Cones plug when there are too many solids to be separated. This can happen when penetration rate is very fast, over 100 feet per hour, and more or larger hydrocyclones are needed.

Plugging also often exhibits when the shale shaker possum belly is emptied into the sand trap during a bit or wiper trip. Shortly after resumption of circulation, the newly dumped but yet not totally settled concentration of large solids moves down the system. On reaching the apex of the desander and/or desilter, hydrocyclone plugging can occur.

20.1.9 Mud Cleaners

The principle use of mud cleaners has always been the removal of drilled solids larger than barite. The mud cleaner is a combination hydrocyclone and fine shaker screen. Hydrocyclone underflow containing a concentration of solids and drilling fluid is sieved through an API 200 or API 150 screen. Because barite is ground so that the majority is smaller than 74 microns (API 200), most barite should pass through a mud cleaner screen. Most of the weight material passes through the screen and is returned to the active drilling fluid system. Solids—a high concentration of drilled solids—are discarded off the screen. Some barite will be discarded and some drilled solids will be retained in the drilling fluid. Solids removed will decrease mud weight.

Mud cleaners are also used to retain expensive liquid phases, even in unweighted drilling fluids such as synthetic or KCl muds. The mud cleaner concept is also frequently used in microtunneling in that manner. It removes solids from water used in tunneling under roads, etc., when water acquisition or disposal is difficult or expensive. *Note*: Even when linear motion or balanced elliptical motion shale shakers are properly operated with API 200 screens, mud cleaners have been found to remove significant quantities of drilled solids.

Mud cleaners are normally located on the surface tanks in the same position as desilters. They are frequently the same piece of equipment, with only the hydrocyclones being used while the drilling fluid is unweighted. When weighting material is added, screens are placed on the mud cleaners: "Barite in, screens on."

Newer rigs, especially those found offshore, are outfitted with several linear or elliptical motion shale shakers. After surface and/or intermediate casing strings have been set, fewer shale shakers are needed to handle the correspondingly lower flow rates for smaller hole sizes. Some rigs have been modified with as many as twenty 4-inch hydrocyclones mounted above one of the main shale shakers. A valve is installed on the flowline to convert a main shaker into a mud cleaner. When drilling smaller-hole sizes, this shaker is taken out of primary service, and with proper plumbing the arrangement can function as a mud cleaner.

When drilling fluid is initially passed through a mud cleaner, the mud weight decreases, and more barite is needed to maintain density. Remember, the principle use of mud cleaners has always been the removal of drill solids larger than barite. When solids—either barite or drilled solids—are removed, mud weight decreases, and to maintain fluid density, some barite must be added to compensate for the discarded drill solids. Overall, any removal of solids larger than 74 microns is beneficial to drilling a trouble free hole. "Bad solids out, barite in."

If fluid passes too quickly through the mud cleaner screen and the materials on the screen become too dry, solid particle separations will be inefficient. Barite will clump together with other solids and be carried over to discard. A light spray of drilling fluid (taken from the desilter overflow) onto the mud cleaner screens promotes efficient separations and prevents discard of too much barite.

Direct screen underflow discharge into a well-agitated section of the surface tanks so weight material will not settle.

The mud cleaner is meant to continually process drilling fluid just like the main shale shakers. The mud cleaner screen removes larger drilled-solid particles from the system. When mud cleaners—or hydrocyclones—are operated only part of the time, solids remain in the drilling fluid system. Particularly on passing through drill-bit jet nozzles, particles degrade into smaller size. Smaller-size particles are progressively more difficult to remove and more damaging to the drilling fluid system. A centrifuge (see the following section) can remove some of these smaller particles, but centrifuges do not process all of the rig flow. The mud cleaner is effective in removing solids before they grind to smaller size, and should be used whenever circulating.

20.1.10 Centrifuges

A decanting centrifuge is a tool used in viscosity control and it is also solids-removal equipment. It is a machine with an internal cylinder, or bowl, that rotates at high speeds (± 1800 rpm). As drilling fluid is pumped into and conveyed through the rotating bowl, it is subjected to a large centrifugal force that increases the separation or settling rate of the suspended solids. The higher-mass particles separate fastest along the wall of the rotating bowl.

An important distinction is that while screens and hydrocyclones are used to separate and discard larger drilled-solids particles, centrifuges are used to separate and discard smaller, ultra-fine particles, both drilled solids and barite.

Why use a centrifuge? Short answer: To remove drilled solids.

During drilling operations, drilled solids accumulate and degrade in size, which causes viscosities and gel strengths to increase, especially in weighted drilling fluids. These fluids necessarily contain concentrations of weight material solids. A centrifuge will separate ultra-fine (less than 2 microns barite and 3 microns drilled solids) from larger, desirable-size barite particles and drilled solids, which, while undesirable, are not as harmful to fluid flow and wall building properties. Expensive treatment additives are also discarded with the ultra-fines stream. However, most chemical treatment expense is directed to combat actions of ultra-fine drilled solids. While some additions will be necessary, the ultimate result is lessened additive and barite usage, and the removal of drilled solids will be beneficial.

Operating Reminders

1. Before startup, rotate bowl or cylinder by hand to make sure it rotates freely.
2. Start centrifuge first, before starting the drilling-fluid feed pump or dilution water.
3. Set drilling-fluid mud and dilution rates according to manufacturer's recommendations, which usually vary according to mud weight.

4. Shutdown:

- Turn off drilling fluid feed
- Turn off dilution water
- Turn off machine.

Reduced viscosity of the fluid within the bowl with dilution enhances separation. Dilution fluid should be added sufficient to maintain an effluent with a funnel viscosity between 35 and 37 seconds.

While centrifuging a water-based system, beneficial bentonite will be discarded along with drilled solids and must be replaced. Usually one or two sacks of bentonite per hour of centrifuge operation over and above normal treatment levels will benefit especially filtration and wall building characteristics. Centrifuge adjustments, other than varying dilution fluid rate, are seldom needed and should be left to a centrifuge mechanic.

Two final notes about solids control and weighted drilling fluids.

1. Ultra-fine particles, both barite and drilled solids, that increase viscosities and gel rates are discarded by a centrifuge. The more valuable larger-size weighting material is retained. Without centrifuging, reduction of ultra-fine concentration would require discard of whole mud, which includes valuable weight material.
2. Mud cleaners do not compete with centrifuges. The mud cleaner and centrifuge are complementary to each other—not competitive with each other. The mud cleaner removes particles larger than barite, the centrifuge removes ultra-fine particles smaller than most barite. These very small particles with much greater relative surface area can cause dramatic viscosity increases.

20.1.11 Piping to Materials Additions (Mixing) Section

The equalizing line leading from the drilled-solids removal section into the materials additions, or mud makeup pit, compartment should be through an L-shaped pipe that allows the discharge end to be raised and lowered. Normally the discharge end will be raised or lowered so that the removal section maintains a constant fluid level.

Pipe diameter should be 8 to 10 inches to provide sufficient flow rate and retard solids settling in the line. Pipes that are too large will plug with settled solids until the drilling-fluid speed is sufficient to keep the pipe open. Any drilling fluid hauled to the location from another source, or drilling fluid from a reserve pit that is added to the system, should be added through the shale shaker.

20.2 EQUIPMENT GUIDELINES

This section provides additional thoughts and considerations concerning solids-removal equipment. The practical operational guidelines for equipment discussed here may not apply to all drilling applications. These guidelines (in italics) were developed as part of API RP 13C. The discussion beneath each captures some of the comments by committee members as they debated the guideline before approval.

20.2.1 Surface Systems

1. *The surface system should be divided into three sections each having a distinct function: removal section, additions section, and check/suction section. Undesirable drilled solids should be removed in the removal section. All mud material and liquid additions should be made in the additions section. The check/suction section provides volume for blending of new mud materials and verification of desired mud properties.*

    This is a simple concept that requires each surface system to have three easily identifiable sections. If not, changes will quickly pay for themselves.

2. *Minimum recommended "usable" surface mud volume is 100 barrels (less for slim holes) plus enough to fill the hole when the largest drill string the rig can handle is pulled wet and all the mud inside the string is lost. In order to maintain fluid properties in large diameter, soft, fast-drilling holes, the minimum surface volume should be at least five or six times the volume of the hole drilled per day.*

    For safety reasons a rig must have enough drilling fluid to fill the hole at all times. This is the situation described above. A second consideration is not a safety feature but a recognition of practicality. When rapidly drilling a large-diameter hole and removing drilled solids from the system, new fluid must be built quickly. The volume of new fluid is the sum of the solids removed and the drilling fluid clinging to them. Mixing equipment on rigs is not usually geared to rapid additions of drilling-fluid products. Drilling operations experience fewer problems if the drilling fluid properties are controlled, which is difficult with large, rapid additions of drilling fluid products.

3. *All removal compartments, except the sand trap, should be well stirred or agitated to ensure even loading of solids-removal equipment.*

   Solids-control equipment works best when the solids loading remains constant. Slugs of large quantity of solids tend to plug the lower discharge opening in desilters. When this occurs, drilled solids will not be removed until the plugged cones are cleaned. On a rig, even with diligent crews, some cones will usually remain plugged, which leads to increased drilled solids in the drilling fluid.

4. *The ideal tank depth would be approximately equal to the width, or the diameter, the tanks. If deeper, special considerations may be necessary for stirring; if shallower, adequate stirring without vortexing will be difficult or impossible.*

   Baffles will help prevent vortexing. When using vertical blade stirrers in circular tanks, baffles are a necessity.

5. *Use top equalization for the sand trap.*

   To take advantage of the maximum settling time, fluid should enter the upstream end of the sand trap. After the fluid moves through the compartment, it exits through an overflow weir into the next compartment. An underflow arrangement would carry settled solids into the next compartment.

   The ability to use API 200 screens on shale shakers means most of the sand-sized particles are removed and few drilled solids are available to settle. (API defines anything larger than 74 microns, or API 200, as sand.) With the introduction of linear motion shale shakers (using API 200 screens) and the emphasis on minimizing rig discharges, fewer rigs are including sand traps in their removal systems. Many rigs place their desander pump suctions in the former sand traps. This requires that these compartments be well agitated.

6. *Use top equalization between the degasser suction and discharge compartments.*

   The degasser should process more fluid than is entering its suction compartment. This will create a backflow from the discharge compartment into the suction compartment. Only processed drilling fluid can flow over the top weir. Any drilling fluid still containing gas will overflow over the top weir back to the degasser suction compartment.

7. *Use bottom equalization between the suction and discharge compartments of desanders, desilters, mud cleaners, and centrifuges.*

Openings in the partitions between the suction and discharges are needed primarily to allow a small backflow. Since the level can vary in these compartments, an underflow is needed so that no adjustment is necessary while drilling. Also, an overflow tends to entrain air as the fluid cascades into the upstream tank.

8. *Use an adjustable equalizer between the removal and additions sections when cyclones and/or centrifuges are being used. Run with the high position on the downstream side.*

    An adjustable equalizer is usually a curved pipe that can swivel in the equalizing line. The upper end is in the downstream compartment. Fluid exits the bottom of the removal section and flows through the adjustable equalizer. The downstream end of the equalizer is elevated so that the liquid level in the removal section remains constant.

9. *Use bottom equalization in the additions section and in the check/ suction section.*

    This will allow the maximum use of the fluid contained in these sections.

10. *For removal devices processing flow rates greater than the rig circulating rate, equalizing flows should always be in the reverse (or upstream) direction.*

    A centrifuge usually processes only a small portion of the total rig flow. Backflow should not be expected between the centrifuge suction and discharge compartments. All other removal equipment (degasser, desander, desilter, and mud cleaner) should process all of the fluid entering the suction compartment. This may exceed the rig flow if drilling fluid enters upstream from another process or from mud guns.

11. *Based on experience, a rule of thumb for the minimum square feet of horizontal area for a compartment is as follows:*

    horizontal surface area (sq ft) = max circulating rate (gpm)/40.

    It has been found from experience that this rule of thumb provides fluid velocities low enough to allow entrained air bubbles to rise to the surface and break out. Note: This rule-of-thumb was developed by George Ormsby and was included in the IADC Mud Equipment Manual, Handbook 2: *"Mud System Arrangements," pp. 2–17.*

    This is strictly an empirical guideline. Following this guideline usually results in most of the air leaving the drilling fluid, although many types of drilling fluids (polymer, synthetics, relaxed fluid-loss oil muds, mineral oil muds, etc.) were not in existence when this rule was developed. Since the air breakout is a function of the interfacial tension and the low-shear-rate viscosity of the fluid, this formula should be used with caution. This is also the reason that all tanks should be adequately stirred. The entrained air must be brought to the surface so that it can leave the system.

12. *Mechanical stirrers are preferred for stirring removal compartments.*

    All fluid entering a suction compartment should be processed. Mud guns suctioning from a downstream compartment will increase the quantity of fluid that must be processed. Mechanical stirrers eliminate this problem. Mud guns can be used in removal compartments if each centrifugal pump stirs its own suction.

13. *Mechanical stirrers should be properly sized and installed according to manufacturer's recommendations.*

    Mechanical stirrers are available in two stirring blade shapes. One type is canted to actually pump fluid vertically downward. The second type has vertical blades, similar to the impeller blades in a centrifugal pump, which propel the fluid outward. Both types are designed to move all of the fluid within a certain volume. The turnover rate (TOR) depends on how much fluid is moving. The TOR must be large enough to adequately blend the fluid within the compartment. TOR is calculated as 60 times the tank volume divided by displacement. Displacement or flow rate associated with each type of blade, based on projected area of blade, is available from manufacturers.

14. *Baffles may be installed around each mechanical stirrer to prevent air vortices and settling in corners. A typical baffle can be 1 inch thick by 12 inches wide and extend from the tank bottom to 6 inches above the top agitator blade. Four baffles are installed around each agitator. They are installed 6 inches past the tips of the agitator blades, along lines connecting the center of the agitator blade with the four actual corners of a square pit or compartment. For a long rectangular pit, with two or more agitators, the tank is divided into imaginary square compartments and a baffle is pointed at each corner (either actual or imaginary).*

    The purpose of installing baffles is to prevent the drilling fluid from swirling in a manner that creates a vortex, which pulls air into the drilling fluid. The baffles can be created in a variety of shapes and positions and still function properly.

15. *Mud guns should not be used in the removal section except where the feed mud to the mud gun(s) comes from the compartment being stirred by the mud gun(s).*

See item 12.
16. *Mud guns can be used in the additions and check/suction sections of the surface system and provide the benefits of shear and dividing and reblending newly added mud materials.*

    Mud guns do an excellent job of shearing new material, enabling it to disperse, and are effective at blending new material into the surface system. Agitators may aid and assist mud guns in these activities. The flow rate through a mud gun can be calculated from the following equation:

    $$\text{flow rate, gpm} = 19.4(D)^2\sqrt{H}$$

    where D = nozzle diameter, in.; H = head, ft.

    For example, if a 1-inch swedge is attached to the end of a mud gun line and 85 feet of head is applied, the flow rate would be = $19.4 \times 1 \text{ sq.in.} \times \sqrt{85 \text{ ft}} = 179$ gpm.

17. *The sand trap is the only settling compartment in the surface mud system. It should not be stirred, nor should any pump take its suction from the sand trap.*

    Sand traps are becoming obsolete except for fast drilling surface holes, where seawater is used as a drilling fluid and coarse mesh screens are used on the shale shakers.

18. *If a sand trap is used, the bottom should slope to its outlet at 45° or steeper. The outlet valve should be large, nonplugging, and quick opening and closing.*

    The bottom slope is needed so that the settled solids can be easily removed from the compartment. The quick opening and closing valve is needed to reduce drilling fluid loss. The valve is normally shaped like a plate or rectangular piece of metal.

19. *The degasser (if needed) should be installed immediately downstream of the shaker and upstream of any piece of equipment requiring feed from a centrifugal pump.*

    Degassers are used to remove entrained hydrocarbon gasses from drilling fluid. Another benefit of degassers is to prevent air or gas from entering centrifugal pumps, where even small quantities significantly reduce pump effectiveness. As liquid (or drilling fluid) is thrust to the outside of the impeller chamber, air or gas collects at the center. Eventually, enough air or gas will collect to completely block the suction and no liquid will be able to enter the pump.

20. *The solids-removal equipment should be arranged sequentially so that each piece of equipment removes successively finer solids. Although every piece of equipment may not be used or needed, general arrangements are as follows:*

Unweighted Mud	Weighted Mud
Gumbo remover	Gumbo remover
Shale shaker	Shale shaker
Degasser	Degasser
Desilter	Centrifuge
Centrifuge	
Dewatering units	

Including the degasser in the unweighted mud list was not agreed upon by the entire committee. Influx of gas into the drilling fluid normally requires the addition of a weighting agent to the drilling fluid. Very few unweighted drilling fluids are degassed.

21. *The overflow for each piece of solids-control equipment should discharge to the compartment downstream from the suction compartment for that piece of equipment. This is termed proper piping, plumbing, or fluid routing.*

    The compartments do not need to be large. Simple partitions in a larger tank can frequently be added to improper systems to improve their performance. Since a backflow is desired between compartments, the partitions do not require a complete seal. A tank can be divided into several compartments by building a simple wall of boards—as long as each compartment is agitated.

22. *Improper fluid routing always leads to solids-laden fluid bypassing the removal device.*

    Unfortunately, improper routing is all too common on drilling rigs. In the 1980s, approximately 90% of the rigs had flaws in their removal system, and the situation did not improve much in the 1990s. Poor tank arrangements, especially on jackup rigs, cost more money than almost any other problem associated with the drilling fluid system.

23. *Two different pieces of solids equipment should not simultaneously operate out of the same suction compartment. Note: Different means, for example, degasser and desander or desander and desilter.*

    If a desander and a desilter take suction from the same compartment and then discharge into the next compartment, some fluid will be desanded and some will be desilted. This is referred to as connecting the equipment in parallel. Desanders are usually used to decrease solids loading in the desilters. If these two pieces of equipment operate in parallel and a significant amount of solids are being processed, the desilters will probably plug.

In some poorly designed systems, where an insufficient number of compartments are available, the degasser and the desander may be connected in parallel. This assumes that only the degasser, and not the desander, will be used on weighted drilling fluid. Obviously, if they are connected in this manner (parallel), both cannot operate simultaneously.

24. *If two of the same piece of solids-control equipment are used simultaneously, the same suction and discharge compartment should be used for both. Example: If two desilter units are used, both should be properly rigged up and have the same suction and discharge compartments.*

This situation is frequently encountered with desilter banks. In the upper part of a borehole, many hydrocyclones are needed to handle the volume (based on 50 gpm per 4-inch hydrocyclone, 1200 gpm will require 24 cones). While some of these cones can be mounted on mud cleaners (the screen is blanked to discard all underflow), all cones should use the same suction compartment and discharge downstream to the next compartment.

Another problem can arise when a separate rig pump is used to increase annular velocity in risers. The additional flow rate onto the shale shakers may require adding additional units in parallel. The degasser capacity requirements may demand additional degassers be added to the mud tank system. These degassers should also be connected in parallel

25. *The degassers, desanders, desilters, and mud cleaners should process 100% of the mud entering their individual suction compartments. In a properly designed system, the processing rate should be at least 10–25% more than the rig circulating rate.*

The intent of this rule is to provide some guideline so that a backflow will exist. As long as there is more fluid processed than is entering the compartment, except for the backflow, all of the fluid entering the compartment will be processed. If a desilter overflow, or cleaned drilling fluid, is returned to an upstream compartment or back into its own suction compartment, less than 50% of the drilling fluid coming from the well will be processed.

26. *If rules 21, 24, and 25 are applicable and are followed, equalizing flow between compartments will be in the reverse (or upstream) direction. Backflow confirms that all mud entering the compartment is being processed.*

    In normal drilling operations rules 21, 24, and 25 are applicable.

27. *Mud should never be pumped from one removal compartment to another except through solids-removal equipment.*

    The intent of this rule is to ensure that all drilling fluid is processed in an orderly fashion. Drilling fluid should not bypass any solids-removal equipment or be pumped upstream from a suction compartment.

28. *Mud should never enter any removal compartment from outside the removal section to feed mud guns, mixers, or the eductor jet of a vacuum degasser.*

    When a system is designed, the equipment is set up to treat a certain quantity of drilling fluid. Increasing the flow rate demand by returning drilling fluid from downstream usually results in an inefficient removal system.

29. *The power mud of the eductor jet for a vacuum degasser should come from the degasser discharge compartment.*

    The drilling fluid to pull the fluid from the vacuum degasser usually comes form a centrifugal pump. This fluid must be degassed. The fluid in the degasser discharge compartment is pumped through the eductor and returns to the same compartment. This will not interfere with solids-removal efficiency.

30. *Single-purpose pumps are necessary in the removal section to ensure proper fluid routing. One suction and one discharge should be used. Suction and discharges should not be manifolded.*

    Manifolding ruins more good drilling fluid systems than just about any other single design. For this reason, dedicated pumps should be used. Since multiple leaky valves will confuse proper fluid routing, a standby pump should be purchased instead of valves. When the desilter pump is on, the system is being desilted and there should be no question concerning routing. All systems, whether dispersed, non-dispersed, polymer, water-based, oil-based, synthetic, or saltwater, will need the same sequential treatment by the removal equipment.

31. *In a properly designed system, solids-control devices should not overflow into mud ditches.*

    Many drilling fluid systems have a square channel (approximately 2×2 ft) along the top of one side of the tanks. Metal plate openings (ditches) are provided so that drilling fluid can bypass compartments. As drilling fluid drops from a ditch opening into the top of the fluid in the tank, air is entrained. Theses ditches can be useful for completion fluids and other specialty conditions; however, their use in drilling operations may result in drilling fluid bypassing solids removal equipment.

32. *Exception to rule 31: Based on field experience, mud foaming problems can be reduced by routing the overflow of a desander or desilter into a mud ditch for a horizontal distance of about 10 feet before the fluid enters its discharge compartment to allow entrained air to break out. If this is done, ensure that the fluid routing is correct.*

    Flowing drilling fluid down a ditch allows the fluid/air interface to expose more of the entrained air to the surface. Sometimes, however, air does not breakout in this distance and more air is entrained as the fluid drops into the active system. Obviously, experts have many differing opinions concerning ditches.

33. *All mud material additions should be made after the removal section. All removal, including all centrifuging, must be finished before the mud material addition begins.*

    This rule is a result of problems incurred when adding a weighting agent upstream from mud cleaners (or pumping fresh weighting agent upstream through mud guns). API barite allows 3% weight larger than API 200, or 75 microns, most of which will be removed by a mud cleaner. When mud cleaners were first introduced, they had to be turned off during weight-up to prevent discarding too much of the newly added barite. This happens because the fresh drilling fluid was pumped back upstream to the removal section before it went downhole, or the mud cleaner was located downstream from the additions tank. All undesirables must be removed before new products are added. One exception to this rule is the addition of flocculants, or other materials, to aid removal of drilled solids.

34. *Mud foaming problems can also be reduced by using a non–air-entraining mud mixing hopper. Jet and venturi hoppers suck air into the mud during mixing.*

    Hoppers should be turned off when they are not being used. At the discharge end of the additions line, an inexpensive air removal cylinder can be added without creating much backpressure. A welder can fabricate it from a piece of 13 3/8-inch to 20-inch casing approximately 1 1/2 feet tall, welded vertically to the end of the hopper discharge line. A plate with an 8- to 10-inch diameter hole is welded on the top of the casing. Fluid enters tangentially and is swirled as it encounters the piece of casing. This swirling action causes drilling fluid to move to the outside wall, and the air moves to the inside. This acts as a centrifugal separator. Air exits through the hole at the top and the drilling fluid drops freely into the pits.

35. *Jet hoppers should include venturi for better mixing.*

    A venturi is needed if the flow line rises to an elevated position. The device converts a velocity head to a pressure head. Without it, fluid does not have enough pressure to rise over the tank wall.

36. *The check/suction section of the surface system should contain a 20–50 barrel slugging tank, which includes a mud gun system for stirring and mixing.*

    An agitator may be used in addition to the mud gun. The mud gun system can be connected to the pump that is used to fill the slug tank. Usually the slug tank is used to prepare a drilling fluid with a higher density. This "slug" is pumped into the drill string. When tripping drill pipe, the fluid level inside the drill pipe will remain below the surface. This prevents spilling drilling fluid when a stand is removed from the drill string. Failure to slug the pipe, or get a good "slug," results in drilling fluid splashing the rig crew as the pipe is pulled and racked.

37. *Mud premix systems should be used on any mud system whose additives require time and shear for proper mixing. Premix systems should especially be used on systems requiring the addition of bentonite, or hard-to-mix polymers, such as CMC, PHPA, XC, etc. Do not add dry bentonite to a drilling fluid.*

    To be effective, bentonite must be prehydrated and dispersed into platelets as small as possible. It should be added to a well-agitated tank of freshwater. No other additives are required. The addition of lignosulfonate will inhibit dispersion as it thins the slurry. Bentonite should be allowed to hydrate for 24 hours (8 hours minimum). Polymers, such as HP007, require many hours of prehydration and shear before use.

38. *Special shear and mixing devices are recommended for premix systems for mixing polymers (especially PHPA, spotting fluids, specialized coring fluids, and for hydrating bentonite).*

    Centrifugal pumps are available that have modified impellers with holes or nozzles through which the fluid shears. These systems are very effective for shearing polymers.

39. *High-shear devices should not be used on the active system because they will rapidly reduce mud solids to colloidal size.*

    Drilled solids are not processed in the same manner as bentonite. The purpose of dispersing bentonite is to take advantage of the very thin clay platelets and their electric charges. Drilled solids usually will not grind as thin as bentonite can disperse. Although they become colloidal, they are still 1000 times larger than bentonite platelets. Increasing the colloidal content will increase the plastic viscosity, which needs to be as low as possible. Bentonite is the ideal clay because only a small amount is necessary to build yield point or control filtration. The capability of hydrated bentonite to disperse is much greater than drilled solids.

40. *The surface system should include a trip tank.*

    Trip tanks are needed to ensure that the well bore receives the correct amount of fluid as the drill string is pulled. For example, after a 10-bbl volume of steel is removed during a trip, the liquid level in the well bore should drop. This might result in an influx of formation fluid at the bottom of the hole. The trip tank continuously supplies fluid to the bell nipple to keep the well full of drilling fluid. If only 3 bbl of drilling fluid are needed to fill the hole after 10 bbl of steel are removed, the additional fluid must be entering the wellbore from the formations. A blowout is imminent. The pipe is run back to the bottom and the situation corrected.

20.2.2 Centrifugal Pumps

1. *Select a pump to handle the highest anticipated flow. Select an impeller size to provide sufficient discharge head to overcome friction in the lines, lift the fluid as required, and have sufficient head remaining to operate the equipment being fed.*

    Initially this guideline suggested that the pump flange size be selected to provide the highest anticipated flow, even though the flange size has nothing to do with the flow rate. Most pump curves are listed in terms of the flange sizes. The size of the pump impeller housing increases as the flange size increases. An impeller rotating at constant speed will create a constant head independent of the size of the housing or the flanges. An impeller that fits inside a 2×3, 3×4, 4×5, or 5×6 pump will produce the same head in each pump if it is rotated at the same speed. Because the housing of a pump with a 2-inch and 3-inch flange is smaller, the internal friction at a high flow rate will be greater than a 5 × 6 pump. This means that the capacity of the various pump sizes will be indicated by their flange sizes. The committee decided to only indicate that the pump should be selected to handle the highest anticipated flow rate, instead of indicating that the flange size is commonly used to specify pump size.

2. *Install the centrifugal pump with a flooded suction that is sumped so that sufficient submergence is available to prevent vortexing or airlocking. Foot valves are not needed or recommended with flooded suctions.*

A small influx of air into the suction of a centrifugal pump can create cavitation problems and diminished flow. As the air enters the chamber with the impeller, it tends to concentrate in the center of the impeller because of the centripetal acceleration of the drilling fluid. The liquid continues to move through the pump. The air does not always continue to the impeller tip, but tends to remain in the center of the impeller. This bubble of air forms a barrier for the incoming fluid, which diminishes the flow rate into the pump. The air also experiences a significant decrease in pressure—possibly even below atmospheric pressure. This causes implosions of vapor bubbles that can remove metal from the impeller. The pump will sound as if it is pumping gravel. If it continues in this mode for a long period of time, the impeller will be severely damaged.

Flooded suctions tend to eliminate most of the air influx problems but sometimes a small vortex will form in the mud tank. These small vortexes can entrain a significant amount of area. Increasing agitation in the tanks may prevent a coherent cylinder of air from reaching the suction line. Alternatively, a plate can be installed in the tank to interrupt the formation of a vortex.

In some cases, a centrifugal pump is placed on the ground above a pond or buried tank. Foot valves are needed if the centrifugal pump is operated above the liquid level of the suction tank. Foot valves are check valves that prevent the suction line from draining when the pump is turned off. Care must be taken to eliminate tiny air leaks in the suction line because the absolute pressure will be below atmospheric pressure. The pump and suction line should be filled with fluid before the pump motor is started. Centrifugal pumps do not move air very well.

A centrifugal pump suction can only lift fluid a certain height above a liquid level. These heights are determined by observing the NPSH (negative pressure suction head) values listed on the centrifugal pump curves. If the NPSH is exceeded, cavitation can destroy the impeller.

3. *Install a removable screen over the suction to keep out large solids and trash. It can be made out of half-inch expanded metal and should have a total screen area at least five times the cross-sectional area of the suction line so it will not restrict flow. An extended handle arrangement reaching to the tank surface is desirable to allow the screen to be pulled during service and cleaned.*

   An expanded metal screen prevents objects (such as gloves, buckets, pieces of clothing, chunks of rubber, etc.) from plugging the suction line or fouling the impeller. A bucket, turned so that the bottom fits into the suction line, can be difficult to diagnose and locate. A box made from expanded metal that covers the suction can prevent these disasters.

   If two alignment yokes are welded to the tank walls to hold a 1-inch pipe handle, the screen can be removed, cleaned, and easily returned to the suction opening. Without these alignment yokes, reseating the expanded metal box is difficult.

4. *Suction and discharge lines should be properly sized and as short as practical. Flow velocities should be in the range of 5 to 10 ft/sec. Less than 5 ft/sec causes solids to form a tight layer obstructing the bottom of horizontal lines. At velocities at or exceeding 10 ft/sec, pipe-turns tend to erode, headers do not distribute properly, and usually there will be cavitation in the suction lines. To calculate the velocity inside the pipe, use the following equation:*

   $$velocity,\ ft/sec = flow\ rate,\ gpm/3.48\ (inside\ diameter,\ in.)^2.$$

   Suction lines should contain no elbows, swages, or reducers closer than three (3) pipe diameters to the pump suction flange.

   Horizontal pipes will fill with solids until the flow rate reaches 5 ft/sec. Barite in equalizing lines between mud tanks is normally settled until the velocity between the tanks reaches 5 ft/sec. Increasing the diameter of connection lines only causes more barite to settle. Above 10 ft/sec, pressure losses in the pipe become too great. Elbows and swages tend to cause turbulence in the flow stream, which can lead to cavitation.

5. *Eliminate manifolding. One suction and one discharge per pump is most cost effective over time. Do not manifold two pumps on the same suction line. Do not pump into the same discharge line with two or more pumps.*

   Flexibility of piping so fluid can be pumped from any tank through any equipment to any other tank has created more problems over the years than just about any other concept. A properly plumbed system should require only one suction and one discharge for each piece of solids-removal equipment. Ignoring this rule allows rig hands the opportunity to open or close the wrong valves. A leaky or incorrectly opened valve can reduce drilled-solids removal efficiency by up to 50%. This translates to an expensive drilling-fluid system. This problem can be eliminated by storing an extra pump and motor. Arrange the centrifugal pumps and motors so that they may be easily replaced. If a pump or motor fails, simply replace the unit. The damaged unit can be replaced during routine maintenance.

Two centrifugal pumps in parallel will not double the head available to equipment because a centrifugal pump is a constant head device. For example, visualize a standpipe that is constantly filled with fluid. If two standpipes of approximately the same height are connected, the flow from both pipes will almost equal the flow from one standpipe. If fluid stands lower in one standpipe than the other. fluid will flow from the highest standpipe to the lowest standpipe. This same flow occurs when two pumps are connected in parallel—fluid will flow backward through one of the pumps.

6. *Install a pressure gauge between the pump discharge and the first valve. When the valve is closed briefly, the pressure reading may be used for diagnostic evaluation of the pump performance.*

A centrifugal pump uses the smallest amount of power when no fluid is moving through the pump (that is, when the discharge valve is completely closed). If the valve remains closed for longer than 5 minutes, the fluid within the pump will become hot from the impeller agitation. This hot fluid may damage the seals. Closing the valve for a short time allows a good reading of the no-flow head produced by the pump. This reading should be compared with the pump manufacturer's charts. The diameter of the impeller can then be determined. (A pump may be stamped 5X6X14. This means that it could house a 14-inch impeller but it does not mean that it has a 14-inch impeller. The impeller size is adjusted so the pump will deliver the proper head.) After the pump has been in service for a period of time, the pressure reading will assess the condition of the impeller. This eliminates the need to dismantle the pump for inspection. If the manifold pressure is incorrect, reading the pump no-flow discharge head will assist in troubleshooting.

7. *Keep air out of the pump by degassing the mud, having adequate suction line submergence, and installing baffles to break mixer vortices.*

Properly sized, baffled, and agitated compartments will not vortex unless the drilling fluid level becomes extremely low.

Centrifugal pumps cannot pump aerated fluid. The air tends to gravitate toward the center of the impeller while the liquid moves toward the outside. This creates an air bubble at the center of the impeller. When the air bubble becomes as large as the suction line diameter, fluid will no longer enter the pump. This is called airlock.

Only a small cylinder of air vortexing into the pump is sufficient to prevent the pump form moving liquid. Since the air accumulates over a period of time, a small vortex the size of a pencil is sufficient to eventually shut down a 6×8 pump.

Baffles are inexpensive and easily installed in an empty tank. Any vertical surface that disrupts the swirling motion of the fluid in a compartment is usually sufficient to destroy a vortex. Rig pump efficiency can decrease from 99 to 85% efficiency if the drilling fluid content rises to 6% volume. Air in the drilling fluid may be calculated by measuring the pressurized and unpressurized mud weight.

8. *Do not restrict the flow to the suction side of the pump. Starving the pump suction causes cavitation and this will rapidly damage the pump.*

    When a pump begins cavitating, small vacuum bubbles adjacent to the impeller surface start imploding. The pump sounds as if it is pumping gravel. The implosions quickly remove metal from the surface to the impeller blade. In a very short period of time, holes will appear in the metal. IMPORTANT: Do not close a valve on the suction line while the pump is running!

    Starving the suction will decrease the output head. If the head, or pressure produced by the pump, is too high, change to a smaller diameter impeller. On a temporary basis, a discharge valve can be partially closed. On a long-term basis, however, considerable valve erosion will occur so a new, properly sized impeller is necessary. Even a large centrifugal pump is not damaged if only 10 to 20 gpm is discharged from the pump. In fact, the lower flow rates will require less horsepower to the motor than pumping fluid at a much higher flow rate.

9. *Make sure the impeller rotation is correct.*

    Centrifugal pumps will pump fluid even if running in reverse. The head produced by the pump will be lower than it should be. The pressure gauge installed between the pump and the first valve will assist with the diagnosis. Usually, switching two wires in the lead-in panel box will correct the rotation.

10. *Startup procedure for an electric motor–driven centrifugal pump with a valve on the discharge side between the pump and the equipment being operated is to start the pump with the valve just slightly open. Once the pump is up to speed, open the valves slowly to full open. This approach will reduce the startup load on the electric motor and will reduce the shock loading on equipment such as pressure gauges and hydrocyclones. An alternative startup procedure is to completely close the discharge valve before startup and then open the valve slowly immediately after startup to prevent overheating and possible damage to the pump seals.*

An electric motor–driven centrifugal pump will immediately try to produce a constant head when it is turned on. If the pump is pumping into an empty line, the flow rate is enormous. Very high flow rates require very high currents to the electric motor. Circuit breakers can stop the pump and avoid motor burnout. Lower horsepower is required if the pump is started with the discharge valve closed.

20.3 SOLIDS MANAGEMENT CHECKLIST

This section contains questions that should be considered before drilling a well. The answers to these questions will assist in the proper selection, sizing, and operation of a solids management system.

20.3.1 Well Parameters/Deepwater Considerations

Where will the well be drilled?
What is the objective (e.g., oil/gas, geothermal, reentry, etc.)?
What formations and geological features are expected?
What type of well (straight hole, directional, horizontal) will be drilled?
What problems are anticipated?

Additional considerations are required when the hole location is in deep water. Among the first problems encountered drilling in waters of any appreciable depths is shallow formations insufficiently consolidated to support weight imposed by a riser annulus loaded with even a low weight drilling fluid and cuttings. Routinely such upper hole segments are drilled without a riser, using seawater as the drilling fluid with intermittent viscous flushes to assist cuttings removal. Returns simply spill out onto the seafloor.

Some of these shallow formation (sands) hold gas or water under pressure that will flow into the well bore when formations are penetrated.

Even when small in magnitude and of brief duration, these occurrences are correctly called blowouts, that is, uncontrolled flows of formation fluid into the well bore. Different approaches are taken in face of these shallow formation flows. A weighted fluid, sufficient to control the shallow formation pressure, can be spotted in the hole and casing run. Sometimes drilling proceeds to some depth below the problem formation before casing is run. Sometimes flow will cease as the formation bridges over or the pocket exhausts itself.

Two of nature's phenomena have combined to cause extensive occurrences of such shallow formation with deepwater flows in portions of the United States Gulf of Mexico. The currents of the Gulf Stream have created a sharp water depth drop off from the outer continental shelf to what is known as deepwater. In severe instances, water depths increase from 600 to 10,000 feet over just a few miles farther distance from shore. This regional current pattern has existed over geological ages whether worldwide seawater depths were rising or falling, the shoreline advancing or receding.

Over the same geological ages, massive deposits of sediment were made by the Mississippi River when it reached the open waters of the Gulf of Mexico. Water depth is an excellent classifier of sediment size. With fluctuations of seawater depth, an alternating deposition pattern has existed with nonporous silt deposited atop porous sand throughout the region. This deposition has often occurred at a pace at which subsidence due to increased weight of newer overlying sediments has transpired with only little consolidation of the formations.

As a result of the actions of the Gulf Stream and the Mississippi River, the Gulf of Mexico—particularly offshore Southeast Louisiana—has become known for shallow gas flows from unconsolidated formations in water depths where drilling with returns through a riser are not possible. Additionally, the magnitude of the two natural phenomena have combined such that these shallow formation with deepwater flows are not only more frequent, but also tend to be much, much larger.

Early attempts to drill these potentially large shallow formation flow zones were made using seawater and viscous flushes, with returns let to seafloor. However, the influx magnitude and force on the unconsolidated near surface formations often created craters so large that reentry, even with casing, was difficult if not impossible. Massive amounts of cement were used in sometimes futile efforts to establish a good surface seat. During the cement setting period, flow can resume.

Drilling with fluid weighted to contain potential influx is one solution to this problem. As returns are still directed to the seafloor while drilling continues, it is has been termed "pump and dump." Admittedly this is an expensive approach, and only undertaken in an expensive drilling environment. Brine of viscosified fluid weighted sufficiently to contain the shallow formation pressures is continuously pumped downhole. The unconsolidated, slightly overpressured formations are rapidly penetrated and a gauge or close-to-gauge hole obtained. Casing is run and with no formation fluid influx routinely cemented.

Because huge volumes are needed, frequently the fluid or brine is premixed at shore-based facilities and transported in bulk to the drill site. When mixed at rig site, "big bags" of barite or salts (usually calcium chloride) are commonly used to facilitate the rapid mixing necessary to continuously fill the large diameter hole being rapidly generated.

20.3.2 Drilling Program

What is the expected total depth?
Where are the casing points?
Are there other drilling parameters (hole size, bit type, ROP)?
What type of drilling fluid will be used?
What is the low-gravity-solids tolerance level?
What flow rate is planned?
What is the annular velocity in all sections of the borehole?
What is the hole cleaning capability?
What nozzle selection optimization will ensure immediate cuttings removal from the bottom of the hole?
What are other desired drilling fluid properties (mud weight, plastic viscosity, yield point, electrical stability)?

20.3.3 Equipment Capability

What type and size solids need removal?
What type of solids-removal equipment is recommended?
Is this solids-removal equipment available? From whom and where?
What are the weights and dimensions of the equipment?
What process rates can be provided?
What is the expected removal efficiency?
What drilling-fluid losses are expected (downhole and surface)?
How much power/fuel is required?

What experience does the vendor have in this geographical region (number of units and references)?
What is the expected downtime?
What is the vendor's safety record?
Is a health and safety plan available? What pit volumes are required?

20.3.4 Rig Design and Availability

What rig will be used?
What equipment is already installed? Is it installed correctly?
What repairs are needed?
Is the tank arrangement correct?
What modifications will be required to make it correct? Are additional tanks or piping needed?
How are removal compartments agitated?
Where are mud guns located?
Where are mud gun suctions located?
Where are additions made to the drilling-fluid system?
What size centrifugal pumps are available?
How are additions and check/suction sections blended and agitated?
What is the residence time on the surface for the drilling fluid in each section of the borehole?
What modifications will be required?
Is space available for required modifications?
Is additional power available?

20.3.5 Logistics

Where is the base of operations?
Where is the stock/service facility?
How many additional people will be required?
Do they need accommodations/meals?
Is additional personal protective equipment required, needed, or available?

20.3.6 Environmental Issues

Can cuttings be buried or discharged without further treatment?
What treatment and disposal options are available?
What determines when cuttings are "clean"?

What analytical testing is required? Where is it located?
How much time is required to reach required standards for "clean"?
For a specific treatment and disposal option: Is it onsite or offsite? What additional equipment is required? Where will it be located? What are fuel and utility requirements?
Are there weather or site constraints?
What permits are required? Who is responsible for obtaining them?

20.3.7 Economics

What is the drilling fluid cost per barrel?
Which is more expensive, commercial solids or the liquid phase?
What does required equipment cost to acquire?
What are installation and modification costs?
What are treatment and disposal costs?
What are expected savings?

APPENDIX

Effect of Temperature on the Density of Water

°F	°C	Density, gm/cc
59.0	15.0	0.9991
59.9	15.5	0.99905
60.8	16.0	0.9990
61.7	16.5	0.9989
62.6	17.0	0.9988
63.5	17.5	0.9987
64.4	18.0	0.9986
65.3	18.5	0.9985
66.2	19.0	0.9984
67.1	19.5	0.9983
68.0	20.0	0.9982
68.9	20.5	0.9981
69.8	21.0	0.9980
70.7	21.5	0.9979
71.6	22.0	0.9977
72.5	22.5	0.9976
73.4	23.0	0.9975
74.3	23.5	0.9974
75.2	24.0	0.9973
76.1	24.5	0.9971
77.0	25.0	0.9970
77.9	25.5	0.9969
78.8	26.0	0.9968
79.7	26.5	0.9966
80.6	27.0	0.9965
81.5	27.5	0.9964
82.4	28.0	0.9962
83.3	28.5	0.9961
84.2	29.0	0.9959
85.1	29.5	0.9958

(continued)

Effect of Temperature on the Density of Water (continued)

°F	°C	Density, gm/cc
86.0	30.0	0.9956
86.9	30.5	0.9955
87.8	31.0	0.9953
88.7	31.5	0.9952
89.6	32.0	0.9950
90.5	32.5	0.9949
91.4	33.0	0.9947
92.3	33.5	0.9945
93.2	34.0	0.9944
94.1	34.5	0.9942
95.0	35.0	0.9940

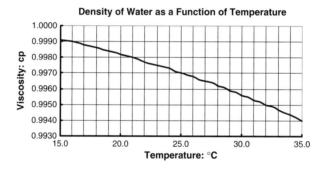

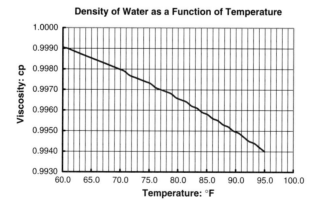

GLOSSARY

Bob De Wolfe
Consultant

Abnormal pressure. A formation pore pressure that is higher than that resulting from a water gradient.

Absolute temperature. Temperature related to absolute zero, the temperature at which all molecular activity ceases. Calculated by adding 460°F to the temperature in Fahrenheit to obtain the absolute temperature in degrees Rankine or by adding 273°C to the temperature in degrees Celsius to obtain the absolute temperature in degrees Kelvin.

Absorb. To take in and make part of an existing whole. See: *absorption, adsorption, adsorb, adsorbed liquid, bound liquid.*

Absorption. The penetration or apparent disappearance of molecules or ions of one or more substances into the interior of a solid or liquid. For example, in hydrated bentonite, the planar water that is held between the mica-like layers is the result of absorption. See: *absorb, adsorption, adsorb, adsorbed liquid, bound liquid.*

Acid. Any chemical compound containing hydrogen capable of being replaced by elements or radicals to form salts. In terms of the dissociation theory, it is a compound, which, on dissociation in solution, yields excess hydrogen ions. Acids lower the pH. Examples of acids or acidic substances are hydrochloric acid (HCl), sodium acid pyrophosphate (SAPP), and sulfuric acid (H_2SO_4). See: *pH, acidity.*

Acidity. The relative acid strength of liquid as measured by pH. A pH value below 7. See: *pH, acid.*

Across-the-line-start. A motor startup method that provides full line voltage to the motor windings.

Active system. The volume of drilling fluid being circulated to drill a hole. It consists of the volume of drilling fluid in the hole plus the volume of drilling fluid in the surface tanks through which the fluid circulates.

Additions section. A (or the) compartment(s) in a drilling-fluid system between the removal section and the suction section that provide(s) a well-agitated location within the fluid circulation system for the addition of commercial materials.

Adhesion. The force that holds unlike molecules together.

Adsorb. (1) The liquid on the surface of a solid particle that cannot be removed by draining or centrifugal force. (2) To hold a liquid on the surface of a solid particle that cannot be removed by draining or centrifugal force. See: *absorption, adsorption, adsorb, adsorbed liquid, bound liquid.*

Adsorbed liquid. The liquid film adhering to the surfaces of solids particles that cannot be removed by draining, even with centrifugal force. See: *absorb, absorption, adsorption, adsorb, bound liquid.*

Adsorption. A surface phenomenon exhibited by a solid (adsorbent) to hold or concentrate gases, liquids, or dissolved substances (adsorptive) upon its surface, a property due to adhesion. For example, water, held to the outside surface of hydrated bentonite, is adsorbed water. *Ad*sorption refers to liquid that is on the outside of some material, and *ab*sorbed refers to the liquid that becomes part of the material. See: *absorb, absorption, adsorb, adsorbed liquid, bound liquid.*

Aerated fluid. Drilling fluid to which air or gas has been deliberately added to lighten the fluid column.

Aeration. (1) The technique of injecting air or gas in varying amounts into a drilling fluid for the purpose of reducing hydrostatic head. (2) The inadvertent mechanical incorporation and dispersion of air or gas into a drilling fluid. If not selectively controlled, it can be very harmful. See: *air cutting, gas cut.*

Agglomerate: The larger groups of individual particles usually originating in sieving or drying operations.

Agglomeration. A group of two or more individual particles held together by strong forces. Agglomerates are stable to normal stirring, shaking, or handling as powder or a suspension. They may be broken by drastic treatment such as the ball milling of a powder or the shearing of a suspension.

Aggregate. To gather together, to clump together. A flocculated drilling fluid will aggregate if flocculent is added.

Aggregation. (1) Formation of aggregates. (2) In drilling fluids, aggregation results in the stacking of the clay platelets face to face. As a consequence, the viscosity and gel strength of the fluid decreases.

Agitation. The process of rapidly moving a slurry within a tank to obtain and maintain a uniform mixture.

Agitator. A mechanically driven impeller used to stir the drilling fluid to assist in the suspension of solids, blending of additives, and maintenance of uniform consistency.

Air cutting. The inadvertent mechanical incorporation and dispersion of air into a drilling fluid system. See: *aeration, gas cut.*

Airlock. A condition causing a centrifugal pump to stop pumping because of a large bubble of air or gas in the center of the pump impeller. This prevents the liquid from entering the pump suction.

Airlocking. See: *airlock.*

Alkali. Any compound having pH properties higher than the neutral state. See: *base.*

Alkalinity. The combining power of a base measured by the maximum number of equivalents an acid with which it can react to form a salt. In water analyses, it represents the carbonates, bicarbonates, hydroxides, and occasionally the borate, silicates, and phosphates in the water. It is determined by titration with standard acid to certain datum points. See API RP 13B for specific directions for determination of phenolphthalein (Pf) and methyl orange (Mf) alkalinities of the filtrate in drilling fluids and the (Pm) alkalinity of the drilling fluid itself. See: *alkali, base, Pf, Mf,* and *Pm.*

Alum. Aluminum sulfate, $Al_2(SO_4)_3$, a common inorganic coagulant.

Aluminum stearate. An aluminum salt of stearic acid used as a defoamer. See: *stearate.*

Amorphous. The property of a solid substance that does not crystallize and is without any definite characteristic shape.

Ampere. The measurement of electric flow per second.

Amplitude. The distance from the mean position to the point of maximum displacement. In the case of a vibrating screen with circular motion, amplitude would be the radius of the circle. In the case of straight-line motion or elliptical motion, amplitude would be one half of the total movement of the major axis of the ellipse; thus, one-half stroke. See: *stroke.*

Anhydrite. A mineral compound, $CaSO_4$, that is often encountered while drilling. It may occur as thin stringers or massive formations. See: *calcium sulfate, gypsum.*

Anhydrous. Without water.

Aniline point. The lowest temperature at which equal volumes of freshly distilled aniline and an oil sample that is being tested are completely miscible. This test gives an indication of the characteristics (paraffinic, naphthenic, asphaltic, aromatic, etc.) of the oil. The aniline point of diesels or crude oils used in drilling fluid is also an indication of the deteriorating effect that these materials may have on natural or rubber. The lower the aniline point of a particular oil, the greater its propensity for damaging rubber parts.

Anion. A negatively charged atom or radical, such as Cl^-, OH^-, SO^{-4}, etc., in solution of an electrolyte. Anions move toward the anode (positive electrode) under the influence of an electrical potential.

Annular pressure loss. The pressure on the annulus required to pump the drilling fluid from the bottom of the hole to the top of the hole in the annular space. See: *ECD*.

Annular velocity. The velocity of a fluid moving in the annulus, usually expressed in ft/min or m/min.

Annulus. The space between the drill string and the wall of the hole or the inside surface of the casing. Also called **annular space.**

Antifoam. A substance used to prevent foam by increasing the surface tension of a liquid. See: *defoamer*.

Aperture. (1) An opening in a screen surface. (2) The opening between the wires in a screen cloth. See: *mesh*.

Apex. The lower end (conical tip) of a hydrocyclone. See: *underflow opening*.

Apex valve. See: *apex underflow opening*.

API Bulletin RP 13E. Recommended practice for shaker screen cloth design. Published by the American Petroleum Institute. This is an alternative method for screen description, which is no longer used by the industry.

API RP 10B. Recommended Practice (RP) for Testing Well Cement. Published by the Petroleum Institute (API).

API RP 13B. Recommended Practice for Standard Procedure for Testing Drilling Fluids at the rig. Published by the American Petroleum Institute.

API RP 13C. Recommended Practice for Drilling Fluid Systems Process Evaluation. Published by the American Petroleum Institute.

API filter press. A device used to measure API fluid loss conditions. See: *API fluid loss*.

API fluid loss. This fluid loss is measured under ambient conditions. Usually these are room temperature and 100 psi differential pressure.

API gravity. The gravity (weight per unit volume) of crude oil or other related fluids as measured by a system recommended by the American

Petroleum Institute (API). It is related to specific gravity by the following formula:

$$\text{degree API} = [141.5/\text{specific gravity}] - 131.5$$

API sand. Solids particles that are too large to pass through a U.S. Standard No. 200 screen (74-micron openings). Note that particle size is the only standard. Particles larger than 74 microns are classified as *sand* even though they may be shale, limestone, wood, or any other material. See: *API RP 13B, sand, sand content.*

Apparent viscosity. The apparent viscosity in centipoise, as determined by the direct-indicating viscometer is equal to one-half the 600-rpm reading. It is the viscosity of a fluid at a shear rate of 1022 $\sec^{-1}$. See: *viscosity, plastic viscosity, yield point, API RP 13B.*

Aromatic hydrocarbons. Hydrocarbons that include compounds containing aliphatic or aromatic groups attached to aromatic rings. Benzene is the simplest example. See: *live oil.*

Asphalt. A natural or mixed blend of solid or viscous bitumen found in natural beds or obtained as a residue from petroleum distillation. Asphalt, blends containing asphalt, and altered air-blown, chemically modified, etc.) asphaltic materials have been added to drilling fluids for purposes such as lost circulation, emulsification, fluid loss control, lubrication, seepage loss, shale stability, etc.

Atom. The smallest quantity of an element capable of entering into chemical combination or that can exist alone.

Atomic weight. The relative weight of an atom of any element as compared with the weight of 1 atom of oxygen. The atomic weight of oxygen is 16.

Attapulgite clay. A colloidal, viscosity-building clay used principally in saltwater drilling fluids to the low shear viscosity. Attapulgite, a special fuller's earth, is a hydrous magnesium aluminum silicate that has long, needle-like platelets, as opposed to the broader, more symmetrical platelets of bentonite.

Axial flow. Flow from a mechanical agitator in which the fluid first moves along the axis of the impeller shaft (usually down toward the bottom of a tank) and then away from the impeller. See: *radial flow.*

Backpressure. The frictional or blocking pressure opposing fluid flow in a conduit. See: *differential pressure.*

Back tank. The compartment on a shale shaker that receives drilling fluid from the flowline. See: *possum belly, mud box.*

Backing plate. The plate attached to the back of screen cloth(s) for support.
Backup screen. See: *support screen.*
Baffles. Plates or obstructions built into a compartment to change the direction of fluid flow.
Balanced design hydrocyclone. A hydrocyclone that has the lower apex adjusted to the diameter of the cylinder of air formed within the cone by the cyclonic forces of drilling fluid spinning within cone. This tends to minimize liquid discharge when there are no separable solids.
Balanced elliptical motion. An elliptical motion of a shale shaker screen such that all ellipse axes are tilted at the same angle toward the discharge end of the shale shaker.
Ball valve. A valve that uses a spherical closure with a hole through its center and rotates 90° to open and close.
Barite. Natural barium sulfate, $BaSO_4$, is used for increasing the density of drilling fluids. The API standard requires a minimum of 4.20 specific gravity. Commercial barium sulfate ore can be produced from a single ore or a blend of ores and may be a straight-mined product or processed by flotation methods. It may contain minerals other than barium sulfate. Because of mineral impurities, commercial barite may vary in color from off-white to gray to red or brown. Common accessory minerals are silicates such as quartz and chert, carbonate compounds such as siderite and dolomite, and metallic oxide and sulfide compounds.
Barite recovery efficiency. Barite recovery efficiency is the ratio of the mass flow rate of barite returning to a drilling fluid from a solids-control device divided by the mass flow rate of barite in the feed to the solids-control device.
Barium sulfate. $BaSO_4$. See: *barite.*
Barrel (bbl). A volumetric unit of measure used in the petroleum industry consisting of 42 U.S. gallons.
Barrel equivalent. One gram of material in 350 ml of fluid is equivalent to a concentration of 1 lb of that material in an oilfield barrel of fluid. See: *barrel, pound equivalent.*
Base. A compound of a metal, or a metal-like group, with hydrogen and oxygen in the proportions that form an OH^- radical, when ionized in an aqueous solution, yielding excess hydroxyl ions. Bases are formed when metallic oxides react with water. Bases increase the pH. Examples of bases are caustic soda, $NaOH$; and lime, $Ca(OH)_2$.
Base exchange. The replacement of the cations associated with the surface of a clay particle by another species of cation, for example,

the substitution of sodium cations by calcium cations on the surface of a clay particle. See: *methylene blue titration, methylene blue test, MBT, cation exchange capacity, CEC.*

Basicity. pH value above 7. Ability to neutralize or accept protons from acids. See: *pH.*

Basket. That portion of a shale shaker containing the deck upon which the screen(s) is mounted; supported by vibration isolation members connected to the bed.

Beach. Area between the liquid pool and the solids discharge ports in a decanting centrifuge or hydrocyclone.

Bed. Shale shaker support member, consisting of mounting skid or frame with or without bottom, flow diverters to direct screen underflow to either side of the skid, and mountings for vibration isolation members.

Bentonite. A colloidal clay, largely made up of the mineral sodium montmorillonite, a hydrated aluminum silicate. Used for developing a low shear rate viscosity and/or good filtration characteristics in water-based drilling fluids. The generic term "bentonite" is not an exact mineralogical name, nor is the clay of definite mineralogical composition. See: *gel, montmorillonite.*

Bentonite (clay) extender. Additive that interacts with clay in a drilling fluid to boost viscosity; usually this at a low to moderate concentration of polymer and depends on the polymer/clay ratio.

Bernoulli Principle. One means of expressing Newton's Second Law of Physics, that is, concerning conservation of energy. Roughly stated, this principle demonstrates that the sum of pressure and velocity through or over a device represents equal quantities, neglecting the effects of losses due to friction and/or increases by adding energy with external devices such as pumps.

Bicarb. See: *sodium bicarbonate.*

Bingham model. A mathematical description that relates shear stress to shear rate in a linear manner. This model requires only two constants (plastic viscosity and yield point) and is the simplest rheological model possible to describe a non-Newtonian liquid. It is very useful for analyzing drilling fluid problems and treatment. See: *viscosity, pseudoplastic fluid, plastic viscosity, yield point, gel strength.*

Blade. See: *flight, flute.*

Blinding: A reduction of open area in a screening surface caused by coating or plugging. See: *coating, plugging.*

Blooie line. The flowline for air or gas drilling.

Blowout. An uncontrolled escape of drilling fluid, gas, oil, or water from the well caused by the formation pressure being greater than the hydrostatic head of the fluid being circulated in the well bore. See: *kick, kill fluid.*

Bonded screens. Multiple screens bonded together with plastic to form a multilayered screen or screens bonded to a metal support plate.

Bonding material. Material used to secure screen cloth to a backing plate or support screen.

Bottom flooding. The behavior of a hydrocyclone when the underflow discharges whole drilling fluid rather than separated solids.

Bound liquid. Adsorbed liquid. See: *absorb, absorption, adsorb, adsorption, adsorbed liquid.*

Bow. See: *crown.*

Bowl. The outer rotating chamber of a decanting centrifuge.

Brackish water. Water containing low concentrations of any soluble salts.

Break circulation. To start movement of the drilling fluid after it has been quiescent in a borehole.

Bridge. An obstruction in a well formed by the intrusion of subsurface formations and/or cuttings or material, which prevents a tubular string from moving down a borehole.

Brine. Water containing a high concentration of common salts such as sodium chloride, calcium chloride, calcium bromide, zinc bromide, etc.

Bromine value. The number of centigrams of bromine that are absorbed by 1 gram of oil under certain conditions. The bromine check is a test for the degree of unsaturation of a given oil.

Brownian movement. Continuous, irregular motion exhibited by particles suspended in a liquid or gaseous medium, usually as a colloidal dispersion.

BS&W. Base sediment and water.

Buffer. Any substance or combination of substances that, when dissolved in water, produces a solution that resists a change in its hydrogen ion concentration upon the addition of an acid or base.

Cable tool drilling. A method of drilling a well by allowing a weighted bit (or chisel) at the bottom of a cable to fall against the formation being penetrated. The cuttings are then bailed from the bottom of the well bore with a bailer. See: *rotary drilling.*

Cake consistency. According to API RP 13B, can be described as hard, soft, tough, rubbery, firm, etc.

Cake thickness. (1) A measurement of the thickness of the filter cake deposited by a drilling fluid against a porous medium, usually filter paper, according to the standard API filtration test. Cake thickness is usually reported in 32nds of an inch or millimeters. (2) A parameter of the filter cake deposited on the wall of the borehole. See: *filter cake, wall cake*.

Calcium. One of the alkaline earth elements with a valence of 2 and an atomic weight of about 40. Calcium compounds are a common cause of the hardness of water. Calcium is also a component of lime, gypsum, limestone, etc.

Calcium carbonate. (1) $CaCO_3$. An acid soluble calcium salt sometimes used as a weighting material (limestone, oyster shell, etc.) in specialized drilling fluids. (2) A term used to denote a unit and/or standard to report hardness. See: *limestone*.

Calcium chloride. $CaCl_2$. A very soluble calcium salt sometimes added to drilling fluids to impart special inhibitive properties, but used primarily to increase the density of the liquid phase (water) in completion fluids and as an inhibitor to the water phase of invert oil emulsion drilling fluids.

Calcium contamination. Dissolved calcium ions in sufficient concentration to impart undesirable properties in a drilling fluid, such as flocculation, reduction in yield of bentonite, increase in fluid loss, etc. See: *calcium sulfate, gyp, anhydrite, lime, calcium carbonate*.

Calcium hydroxide. $Ca(OH)_2$. The active ingredient of slaked lime. It is also the main constituent in cement (when wet) and is referred to as "lime" in field terminology. See: *lime*.

Calcium sulfate. Anhydrite, $CaSO_4$, plaster of Paris, $CaSO_4 \cdot \frac{1}{2}H_2O$, and gypsum, $CaSO_4 \cdot 2H_2O$. Calcium sulfate occurs in drilling fluids as a contaminant or may be added as a commercial product to certain drilling fluids to impart special inhibitive properties. See: *gypsum, anhydrite*.

Calcium-treated drilling fluids. Drilling fluids to which quantities of soluble calcium compounds have been added or allowed to remain from the formation drilled in order to impart special inhibitive properties to the drilling fluid.

Calendered wire cloth. Wire cloth that has been passed through a pair of heavy rollers to reduce the thickness of the cloth or flatten the intersections of the wire and produce a smooth surface. This process is usually done to the coarser backing cloths. See: *market grade cloth, mill grade cloth*.

Capacity. The maximum volume flow rate at which a solids-control device is designed to operate without detriment to separation. See: *feed capacity, solids discharge capacity*.

Cascade. Gravity-induced flow of fluid from one unit to another.

Cascade shaker arrangement. System that processes the drilling fluid through two or more shakers in series.

Casing. Steel pipe placed in an oil or gas well to prevent the wall of the hole in a drilled interval from caving in, as well as to prevent movement of fluids from one formation to another.

Cation. The positively charged particle in the solution of an electrolyte, which, under the influence of an electrical potential, moves toward the cathode (negative electrode). Examples are Na^+, H^+, NH^{4+}, Ca^+, Mg^{++}, and Al^{+++}.

Cation exchange capacity. The total amount of cations adsorbed on the basal surfaces or broken bond edges of a clay sample, expressed in milliequivalents per 100 grams of dry clay. See: *base exchange, methylene blue titration, methylene blue test, MBT, CEC*.

Caustic. See: *sodium hydroxide*.

Caustic soda. See: *sodium hydroxide*.

Cave in. A severe form of sloughing. See: *sloughing*.

Cavernous formation. A formation having voluminous voids, usually the result of dissolution by formation waters that or may not be still present.

Caving. Caving is a severe form of sloughing. See: *sloughing, heaving*.

Cavitation. Cavitation is the formation and collapse of low-pressure bubbles in a liquid. In centrifugal pumps it occurs when the pressure within the impeller chamber decreases below the vapor pressure of the liquid. As these vapor bubbles move to the impeller tip and into a higher-pressure region, they implode or collapse. The pressure at the suction entry may be considerably below atmospheric pressure if the pressure loss in the suction line is too large, if the flow rate from the pump is too large for the inlet size, or if the fluid must be lifted to excessive heights. As the bubbles move out to the tips of the impeller, they implode, releasing a large amount of energy that can actually chip metal pieces from the impeller blade. Cavitation frequently sounds like the centrifugal pump is pumping gravel. See: *centrifugal pump*.

CEC. See: *cation exchange capacity*.

Cement. A mixture of calcium aluminates and silicates made by combining lime and clay while heating. Slaked cement contains about 62.5% calcium hydroxide, which can cause a major problem when cement contaminates drilling fluid.

Centipoise (cP). Unit of viscosity equal to 0.01 Poise. Poise equals 1 dyne-second per square centimeter. The viscosity of water at 20°C is 1.005 cP (1 cP = 0.000672 lb/ft sec).

Centrifugal force. That force which tends to impel matter outward from the center of rotation. See: g *force*.

Centrifugal pump. A machine for moving fluid by spinning it using a rotating impeller in a pump casing with a central inlet and a tangential outlet. The path of the fluid is an increasing spiral from the inlet at the center to the outlet, tangent to the annulus. In the annular space between the impeller vane tips and the casing wall, the fluid velocity is roughly that of the impeller vane tips. Useful work is produced by the pump when some of the spinning fluid flows out of the casing tangential outlet into the pipe system. Power from the motor is used to accelerate the fluid coming into the inlet up to the speed of the fluid in the annulus. (Some of the motor power is expended as friction of the fluid in the casing and impeller.)

Centrifugal separator. A general term applicable to any device using centrifugal force to shorten and/or control the settling time required to separate a heavier mass from a lighter mass.

Centrifuge. A centrifugal separator, specifically a device rotated by an external force for the purpose of separating materials of different masses. This device is used for the mechanical separation of solids from a drilling fluid. Usually in a weighted drilling fluid, it is used to eliminate colloidal solids. In an unweighted drilling fluid, it is used to remove solids larger than colliods. The centrifuge uses high-speed mechanical rotation to achieve this separation, as distinguished from the cyclone type of separator, in which the fluid energy alone provides the separating force. See: *hydrocyclone, desander, desilter*.

Ceramics. A general term for heat-hardened clay products, which resist abrasion; used to extend the useful life of wear parts in pumps and hydrocyclones.

Check/suction section. The last active section in the surface system. It provides a location for the rig pump and drilling-fluid hopper suction. This section should be large enough to check and adjust drilling-fluid properties before the drilling fluid is pumped downhole.

Chemical barrel. A container in which soluble chemicals can be mixed with a limited amount of fluid prior to addition to the circulating system.

Chemical treatment. The addition of chemicals (such as caustic, thinners, or viscosifiers) to the drilling fluid to adjust the drilling-fluid properties.

Chemicals. In drilling-fluid terminology, a chemical is any material that produces changes in the low-shear-rate viscosity, yield point, gel strength, fluid loss, pH, or surface tension.

Choke. An opening, aperture, or orifice used to restrict a rate of flow or discharge.

Chromate. A compound in which chromium has a valence of 6, for example, sodium dichromate. Chromate may be added to drilling fluids either directly or as a constituent of chrome lignites or chrome lignosulfonates to assist with rheology stabilization. In certain areas, chromate is widely used as an anodic corrosion inhibitor, often in conjunction with lime.

Chrome lignite. Mined lignite, usually leonardite, to which chromate has been added and/or reacted. The lignite can also be causticized with either sodium or potassium hydroxide. The chrome lignite is used for rheology stabilization and filtration control of the drilling fluid.

Circular motion. A shale shaker screen moves in a uniform circular motion when the vibrator is located at the center of gravity of the vibrating basket.

Circulation. The movement of drilling fluid through the flow system on a drilling or workover rig. This circulation starts at the suction pit and goes through the mud pump, drill pipe, bit, annular space in the hole, flowline, fluid pits, and back again to the suction pit. The time involved is usually referred to as **circulation time.** See: *reverse circulation*.

Circulation rate. The volume flow rate of the circulating drilling fluid, usually expressed in gallons per minute or barrels per minute. See: *flow rate*.

Clabbered. A slang term used to describe moderate to severe flocculation of drilling fluid due to various contaminants. See: *gelled up*.

Clarification. Any process or combination of processes the primary purpose of which is to reduce the concentration of suspended matter in liquid.

Clay. (1) A soft, variously colored earth, commonly hydrous silicates of alumina, formed by the decomposition of feldspar and other aluminum silicates. Clay minerals are essentially insoluble in water but disperse under hydration, grinding, or velocity effects. Shearing forces break down the clay particles to sizes varying from submicron particles to particles 100 microns or larger. (2) Solids particles of less than 2 micrometer equivalent spherical diameter. See: *attapulgite clay, bentonite, high yield clay, low yield clay*, and *natural clays*.

Clay extender. Substances, usually high-molecular-weight organic compounds, that when added in low concentrations to a bentonite or to other specific clay slurries, will increase the low-shear-rate viscosity of the system. An example would be polyvinyl acetate-maleic anhydride copolymer. See: *low solids drilling fluids*.

Clay-size particles. See: *clay*.

Closed loop mud systems. A drilling-fluid processing system that minimizes the liquid discard. Usually as much as possible of the liquid phase normally separated with drilled solids is returned to the active system.

Closed loop systems (pressurized). In underbalanced drilling, this refers to a system in which formation fluid is contained in tanks and not exposed to the atmosphere until sent to the flare line or the holding tank.

CMC. Ceramic matrix compound(s). See: *sodium carboxymethylcelluose*.

Coagulation. The destabilization and initial aggregation of colloidal and finely divided suspended matter by the addition of a floc-forming agent. See: *floc*.

Coalescence. (1) The change from a liquid to a thickened curdlike state by chemical reaction. (2) The combination of globules in an emulsion caused by molecular attraction of the surfaces.

Coarse solids. Solids larger than 2000 microns in diameter.

Coating. (1) A material adhering to a surface to change the properties of the surface. (2) A condition in which material forms a film that covers the apertures of the screening surface. See: *blinding, plugging*.

Cohesion. The attractive forces between molecules of the same kind, that is, the force that holds the molecules of a substance together.

Colloid. A particle smaller than 2 microns. The size and electrical charge of these particles determine the different phenomena observed with colloids, for example, Brownian movement. See: *clay, colloidal solids*.

Colloidal composition. A colloidal suspension containing one or more colloidal constituents.

Colloidal matter. Finely divided solids that will not settle but may be removed by coagulation.

Colloidal solids. Particles smaller than 2 microns. These are so small that they do not settle out when suspended in a drilling fluid. Commonly used as a synonym for clay. See: *clay, colloid*.

Colloidal suspension. Finely divided particles that are so small that they remain suspended in a liquid by Brownian movement.
Combining weight. See: *equivalent weight.*
Conductance. The permeability of a shaker screen per unit thickness of the screen, measured in units of kilodarcys/millimeter, while the screen is stationary.
Conductivity. Measure of the quantity of electricity transferred across unit area per unit potential per unit time. It is the reciprocal of resistivity. Electrolytes may be added to a fluid to alter its conductivity. See: *resistivity.*
Cone. See: *hydrocyclone, hydroclone.*
Connate water. Water trapped within sedimentary deposits, particularly as hydrocarbons displaced most of the water from a reservoir.
Consistometer. A thickening time tester having a stirring apparatus to measure the relative thickening time for drilling fluid or cement slurries under predetermined temperatures and pressures. See: *API RP 10B.*
Contamination. In a drilling fluid, the presence of any material that may tend to harm the desired properties of the drilling fluid.
Continuous phase. (1) The fluid phase that completely surrounds the dispersed phase. (2) The fluid phase of a drilling fluid: either water, oil, or synthetic oil. The dispersed (noncontinuous) phase may be solids or liquid.
Controlled aggregation. The condition in which the clay platelets are maintained stacked by a polyvalent cation, such as calcium.
Conventional drilling fluid. A drilling fluid containing essentially clay and water. Also called **conventional mud.**
Conventional shale shakers. Usually refers to device that vibrates screens with a circular or an unbalanced elliptical motion. These shale shakers are usually limited to processing drilling fluid through screens up to 100 mesh.
Conveyor. A mechanical device for moving material from one place to another. In a decanting centrifuge, this is a hollow hub fitted with flights rotating in the same direction but at a different speed than the centrifuge bowl. These flights are designed to move the coarse solids out of the bowl and are part of the conveyor.
Copolymer. A substance formed when two or more substances polymerize at the same time to yield a product that is not a mixture of separate polymers but a complex substance having properties different from either of the base polymers. Examples are polyvinyl

acetate-maleic anyhdride copolymer (clay extender and selective flocculant), acrylamide-carboxylic acid copolymer (total flocculant), etc. See: *polymer*.

Corrosion. A chemical degradation of a metal by oxygen in the presence of moisture. An oxide is the by-product of corrosion.

Corrosion inhibitor. An agent that, when added to a system, slows down or prevents a chemical or corrosion. Corrosion inhibitors are used widely in drilling and producing operations to prevent corrosion of metal equipment exposed to hydrogen sulfide, carbon dioxide, oxygen, saltwater, etc. Common inhibitors added to drilling fluids are filming amines, chromates, and oxygen scavengers.

Crater. The formation of a large funnel-shaped cavity at the top of a hole resulting from either a blowout or from caving.

Creaming of emulsions. The settling or rising of particles from the dispersed phase of an emulsion as observed by a difference in color shading of the layers formed. This separation can be either upward or downward, depending on the relative densities of the continuous and dispersed phases.

Created fractures. Induced fractures by means of hydraulic or mechanical pressure exerted on the formation by the drill string and/or circulating fluid.

Critical velocity. That velocity at the transitional point between laminar and turbulent types of fluid flow. This point occurs in the transitional range of Reynolds numbers between approximately 2000 to 3000.

Crown. The curvature of a screen deck or the difference in elevation between its high and low points. See: *bow*.

Cryogenic (nitrogen). Nitrogen in its liquid form.

Cubic centimeter (cc). A metric-system unit for the measure of volume. A cube measuring 1 centimeter on each side would have a volume of 1 cubic centimeter. It is essentially equal to the milliliter, with which it is commonly used interchangeably. One cc of water at room temperature weighs approximately 1 gram.

Cut point. Cut point curves are developed by dividing the mass of solids in a certain size range removed by the total mass of solids in that size range that enters the separation device. A cut point usually refers to the size of particle that has a 50% chance of being discarded. See *median cut*.

Cutt points (pronounced "Koot"). The equivalent spherical diameters corresponding to the ellipsoidal volume distribution of a screen's opening sizes, as determined by image analysis. See: *API RP 13E*.

Cuttings. The pieces of formation dislodged by the bit and brought to the surface in the drilling fluid. Field practice is to call all solids removed by the shaker screen "cuttings," although some can be sloughed material from the wall of the borehole. See: *drilled solids, low-gravity solids, samples.*

Cyclone. A device for the separation of solid particles from a drilling fluid. The most common cyclones used for solids separation are a desander or desilter. In a cyclone, fluid is pumped tangentially into a cone, and the fluid rotation provides enough centrifugal force to separate particles by mass (weight). See: *desander, desilter, hydrocyclone, hydroclone.*

Cyclone bottom. See: *apex, apex valve.*

Darcy. A unit of permeability. A porous medium has a permeability of 1 darcy when a pressure of 1 atm on a sample 1 cm long and 1 sq cm in cross section will force a liquid of 1 cP viscosity through the sample at the rate of 1 cc per sec. See: *millidarcy, permeability.*

Decanter. See: *decanting centrifuge.*

Decanting centrifuge. A centrifuge that removes solids from the feed slurry and discharges them as damp underflow. Ultra-fine colloidal solids are discharged with the liquid overflow. The decanting centrifuge has an internal auger that moves the solids that have been settled to the bowl walls, out of a pool of liquid, and to the underflow. See: *centrifuge.*

Deck. The screening surface in a shale shaker basket.

De-duster. A tank at the end of the blooie line in air or gas drilling in which water is injected to settle the dust caused by drilling.

Deflocculant. Chemical that promotes deflocculation. See: *thinner.*

Deflocculation. (1) The process of thinning the drilling fluid by bonding with (neutralizing or covering) the positive electrical charges of drilling-fluid additives to prevent one particle of drilling fluid from being attracted to another particle. (2) Breakup of flocs of gel structures by use of a thinner.

Defoamer. Any substance used to reduce or eliminate foam by reducing the surface tension of a liquid. See: *antifoam.*

Degasser. A device that removes entrained gas from a drilling fluid, especially the very small bubbles that do not float readily in viscous drilling fluid.

Dehydration. Removal of free or combined water from a compound.

Deliquescence. The liquification of a solid substance due to the solution of the solid by absorption of moisture from the air, for example, calcium chloride deliquesces in humid air.

Density. Mass per unit volume expressed in pounds per gallon (ppg), grams per cubic (g/cc), or pounds per cubic ft (lb/cu.ft). Drilling-fluid density is commonly referred to as **mud weight**.

Desand. To remove most API sand (> 74 microns) from drilling fluid.

Desander. A hydrocyclone with an inside diameter of 6 inches or larger that can remove a very high proportion of solids larger than 74 micrometers. Generally, desanders are used on unweighted muds. See: *cyclone, hydrocyclone, hydroclone, desilter*.

Desilt. To remove most silt particles greater than 15–20 microns from an unweighted fluid. The desilter is not normally not used on weighted drilling fluids because it can remove large amounts of barite.

Desilter. A hydrocyclone with an inside diameter less than 6 inches. It can remove a large fraction of solids larger than 15–20 microns. See: *cyclone, hydrocyclone, hydroclone, desander*.

Destabliziation. A condition in which colloidal particles no longer remain separate and discrete, but contact and agglomerate with other particles.

Diatomaceous earth. A very porous natural earth compound composed of siliceous skeletons. Sometimes used for controlling lost circulation and seepage losses and as an additive to cement.

Diesel oil plug. See: *gunk plug*.

Differential angle deck. A screen deck in which successive screening surfaces of the same deck are at different angles.

Differential pressure. The difference in pressure between two points. It is usually the difference in pressure at a given point in the well bore between the hydrostatic pressure of the drilling-fluid column and the formation pressure. Differential pressure can be positive, zero, or negative with respect to the formation pressure. See: *backpressure*.

Differential pressure sticking. Sticking that occurs when a portion of the drill string (usually the drill collars) becomes embedded in the filter cake resulting in a nonuniform distribution of pressure around the circumference of the pipe. The conditions essential for sticking require a permeable formation and a positive pressure (from well bore to formation) differential across a drill string embedded in a poor filter cake. See: *stuck*.

Diffusion. The spreading, scattering, or mixing of material (gas, liquid, or solid).

Dilatant fluid. Opposite of shear thinning. A dilatant or inverted plastic fluid is usually made up of a high concentration of well-dispersed solids that exhibit a nonlinear consistency curve passing through the origin. The apparent viscosity increases instantaneously with increasing shear rate. The yield point, as determined by conventional calculations from the direct-indicating viscometer readings, is negative. See: *apparent viscosity, viscosity, Bingham model, plastic viscosity, yield point, gel strength.*

Diluent. Liquid added to dilute or thin a solution or suspension.

Dilution. (1) Decreasing the percentage of drilled-solids concentration by addition of liquid phase. (2) Increasing the liquid content of a drilling fluid by addition of water or oil. Dilution fluid may be a clean drilling fluid or the liquid phase of a drilling fluid.

Dilution factor. The ratio of the actual volume of drilling fluid required to drill a specified interval of footage using a solids-removal system versus a calculated volume of drilling fluid required to maintain the same drilled-solids fraction over the same specified interval of footage with no drilled-solids removal.

Dilution rate. The rate, in gpm or bbl/hr, at which fluids and/or premix is added to the circulating system for the purpose of solids management.

Dilution ratio. Ratio of volume of dilution liquid to the volume of raw drilling fluid in the feed prior to entering a liquid/solids separator.

Dilution water. Water used for dilution of water-based drilling fluid.

Direct-indicating viscometer. Commonly called a "V-G meter." The direct-indicating viscometer shears fluid between a rotating outer cylinder and a stationary cylindrical bob in the center of the rotating cylinder. The bob is constrained from rotating by a spring. The spring reads the drag force on the bob, which is related to the shear stress. The rotational speed of the outer cylinder and the spacing between the bob and the cylinder the shear rate. Viscosity is the ratio of shear stress to shear rate, so this instrument may be used to determine viscosity of a fluid at a variety of shear rates. Gel strengths may also be determined after a quiescent period of a drilling fluid between the bob and the cylinder. See: *API RP 13B.*

Discharge. Material removed from a system. See: *effluent.*

Discharge spout. Extension at the discharge area of a screen. It may be vibrating or stationary. Also called **discharge lip.**

Dispersant. (1) Any chemical that promotes the subdivision of a material phase. (2) Any chemical that promotes dispersion of particles in a fluid. Frequently, a deflocculant is inaccurately called a dispersant. Caustic soda is a dispersant but not a deflocculant.

Disperse. To separate into component parts. Bentonite disperses by hydration into many smaller pieces.

Dispersed phase. The scattered phase (solid, liquid, or gas) of a dispersion. The particles are finely divided and completely surrounded by the continuous phase.

Dispersion. (1) Process of breaking up, scattering (as in reducing particle size), and causing to spread apart. (2) Subdivision of aggregates. Dispersion increases the specific surface of the particle, which results in an increase in viscosity and gel strength.

Dispersoid. A colloid or finely divided substance.

Disassociation. The splitting of a compound or element into two or more simple molecules, atoms, or ions. Applied usually to the effect of the action of heat or solvents upon dissolved substances. The reaction is reversible and not as permanent as decomposition; that is, when the solvent is removed, the ions recombine.

Distillation. Process of first vaporizing a liquid and then condensing the vapor into a liquid (the distillate), leaving behind nonvolatile solid substances of a drilling fluid. The distillate is the water and/or oil content of a fluid.

Divided deck. A deck having a screening surface longitudinally divided by partition(s).

Dog leg. The elbow caused by a sharp change of drilling direction in the well bore.

Double flute. The flutes or leads advancing simultaneously at the same angle and 180° apart. See: *flute, flight, blade*.

Downstream venturi. See: *venturi*.

Drill bit. The cutting or boring element at the end of the drill string.

Drill stem test (DST). A postdrilling and preproduction test that allows formation fluids to flow into the drill pipe under controlled conditions, to determine whether oil and/or gas in commercial quantities have been encountered in the penetrated formations.

Drill string. The column of drill pipe with attached tool joints that transmits fluid and rotational power from the kelly to the drill collars and bit.

Drilled solids. Formation solids that enter the drilling-fluid system, whether produced by a bit or from the side of the borehole. See *low-gravity solids, cuttings*.

Drilled-solids fraction. The average volume fraction of drilled solids maintained in the drilling fluid over a specified interval of footage.

Drilled-solids removal system. All equipment and processes used while drilling a well that remove the solids generated from the hole and carried by the drilling fluid, that is, settling, screening, desanding, desilting, centrifuging, and dumping.

Drilling fluid. Term applied to any liquid or slurry pumped down the drill string and up the annulus of a hole to facilitate drilling. See: *drilling mud, mud*.

Drilling-fluid additive. Any material added to a drilling fluid to achieve a particular effect.

Drilling-fluid analysis. Examination and testing of the drilling fluid to determine its physical and chemical properties and functional ability. See: *API RP 13B*.

Drilling-fluid cycle time. The time necessary to move a fluid from the kelly bushing to the flowline in a borehole. The cycle, in minutes, equals the barrels of drilling fluid in the hole minus pipe displacement divided by barrels per minute of circulation rate:

$$(\text{Hole}_{bbl} - \text{Pipe Voume}_{bbl})/\text{Circulation Rate}_{bbl/min}.$$

Drilling-fluid engineer. One versed in drilling fluids, rig operations, and solids and waste management, whose duties are to manage and maintain the drilling-fluid program at the well site.

Drilling-fluid program. A proposed plan or procedure for application and properties of drilling fluid(s) used in drilling a well with respect to depth. Some factors that influence the drilling-fluid program are the casing program and formation characteristics such as type, competence, solubility, temperature, pressure, etc.

Drilling in. The drilling operation starting at the point of drilling into the producing formation.

Drilling mud. See: *drilling fluid*, which is the preferred term.

Drilling out. The operation of drilling out of the casing shoe after the cementing of a casing or liner in place. Drilling out of the casing is done before further hole is made or completion attempted.

Drilling rate. The rate at which hole depth progresses, expressed in linear units per unit of time (including connections) as feet/minute or feet/hour. See: *ROP, rate of penetration, penetration rate*.

Dry bottom. An adjustment to the underflow opening of a hydrocyclone that causes a dry beach, usually resulting in severe plugging. See: *dry plug*.

Dry plug. The plugging of the underflow opening of a hydrocyclone caused by operating with a dry bottom.

Dryer. A shale shaker with a fine mesh screen that removes excess fluid and fine solids from discarded material from other shale shakers and hydrocyclones. Typically, this is used to decrease the liquid waste from a drilling fluid to decrease discarded volumes. See *mud cleaner*.

Dual wound motors. Motors that may be connected to either of two voltages and starter configurations.

Dynamic. The state of being active or in motion, as opposed to static.

ECD. Equivalent circulating density. See: *equivalent circulating density, annular pressure loss.*

Eductor. (1) A device utilizing a fluid stream discharging under high pressure from a jet through an annular space to create a vacuum. When properly arranged, it can evacuate degassed drilling fluid from a vacuum-type degasser. (2) A device using a high-velocity jet to create a vacuum that draws in liquid or dry material to be blended with drilling fluid.

Effective screening area. The portion of a screen surface available for solids separation.

Effluent. A discharge of liquid. Generally used to describe a stream of liquid after some attempt at separation or purification has been made. See: *discharge.*

Elastomer. Any rubber or rubber-like material (such as polyurethane).

Electric logging. Logs run on a wire line to obtain information concerning the porosity, permeability, density, and/or fluid content of the formations drilled. The drilling-fluid characteristics may need to be altered to obtain good logs.

Electrolyte. A substance that dissociates into charged positive and negative ions when in solution or a fused state. This electrolyte will then conduct an electric current. Acids, bases, and salts are common electrolytes.

Elevation head. The pressure created by a given height of fluid. See: *hydrostatic pressure head.*

Emulsifier. A substance used to produce a mixing of two liquids that do not solubilize in each other or maintain a stable mixture when agitated in the presence of each other. Emulsifiers may be divided into ionic and nonionic agents, according to their behavior. The ionic types may be further divided into anionic, cationic, and, depending on the nature of the ionic groups.

Emulsion. A substantially permanent heterogeneous mixture of two or more liquids that do not normally dissolve in each other but are held in a dispersed state, one within the other. This dispersion is accomplished by the combination of mechanical agitation and presence of fine solids and/or emulsifiers. Emulsions may be mechanical, chemical, or a combination of the two. Emulsions may be either oil-in-water or water-in-oil. See: *interfacial tension, surface tension.*

Emulsoid. Colloidal particle that takes up water.

Encapsulation. The process of totally enclosing electrical parts or circuits with a polymeric material (usually epoxy).

End point. Indicates the end of a chemical testing operation when a clear and definite change is observed in the test sample. In titration, this change is frequently a change in color of an indicator or marker added to the solution, or the disappearance of a colored reactant.

Enriching. The process of increasing the concentration of a flammable gas or vapor to a point at which the atmosphere has a concentration of that flammable gas or vapor above its upper flammable or explosive limit.

Extreme pressure (EP) additive. See: *extreme pressure lubricant.*

EPL. Extreme pressure lubricant. See: *extreme pressure lubricant.*

epm. Equivalents per million. See: *equivalents per million, parts per million.*

Equalizer. An opening for flow between compartments in a surface fluid holding system.

Equivalent circulating density (ECD). The effective drilling-fluid weight at any point in the annulus of the well bore during fluid circulation. ECD includes drilling-fluid density, cuttings in the annulus, and annular pressure loss. See: *annular pressure loss.*

Equivalent spherical diameter (ESD). The theoretical dimension usually referred to when the sizes of irregularly shaped particles are discussed. These dimensions can be determined by several methods, such as settling velocity, electrical resistance, and light reflection. See: *particle size.*

Equivalent weight. The atomic weight or formula weight of an element, compound, or ion divided by its valence. Elements entering into combination always do so in quantities to their equivalent weights. Also known as **combining weight.**

Equivalents per million (epm). Unit chemical weight of solute per million unit weights of solution. The epm of a solute in solution is equal to the ppm (parts per million) divided by the equivalent weight. See: *parts per million.*

ESD. Equivalent spherical diameter. See: *equivalent spherical diameter, particle size.*

Extreme pressure lubricant (EPL). Additives to the drilling fluid that impart lubrication to bearing surfaces when subjected to extreme pressure conditions.

Fault. Geological term denoting a formation break across the trend of a subsurface strata. Faults can significantly affect the drilling fluid and casing programs due to possibilities for lost circulation, sloughing hole, or kicks.

Feed. A mixture of solids and liquid (including dilution liquid) entering a liquid/solids separation device.

Feed capacity. The maximum volume flow rate at which a solids-control device is designed to operate without detriment to separation efficiency. This capacity will be dependent on particle size, particle concentration, viscosity, and other variables of the feed. See: *capacity, flow capacity, solids discharge capacity.*

Feed chamber. That part of a device that receives the mixture of diluents, drilling fluid, and solids to be separated.

Feed head. The equivalent height, in feet or meters, of a column of fluid at the cyclone feed header.

Feed header. A pipe, tube, or conduit to which two or more hydrocyclones have been connected and from which they receive their feed slurry.

Feed inlet. The opening through which the feed fluid enters a solids-separation device. Also known as **feed opening.**

Feed mud. See: *feed.*

Feed opening. See: *feed inlet.*

Feed pressure. The actual gauge pressure measured as near as possible to, and upstream of, the feed inlet of a device.

Feed slurry. See: *feed.*

Fermentation. Decomposition process of certain substances, for example, starch, in which a chemical change is brought about by enzymes, bacteria, or other microorganisms. Often referred to as "souring."

Fibrous materials. Any tough, stringy material used to prevent loss of circulation or to restore circulation. In field use, "fiber" generally refers to the larger fibers of plant origin.

Filter cake. The suspended solids that are deposited on a porous medium during the process of filtration. See: *wall cake.*

Filter cake texture. The physical properties of a cake as measured by toughness, slickness, and brittleness. See: *cake consistency.*

Filter paper. Porous paper without surface sizing for filtering solids from liquids. The API filtration test specifies 9-cm-diameter filter paper Whatman No. 50, S&S No. 576, or equivalent.

Filtrate. The liquid that is forced through a porous medium during the filtration process. See: *fluid loss.*

Fill-up line. The line through which fluid is added to the annulus to maintain the fluid level in the well bore during the extraction of the drilling assembly.

Filter cake. (1) The soild residue deposited by a drilling fluid against a porous medium, usually filter paper, according to the standard API filtration test. (2) The soild residue deposited on the wall of a borehole during the drilling of permeable formations. See: *wall cake*.
Filter cake thickness. See: *cake thickness*.
Filter press. A device for determining the fluid loss of a drilling fluid having specifications in accordance with API RP 13B. See: *API RP 13B*.
Filter run. The interval between two successive backwashing operations of a filter.
Filterability. The characteristic of a clear fluid that denote both the ease of filtration and the ability to remove solids while filtering.
Filtrate loss. See: *fluid loss*.
Filtration. (1) The process of separation of suspended solids from liquid by forcing the liquid a porous medium while screening back the solids. Two types of fluid filtration occur in a well: dynamic filtration while circulating, and static filtration when the fluid is at rest. (2) The process of drilling fluid losing a portion of the liquid phase to the surrounding formation. See: *water loss*.
Filtration rate. See: *fluid loss*.
Fine-screen shaker. A vibrating screening device designed for screening drilling fluids through screen cloth finer than 80 mesh.
Fine-screen shale shakers. Usually refers to shale shakers that vibrate screens with a balanced elliptical or linear motion. These are usually capable of processing large flow rates of drilling fluid through 120 to 250 mesh screens.
Fine solids. Solids 44–74 microns in diameter, or sieve size 325–200 mesh. See: *API RP 13C*.
Fishing. Operations on the rig for the purpose of retrieving sections of pipe, collars, or other obstructive items that are in the hole and would interfere with drilling or logging operations.
Flat decked. Shaker screens that do not have a crowned, or bowed, surface.
Flat gel. A condition wherein the gel strength does not increase appreciably with time and is essentially equal to the initial gel strength. Opposite of progressive gel. See: *progressive gel, zero-zero gel*.
Flight. On a decanting centrifuge, one full turn of a spiral helix, such as a flute or blade of a screw-type conveyor. See: *blade, flute*.
Flipped. A slang term for an extreme imbalance in a drilling fluid. In a water-in-oil emulsion, the emulsion is identified as "flipped" when the continuous and dispersed phases separate and the solids begin to settle.

Floc. Small gelatinous masses of solids formed in a liquid.

Flocculates. A group of aggregates or particles in a suspension formed by electrostatic attraction forces between negative and positive charges. Bentonite clay particles have negatively charged surfaces that will attract positive charges such as those of other bentonite positive edge charges.

Flocculating agent. Substances, for example, most electrolytes, a few polysaccharides, certain natural or synthetic polymers, that bring about the thickening of a drilling fluid. In Bingham plastic fluids, the yield point and gel strength increase with flocculation.

Flocculation. (1) Loose association of particles in lightly bonded groups, sometimes called "flocs," with nonparallel association of clay platelets. In concentrated suspensions, such as drilling fluids, flocculation results in gelation. In some drilling fluids, flocculation may be followed by irreversible precipitation of colloids and certain other substances from the fluid, such as red beds and polymer flocculation. (2) A process in which dissimilar electrical charges on clay platelets are attracted to each other. This increases the yield point and gel strength of a slurry.

Flooding. (1) The effect created when a screen, hydrocyclone, or centrifuge is fed beyond its capacity. (2) Flooding may also occur on a screen as a result of blinding.

Flowback pan. A pan or surface below a screen that causes fluid passing through one screen to flow back to the feed end of a lower screen.

Flow capacity. The rate at which a shaker can process drilling fluid and solids. This depends on rnany variables, including shaker configuration, design and motion, drilling fluid rheology, solids loading, and blinding by near-size particles. See: *feed capacity*.

Flow drilling. Drilling in which there is a constant flow of formation fluid.

Flowline. The pipe (usually) or trough that conveys drilling fluid from the rotary nipple to the solids-separation section of the drilling fluid tanks on a drilling rig.

Flow rate. The volume of liquid or slurry moved through a pipe in one unit of time, that is, gpm, bbl/min, etc. See: *circulation rate*.

Flow streams. With respect to centrifugal separators, all liquids and slurries entering and leaving a machine, such as feed drilling fluid stream plus dilution stream equals overflow stream plus underflow stream.

Fluid. Any substance that will readily assume the shape of the container in which it is placed. The term includes both liquids and gases. It is a substance in which the application of every system of stress (other than hydrostatic pressure) will produce a continuously increasing deformation without any relation between time rate of deformation at any instant and the magnitude of stress at the instant.

Fluid flow. The state of dynamics of a fluid in motion as determined by the type of fluid (e.g., Newtonian plastic, pseudoplastic, dilatant), the properties of the fluid such as viscosity and density, the geometry of the system, and the velocity. Thus, under a given set of conditions and fluid properties, the fluid flow can be described as plug flow, laminar (called also Newtonian, streamline, parallel, or viscous) flow, or turbulent flow. See: *Reynolds number*.

Fluid loss. Measure of the relative amount of fluid loss (filtrate) through permeable formations or membranes when the drilling fluid is subjected to a pressure differential. See: *filtrate loss, API RP 13B*.

Fluidity. The reciprocal of viscosity. The measure of rate with which a fluid is continuously deformed by a shearing stress. Ease of flowing.

Fluorescence. Instantaneous re-emission of light of a greater wavelength than that of the light originally absorbed.

Flute. A curved metal blade wrapped around a shaft as on a screw conveyor in a centrifuge. See: *blade, flight*.

Foam. (1) A two-phase system, similar to an emulsion, in which the dispersed phase is a gas or air. (2) Bubbles floating on the surface of the drilling fluid. The bubbles are usually air but can be formation gas.

Foaming agent. A substance that produces fairly stable bubbles at the air/liquid interface due to agitation, aeration, or ebullition. In air or gas drilling, foaming agents are added to turn water influx into aerated foam. This is commonly called "mist drilling."

Foot. Unit of length in British (foot-pound-second) system.

Foot-pound. Unit of work or of mechanical energy, which is the capacity to do work. One foot is the work performed by a force of 1 pound acting through a distance of 1 foot; or the work required to lift a 1-pound weight a vertical distance of 1 foot.

Foot valve. A check valve installed at the suction end of a suction line.

Formation. A bed or deposit composed throughout of substantially the same kind of rock.

Formation damage. Damage to the productivity of a well as a result of invasion of the formation by drilling-fluid particles, drilling-fluid filtrates, and/or cement filtrates. Formation damage can also result from changes in pH and a variety of other conditions. Asphalt from crude oil will also damage some formations. See: *mudding off*.

Formation fluid. The fluid—brine, oil, gas—that is in the pores of a formation.

Formation sensitivity. The tendency of certain producing formations to adversely react with the drilling and completion process.

Free liquid. The liquid film that can be removed by gravity draining or centrifugal force. See: *absorb, absorption, adsorption, adsorb, bound liquid.*

Free-water knockout. A water/gas separator ahead of the flare line.

Freshwater drilling fluid. A drilling fluid in which the liquid phase is freshwater.

Freshwater mud. See: *freshwater drilling fluid.*

Friction loss. See: *pressure drop, pressure loss.*

Functions of drilling fluids. Drilling fluids in rotary drilling must remove cuttings from the bottom of the hole, bring those cuttings and any material from the side of the hole to the surface, subsurface formation pressures, cool the drill bit, lubricate the drill string, create an impermeable filter cake, refrain from invading the formations with excessive quantities of drilling-fluid filtrate, and provide a well bore that can be evaluated and produce hydrocarbons.

Funnel viscosity. See: *kinematic viscosity, marsh funnel viscosity.*

Galena. Lead sulfide (PbS). Technical grades (specific gravity about 7.0) are used for increasing the density of drilling fluids to points impractical or impossible with barite. Almost entirely used in preparation of "kill fluids." See: *kill fluid.*

Gas buster. See: *poor boy degasser, mud/gas separator.*

Gas cut. Gas entrained by a drilling fluid. See: *air cutting, aeration.*

Gel. (1) A state of a colloidal suspension in which shearing stresses below a certain finite value fail to produce permanent deformation. The minimum shearing stress that will produce permanent deformation is known as the shear or gel strength of the gel. Gels commonly occur when the dispersed colloidal particles have a great affinity for the dispersing medium, that is, are lyophilic. Thus, gels commonly occur with bentonite in water. (2) A term used to designate highly colloidal, high-yielding, viscosity-building, commercial clays, such as bentonite and attapulgite. See: *gel strength.*

Gelation. Association of particles forming continuous structures at low shear rates.

Gel cement. Cement having a small to moderate percentage of bentonite added as a filler and/or reducer of the slurry weight. The bentonite may be dry-blended into the mixture or added as a prehydrated slurry.

Gel strength. (1) The ability or measure of the ability of a colloid to form gels. Gel strength is a pressure unit usually reported in lb/100 sq ft. It is a measure of the same interparticle forces of a fluid as determined by the yield point, except that gel strength is measured under static conditions, whereas the yield point is measured under dynamic conditions. The common gel strength measurements are initial, 10-minute, and 30-minute gels. (2) The measured initial gel strength of a fluid is the maximum reading (deflection) taken from a direct-reading viscometer after the fluid has been quiescent for 10 seconds. It is reported in lb/100 sq ft. See: *API RP 13B, shear rate, shear stress, thixotropy.*

Gelled up. Oilfield slang usually referring to any fluid with a high gel strength and/or highly viscous properties. Often a state of severe flocculation. See: *clabbered.*

g Factor. The acceleration of an object relative to the acceleration of gravity.

g Force. The centrifugal force exerted on a mass moving in a circular path. See g *factor.*

Glosses. Explanations or comments to elucidate some difficulty or obscurity in the text; or annotations.

Grains per gallon (gpg). Ppm equals gpg × 17.1.

Greasing out. In some cases, certain organic substances, usually fatty acid derivatives, that are added to drilling fluids as emulsifiers, EPLs, etc., may react with ions such as calcium and magnesium to form a water-insoluble, greasy material that separates out from the drilling fluid. This separation process is called greasing out.

Guar gum. A naturally occurring hydrophilic polysaccharide derived from the seed of the guar plant. The gum is chemically classified as a galactomannan. Guar gum slurries made up in clear fresh- or brine water possess pseudoplastic flow properties.

Gum. Any hydrophilic plant polysaccharides or their derivatives that, when dispersed in water, swell to produce a viscous dispersion or solution. Unlike resins, they are soluble in water and insoluble in alcohol.

Gumbo: Small, sticky drilled solids that hydrate as they move up an annulus, forming large of cuttings. Gumbo is characteristically observed with water-based drilling fluids during drilling of shales containing large quantities of smectite clay.

Gunk plug. A volume of bentonite in oil that is pumped in a well to combat lost circulation. the bentonite encounters water, it expands and creates a gunk plug with a very high viscosity and gel structure. The plug may or may not be squeezed. See: *diesel oil plug.*

Gunning the pits. Mechanical agitation of the drilling fluid in a pit by means of a mud gun. See: *mud gun*.

Gyp. Gypsum.

Gypsum. Calcium sulfate, $CaSO_4 \cdot 2H_2O$, frequently encountered while drilling. It may occur as thin stringers or in massive formations. See: *anhydrite, calcium sulfate*.

Hardness (water). The hardness of water is due principally to calcium and magnesium ions. The total hardness is measured in terms of parts per million of calcium carbonate or calcium and sometimes epm of calcium. See: *API RP 13B*.

Head. The height a column of fluid would stand in an open-ended pipe if it was attached to the point of interest. The head at the bottom of a 1000-ft well is 1000 ft, but the pressure would be dependent on the density of the drilling fluid in the well.

Heaving. The partial or complete collapse of the walls of a hole resulting from internal pressures due primarily to swelling from hydration or formation pressures or from internal stresses. See: *sloughing*.

Heavy solids. See: *high-gravity solids*.

Hertz. A unit of frequency: cycles per second.

Heterogeneous. A substance that consists of more than one phase and is not uniform, such as colloids, emulsions, etc. It has different properties in different parts.

High-gravity solids (HGS). Solids purchased and added to a drilling fluid specifically and solely to increase drilling-fluid density. Barite (4.2 specific gravity) and hematite (5.05 specific gravity) are the most common additives used for this purpose. See: *low-gravity solids*.

High-pH drilling fluid. A drilling fluid with a pH range above 10.5. A high-alkalinity drilling fluid. See: *pH*.

High-yield clay. A classification given to a group of commercial drilling-clay preparations having yield of 35 to 50 bbl/ton, an intermediate rating between bentonite and low-yield clays. High-yield drilling clays are usually prepared by peptizing low-yield calcium montmorillonite clays or, in a few cases, by blending some bentonite with the peptized low-yield clay. See: *low-yield clay, bentonite*.

HLB. Hydrophilic-lipophilic balance. See: *hydrophilic-lipophilic balance*.

Homogeneous. Of uniform or similar nature throughout, or a substance or fluid that has at all points the same property or composition.

Hook strips. The hooks on the edges of a screen section of a shale shaker that accept the tension member for screen mounting.

Hook-strip panel. One of the two main screen panel types, which consists of one to three layers of screen bordered by metal strips running parallel to the loom. The metal strips have a U-shaped cross section that allows them to be secured and stretched by the shaker tensioning drawbars. These screens are nonpretensioned. See: *Rigid frame panel*.

Hopper. A large funnel-shaped or cone-shaped device for mixing dry solids or liquids into a drilling-fluid stream in order to uniformly mix these materials into the slurry. The solids are wetted prior to entry into the drilling-fluid system. The system usually consists of a jet nozzle, an open top hopper, and a downstream venturi. See: *mud hopper*.

Horsepower. The rate of doing work or of expending mechanical energy; that is, horsepower is work performed per unit of time.

$$1 \text{ hp} = 550 \text{ ft-lb per sec} = 0.7067 \text{ Btu per sec.}$$
$$= 0.7457 \text{ kilowatt (rated horsepower, converted to kilowatts}$$
$$- \text{ horsepower} \times 0.746 = \text{kilowatts})$$

Motor nameplate horsepower is the maximum steady load that the motor can pull without damage.

Horsepower-hour. Horsepower-hour (hp-hr) and kilowatt-hour (kW-hr) are units of work.

$$1 \text{ hp-hr} = 1,980,000 \text{ ft-lb} = 2545 \text{ Btu}$$
$$1 \text{ hp-hr} = 0.7457 \text{ kW-hr}$$
$$1 \text{ kW-hr} = 1.341 \text{ hp-hr} = 3413 \text{ Btu} = 2,655,000 \text{ ft-lb}$$

Horseshoe effect. The U shape formed by the leading edge of drilling fluid moving down a shale shaker screen. The drilling fluid usually tends to pass through the center of a crowned screen faster than it passes through the edges, creating the U shape.

HTHP. High temperature high pressure.

HTHP filter press. A device used to measure the fluid loss under HTHP conditions. See: *HTHP fluid loss*.

HTHP fluid loss. The fluid loss measured under HTHP conditions, usually 300°F and 500 psi differential pressure. See: *HTHP filter press*.

Humic acid. Organic acids of indefinite composition found in naturally occurring leonardite lignite. The humic acids are the active constituents that assist in the positive adjustment of drilling-fluid properties. See: *lignin*.

Hydrate. A substance containing water combined in molecular form (such as $CaSO_4 \cdot 2H_2O$). A crystalline substance containing water of crystallization.

Hydration. The act of a substance to take up water by means of absorption and/or adsorption; usually results in swelling, dispersion and disintegration into colloidal particles. See: *absorb, absorption, adsorb, adsorbed liquid.*

Hydroclone. See: *cyclone, hydrocyclone.*

Hydrocyclone. A liquid/solids separation device utilizing centrifugal force for settling. Fluid tangentially and spins inside the cone. The heavier solids settle to the walls of the cone and move downward until they are discharged at the cone bottom (cone apex). The spinning fluid travels partway down the cone and back up to exit out the top of the cone through the vortex finder.

Hydrocyclone balance point. (1) That adjustment of the apex that creates an opening about the same diameter as the air cylinder inside of the hydrocyclone. (2) In the field, to adjust a balanced design hydrocyclone during the setup of the solids-control system so that it discharges only a slight drip of water at the underflow opening.

Hydrocyclone size. The maximum inside working diameter of the cone part of a hydrocyclone.

Hydrocyclone underflow. The discharge stream from a hydroclone that contains a higher percentage of solids than does the feed. See: *solids discharge.*

Hydrogen ion concentration. A measure of either the acidity or alkalinity of a solution, normally expressed as pH. See: *pH.*

Hydrolysis. Hydrolysis is the reaction of a salt with water to form an acid or base. For example, soda ash (Na_2CO_3) hydrolyzes basically, and hydrolysis is responsible for the increase in the pH of water when soda ash is added.

Hydrometer. A floating instrument for determining the specific gravity or density of liquids, solutions, and slurries.

Hydrophile. Any substance, usually in the colloidal state or an emulsion, that is wetted by water; that is, it attracts water or water adheres to it. See: *lipophile.*

Hydrophilic. A property of a substance having an affinity for water or one that is wetted by water. See: *lipophilic.*

Hydrophilic lipophilic balance. The relative attraction of an emulsifier for water and for oil. It is determined largely by the chemical composition and ionization characteristics of a given emulsifier. The HLB of an emulsifier is not directly related to its solubility, but it determines the type of an emulsion that tends to be formed. It is an indication of the behavioral characteristics and not an indication of emulsifier efficiency.

Hydrophobe. Any substance, usually in the colloidal state, that is not wetted by water.

Hydrophobic. Any substance, usually in the colloidal state or an emulsion, that is not wetted by water; that is, it repels water or water does not adheres to it.

Hydrostatic pressure head. The pressure exerted by a column of fluid, usually expressed in pounds per square inch. To determine the hydrostatic head in psi at a given depth, multiply the depth in feet by the density in pounds per gallon by the conversion factor, 0.052.

Hydroxide. Designation that is given basic compounds containing the OH^- radical. When these substances are dissolved in water, the pH of the solution is increased. See: *base, pH*.

Hygroscopic. The property of a substance enabling it to absorb water from the air.

ID. Inside diameter of a pipe.

Ideal nozzle. Orifice that will pass fluid without friction loss, theoretically.

Impeller. A spinning disc in a centrifugal pump with protruding vanes used to accelerate the fluid in the pump casing.

Indicator. Substances in acid/base titrations that in solution change color or become colorless as the hydrogen ion concentration reaches a definite value. These values vary with the indicator. In other titrations such as chloride, hardness, and other determinations, these substances change color at the end of the reaction. Common indicators are phenolphthalein, methyl orange, and potassium chromate.

Inertia. Force that makes a moving particle tend to maintain its direction or a particle at rest to remain at rest.

Inhibited drilling fluid. A drilling fluid having an aqueous phase with a chemical composition that tends to retard and even prevent (inhibit) appreciable hydration (swelling) or dispersion formation clays and shales through chemical and/or physical reactions. See: *calcium-treated drilling fluids, saltwater drilling fluid*.

Inhibited mud. See: *inhibited drilling fluid*.

Initial gel. See: *gel strength*.

Inlet. The opening through which the feed mud enters a solids-control device. See: *feed inlet, feed opening*.

Interfacial tension. The force required to break the surface definition between two immiscible liquids. The lower the interfacial tension between the two phases of an emulsion, the greater the ease of emulsification. When the values approach zero, emulsion formation is spontaneous. See: *emulsion, surface tension.*

Intermediate (solids). Particles whose diameter is between 250 and 2000 microns.

Intercalation. A shale stabilization mechanism that involves penetration of a foreign material, such as a glycol, between clay lamellae in a shale to retard interaction of the clay with water.

Interstitial water. Water contained in the interstices or voids of formations.

Intrinsic safety. A feature of an electrical device or circuit in which any spark or thermal effect from the electrical device or circuit is incapable of causing ignition of a mixture of flammable or combustible material in air.

Invert drilling fluid. See: *invert oil emulsion drilling fluid.*

Invert oil emulsion drilling fluid. A water-in-oil emulsion in which water (sometimes containing sodium or calcium chloride) is the dispersed phase, and diesel oil, crude oil, or some other oil is the phase. Water addition increases the emulsion viscosity, and oil reduces the emulsion viscosity. The water content exceeds 5% by volume. See. *oil-based drilling fluid*

Iodine number. The number indicating the amount of iodine absorbed by oils, fats, and waxes, giving a measure of the unsaturated linkages present. Generally, the higher the iodine number, the more severe the destructive action of the oil on rubber.

Ions. Molecular condition due to loss or gain of electrons. Acids, bases, and salts electrolytes), when dissolved in certain solvents, especially water, are more or less dissociated into electrically charged ions or parts of the molecules. Loss of electrons results in positive charges, producing a cation. A gain of electrons in the formation of an anion, with negative charge. The valence of an ion is equal to the number of charges borne by the ion. See: *anion, cation.*

Irreducible fraction. See: *adsorbed liquid, bound liquid.*

Jet. See: *eductor.*

Jet hopper. A device that has a jet that facilitates the addition of drilling-fluid additives to the system. See: *hopper, mud hopper.*

Jetting. The process of periodically removing a portion of the water, drilling fluid, and/or solids from the pits, usually by means of pumping through a jet nozzle to agitate the drilling fluid while simultaneously removing it from the pit.

Jones effect. The net surface tension of all salt solutions first decreases with an increase in concentration, passes through a minimum, and then increases as the concentration is raised. The initial decrease is called the Jones effect.

Kelly. A heavy square or hexagonal pipe that passes through rollers in a bushing on the drill floor to transmit rotational torque to the drill string.

Key seat. A section of a hole, usually of abnormal deviation and relatively soft formation, that has been eroded or worn by drill pipe to a size smaller than the tool joints or collars of the drill string. This keyhole-type configuration resists passage of the shoulders of these pipe upset (box) configurations when pulling out of the hole.

Kick. Situation caused when the annular hydrostatic pressure in a drilling well temporarily (and usually relatively suddenly) becomes less than the formation, or pore, pressure in a permeable downhole section. A kick occurs before control of the fluid intrusion is totally lost. A blowout is an uncontrolled influx of formation fluid into the well bore. See: *blowout, kill fluid*.

Kill fluid. A fluid built with a specific density aimed at controlling a kick or blowout. See: *galena*.

Kill line. A line connected to the annulus below the blowout preventers for the purpose of pumping into the annulus while the preventers are closed.

Killing a well. (1) Bringing a well kick under control. (2) The procedure of circulating a fluid into a well to overbalance formation fluid pressure after the bottom-hole pressure has been less than formation fluid pressure. See: *kick, blowout, kill fluid*.

Kilowatt-hour. Horsepower-hour (hp-hr) and kilowatt-hour (kW-hr) are units of work.

$$1 \text{ hp-hr} = 1{,}980{,}000 \text{ ft-lb} = 2545 \text{ Btu}$$

$$1 \text{ hp-hr} = 0.7457 \text{ kW-hr}$$

$$1 \text{ kW-hour} = 1.341 \text{ hp-hr} = 3413 \text{ Btu} = 2{,}655{,}000 \text{ ft-lb}.$$

Kinematic viscosity. The kinematic viscosity of a fluid is the ratio of the viscosity (e.g., cP in g/cm-sec) to the density (e.g., g/cc) using consistent units. In several common commercial viscometers, the kinematic viscosity is measured in terms of the time of efflux, in seconds, of a fixed volume of liquid through a standard capillary tube or orifice. See: *marsh funnel viscosity*.

In laminar flow, the fluid moves in plates or sections with a differential velocity across the front of the flow profile that varies from zero at the wall to a maximum toward the center for flow. These fluid elements flow along fixed stream lines that are parallel to the walls of the channel of flow. Laminar flow is the first stage of flow in a Newtonian fluid. It is the second stage of flow in a Bingham plastic fluid. This type of motion is also called parallel, streamline, or viscous flow. See: *plug flow, parallel flow, turbulent flow*.

LCM. Circulation material. See: *lost circulation materials*.

Lead. In a decanting centrifuge, the slurry-conducting channel formed by the adjacent walls of the flutes or blades of the screw conveyor.

Leonardite. A naturally occurring oxidized lignite. See: humic acid, lignin.

Light solids. See: *low-gravity solids*.

Lignin. Mined lignin is a naturally occurring special lignite, for example, leonardite, produced by strip mining from special lignite deposits. The active ingredients are the humic acids. Mined lignins are used primarily as thinners, which may or may not be chemically modified. See: *leonardite, humic acid*.

Lignosulfonates. Organic drilling-fluid additives derived from by-products of the sulfite paper manufacturing process from coniferous woods. Some of the common salts, such as ferrochrome, chrome, calcium, and sodium, are used as deflocculants while other lignosulfonates are used selectively for calcium-treated systems. In large quantities, the "heavy metal" ferrochrome and chrome salts are used for fluid loss control and shale inhibition.

Lime. $Ca(OH)_2$. Commercial form of calcium hydroxide.

Lime-treated drilling fluids. Commonly referred to as "lime-based" muds. These high-pH systems contain most of the conventional freshwater drilling-fluid additives to which slaked lime has been added to impart special inhibition properties. The alkalinities and lime contents of the fluids may vary from low to high. See: *calcium-treated drilling fluids*.

Limestone. $Ca(CO)_3$. See: *calcium carbonate*.

Line sizing. Ensuring that the fluid velocity through all piping within the surface system has the proper flow and pipe diameter combination to prevent solids from settling and pipe from eroding. A good rule of thumb is to ensure that fluid flow is between 5 and 9 feet per second, as determined by the following:

$$V_{ft/sec} = [Q_{gpm} \times 0.4087]/d^2$$

where

$V_{ft/sec}$ = velocity of flow, in feet per second
Q_{gpm} = flow rate in gallons per minute
d = ID of the pipe, in inches

Linear motion. Linear motion of a shale shaker screen is produced by two counterrotational motors located above the shaker basket in such a way that a line connecting the two motor axes is perpendicular to a line passing through the center of gravity of the basket. Because the acceleration is applied directly through the center of gravity of the basket, the basket is dynamically balanced; the same pattern of motion will exist at all points along the shaker screen. The resultant screen motion is linear, and the angle of this uniform motion is usually 45° to 60° relative to the shaker screen deck.

Lipophile. Any substance, usually in the colloidal state or an emulsion, that is wetted by oil; that is, it attracts oil or oil adheres to it. See: *hydrophile*.

Lipophilic. A property of a substance having an affinity for oil or one that is wetted by oil. See: *hydrophilic*.

Liquid. Fluid that will flow freely and takes the shape of its container.

Liquid-clay phase. See: *overflow*.

Liquid discharge. See: *underflow*.

Liquid film. The liquid surrounding each particle discharging from the solids discharge of cyclones and screens. See: *bound liquid, free liquid*.

Live oil. Crude oil that contains gas and distillates and has not been stabilized or weathered. This oil can cause gas cutting when added to drilling fluid and is a potential fire hazard. See: *aromatic hydrocarbons*.

Load. A device connected to a motor that is receiving output mechanical power from the motor.

Logging. See: *mud logging, electric logging*.

Loom. See: *warp*.

Loss of circulation. See: *lost circulation*.

Lost circulation. The result of drilling fluid escaping into a formation, usually in fractures, cavernous, fissured, or coarsely permeable beds, evidenced by the complete or partial failure of the drilling fluid to return to the surface as it is being circulated in the hole.

Lost circulation additives. Materials added to the drilling fluid to gain control of or prevent the loss of circulation. These materials are added in varying amounts and are classified as fibrous, flake, or granular.

Lost circulation materials. See: *lost circulation additives.*

Lost returns. See: *lost circulation.*

Low-gravity solids. Salts, drilled solids of every size, commercial colloids, lost circulation materials; that is, all solids in drilling fluid, except barite or other commercial weighting materials. Salt is considered a low–specific gravity solid. See: *heavy solids, high-gravity solids.*

Low-silt drilling fluid. An unweighted drilling fluid that has all the sand and a high proportion of the silts removed and has a substantial content of bentonite or other water–loss–reducing clays.

Low-silt mud. See: *low-silt drilling fluid.*

Low-solids drilling fluids. A drilling fluid that has polymers, such as ceramic matrix compound (CMC) or xanthan gum (XC) polymer, partially or wholly substituted for commercial or natural clays. For comparable viscosity and densities, a low-solids drilling fluid will have a lower volume percentage solids content. In general, the lower the solids content in a mud, the faster a bit can drill.

Low-solids muds. See: *low-solids drilling fluids.*

Low-solids nondispersed (LSND) drilling fluids. A drilling fluid to which polymers have been added to simultaneously extend and flocculate bentonite drilled solids. These fluids contain low concentrations of dispersed bentonite and do not contain deflocculants such as lignites, lignosulfonates, etc.

Low-yield clay. Commercial clay chiefly of the calcium montmorillonite type having a yield of approximately 15 to 30 barrels per ton. See: *high-yield clay, bentonite.*

Lyophilic. Having an affinity for the suspending medium, such as bentonite in water.

Lyophlic colloid. A colloid that is not easily precipitated from a solution and is readily dispersible after precipitation by addition of a solvent.

Lyophobic colloid. A colloid that is readily precipitated from a solution and cannot be redispersed by addition of the solution.

Main shaker. The shale shaker that processes drilling fluid from the flowline through the finest-mesh screen.

Manifold. (1) A length of pipe with multiple connections for collecting or distributing fluid. (2) A piping arrangement through which liquids, solids, or slurries from one or more sources can be fed to or discharged from a solids-separation device.

Market grade cloth. A group of industrial wire cloth specifications selected for general-purpose work, made of high-strength, square mesh cloth in several types of metals. The common metal for oilfield use is

304 or 316 stainless steel. The wire diameters are marginally larger than mill grade cloth, resulting in a lower percentage of open area. Market grade and mill grade cloths are used mostly as support screens for fine-mesh screens. See: *mill grade cloth, tensile bolting cloth, ultrafine wire cloth, support screen,* and *calendered.*

Marsh funnel. An instrument used in determining the Marsh funnel viscosity. The Marsh funnel is a container with a fixed orifice at the bottom so that when filled with 1500 cc freshwater, 1 qt (946 ml) will flow out in 26 ± 0.5 sec. For 1000 cc out, the efflux time for water is $27.5 \pm$ sec. It is used for comparison values only and not to diagnose drilling fluid problems. See: *API Bulletin RP 13B, funnel viscosity, Marsh funnel viscosity, kinematic viscosity.*

Marsh funnel viscosity. Commonly called funnel viscosity. The Marsh funnel viscosity is reported as the time, in seconds, required for 1 qt of fluid to flow through an API standardized funnel. In some areas, the efflux quantity is 1000 cc. See: *API RP 13B, funnel viscosity, kinematic viscosity, Marsh funnel.*

Martin's radii. The distance from the centroid of an object to its outer boundary. The direction of this measurement is specified by the azimuth orientation of the line (the radii in the 0°, 90°, 180°, 270° angle from horizontal).

Mass. The inertial resistance of a body to acceleration, considered in classical physics, to a conserved quantity independent of speed. The weight of a body is the product of the mass of the body and the acceleration of gravity for the specific location. In space the mass would stay constant but the weight would disappear as the gravitational acceleration approaches zero.

MBT. Methylene blue test. See: *methylene blue test.*

Mechanical agitator. A device used to mix, blend, or stir fluids by means of a rotating impeller blade. See: *agitator, mechanical stirrer.*

Mechanical stirrer. See: *agitator, mechanical agitator.*

Median cut. The median cut is the particle size that reports 50% of the weight to the overflow 50% of the weight to the underflow. Frequently identified as the D_{50} point. See: *cut point.*

Medium (solids). Particles whose diameter is between 74 and 250 microns.

Membrane nitrogen. Air from which water and oxygen have been removed by a filter (membrane) system.

Meniscus. The curved upper surface of a liquid column, concave when the containing walls are wetted by the liquid and convex when they are not wetted.

Mesh. (1) The number of openings (and fraction thereof) per linear inch in a screen, counted in both directions from the center of a wire. (2) An indication of the weave of a woven material, screen or sieve. A 200 mesh sieve has 200 openings per linear inch. A 200 mesh screen with a wire diameter of 0.0021 inch (0.0533 mm) has an opening of 0.0029 in. (0.074) mm and will pass a spherical particle of 74 microns diameter. See: *micron*.

Mesh count. Such as 30 × 30, or often 30 mesh, indicating the number of openings per linear inch of screen; and a square mesh. A designation of 70 × 30 mesh indicates rectangular mesh with 70 openings per inch in one direction and 30 openings per inch in a perpendicular direction.

Mesh equivalent. As used in oilfield drilling applications, the U.S. Sieve number that has the same-size opening as the minimum opening of the screen in use.

Methylene blue test. A test that serves to indicate the amount of active clay in a fluid system, clay sample, or shale sample. Methylene blue is titrated into a slurry until all of the negative charge sites are covered with the methylene blue. This indicates the number of active charge sites present in the slurry. See: *base exchange, methylene blue titration, MBT, cation exchange capacity, CEC*.

Methylene blue titration. Methylene blue is a cation that seeks all negative charges on a clay surface after the surface has been properly prepared (see *API RP13B*). By titrating with a known concentration, this test provides an indication of the amount of clay present in the drilling fluid. See: *methylene blue test, MBT, cation exchange capacity, CEC*.

Mf. The methyl orange alkalinity of the filtrate, reported as the number of milliliters of 0.02 normal sulfuric acid required per milliliter of filtrate to decrease the pH to reach the methyl orange endpoint (pH 4.3).

Mica. Naturally occurring mineral flake material of various sizes used in controlling lost circulation. An alkali aluminum silicate.

Micelles. Organic and inorganic molecular aggregates occurring in colloidal solutions. chains of individual structural units chemically joined to one another and deposited side by side to form bundles. When bentonite hydrates, certain sodium, or other metallic ions go into solution, the clay particle plus its complement of ions is technically known as a micelle.

Micron. A unit of length equal to one-thousandth of a millimeter. Used to specify particle sizes in drilling fluids and solids control discussions (25,400 microns = 1 inch).

Mil. A unit of length equal to 1/1000 inch.

Milk emulsion. See: *oil-in-water emulsion drilling fluid.*

Mill grade cloth. A group of industrial wire cloth specifications with lighter wire than market grade cloth. The standard wire diameter of the grade produces a median percentage of open area. Market grade and mill grade cloths are used mostly as support screens for fine-mesh screens. See: *market grade cloth, tensile bolting cloth, ultra-fine wire cloth, support screen,* and *calendered.*

Millidarcy. 1/1000 darcy. See: *darcy.*

Milliliter. A metric system unit for the measurement of volume. Literally 1/1000th of a liter. In drilling-fluid analyses, this term is used interchangeably with cubic centimeter (cc). One quart is equal to approximately 946 ml.

Mini still. An instrument used to distill oil, water, and any other volatile material in a drilling fluid to determine oil, water, and total solids contents as volume percentage. See: *distillation, mud still.*

Mist drilling. A method of rotary drilling whereby water and/or oil is dispersed in air and/or gas as the drilling fluid. See: *foam.*

ml. See: *milliliter.*

Molecule. Atoms combine to form molecules. For elements or compounds, a molecule is the smallest unit that chemically still retains the properties of the substance in mass.

Monovalent. See: *valence.*

Montmorillonite. A clay mineral commonly used as an additive to drilling muds. Sodium montmorillonite is the main constituent of bentonite. Each platelet of the crystalline structure of montmorillonite has two layers of silicon tetrahedra attached to a center layer of alumina octahedra. The platelets are thin and have a broad surface. Exchangeable cations are located on the clay surfaces between the platelets. Calcium montmorillonite is the main constituent in low-yield clays. See: *gel, bentonite.*

Mud. See: *drilling fluid, which is the preferred term.*

Mud analysis. See: *drilling fluid analysis, API RP 13B.*

Mud balance. A beam-type balance used in determining drilling-fluid density (mud weight). It consists primarily of a base, a graduated beam with constant volume cup, lid, rider, knife-edge, and counterweight. See: *API RP 13B.*

Mud box. See: *back tank, possum belly.*

Mud cleaner. A device that places a screen in series with the underflow of hydrocyclones. The hydrocyclone overflow returns to the mud system, and the underflow reports to a vibrating screen. Solids discharged from the screen are discarded and the screen throughput returns to the system.

Mud compartment. A subdivision of the removal, additions, or check/suction sections of a surface system. See: *mud pits, mud tanks.*

Mud ditch. A trough built along the upper edge of many surface systems that is used to direct flow to selected compartments of the surface system. See: *mud pits, mud compartment.*

Mud engineer. See: *drilling fluid engineer.*

Mud gun. A submerged nozzle used to stir the drilling fluid with a high-velocity stream. See *gunning the pits.*

Mud hopper. See: *hopper.*

Mud house. A structure at the rig to store and shelter sacks of materials used in drilling fluids.

Mud inhibitor. Additives such as salt, lime, lignosulfonate, and calcium sulfate that prevent clay dispersion.

Mud logging. A process that helps determine the presence or absence of oil or gas in the various formations penetrated by the drill bit, and assists with a variety of indicators that assist drilling operations. Drilling fluid and cuttings are continuously tested on their return to the surface, and the results of these tests are correlated with the drilling depth for depth of origin.

Mud mixing devices. The most common device for adding solids to the drilling fluid is by means of the jet hopper. Some other devices to assist mixing are eductors, mechanical agitators, paddle mixers, electric stirrers, mud guns, chemical barrels, etc.

Mud pit. See: *mud compartments, mud tanks.*

Mud pump. Pumps at the rig used to circulate drilling fluids.

Mud scales. See: *mud balance.*

Mud still. See: *distillation, mini still.*

Mud tanks. (1) Drilling-fluid system compartments constructed of metal and mounted so they can be moved from location to location, either as a part of the rig (such as on a semisubmersible rig) or separately on unitized skids (as on most land rigs). (2) Earthen or steel storage facilities for the surface system. Mud pits are of two types: circulating and reserve. Drilling-fluid testing and conditioning is normally done in the circulating pit system.

Mud weight. A measurement of density of a slurry usually reported in lb/gal, lb/cu ft, psi/1000 ft or specific gravity. See: *density.*

Mud/gas separator. A vessel into which the choke line discharges when a "kick" is being taken. Gas is separated in the vessel as the drilling fluid flows over baffle plates. The gas flows through a line to a flare. The liquid mud discharges into the shale shaker back tank. See: *gas buster, poor boy degasser.*

Mudding off. A condition promoting reduced production caused by the penetrating, sealing or plastering effect of a drilling fluid. See: *formation damage.*

Mudding up. Process of mixing drilling fluid additives to a simple, native clay water slurry to achieve some properties not possible with the previous fluid.

MW. Abbreviation for mud weight. See: *density, mud weight.*

Natural clays. Natural clays, as opposed to commercial clays, are clays that are encountered when drilling various formations. The yield of these clays varies greatly, and they may or may not be purposely incorporated into the drilling fluid system. See: *attapulgite clay, bentonite, high-yield clay, low-yield clay, clay.*

Near-size plugging. A term used in describing screen plugging, referring to particles with a dimension slightly larger than the screen opening. See: *blinding, plugging.*

Neat cement. A slurry composed only of Portland cement and water.

Negative deck angle. The angle of adjustment to a screen deck that causes the screened solids to travel "downhill" (usual travel) to reach the discharge end of the screen surface. This downhill travel decreases the fluid throughput of a screen but usually lengthens the life of a screen. See: *positive deck angle.*

Neutralization. A reaction in which the hydrogen ion of an acid and the hydroxyl ion of a base unite to form water, the other ionic product being a salt.

Newtonian flow. See: *Newtonian fluid.*

Newtonian fluid. The basic and simplest fluids from the standpoint of viscosity, in which the shear force is directly proportional to the shear rate. These fluids will immediately begin to move when a pressure or force in excess of zero psi is applied. Examples of Newtonian fluids are water, diesel oil, and glycerine. The yield point as determined by direct-indicating viscometer is zero. See: *Newtonian flow.*

Nonconductive drilling fluid. Any drilling fluid, usually oil-based or invert-emulsion drilling fluid, whose continuous phase does not conduct electricity. The spontaneous potential (SP) and normal resistivity cannot be logged, although such other logs as the gamma rays, induction, acoustic velocity, etc., can be run.

Nondispersed. A condition in which the clays do not separate into individual platelets. Dispersion is inhibited.

Normal solution. A solution of such a concentration that it contains 1 gram equivalent of a substance per liter of solution.

Oblong mesh. A screen cloth that has more openings per inch in one direction than in the perpendicular direction. For example, a 70 × 30 mesh has 70 openings per inch in one direction and 30 openings per inch in the perpendicular direction, creating a rectangular opening. The smaller opening dimension controls the sizing of spherical material. See: *rectangular screen.*

Oblong weave. See: *oblong mesh.*

OD. Outside diameter of a pipe.

Ohm. The measurement of resistance or electrical friction.

Oil-based drilling fluid. The term "oil-based mud" is applied to a special type of drilling fluid in which oil is the continuous phase and water the dispersed phase. Oil-based drilling fluid contains from 1 to 5% water emulsified into the system with lime and emulsifiers. Oil-based muds are differentiated from invert emulsion muds (both water-in-oil emulsions) by the amounts of water used, the method of controlling viscosity, the thixotropic properties, wall-building materials, and fluid loss. See: *invert oil emulsion drilling fluid.*

Oil breakout. Oil that has risen to the surface of a drilling fluid. This oil had been previously emulsified in the drilling fluid or may derive from oil-bearing formations that have been penetrated.

Oil content. The oil content of any drilling fluid is the amount of oil in volume percentage.

Oil immersion. An oil-filled construction in which an electrical device has no electrical connections, joints, terminals, or arcing parts at or above the normal oil level.

Oil wet. A surface on which oil easily spreads. If the contact angle of an oil droplet on a surface is less than 90°, the surface is oil wet. See: *lipophilic, water wet.*

Oil-in-water emulsion drilling fluid. Any conventional or special water-base drilling fluid to which oil has been added. A drilling fluid in which the oil content is usually kept between 3 and 7% and seldom over 10% (it can be considerably higher). Commonly called "emulsion mud." The oil becomes the dispersed phase and may be emulsified into the mud either mechanically or chemically. The oil is emulsified into fresh- or saltwater with a chemical emulsifier

Overflow. The discharge stream from a centrifugal separation that normally contains a higher percentage of liquids than does the feed.

Overflow header. A pipe into which two or more hydrocyclones discharge their overflow.

Overslung. Field terminology denoting that the support ribs for the shaker screen are located below the screen surface. See: *underslung*.

Packer fluid. A fluid placed in the annulus between the tubing and casing above a packer. The hydrostatic pressure of the packer fluid is utilized to reduce the pressure differentials between the formation and the inside of the casing and across the packer.

Panel-mounted units. Shale shaker screens mounted to a rigid frame.

Parallel flow. See: *laminar flow*.

Particle. A discrete unit of solid material that may consist of a single grain or of any number of grains stuck together.

Particle size. Particle diameter expressed in microns. See: *ESD, equivalent spherical diameter*.

Particle size distribution. The volume classification of solid particles into each of the various size ranges as a percentage of the total solids of all sizes in a fluid sample.

Parts per million. The unit weight of solute per million unit weights of solution (solute plus solvent), corresponding to weight percentage. The results of standard API titration of chloride hardness, etc., are correctly expressed in milligrams (mg) per liter but not in ppm. At low concentrations mg/L is about numerically equal to ppm. A correction for the solution specific gravity or density in g/ml must be made as follows:

ppm = [milligrams/liter]/solution density(grams/liter)

weight% = [milligrams/liter]/[10,000 × solution density (grams/liter)]

weight% = [ppm]/[10,000]

Thus, 316,000 mg/L salt is commonly called 316,000 ppm, or 31.6%, which correctly should be 264,000 pprn and 26.4%, respectively.

Pay zone. A formation that contains oil and/or gas in commercial quantities.

Penetration rate. The rate at which the drill bit penetrates the formation, usually expressed in feet per hour or meters per hour. See: *rate of penetration, ROP*.

Peptization. An increased flocculation of clays caused by the addition of electrolytes or other chemical substances. See: *deflocculation dispersion*, *high-yield clay*.

Peptized clay. A clay to which an agent has been added to increase its initial yield. For example, soda ash is frequently added to calcium montmorillonite clay to increase the yield. See: *high-yield clay*.

Percent open area. Ratio of the area of the screen openings to the total area of the screen surface.

Percent separated curve. A plot of mass distributions of solids sizes discarded from a solids-separation device divided by the mass distributions of each size of solids fed to the device.

Perforated cylinder centrifuge. A mechanical centrifugal separator in which the rotating element is a perforated cylinder (the rotor) inside of and concentric with an outer stationary cylindrical case.

Perforated panel screen. A screen in which the backing plate used to provide support to the screen cloths is a metal sheet with openings.

Perforated plate screen. Shale shaker screens mounted on metal plates that have holes punched through.

Perforated rotor. The rotating inner cylinder of the perforated cylinder centrifuge. See: *perforated cylinder centrifuge*.

Permeability. Permeability is a measure of the ability of a formation to allow passage of a fluid. Unit of permeability is the darcy. See: *darcy*, *porosity*.

Pf. The phenolphthalein alkalinity of the filtrate is reported as the number of milliliters of 0.02 normal sulfuric acid required per milliliter of filtrate for the pH to reach the phenolphthalein endpoint, which is a pH of 8.3.

pH. The negative logarithm of the hydrogen ion concentration in gram ionic weights per liter. The pH range is numbered from 0 to 14, with 7 being neutral, and is an index of the acidity (below 7) or alkalinity (above 7) of the fluid. At a temperature of 70°F, a neutral pH is 7 or a hydrogen ion concentration of 10^{-7}. The neutral pH is a function of temperature. At higher elevated temperatures the neutral pH is lower. The pH of a solution offers valuable information as to the immediate acidity or alkalinity, in contrast to the total acidity or alkalinity, which may be determined by titratration.

Phosphate. Certain complex phosphates, commonly sodium tetraphosphate ($Na_6P_4O_{13}$) and sodium acid pyrophosphate (SAPP, $Na_2H_2P_2O_4$), are used either as drilling-fluid thinners or for treatment of various forms of calcium and magnesium contamination.

Piggyback, -ing. The attachment of fine solids particles to the surface of larger solids particles due to surface attraction, fluid consistency, and particle concentration. This attachment phenomenon causes fine solids to be discharged from the screen that would normally pass through the screen.

Pill. A small volume of a special fluid slurry pumped through the drill string and normally placed in the annulus. See: *slug*.

Pilot testing. A method of predicting behavior of drilling-fluid systems by adding various chemicals to a small quantity of drilling fluid (usually 350 cc), then examining the results. One gram of an additive in 350 cc is equivalent to 1 lb/bbl.

Plastic flow. See: *plastic fluid*.

Plastic fluid. A complex, non-Newtonian fluid in which shear force is not proportional to shear rate. A definite pressure is required to start and maintain fluid movement. Plug flow is the initial flow type and only occurs in plastic fluids. Most drilling fluids are plastic fluids. The yield point, as determined by a direct-indicating viscometer, is in excess of zero.

Plastic viscosity. This is a measure of the internal resistance to fluid flow attributable to the concentration, type, and size of solids present in a given fluid and the viscosity of the continuous phase. This value, expressed in centipoise, is proportional to the slope of the shear stress/shear rate curve determined in the region of laminar flow for materials whose properties are described by Bingham's law of plastic flow. When using the direct-indicating viscometer, plastic viscosity is found by subtracting the 300-rpm reading from the 600-rpm reading. See: *viscosity, yield point, API RP 13B*.

Plasticity. The property possessed by some solids, particularly clays and clay slurries, of changing shape or flowing under applied stress without developing shear planes or fractures; that is, it deforms without breaking. Such bodies have yield points, and stress must be applied before movement begins. Beyond the yield point, the rate of movement is proportional to the stress applied, but movement ceases when the stress is removed. See: *fluid*.

Plug flow. The movement of material as a unit without shearing within the mass. Plug flow is the first type of flow exhibited by a plastic fluid after overcoming the initial force required to produce flow. See: *Bingham model, Newtonian fluid, laminar flow, turbulent flow*.

Plugging. The wedging or jamming of openings in a screening surface by near-size particles, preventing passage of undersize particles and leading to the blinding of the screen. See: *blinding, coating*.

Pm. The phenolphthalein alkalinity of drilling fluid is reported as the number of milliliters of 0.02 normal (N/50) sulfuric acid required per milliliter of drilling fluid for the pH to reach the phenolphthalein endpoint of 8.3.

Polyelectrolytes. Long-chain organic molecules possessing ionizable sites that when dissolved in water become charges.

Polymer. A substance formed when two or more molecules of the same kind are linked end to end into another compound having the same elements in the same proportion but higher molecular weight and different physical properties, for example, paraformaldehyde. Polymers are used in drilling fluids to maintain viscosity and control fluid loss. See: *copolymer*.

Polyurethane. A high-performance elastomer polymer used in construction of hydrocyclones for its unique combination of physical properties, especially abrasion, toughness, and resiliency.

Pool. (1) The reservoir or pond of fluid, or slurry, formed inside the wall of hydrocyclones and centrifuges and in which classification or separation of solids occurs due to the settling effect of centrifugal force. (2) The reservoir or pond of fluid that can form on the feed end of an uphill shaker basket, a shaker basket with positive deck angle.

Poor boy degasser. See: *gas buster, mud/gas separator*.

Porosity. The volume of void space in a formation rock usually expressed as percentage of void volume per bulk volume.

Ports. The openings in a centrifuge for entry or exit of materials. Usually applied in connection with a descriptive term, that is, feed ports, overflow ports, etc.

Positive deck angle. The angle of adjustment to a screen deck that causes the screened solids to travel "uphill" to reach the discharge end of the screen surface. This so-called uphill travel increases the fluid throughput of a screen but also shortens the life of a screen. See: *negative deck angle*.

Possum belly. The compartment on a shale shaker into which the flow-line discharges, and from the drilling fluid is fed, either to the screens or to a succeeding tank. See: *back tank, mud box*.

Potassium. One of the alkali metal elements with a valence of 1 and an atomic weight of approximately 39. Potassium compounds, most commonly potassium hydroxide (KOH), are sometimes added to drilling fluids to impart special properties, usually inhibition.

Potential separation curve. A distribution curve of sizes determined by the optical image analysis for separation potential.

Pound equivalent. A laboratory unit used in pilot testing. One gram of a material added to 350 ml of fluid is equivalent to 1 lb of material added to one barrel. See: *barrel, barrel equivalent.*

ppm: Parts per million. See: *parts per million.*

Precipitate. Material that separates out of solution or slurry as a solid. Precipitation of solids in a drilling fluid may follow flocculation or coagulation.

Preformed foam. Foam formed at the drill bit (obsolete).

Prehydration tank. A tank used to hydrate materials (such as bentonite, polymers, etc.) that require a long time (hours to days) to hydrate fully and disperse before being added to the drilling fluid. See: *premix system.*

Premix system. A compartment used to mix materials (such as bentonite, polymers, etc.) that require time to hydrate or disperse fully before they are added to the drilling fluid. See: *prehydration tank.*

Preservative. Any material used to prevent starch or any other organic substance from fermenting via bacterial action. A common preservative is paraformaldehyde. See: *fermentation.*

Pressure drop. See: *friction loss, pressure loss.*

Pressure head. Pressure within a system equal to the pressure exerted by an equivalent height of fluid (expressed in feet or meters). See: *head, hydrostatic head, centrifugal pump.*

Pressure loss. The pressure lost in a pipeline or annulus due to the velocity of the liquid in the pipeline, the properties of the fluid, the condition of the pipe wall, and the configuration of the pipe. See: *friction loss, pressure drop.*

Pressure surge. A sudden, usually brief increase in pressure. When pipe or casing is run into a hole too rapidly or the drill string is set in the slips too quickly, an increase in the hydrostatic pressure results due to pressure surge which may be great enough to create lost circulation. See: *ECD, annular pressure loss.*

Pressurization. The process of supplying an enclosure with a protective gas with or without continuous flow at sufficient pressure to prevent the entrance of a flammable gas or vapor, a combustible dust, or an ignitable fiber.

Pretensioned screen. A screen cloth that is bonded to a frame or backing plate with proper tension applied prior to its installation on a shaker. See: *backing plate, perforated panel screen.*

Progressive gel. A condition wherein the 10-min gel strength is greater than the initial gel strength. Opposite of flat gel. See: *flat gel, zero-zero gel.*

Pseudoplastic fluid. A complex non-Newtonian fluid that does not possess thixotropy. A pressure or force in excess of zero will start fluid flow. The apparent viscosity or consistency decreases instantaneously with increasing rate of shear until at a given point the viscosity becomes constant. The yield point as determined by direct-indicating viscometer is positive, the same as in Bingham plastic fluids. However, the true yield point is zero. An example pseudoplastic fluid is guar gum in fresh- or saltwater. See: *viscosity, Bingham model, plastic viscosity, yield point, gel strength.*

Purging. The process of supplying an enclosure with a protective gas at a sufficient flow and positive pressure to reduce the concentration of any flammable gas or vapor initially present to an acceptable level.

Quebracho. An additive used extensively for thinning/dispersing to control low-shear-rate viscosity and thixotropy. It is a crystalline extract of the quebracho tree consisting mainly of tannic acid. See: *thinner.*

Quicklime. Calcium oxide, CaO. Used in certain oil-based drilling fluids to neutralize the organic acid.

Quiescence. The state of being quiet or at rest, being still. See: *static.*

Radial flow. Flow of a fluid outwardly in a 360° pattern. This describes the flow from a mechanical agitator in which fluid moves away from the axis of the impeller shaft (usually horizontally toward a mud tank wall). See: *axial flow.*

Radical. Two or more atoms behaving as a single chemical unit, that is, as an atom; for instance, sulfate and phosphate are nitrate are radicals.

Rate of penetration. The rate at which the drill bit penetrates the formation, expressed in lineal units of feet/minute. See: *penetration rate.*

Rate of shear. The change in velocity between two parallel layers divided by the distance between the layers. Shear rate has the units of reciprocal seconds (sec^{-1}). See: *shear rate.*

Raw drilling fluid. Drilling fluid, before dilution, that is to be processed by solids-removal equipment.

Rectangular screen. See: *oblong mesh.*

Reduced port. A valve whose bore size is less than the area of the pipe to which it is attached.

Removal section. The first section in the drilling-fluid system, consisting of a series of compartments and solids-removal equipment to remove gas and undesirable solids.

Reserve pit. (1) An earthen pit used to store drilling waste in land drilling operations. (2) A section of a surface system used to store drilling fluid.

Resin. A semisolid or solid complex or amorphous mixture of organic compounds having no definite melting point or tendency to crystallize. Resin may be a component of compounded materials that can be added to drilling fluids to impart special properties to the system, that is, wall cake, fluid loss, etc.

Resistivity. Resistivity is a characteristic electrical property of a material and is equal to the electrical resistance of a 1-meter cube of the material to passage of a 1-ampere electric current perpendicular to two parallel faces. The electrical resistance offered to the passage of a current is expressed in ohm-meters. It is the reciprocal conductivity. Freshwater muds are usually characterized by high resistivity; saltwater muds by low resistivity. See: *conductivity*.

Resistivity meter. An instrument for measuring the electrical resistivity of drilling fluids.

Retention time. The time any given particle of material is retained in a region, for example, the time a particle is actually on a screening surface, within a hydroclone, or within the bowl of a centrifuge.

Retort. An instrument used to distill oil, water, and other volatile material in a drilling fluid to determine oil, water, salt, and total solids contents in volume percentage. See: *mud still, mini still, API RP 13B*.

Reverse circulation. The method by which the normal flow of a drilling fluid is reversed by circulating down the annulus, then up and out the drill string. See: *circulation*.

Reynolds number. A dimensionless number, Re, that occurs in the theory of fluid dynamics. The Reynolds number for a fluid flowing through a cylindrical conductor is determined by the equation:

$$R_e = DV\rho/\mu.$$

where

$D =$ diameter
$V =$ velocity
$\rho =$ density
$\mu =$ viscosity

The number is important in fluid-hydraulics calculations for determining the type of fluid flow, that is, whether laminar or turbulent. The transitional range occurs approximately from 2000 to 3000. Below 2000, the flow is laminar; and above 3000, the flow is turbulent. See: *fluid flow*.

Rheology. The science that deals with deformation and flow of matter. See: *viscosity, Bingham model, plastic viscosity, yield point, gel strength*.

Rig pump. The reciprocating, positive displacement, high-pressure pump on a drilling rig used to circulate the hole. See: *mud pump*.

Rig shaker. Slang term for a shale shaker.

Rigid frame panel. One of the two main screen panel types, consisting of a rigid panel to which the screen or layers of screen are attached. The screen panel fastening device can be designed for fast panel replacement. See: *hook-strip panel*.

ROP. See: *rate of penetration, penetration rate*.

Rope discharge. The characteristic underflow of a hydrocyclone so overloaded with separable solids that not all the separated solids can crowd out through the underflow opening (apex), causing those solids that can exit to form a slow moving, heavy, ropelike stream. Also referred to as "rope" or "rope underflow."

Rotary drilling. The method of drilling wells in which a drill bit attached to a drill string is rotated on the formation to be drilled. A fluid is circulated through the drill pipe to remove cuttings from the bottom of the hole, bring cuttings to the surface, and perform other functions. See: *cable tool drilling*.

Rotary mud separator (RMS). A centrifuge consisting of a perforated cylinder rotating inside of an outer cylinder housing. As drilling fluid flows outside of the perforated cylinder, only the very small particles pass through the perforations.

Round trip. See: *trip*.

rpm. Revolutions per minute.

Salt. A class of compounds formed when the hydrogen of an acid is partially or wholly replaced by a metal or a metallic radical. Salts are formed by the action of acids on metals, or oxides and hydroxides, directly with ammonia and by other methods. See: *sodium chloride*.

Saltwater drilling fluid. A water-based drilling fluid whose external liquid phase contains sodium chloride or calcium chloride.

Saltwater mud. See: *saltwater drilling fluid*.

Samples. Cuttings obtained for geological information from the drilling fluid as it emerges from the hole. They are washed, dried, and labeled as to the depth.

Sand. (1) Particle-size classification for solids larger than 74 microns. (2) A loose, granular material resulting from the disintegration of rocks with a high silica content. See: *API RP 13B, API sand.*

Sand content. The solids particles retained on a U.S. Standard No. 200 test screen, expressed as the bulk percentage by volume of the drilling fluid slurry sample. The opening in this screen is 74 microns. The retained solids may be of any mineral or chemical composition and characteristic. For example, barite, shale, mica, silica, steel, chert, etc., larger than 74 microns are called API sand. See. *API sand.*

Sand trap. The first compartment and the only unstirred compartment in a well-designed drilling-fluid system intended as a settling compartment.

Scalping shakers. The first set of shale shakers after the flowline in a cascade shaker arrangement. These shakers are usually circular or elliptical motion shakers with coarse mesh screens that are used to remove the bulk of the large-diameter drilled solids or gumbo. This initial fluid preparation allows the second set of fine-screen shale shakers in the series to operate more efficiently with less possibility of flooding. See: *fine-screen shale shakers, flooding, blinding.*

Screen cloth. A type of screening surface, woven in square, rectangular, or slotted openings. See: *wire cloth.*

Screen support rubbers. Elastomers that cushion the contact between screens and shale shaker frames.

Screen underflow. The discharge stream from a screening device that contains a greater percentage of liquids than does the feed. See: *liquid discharge.*

Screening. A mechanical process resulting in a division of particles on the basis of size by their acceptance or rejection by a screening surface.

Screening surface. The medium containing the openings for passage of undersize material.

Scroll. See: *flute.*

Self-lubricating. Units that provide their own means of lubrication.

Separation potential. Separation potential of a shale shaker screen is the size distribution of equivalent spherical volumes calculated by determining the equivalent ellipsoidal volumes of at least 1500 openings in a screen as determined by image analysis. Also called the Cutt point distribution. See: *Cutt point.*

Separator. A tank in which mixed water, oil, and gas are allowed to separate by gravity or enhanced force.

Separator (open/atmospheric). A separator for drilling fluid/formation fluid that is open to atmospheric pressure.

Separator (closed/pressurized). A separator for drilling fluid/formation fluid that is closed and pressurized.

Separator (West Texas). A type of open separator. A large tank at atmospheric or 1–3 psi gauge that is used to separate gas from drilling fluid at the flowline.

Settling velocity. The velocity a particle achieves in a given fluid when gravity forces equal friction forces of the moving particle, that is, when the particle achieves its maximum velocity.

Shale. Stone of widely varying hardness, color, and compaction that is formed of clay-sized grains (less than two microns). See: *natural clay*.

Shale shaker. Any of several mechanical devices for removing cuttings and other large solids from drilling fluid. Common examples are the vibrating screen and rotating cylindrical screen.

Shale stabilizer. Drilling-fluid additive that reduces the rate of interaction of water with the clays in shale. Also known as **shale shaker inhibitor.**

Sharpness of cut. The slope of a straight line drawn between the solids separated at the 84% point and the 16% point on a graph of the percentage of solids separated versus particle size. The more vertical the slope, the sharper the cut. Also known as **sharpness of separation.**

Shear rate. The change of velocity with respect to the distance perpendicular to the velocity changes. See: *rate of shear*.

Shear stress. The force per unit of an area parallel to the force that tends to slide one surface past another. See: *viscosity, Bingham model, plastic viscosity, yield point, gel strength*.

Shear thinning. Opposite of dilatant. The apparent viscosity decreases instantaneously with increasing shear rate. See: *apparent viscosity, viscosity, Bingham model, plastic viscosity, yield point, gel strength*.

Short circuiting. A hydraulic condition existing in parts of the tank basin, reservoir, or hydrocyclone in which the time of travel of liquid/ solids is less than the normal flowthrough time—for instance, if the surface tanks contain very viscous fluid, but the returns from the flowline have a very low viscosity; the flowline returns might tend to channel across the top of the surface system toward the pump suction. In this case the flowline returns would be "short circuiting" or bypassing the solids-separation equipment. In hydrocyclones, separable solids that pass directly from the feed inlet and out through the vortex finder without passing through the cone section of the hydrocyclone have "short circuited" the hydrocyclone processing system.

Shute. In a woven cloth, the direction of the wires running perpendicular to the loom or running across the roll of cloth. In wire cloth production, these are the short or transverse wires. See: *weft*.

Sieve. See: *testing sieve*.

Sieve analysis. The mass classification of solid particles passing through or retained on a sequence of screens of increasing mesh count. Analysis may be by wet or dry methods. See: *particle size distribution*.

Silencer. A tank or pit used to muffle the sound at the blooie line.

Silt. Materials whose particle size generally falls between 2 and 74 microns. A certain portion of dispersed clays and barite fall into this particle size range, as well as drilled solids.

Size distribution. See: *particle size distribution*.

Slip. The difference between synchronous speed and operating speed compared with synchronous speed, expressed as a percentage. If expressed in rpm, slip is the difference between synchronous speed and operating speed.

Sloughed solid. A solid entering the well bore from the exposed formation; not a drilled solid.

Sloughing. A situation in which portions of a formation fall away from the walls of a hole, as a result of incompetent unconsolidated formations, tectonic stresses, high angle of repose, wetting along internal bedding planes, or swelling of formations. See: *caving, cuttings, heaving*.

Slug. A small volume of weighted fluid pumped into the drill string to keep the drilling fluid liquid level below the rig floor while pulling drill pipe during a trip. This prevents drilling fluid from spilling on the rig floor as the pipe is pulled. See: *pill*.

Slug tank. A small compartment (normally adjacent to the suction compartment) used to mix special fluids to pump downhole. The most common use is to prepare a slug or a small volume of weighted mud before a trip. See: *pill tank*.

Slurry. A mixture or suspension of solid particles in one or more liquids.

Sodium bicarbonate. $NaHCO_3$. A material used extensively for treating cement contamination and occasionally other calcium contamination of drilling fluids. It is the half-neutralized salt of carbonic acid. See: *bicarb*.

Sodium carboxymethylcelluose. An organic polymer, available in various grades of purity, used to control filtration, suspend weight material, and build low-shear-rate viscosity in drilling fluids. It can be used in conjunction with bentonite where low-solids drilling fluids (muds) are desired. See: *CMC, low-solids drilling fluids*.

Sodium chloride. NaCl. Commonly known as salt. Salt may be present in the drilling fluid as a contaminant or may be added purposely for inhibition. See: *salt*.

Sodium chromate. Na_2CRO_4. See: *chromate*.

Sodium hydroxide. NaOH. Commonly referred to as "caustic" or "caustic soda." A chemical used primarily to raise pH.

Sodium polyacrylate. A synthetic, high-molecular-weight polymer of acrylonitrile used primarily for fluid loss control.

Sodium silicate drilling fluids. Special class of inhibited chemical drilling fluid using sodium silicate, saltwater, and clay.

Solid. A firm substance that holds its form; not gaseous or liquid.

Solids. All particles of matter in the drilling fluid, that is, drilled formation cuttings, barite, bentonite, etc.

Solids content. The total amount of solids in a drilling fluid. This is usually determined by distillation that measures the volume fraction of both the dissolved and the suspended, or undissolved, solids. The suspended solids content may be a combination of high– and low–specific gravity solids and native or commercial solids. Examples of dissolved solids are the soluble salts of sodium, calcium, and magnesium. Suspended solids make up the wall cake; dissolved solids remain in the filtrate. The total suspended and dissolved solids contents are commonly expressed as percentage by volume and less commonly as percentage by weight. See: *retort*.

Solids discharge. That stream from a liquid/solids separator containing a higher percentage of solids than does the feed.

Solids discharge capacity. The maximum rate at which a liquid/solids separation device can discharge solids without overloading.

Solids removal equipment efficiency. A measure of the performance of surface equipment in removing drilled solids from the drilling fluid. It is a calculation based on a comparison of the dilution required to maintain the desired drilled-solids content with that which would have been required if no drilled solids were removed. Also called **solids removal equipment performance** and **drilled solids removal system performance.**

Solids separation equipment. Any and all of the devices used to remove solids from liquids in drilling, that is, shale shaker, desander, desilter, mud cleaner, and centrifuge.

Solubility. The degree to which a substance will dissolve in a specific solvent.

Solute. A substance that is dissolved in another (the solvent).

Solution. A mixture of two or more components that form a homogeneous single phase. An example of a solution is salt dissolved in water.

Solvent. Liquid used to dissolve a substance (the solute).

Souring. A term commonly used to mean fermentation.

Specific gravity (SG). The weight of a specific volume of a liquid, solid, or slurry in reference to the weight of an equal volume of water at a reference temperature of 3.89°C (water has a density of 1.0 g/cc at this temperature).

Specific heat capacity. The number of calories required to raise 1 g of a substance one degree Celsius.

Spray bar. A pipe located over the bed of a shale shaker through which dilution fluid is sprayed onto the screen surface during separation of the drilled solids. In practice, spray bars may supply a mist or small amount of liquid, not a hard spray, to prevent washing fine solids through the screen panels and back into the circulating system.

Spray discharge. See: *spray underflow*.

Spray underflow. The characteristic underflow of certain balanced hydrocyclones discharging to the atmosphere and not overloaded with separable solids.

Spud mud. The drilling fluid used when drilling starts at the surface, often a thick bentonite-lime slurry.

Spudding in. The initiating of the drilling operations in the first top-hole section of a new well.

Spurt loss. The flux of fluids and solids that occurs in the initial stages of any filtration before pore openings are bridged and a filter cake is formed. See: *surge loss*.

Square mesh. Screen cloth with the same mesh count in both directions.

Square weave. See: *square mesh*.

Squeeze. A procedure whereby slurries of cement, drilling fluid, gunk plug, etc., are forced into the formation by pumping into the hole while maintaining a backpressure. This is usually achieved by closing the blowout preventers or by using a retrievable downhole packer.

Squirrel-cage motor. An induction motor that gets its name from the rotor assembly that looks like a squirrel cage, typical of those used earlier in the twentieth century. The cage consists of rotor bars secured at each end to the shorting rings. An induction motor is one in which there is no physical electrical connection to the rotor. Current in the rotor is induced by the magnetic field of the stator.

Stability meter. An instrument to measure the breakdown voltage of oil-based drilling fluids. This gives an indication of the emulsion stability.

Stacking a rig. Storing a drilling rig upon completion of a job when the rig is to be withdrawn from service for a period of time.

Starch. A group of carbohydrates occurring in plant cells. Starch is specially processed (pregelatinized) for use in drilling fluids to reduce filtration rate and occasionally to increase the viscosity. Without proper preservative, starch can ferment.

Static. Not moving, or at rest. Opposite of dynamic. See: *quiescence*.

Stearate. Salt of stearic acid, which is a saturated, 18-carbon fatty acid. Certain compounds, such as aluminum stearate, calcium stearate, and zinc stearate, have been used in drilling fluids for defoaming, lubrication, air drilling in which a small amount of water is encountered, etc.

Stiff foam. A foam in which a bentonite or long-chain polymer has been added.

Stirrer. See: *agitator, mechanical agitator*.

Stokes' law. Stokes' law states that the terminal settling velocity of a spherical particle is proportional to the square of the particle diameter, the acceleration of gravity, and the density difference between the density of the particle and the density of the liquid medium; the terminal settling velocity is inversely proportional to the viscosity of the liquid medium:

$$V_T = [gD_P^2(\rho S - \rho L)(10^{-6})]/116\rho$$

where

V_T = terminal settling velocity, in in./sec
D_P = particle diameter, in microns
ρS = density of the solids, in g/cm^3
ρL = density of the liquid, in g/cm^3
μ = viscosity of the feed slurry, in centipoise

Stormer viscometer. A rotational shear viscometer used for measuring the viscosity and gel strength of drilling fluids. This instrument has been largely replaced by the direct-indicating viscometer.

Streaming potential. The electrokinetic portion of the spontaneous potential (SP) electric-log curve that can be significantly influenced by the characteristics of the filtrate and filter cake of the drilling fluid.

Streamline flow. See: *laminar flow*.

Stroke. The distance between the extremities of motion or total displacement normal to the screen; that is, the diameter of a circular motion or twice the amplitude. See: *amplitude*.

Stuck. A condition whereby the drill pipe, casing or any other device inserted into the well bore inadvertently becomes lodged in the hole. Sticking may occur while drilling is in progress, while casing is being run in the hole, or while the drill pipe is being tripped. Frequently a fishing job results.

Stuck pipe. See: *differential pressure sticking, stuck*.

Suction compartment. (1) The area of the check/suction section from which drilling fluid is picked up by the suction of the mud pumps. (2) Any compartment from which a pump moves fluids.

Sump. (1) A disposal compartment or earthen pit for holding discarded liquids and solids. (2) The pan or compartment below the lowest shale shaker screen.

Supersaturation. If a solution contains a higher concentration of a solute in a solvent than would normally correspond to its solubility at a given temperature, a state of supersaturation exists. This is an unstable condition, because the excess solute separates when the solution is seeded by introducing a crystal of the solute. The term is frequently used erroneously for hot salt drilling fluids.

Support screen. A heavy, wire mesh either plain or calendered that supports a finer mesh screen for use in filtering or screen separation. See: *backup screen*.

Surface active materials. See: *surfactant*.

Surface tension. Generally the cohesive forces acting on surface molecules at the interface between a liquid and its own vapor. This force appears as a tensile force per unit length along the interface surface and is usually expressed in units of dynes per centimeter. Since the surface tension is between the liquid and the air, it is common practice to refer to values measured against air as surface tension, and to use the term "interfacial tension" for measurements at an interface between two liquids or a liquid and a solid. See: *interfacial tension, emulsion*.

Surfactant. Material that tends to concentrate at an interface of an emulsion or a solid/liquid interface. Used in drilling fluids to control the degree of emulsification, aggregation, dispersion, interfacial tension, foaming, defoaming, wetting, etc.

Surfactant drilling fluid. A drilling fluid that contains a surfactant, usually to effect control over the degree of aggregation and dispersion or emulsification.

Surge. The pressure increase in a well bore caused by lowering tubulars. Viscous drilling fluid flowing up the annulus, displaced by drill pipe, tubing, or casing, creates the pressure surge.

Surge loss. This is a colloquial term used to describe a spurt of filtrate and solids that occurs in the initial stages of any filtration before pore openings are bridged and a filter cake is formed. The preferred term is "spurt loss." See: *spurt loss*.

Suspensoid. A mixture that consists of finely divided colloidal particles floating in a liquid. The particles are so small that they do not settle but are kept in motion by the moving molecules of the liquid (Brownian movement).

Swabbing. When pipe is withdrawn from the hole in a viscous drilling fluid or if the bit is balled, a decrease in pressure in the well bore can cause formation fluid to flow into the well.

Swelling. See: *hydration*.

Synergism. Term describing an effect obtained when two or more products are used simultaneously to obtain a certain result. Rather than the result of each product being additive to the other, the result is a multiple of the effects.

Synergistic properties. See: *synergism*.

Tannic acid. The active ingredient of quebracho and other quebracho substitutes such as mangrove bark, chestnut extract, hemlock, etc.

Temperature survey. An operation to determine temperatures at various depths in the well bore. This survey is used to find the location of inflows of water into the borehole or where proper cementing of the casing has taken place.

Ten-minute gel. See: *gel strength*.

Tensile bolting cloth. A group of industrial wire cloth specifications woven of extremely smooth and durable stainless steel in a square mesh pattern. The wire diameter is lighter than mill grade cloth, producing a higher percentage of open area. See: *market grad cloth*, *mill grade cloth*, *ultra-fine wire cloth*, and *calendered*.

Tensioning. The stretching of a screening surface of a shale shaker within the vibrating frame, to the proper tension.

Testing sieve. A cylindrical or traylike container with a screening surface bottom of standardized apertures. See: *sieve*.

Thermal decomposition. Chemical breakdown of a compound or substance by temperature into simple substances or into its constituent elements. Starch thermally decomposes in drilling fluids as the temperature approaches 300°F.

Thinner. Any of the various organic agents (e.g., tannins, lignins, lignosulfonates, etc.) and inorganic agents (pyrophosphates, tetraphosphates, etc.) that are added to a water-based drilling fluid to reduce the low-shear-rate viscosity and/or thixotropic properties by deflocculation.

Thixotropy. The ability of a fluid to develop gel strength with time. That property of a fluid at rest that causes it to build up a rigid or semirigid gel structure if allowed to remain at rest. The fluid can be returned to a liquid state by mechanical agitation. This change is reversible. See: *gel strength*.

Thrust. A force that pushes; for example. as solids experience a thrust on a shale shaker screen.

Tighten-up emulsion. Jargon describing condition in oil-based drilling fluids in which either chemicals or shear or both are used to emulsify water in oil into smaller droplets to prevent the emulsion from breaking, or coming apart. Also known as **tighten-up emulsion mud.**

Titration. The process of using a standard solution in order to determine of the amount of some substance in another solution. The known solution is usually added in a definite quantity to the unknown until a reaction is complete.

Tool joint. A drill-pipe coupler consisting of a threaded pin and a box of various designs and sizes.

Torque. (1) The turning effort caused by a force acting normal to the radius at a specified distance from the axis of rotation. Torque is expressed in pound-feet (pounds at a radius of one foot). Torque, lb-ft = force, lbs × lever arm, ft. (2) Drill string connections require a specific torque to be properly tightened. The drill string in a borehole experiences a frictional force as it is rotated. This causes a torque in the drill string. Torque reduction can usually be accomplished by the addition of various drilling-fluid additives.

Total depth (TD). The greatest depth reached by the drill bit in a particular well.

Total dilution. The volume of drilling fluid that would be built to maintain a specified fraction of drilled solids over a specified interval of footage if there were no solids-removal system.

Total hardness. See: *hardness (water)*

Total head. The sum of all the heads within a system (total head = velocity head + pressure head + elevation head).

Total nonblanked area. The net unblocked area, in square feet, that will permit the passage of fluid through a screen. Some screen designs can eliminate as much as 40% of the gross screen panel area from fluid flow due to backing plate and bonding material blockage.

Tour. Pronounced like "tower." A person's turn in an orderly schedule, designating the shift of a drilling crew.

Trenchless drilling. Excavating material near the surface for tunnels, cables, pipelines, etc., by drilling instead of digging ditches.

Trip. The process of pulling the drill string from the hole and running it back to the bottom again. One way (either in or out) is referred to as a half-trip. See: *round trip*.

Trip tank. A gauged and calibrated vessel used to account for fill and displacement volumes as pipe is pulled from and run into the hole. Close observation allows early detection of formation fluid entering the well bore and of drilling fluid loss to a formation.

Turbidity. A condition in a clear fluid that causes a lack of clarity caused by the presence of suspended matter, resulting in the scattering and absorption of light rays.

Turbine. See: *impeller*.

Turbulent flow. Fluid flow in which the velocity at a given point changes constantly in magnitude and the direction of flow; pursues erratic and continually varying courses. See: *critical velocity*, *Reynolds number*.

Twist-off. The severing or failure of a joint of drill pipe caused by excessive torque.

Ultra-fine solids. Particles whose diameter is between 2 and 44 microns.

Ultra-fine wire cloth. A group of industrial wire cloth specifications with lighter than normal wire. The wire diameter of this grade produces the highest percentage of open area of all other grades for any specific mesh size. This cloth is used in multiple layer screens. See: *market grade cloth*, *mill grade cloth*, *tensile bolting cloth*, and *calendered*.

Ultraviolet light. Light waves shorter than the visible blue and violet waves of the spectrum. Crude oil, colored distillates, residium, a few drilling-fluid additives, and certain minerals and chemicals fluoresce in the presence of ultraviolet light. These substances, when present in drilling fluid, may cause the drilling fluid to fluoresce.

Unbalanced elliptical motion. An elliptical motion of a shale shaker screen such that the ellipse axes at the feed end are tilted toward the discharge end of the screen and the ellipse axes at the discharge end are tilted toward the feed end. Usually these screens are tilted downward to assist solids removal from the end of the screen. The vibrator is usually located above the center of gravity of the shaker basket.

Underflow. (1) Centrifugal separators: the discharge stream from centrifugal separators that contains a higher percentage of solids than does the feed. (2) Screen separators: the discharge stream from screen separators that contains a lower percentage of solids than does the feed.

Underflow header. A pipe, tube, or conduit into which two or more hydrocyclones discharge their underflow.

Underflow opening. See: *apex, apex valve.*

Undersize solids particles. (1) Particles, in a given situation, that will pass through the mesh of the screen in use. (2) Particles, in a given situation, that will remain with the liquid phase when subjected to centrifugal force.

Underslung. Field terminology denoting that the support ribs for the shaker screen are located above the screen surface. See: *overslung.*

Unoccluded area. Unobstructed area of a screen opening.

Unweighted drilling fluid. A drilling fluid that does not contain commercial suspended solids added for the purpose of increasing the density of the drilling fluid.

V-G meter. See: *direct-indicating viscometer.*

VAC. Alternating current voltage.

Valence. A number representing the combining power of an atom, that is, the number of electrons lost, gained, or shared by an atom in a compound. It is also a measure of the number of hydrogen atoms with which an atom will combine or replace, for example, an oxygen atom combines with two hydrogens, hence has a valence of 2. Thus, there are mono-, di-, tri-, etc., valent ions.

Valence effect. In general, the higher the valence of an ion, the greater the loss of stability to emulsions, colloidal suspensions, etc., these polyvalent ions will impart.

Velocity. Time rate of motion in a given direction and sense. It is used as a measure of fluid flow and may be expressed in terms of linear velocity, mass velocity, volumetric velocity, etc. Velocity is one of the factors that contribute to the carrying capacity of a drilling fluid.

Velocity head. Head (relating to pressure when divided by the density of the fluid) created by movement of a fluid, equal to an equivalent height of static fluid.

Venturi. Streamlining up to given pipe size following a restriction (as in a jet in a mud hopper) to minimize turbulence and pressure drop.

Vibrating screen. A screen with motion induced as an aid to solids separation. See: *shale shaker*.

Vibration isolaters. Elastomers ranging from solid to air-pressured or springs that allow the shale shaker screens to vibrate but do not transmit the vibratory motion to the rest of the machine.

Vibrators. Weights rotated about an axis that does not pass through the center of mass.

Viscometer. An apparatus to determine the viscosity of a fluid or suspension. Viscometers vary considerably in design and methods of testing.

Viscosifiers. Material added to a drilling fluid to increase the low-shear-rate viscosity.

Viscosity. The ratio of shear stress to shear rate in a fluid. If the shear stress is measured in dynes/cm^2 and the shear rate in reciprocal seconds, the ratio is the viscosity, in Poise. Viscosity may be viewed as the internal resistance offered by a fluid to flow. This phenomenon is attributable to the attractions between molecules of a liquid and is a measure of the combined effects of adhesion and cohesion to the effects of suspended particles and to the liquid environment. The greater this resistance, the greater the viscosity. (2) A characteristic property of a fluid, liquid, or slurry crudely defined as resistance to flow (by accurate definition the ratio of shear stress to shear rate). See: *apparent viscosity, plastic viscosity, API RP 13B*.

Viscosity gravity (V-G) meter. The name more commonly used for the direct-indicating viscometer. See: *viscometer*.

Viscous flow. See: *laminar flow*.

Volatile matter. Normally gaseous products given off by a substance, such as gas breaking out of live crude oil that has been added to a drilling fluid. In distillation of drilling fluids, the volatile matter is the water, oil, gas, etc., that are vaporized, leaving behind the total solids, which can consist of both dissolved and suspended solids.

Volt. The unit of electrical pressure or electromotive force. One volt produces a current flow of 1 ampere through a resistance of 1 ohm.

Volume percentage. Volume percentage is the number of volumetric parts of the total volume. Volume percentage is the most common method of reporting solids, oil, and water contents of drilling fluids. See: *weight percentage, ppm.*

Vortex. A cylindrical or conical-shaped core of air or vapor lying along the central axis of the rotating slurry inside a hydrocyclone.

Vortex finder. A cylinder extending into the upper end of a hydrocyclone to cause drilling fluid to move in a circular spiral direction within the cone and prevent the entering fluid from short circuiting directly to the hydrocyclone overflow.

Wall cake. The solid material deposited along the wall of the hole resulting from filtration of the fluid part of the drilling fluid into the formation. See: *cake thickness, filter cake.*

Wall sticking. See: *differential pressure sticking.*

Warp. In a woven cloth, the direction of the wires running parallel with the loom or running the length of a roll of cloth. In wire cloth production, these are the long or longitudinal wires. See: *loom.*

Water-based drilling fluid. Common, conventional drilling fluid. Water is the suspending medium for solids and is the continuous phase, whether or not oil is present. See: *water loss, filtration.*

Water-based mud. See: *water-based drilling fluid.*

Water block. A reduction in the permeability of the formation caused by the invasion of water into the pores (capillaries). The decrease in permeability can be caused by swelling of clays, thereby shutting off the pores, or in some cases by a capillary block of wetted pores due to surface tension phenomena.

Water feed. Water to be added for dilution of the mud feed into a centrifugal separator. See: *dilution water.*

Water loss. See: *filtration, fluid loss.*

Water wet. Not oil wet. A surface on which water easily spreads. If the contact angle of a water droplet on a surface is less than 90°, the surface is water wet. See: *hydrophilic, oil wet.*

Water-in-oil emulsion. See: *invert oil-emulsion drilling fluid.*

Weft: See: *shute.*

Weight. In drilling fluid terminology, the density of a drilling fluid. This is normally expressed in either lb/gal, lb/cu ft, psi hydrostatic pressure per 1000 ft of depth, or specific gravity related to water. See: *density.*

Weight material. Any of the high–specific gravity materials used to increase the density of drilling fluids. This material is most commonly barite or hematite. In special applications, limestone is also called a weight material.

Weight percentage. The number of weighted parts of the total weight. Weight percentage is the most common method of reporting oil in solids discharges and mass balance calculations. See: *volume percentage, ppm.*

Weight up. To increase the weight of a drilling fluid, usually by the addition of weight material.

Weighted drilling fluid. A drilling fluid to which commercial solids have been added to increase the slurry weight.

Weighted mud. See: *weighted drilling fluid.*

Well bore. The hole drilled by the bit, also known as the borehole.

Well-bore stabilization. Maintenance of well-bore integrity, which generally requires manipulating the properties of the drilling fluid to simulate the physicochemical environment of the rock before it was drilled.

Well logging. See: *electric logging, mud logging.*

Wetting. The adhesion of a liquid to the surface of a solid.

Wetting agent. A substance that, when added to a liquid, increases the spreading of the liquid on a surface or the penetration of the liquid into a material.

Whipstock. A device inserted into a well bore to cause the drill bit to exit the established path of the existing well bore. The whipstock is the tool used for the initiation of directional drilling.

Wildcat. A well in unproved territory.

Windage loss. (1) The resisting power of air or air friction acting against a rapidly rotating armature or cooling fan to create a power loss. (2) The resisting power of air or air friction against the rotating bowl of a centrifuge.

Wire cloth. Screen cloth of woven wire. See: *screen cloth.*

Working pressure (WP). The maximum pressure to which equipment should be exposed in order to comply with manufacturer's warranty and be within industry codes and safety standards.

Workover fluid. Any type of fluid used in the workover operation of a well.

Yield. The quality of a clay in terms of the number of barrels of a given viscosity (usually 15 cP) slurry that can be made from a ton of the clay. Based on the yield, clays are classified as bentonite, high-yield, low-yield, etc. Not related to yield point. See *API RP 13B.*

Yield point. (1) A term derived from a direct-reading viscosimeter (Fann V-G or equivalent) based on subtracting the plastic viscosity from the 300-rpm reading. (2) An extrapolated shear stress at zero shear rate created by assuming a linear relationship between shear stress and shear rate and determining the intercept on the shear stress axis. The linear relationship between shear stress and shear rate that results in a yield point is called a Bingham Plastic model. See: *viscosity, plastic viscosity, API RP 13B*.

Yield value. See: *yield point*.

Zero-zero gels. A condition wherein the drilling fluid fails to form measurable gels during a quiescent time interval (usually 10 minutes). The measurements of gel are made with a direct-reading viscometer at intervals of 10 seconds and 10 minutes. See: *progressive gel, flat gel*.

Zeta potential. The electrokinetic potential of a particle as determined by its electrophoretic mobility. This electric potential causes colloidal particles to repel each other and stay in suspension.

Zinc bromide. $ZnBr_2$. A very soluble salt used to increase the density of water or brine to more than double that of water. Normally added to calcium chloride/calcium bromide mixed brines.

INDEX

A

AC, *see* Alternating current
Adjustable speed drive
  benefits and disadvantages, 431–432
  components, 430
  functions, 429–430
  torque versus rpm load characteristics, 430–431
  types used with induction motors, 431
Agitators
  baffles
    American Petroleum Institute guidelines, 565
    round tank baffling, 227
    square tank baffling, 227
  components, 214–215
  design parameters
    compartment shape, 226
    impeller selection, 223–225
    internal piping, 226
    overview, 223
    tank and compartment dimensions, 226
  gearbox, 222
  impellers, *see* Impeller
  pros and cons, 237–238
  purpose, 213–214
  shafts, 222–223
  sizing, 227–232
Air pycnometer, density of weighting material measurement, 29
Alternating current, direct current comparison, 414
American Petroleum Institute
  dilution calculation, 361–362
  equipment guidelines
    centrifugal pumps, 572–577
    surface systems, 562–572
  Fluid Loss Test, 46
  shaker screen designation system
    API number, 168–171
    flow capacity, 171–173
    identification tag contents, 173–174
    manufacturer's designation, 167
    nonblanked area, 173
Ampere-turn definition, 423
Annular velocity, hole cleaning effects, 40–41
API, *see* American Petroleum Institute
Apparent power, definition, 419–420
ASD, *see* Adjustable speed drive

Augers, waste handling, 400–402
AV, see Annular velocity
Average particle density,
    measurement, 28–29

B

Bailing, see Shale inhibition
Balanced elliptical motion shale
    shaker, principles,
    132–133
Barite
  cost analysis, 11
  mud cleaner low-gravity solids
    volume/barite volume
    ratio estimation in screen
    discard, 293–294
  recovery via centrifugation, 314
  settling rate, 186–187
  shale shaker discard calculation,
    75–76
  size distribution in drilling fluid,
    284–285
Bentonite, viscosity control, 55
Bernoulli's principle, 239
Bingham Plastic model
  overview, 33–34
  rotary viscometer data
    application, 37–38
  yield point conversion, 43–44
Burial, see Land disposal, drilling
    waste

C

Capacitance, calculation, 416, 421
Capture
  analysis
    calculations, 331–332
    data collection, 330
    economics
      unweighted fluids, 333
      weighted fluids, 333–334
    laboratory work, 330
    removed solid characterization
      particle size, 332
      specific gravity, 331–332
    supplementary information,
      334
  definition, 327
  equation, 327–329
Carrying capacity index,
    calculation, 43
Cascade shale shaker
  advantages, 148–150
  design
    integral unit
      multiple vibratory motions,
        150
      single vibratory motion, 152
    separate unit system, 150
  high solids loading, 149, 152
  historical perspective, 148
  screen mesh, 153
Casson model, 35
CCI, see Carrying capacity index
Centrifuges
  applications
    unweighted drilling fluids,
      317
    weighted drilling fluids,
      317–318
  barite recovery, 314
  bowl shape, 315
  capture analysis, see Capture
  costs, 320, 322–326
  cut points, 308–310
  decanting centrifuge
    components and
      principles, 303, 305, 307,
      314–315

Derrickman's guidelines, 560–561
drilled solids effects on drilling fluids, 311–313
flocculation units, 320
gas cutting problems, 193
g force calculation, 315–316
hydrocyclone underflow centrifugation, 321
installation, 316
motors, 459
operating guidelines, 321
overflow, 313–314
pump, see Pumps
rotary mud separator, 321–322
separation limits, 308
series centrifugation, 318–319
suction and discharge pits, 103–104
traditional centrifuging, 314–315
underflow, 313–314
Chip hold-down pressure, rate of penetration relationship, 50
Circular motion shale shaker, principles, 127–128
Conductance, screens, 167, 171–173
Corrosion
control, 53–54
mechanisms, 53
Costs
capture analysis
unweighted fluids, 333
weighted fluids, 333–334
centrifuging drilling fluids, 320, 322–326
dilution, 364
drilled solids removal, 145–146
estimation, 11–12
mud cleaner use, 297–299
solids management checklist, 581
waste management, 13
Current
properties, 413
wire capacity by gauge, 142–143
Cut point
centrifuges, 308–310
curve generation
discard, 88–89
feed, 85–86
plotting, 89–90
shale shaker example, 90–92
underflow, 86
hydrocyclones, 261, 269–276, 282, 283–284
overview, 81–84
Cuttings boxes, waste handling, 403–404
Cuttings dryers
installation, 411–412
legislation, 408–409
oil retention, 406–409
operation, 411
removed fluid processing, 410–411
volume reduction, 406
Cuttings, see Drilled solids

D

DC, see Direct current
Degasser
American Petroleum Institute guidelines, 566
combination separator and degasser, 202
Derrickman's guidelines, 553–554
Magna-Vac degasser, 207
mechanisms, 193, 195–196, 203

Degasser *(continued)*
    operation variables, 203, 205
    pump degassers, 207
    purpose, 202
    suction and discharge pit, 102
    top equalization, 563
    treatment calculations, 208
    vacuum effects on entrained gas, 205
    vacuum-tank degassers, 205–207
Delta connection, three-phase power, 417–418
Density equation, 69
Derrickman's guidelines
    centrifuges, 560–561
    degassers, 553–554
    equipment checklist, 552
    hydrocyclones, 554–558
    mud cleaners, 558–560
    piping to mixing section, 561–562
    sand traps, 552–553
    shale shakers, 550–551
    tank and equipment arrangements, 549–550
Desander
    American Petroleum Institute guidelines, 568
    gas cutting problems, 192–193
    hydrocyclone arrangement, 267–269
    suction and discharge pit, 102–103
Desilter
    American Petroleum Institute guidelines, 568
    gas cutting problems, 192–193
    historical perspective, 8
    hydrocyclone arrangement, 268–269
    suction and discharge pit, 103
Dewatering, waste treatment, 391–394
Dilution
    American Petroleum Institute method for calculation, 361–362
    calculation examples, 362–366
    cost analysis, 364
    definition, 335
    examples, 335–336
    porosity effects, 337–338
    rationale, 339–341
    solids removal equipment efficiency, *see* Solids removal equipment efficiency
    volume increase factor calculation, 361
Dilution volume, calculation, 23–24
Direct current, alternating current comparison, 414
Disaggregation, definition, 5
Drilled solids
    associated problems, 2, 548
    centrifugation, *see* Centrifuges
    characteristics
        overview, 25–26
        physical properties, 26–31
    checklist for management, *see* Solids management checklist
    commercial solids, 310
    definition, 3
    economic impact, 2–3
    effects on drilling fluids, 311–313
    history of management, 4–11
    removal
        overview, 3–4, 20–25

rationale, 339–341, 548–549
safety in handling of cuttings, 58
Drilled solids removal factor,
    calculation, 361–362
Drilling fluid
  circulating system, 22–23
  dilution, see Dilution
  drilled solids removal overview,
    3–4, 20–25
  functions, 15–16
  rheology, 32–38
  selection considerations, 17, 20
  stability and maintenance, 54
  types and classification, 16–19
  viscosity maintenance, 30
Drilling fluid products
  colloidal and fine solids, 54–55
  conventional polymers, 56–57
  hazard classification, 59–61
  macropolymers, 55–56
  safety in handling, 58
  storage, 58
  surface-active materials, 57–58
  waste management and disposal,
    62–65
Drilling waste, see Waste
    management
Dryer shaker, principles,
    153–154
DSRF, see Drilled solids removal
    factor

E
Einstein equation, particle effects
    on effective velocity, 30
EIR, see Environmental impact
    reduction
Electromagnetic theory, 421–423
Environmental impact reduction,
    waste minimization, 377

F
Fann Reading, calculation,
    35–38
Flame propagation, definition,
    448
Flashpoint, definition, 447
Flocculation, applications, 320
Flow rate
  centrifugal pump selection with
    standard drilling
    equipment, 491
  friction losses, 472–479,
    490–491
  hydrocyclones, 260–261
  shale shaker selection,
    146–148
  velocity calculation, 98
Fluid limit, shale shaker,
    118–119

G
Gas buster
  design, 196–197
  mechanisms, 193, 195–196
Gas cutting
  bottom-hole pressure loss,
    189–191
  equipment effects
    centrifuge, 193
    desander, 192–193
    desilter, 192–193
    shale shaker, 192
  mud density adjustment, 191
  mud handling equipment, see
    Degasser; Gas buster;
    Separators
  problems, 189
  pump output reduction
    calculation, 194
  separation guidelines, 209–210

g factor
  calculation, 136–139
  definition, 136
  relationship to stroke and speed of rotation, 140
  shale shaker design considerations, 137–140
g force, calculation for centrifuges, 315–316
Gumbo
  conveyor, historical perspective, 11
  definition, 31
  emergency removal, 107
  formation, 107
  scalping shakers, 107–109
  transport, 31

H

Herschel-Bulkley model, 34
Hole cleaning
  decision algorithm, 39
  drilling element effects
    carrying capacity, 42–44
    cuttings characteristics, 44
    drill string eccentricity, 45
    flow rate/annular velocity, 40–41
    hole angle, 50
    overview, 40
    pipe rotation, 45
    rate of penetration, 44
    rheology, 41–42
  filtration, 45–47, 50
  problem detection, 38
Hydrocyclones, *see also* Mud cleaners
  advantages and limitations, 279–281
  arrangements
    desanders, 267–269
    desilters, 268–269
  capacity, 265–266
  capture analysis, *see* Capture
  centrifugal forces, 257
  components, 257–258
  countercurrent spiraling streams, 260
  cut points, 261, 270–276, 282, 283–284
  Derrickman's guidelines, 554–558
  discharge
    rope discharge, 264–265
    spray discharge, 261–264
  feed header problems, 269
  flow rates, 260–261
  installation, 278–279
  motors, 459
  operating guidelines, 276–278
  plugging, 558
  pressure relationship with mud weight, 258–259, 555
  principles, 257–259
  siphon breaker, 261
  sizing, 260, 281–282
  tanks, 266
  troubleshooting, 280, 557–558
  underflow centrifugation, 321
Hydrogen, burns, 447
Hydrogen sulfide, control in underbalanced drilling, 540

I

Ignitable mixture, definition, 447–448
Impeller
  axial flow impellers, 221

contour impellers, 222
design, 215–217, 466
diameter formulas, 507
displacement values, 229
head equation, 465, 467
power transmission, 215
prerotation of fluid in suction piping, 466
radial flow impellers, 217–220
rotational velocity determination, 467
turnover rate determination for sizing, 228–232
Inductance, definition, 415
Induction motors, *see* Motors
Inductive reactance, calculation, 416
Ingress protection code, motor enclosures, 443–444, 446
IP, *see* Ingress protection code

J

Jet hopper, American Petroleum Institute guidelines, 570–571

K

Kindling temperature, definition, 447

L

Land disposal, drilling waste
  burial
    cells, 386
      chemical content limits, 387
      depth or placement, 387–388
      leakage and leaching, 389–391
      moisture content, 388–389
  concerns, 374
  land application, 382–386

Laser granulometry, particle size distribution measurement, 27
LCM, *see* Lost circulation material
LGS, *see* Low-gravity solid
Linear motion shale shaker, principles, 9–10, 128–132
Lost circulation material, mud treatment, 55
Low-gravity solid
  barite discarded by shale shaker, 75–76
  measurement, 29, 70–77
  mud cleaner low-gravity solids volume/barite volume ratio estimation in screen discard, 293–294
  settling rate, 186–187
  volume calculation, 69–70, 77–78
Lubricity
  drilling solids removal advantages, 52–53
  rate of penetration effects, 51

M

Magna-Vac degasser, 207
MBT, *see* Methylene blue test
Mesh, counting, 160
Meter model, 34–35
Methylene blue, clay test, 21
Motors
  adjustable speed drive, *see* Adjustable speed drive
  alternating current induction motor advantages, 424, 429–430
  ambient temperature effects on performance, 435–437
  centrifuges, 459

Motors *(continued)*
  electromagnetic theory, 421–423
  enclosures, 441–443
  energy losses, 433–434
  frame dimension nomenclature, 442
  hazardous duty
    European Community regulations, 450, 453–454
    explosion risks, 444–448
    international nomenclature, 451–452
    location designations, 449–451
  horsepower calculation, 424
  hydrocyclones, 459
  induction motor performance characteristics, 423
  ingress protection code, 443–444, 446
  installation and troubleshooting, 438
  ratings, 432–433
  rotor, 423
  rotor circuits, 424–425
  shale shakers, 454–457, 459
  standards, 439–441
  stator, 423
  stator circuits, 425, 427
  temperature rise, 434–435
  voltage imbalance, 435–436
Mud, *see* Drilling fluid
Mud cleaners
  applications, 288–291
  arrangement, 291–292
  cut point curves, 284
  Derrickman's guidelines, 558–560
  economics, 297–299
  heavy drilling fluids, 301–302
  historical perspective, 9, 283, 286–288
  low-gravity solids volume/ barite volume ratio estimation in screen discard, 293–294
  operation, 292–293
  performance, 295–297
  specific gravity accuracy requirements, 300–301
Mud ditch, American Petroleum Institute guidelines, 569–570
Mud guns
  American Petroleum Institute guidelines, 565–566
  eductors, 234
  high-pressure mud guns, 233
  low-pressure mud guns, 233–234
  placement, 234–235
  pros and cons, 237–239
  pump suction sites, 232–233
  purpose, 213–214
  sizing, 235–237, 254
Mud hoppers
  eductor, 246
  guidelines for use, 248–250, 570–571
  installation and operation, 246–248
  low-pressure mud hoppers, 244–245
  venturi utilization, 245
Mud premix systems, American Petroleum Institute guidelines, 571
Mud processing circle, 31
Mud pump, supercharging mud pumps, 510–512

Mud tank separator, 197–198

N
Net positive suction head, calculation, 503–506
NPSH, *see* Net positive suction head

O
Offshore disposal, drilling waste
  collection and transport to shore, 380
  commercial services, 380–382
  concerns, 373–374
  direct discharge, 378
  injection, 378–380
Ohm's law, 414
Opening size
  determination for screens, 160–161
  screen performance correlation, 161

P
Partially hydrolyzed polyacrylamide
  shale encapsulation, 56–57
  shale shaker interactions, 118–119
Particle size
  capture analysis, 332
  distribution measurement, 27
PHPA, *see* Partially hydrolyzed polyacrylamide
Piping
  agitator tanks, 226
  Bernoulli's principle, 239
  Derrickman's guidelines for piping to mixing section, 561–562
  friction losses, 472–479, 489–491
  pressure and velocity relationship, 240–243
  suction pipe configurations, 509–511
  surface circulation system, 96, 98
Plastic viscosity, mud density relationship, 21–22
Possum belly, dumping, 188
Power
  definition, 414
  power triangle, 420
  three-phase power, 416–419
Power factor, definition, 420
Power Law model
  overview, 34
  rotary viscometer data application, 36–37
  yield point conversion, 43–44
Power mud, pumping, 102
Power supply
  current capacity by wire gauge, 142–143
  motor current requirements by horsepower rating, 142–143
  shale shakers, 140–143
Prehydration, clay, 250
Pressure
  head relationship, 258
  mud weight relationship in hydrocyclones, 258–259, 555
  velocity relationship in piping, 240–243
Pressure tank, functions, 250
PSD, *see* Particle size distribution
P-tank, *see* Pressure tank

Pumps
  casing
    concentric versus solute casings, 488–489
    cutwater, 469–470
    design, 468–469
    functions, 467–468
    gap size, 469
  centrifugal pumps
    affinity laws, 506–507
    American Petroleum Institute guidelines, 572–577
    cavitation, 485–486
    entrained air, 486–488
    friction losses
      formulas, 507–508
      piping, 472–479, 489–491
    guidelines, 513–514
    head pressure and flow, 479–480
    net positive suction head calculation, 503–506
    nomenclature, 471, 479
    priming, 484–485
    pump curve interpretation, 480–484
    selection factors
      flow rate needed for specific equipment, 491
      friction loss and elevation considerations, 491–503
    sizing, 470, 491
    speed formulas, 507
    system head requirement worksheet, 506, 515–519
  degassers, 207
  gas cutting and output reduction, 194
  hydraulic-driven submersible pumps, 405
  impeller, *see* Impeller
  mud gun suction sites, 232–233
  suction pipe configurations
    baffle plate, 509
    duplicity, 513
    parallel operation, 513
    piping practices, 509–511
    series operation, 512–513
    submergence levels, 508–509
    supercharging mud pumps, 510–512
PV, *see* Plastic viscosity
Pycnometer, low-gravity solid measurement, 70–77

R

Rate of penetration
  drilling fluid parameter effects
    density, 48–49
    filtration, 50
    lubricity, 51
    overview, 47–48
    rheological profile, 50
    shale inhibition, 51
    solids content, 49–50
  hole cleaning effects, 44
Reactive power, definition, 420–421
Real power, definition, 420
Reserve tanks
  agitation, 254
  functions, 105–106
Resistance
  properties, 413–414
  temperature relationship, 460
RMS, *see* Rotary mud separator
ROP, *see* Rate of penetration
Rotor, motors, 423

Rotary mud separator, principles and uses, 321–322

S

Sacks, lifting and handling systems, 251
Sand trap
  American Petroleum Institute guidelines, 566
  applications, 187–188
  Derrickman's guidelines, 552–553
  design, 100–102, 183
  top equalization, 563
Scalping shakers, gumbo removal, 107–109
Screens, *see* Shale shaker; Wire cloth
Separators
  atmospheric separators
    mud tanks, 197–198
    West Texas separator, 198–199
  mechanisms, 193, 195–196
  pressurized separators
    closed separators, 200–202
    combination separator and degasser, 202
    flare systems, 199–200
  separation guidelines, 209–210
Settling rate
  barite, 186–187
  calculation, 184–186
  forces affecting particles, 184–185
  low-gravity solids, 186–187
Shaker, *see also* Shale shaker
  historical perspective, 6–7, 9
  mesh size, 6–7, 9–10

Shale barge, waste handling, 404–405
Shale encapsulators
  high-molecular-weight polymers, 56–57
  mixing guidelines, 56–57
  types, 56
Shale inhibition
  definition, 51
  rate of penetration effects, 51
  wetting characteristics, 51–52
Shale inhibitors, mechanisms of action, 57–58
Shale shaker
  applications
    fiber-optic cables, 182
    microtunneling, 181–182
    river crossing, 182
    road crossing, 182
  bypassing, 188
  cascade systems, *see* Cascade shale shaker
  configurations, 111
  cut point curve, *see* Cut point
  definition, 111
  Derrickman's guidelines, 550–551
  description, 116–117
  design elements
    g factor, 136–140
    overview, 122–123
    power systems, 140–143
    screen deck design, 134–136
    shape of motion
      balanced elliptical motion, 132–133
      circular motion, 127–128
      classification, 123–124
      linear motion, 9–10, 128–131

Shale shaker *(continued)*
  unbalanced elliptical
    motion, 124–127
  vibrating systems, 133–134
 dryer shaker, 153–154
 flow rate, charts and factors
   affecting, 112
 gas cutting problems, 192
 historical perspective, 6, 10,
   121–122
 importance in drilling fluid
   system, 111–112
 limits
   factors affecting
     density of fluid, 120
     hole cleaning, 121
     plastic viscosity, 119–120
     solid quantity, 121
     solid types, sizes, and
       shapes, 120–121
     surface tension of fluids,
       120
     wire wettability, 120
   fluid limit, 118–119
   solids limit, 118–119
 mechanisms, 113–115
 motors, 454–457, 459
 percentage separated curve
   generation, 174–176
 screens, *see also* Wire cloth
   American Petroleum Institute
     designation system
     API number, 168–171
     flow capacity, 171–173
     identification tag contents,
       173–174
     manufacturer's designation,
       167
     nonblanked area, 173
   cloth weaves, 160–167

   conductance, 167, 171–173
   deck design, 134–136
   desirable characteristics,
     159–160
   factors affecting performance
     blinding, 176–177
     bonded screens, 180
     hook-strip screens, 180
     metal screens, 177
     plastic screens, 178
     pretensioned panels,
       179–180
     three-dimensional screen
       panels, 180–181
   open area calculation, 162,
     166–167
   requirements, 178
   selection, 112–113, 145
 selection factors
   costs, 145–146
   discharge dryness, 148
   flow rate, 146–148
   overview, 143–145
   rig configuration, 148
   screen selection, 112–113,
     145
 stroke, 463–464
 users guidelines
   installation, 155–156
   maintenance, 157–158
   operating precautions,
     158–159
   operation, 156–157
   vibrator speed, 463–464
Slip, calculation, 427
Slug tank, functions, 105
Solidification, waste treatment,
 397–399
Solids limit, shale shaker,
 118–119

Solids management checklist
  drilling program, 579
  economics, 581
  environmental issues, 580–581
  equipment capability, 579–580
  logistics, 580
  rig design and availability, 580
  well parameters/deepwater considerations, 577–579
Solids removal equipment efficiency
  calculation
    formulas, 338–339, 341–342
    unweighted drilling fluid, 354–357
    weighted drilling fluid
      discard volume calculation, 360
      excess drilling fluid generated, 360
      overview, 357–358
      volume of new drilling fluid built, 358–359
  definition, 338
  effects on drilling performance
    70% efficiency, 347–348
    80% efficiency, 346
    90% efficiency, 344–346
    100% efficiency, 343–344
    overview, 341–343
  minimum volume of drilling fluid to dilute drilled solids determination
    discarded solids, 350–351
    equation derivation, 349–350
    optimum solids-removal efficiency equation, 349
    optimum value, 351–354

Specific gravity
  accuracy requirements for mud cleaners, 300–301
  average specific gravity calculation, 77–78
  capture analysis, 331–332
  definition, 28
  rig-site determination for drilled solids, 78–79
SREE, see Solids removal equipment efficiency
Stabilization, see Solidification
Stator, motors, 423
Stereopycnometer, density of weighting material measurement, 29
Stokes' law, settling rate calculation, 184–186, 271–276, 307–308
Surface circulation system
  active system
    additions section, 95
    centrifuge suction and discharge pits, 103–104
    degasser suction and discharge pit, 102
    desander suction and discharge pit, 102–103
    desilter suction and discharge pit, 103
    equalization, 98–99
    piping and equipment arrangement, 96, 98
    removal section, 95–96
    sand traps, 100–102
    suction and testing section, 94–95
    surface tanks, 99
  overview, 93, 253
  reserve tanks, 105–106

Surface circulation system
  (*continued*)
  slug tank, 105
  trip tank, 104–105
Surface systems, American Petroleum Institute guidelines, 562–572
Suspended solids, calculation, 70
System head requirement worksheet, 506, 515–519

T
Tanks
  agitators, 226–227
  American Petroleum Institute guidelines, 572
  Derrickman's guidelines, 549–550
  hydrocyclones, 266
  pressure tanks, 250
  reserve tanks, 105–106, 255
  slug tank, 105
  trip tank, 104–105, 255, 572
TFM, *see* Total fluid management
Thermal desorption, waste treatment, 395–397
Three-phase circuit
  features, 414–415
  power, 416–419
TOR, *see* Turnover rate
Total fluid management, waste minimization, 375–377
Transformers
  constant-potential transformers, 428
  counter-electromotive force, 428
  functions, 427–428
  ideal properties, 428
  stepdown versus stepup transformers, 428
  turn ratio, 428
Trip tank, functions, 104–105, 255
Triplex mud pumps, features, 510–512
Turnover rate
  American Petroleum Institute guidelines, 565
  determination for agitator impeller sizing, 228–232
Turn ratio, transformers, 428

U
UBD, *see* Underbalanced drilling
Unbalanced elliptical motion shale shaker, principles, 124–127
Underbalanced drilling
  definition, 521
  solids control
    air/gas drilling
      recycling versus flaming, 523–524
      environmental contamination, 524–525
      natural gas, 525–526
      sample collection, 526–527
    mist systems, 527–528
    conventional or weighted drilling fluids, 536–537
    fluid types, 523
    foam drilling
      disposable foam systems, 529–530
      recyclable foam systems, 530–532
      sample collection, 532
    liquid/gas systems
      oil systems, 535

overview, 532–534
sample collection, 535–536
overview, 522–523
pressurized closed separator
system, 538–539
problems
corrosion control, 542
downhole fires and
explosions, 540–541
excess formation water, 540
fluid surges, 541
foam control, 542
hydrogen sulfide, 540
shale, 539–540
small cuttings, 541
waste management, 537–538

V

Vacuum transfer sysems, 402–403
VIF, see Volume increase factor
Viscoelasticity
definition, 32
types, 32
Viscosity
drilling fluid maintenance, 30
equation, 33
measurement, 36–37
shear rate relationship, 32–33
Voltage, properties, 413
Volume increase factor,
calculation, 361

W

Waste management
contents of drilling waste,
372–373
drilling fluid products, 62–65
drilling waste contaminants,
383–385
equipment

augers, 400–402
cuttings boxes, 403–404
cuttings dryers
installation, 411–412
legislation, 408–409
oil retention, 406–409
operation, 411
removed fluid processing,
410–411
volume reduction, 406
hydraulic-driven submersible
pumps, 405
pneumatic system, 405–406
shale barge, 404–405
vacuums, 402–403
land disposal
burial
cells, 386
chemical content limits,
387
depth or placement,
387–388
leakage and leaching,
389–391
moisture content, 388–389
concerns, 374
land application, 382–386
minimization of drilling waste
environmental impact
reduction, 377
total fluid management,
375–377
offshore disposal
collection and transport to
shore, 380
commercial services,
380–382
concerns, 373–374
direct discharge, 378
injection, 378–380

Waste management *(continued)*
  quantification of drilling waste, 367–372
  treatment
    dewatering, 391–394
    solidification, 397–399
    thermal desorption, 395–397
  underbalanced drilling waste, 537–538
Water
  dewatering for waste treatment, 391–394
  vapor pressure, 504–505
Weighting agents, discard costs, 12
West Texas separator, 198–199
Wire
  copper wire size required to limit line voltage drop, 142–143
  current capacity by wire gauge, 142–143
Wire cloth
  conductance, 167, 171–173
  market grade and tensile bolting cloth shaker screen characteristics, 166
  mesh counting, 160
  opening size determination, 160–161
  sieve designations of National Bureau of Standards, 162–165
Work, definition, 414
Wye connection, three-phase power, 417–418

Y

Yield point, conversion between models, 43–44